Microbial Cultures and Enzymes in Dairy Technology

Şebnem Öztürkoğlu Budak
Ankara University, Turkey

H. Ceren Akal
Ankara University, Turkey

A volume in the Advances in Medical Technologies and Clinical Practice (AMTCP) Book Series

IGI Global
DISSEMINATOR OF KNOWLEDGE

Published in the United States of America by
IGI Global
Medical Information Science Reference (an imprint of IGI Global)
701 E. Chocolate Avenue
Hershey PA, USA 17033
Tel: 717-533-8845
Fax: 717-533-8661
E-mail: cust@igi-global.com
Web site: http://www.igi-global.com

Library of Congress Cataloging-in-Publication Data

Names: Budak, Sebnem Ozturkoglu, 1978- editor. | Akal, H. Ceren, 1986- editor.
Title: Microbial cultures and enzymes in dairy technology / Sebnem Ozturkoglu Budak and H. Ceren Akal, editors.
Description: Hershey PA : Medical Information Science Reference, [2018] | Includes bibliographical references.
Identifiers: LCCN 2017040556| ISBN 9781522553632 (hardcover) | ISBN 9781522553649 (ebook)
Subjects: | MESH: Dairy Products--microbiology | Microbiological Techniques--methods | Probiotics | Dairying--methods
Classification: LCC SF250.5 | NLM QW 85 | DDC 637--dc23 LC record available at https://lccn.loc.gov/2017040556

This book is published in the IGI Global book series Advances in Medical Technologies and Clinical Practice (AMTCP) (ISSN: 2327-9354; eISSN: 2327-9370)

British Cataloguing in Publication Data
A Cataloguing in Publication record for this book is available from the British Library.

All work contributed to this book is new, previously-unpublished material. The views expressed in this book are those of the authors, but not necessarily of the publisher.

For electronic access to this publication, please contact: eresources@igi-global.com.

Advances in Medical Technologies and Clinical Practice (AMTCP) Book Series

Srikanta Patnaik
SOA University, India
Priti Das
S.C.B. Medical College, India

ISSN:2327-9354
EISSN:2327-9370

Mission

Medical technological innovation continues to provide avenues of research for faster and safer diagnosis and treatments for patients. Practitioners must stay up to date with these latest advancements to provide the best care for nursing and clinical practices.

The **Advances in Medical Technologies and Clinical Practice (AMTCP) Book Series** brings together the most recent research on the latest technology used in areas of nursing informatics, clinical technology, biomedicine, diagnostic technologies, and more. Researchers, students, and practitioners in this field will benefit from this fundamental coverage on the use of technology in clinical practices.

Coverage

- Biomechanics
- Medical informatics
- Diagnostic Technologies
- Clinical Studies
- Neural Engineering
- Biometrics
- Telemedicine
- Patient-Centered Care
- Medical Imaging
- Clinical High-Performance Computing

IGI Global is currently accepting manuscripts for publication within this series. To submit a proposal for a volume in this series, please contact our Acquisition Editors at Acquisitions@igi-global.com or visit: http://www.igi-global.com/publish/.

The Advances in Medical Technologies and Clinical Practice (AMTCP) Book Series (ISSN 2327-9354) is published by IGI Global, 701 E. Chocolate Avenue, Hershey, PA 17033-1240, USA, www.igi-global.com. This series is composed of titles available for purchase individually; each title is edited to be contextually exclusive from any other title within the series. For pricing and ordering information please visit http://www.igi-global.com/book-series/advances-medical-technologies-clinical-practice/73682. Postmaster: Send all address changes to above address.

Titles in this Series

For a list of additional titles in this series, please visit: www.igi-global.com/book-series

Multifunctional Nanocarriers for Contemporary Healthcare Applications
Md. Abul Barkat (K.R. Mangalam University, India) Harshita A.B. (K.R. Mangalam University, India) Sarwar Beg (Jubilant Generics, India) and Farhan J. Ahmad (Jamia Hamdard, India)
Medical Information Science Reference • copyright 2018 • 335pp • H/C (ISBN: 9781522547815) • US $265.00 (our price)

Biomedical Signal and Image Processing in Patient Care
Maheshkumar H. Kolekar (Indian Institute of Technology Patna, India) and Vinod Kumar (Indian Institute of Technology Roorkee, India)
Medical Information Science Reference • copyright 2018 • 312pp • H/C (ISBN: 9781522528296) • US $265.00 (our price)

Next-Generation Mobile and Pervasive Healthcare Solutions
Jose Machado (University of Minho, Portugal) António Abelha (University of Minho, Portugal) Manuel Filipe Santos (University of Minho, Portugal) and Filipe Portela (University of Minho, Portugal)
Medical Information Science Reference • copyright 2018 • 286pp • H/C (ISBN: 9781522528517) • US $245.00 (our price)

Innovative Research in Thermal Imaging for Biology and Medicine
Ricardo Vardasca (University of Porto, Portugal) and Joaquim Gabriel Mendes (University of Porto, Portugal)
Medical Information Science Reference • copyright 2017 • 340pp • H/C (ISBN: 9781522520726) • US $210.00 (our price)

Internet of Things and Advanced Application in Healthcare
Catarina I. Reis (Polytechnic Institute of Leiria, Portugal) and Marisa da Silva Maximiano (Polytechnic Institute of Leiria, Portugal)
Medical Information Science Reference • copyright 2017 • 349pp • H/C (ISBN: 9781522518204) • US $210.00 (our price)

Integrating Biologically-Inspired Nanotechnology into Medical Practice
B.K. Nayak (K.M. Centre for Post Graduate Studies, India) Anima Nanda (Sathyabama University, India) and M. Amin Bhat (Sathyabama University, India)
Medical Information Science Reference • copyright 2017 • 394pp • H/C (ISBN: 9781522506102) • US $190.00 (our price)

Recent Advances in Drug Delivery Technology
Raj K. Keservani (Rajiv Gandhi Proudyogiki Vishwavidyalaya, India) Anil K. Sharma (Delhi Institute of Pharmaceutical Sciences and Research, India) and Rajesh Kumar Kesharwani (National Institute of Technology, Warangal, India)
Medical Information Science Reference • copyright 2017 • 510pp • H/C (ISBN: 9781522507543) • US $215.00 (our price)

IGI Global
DISSEMINATOR OF KNOWLEDGE

701 East Chocolate Avenue, Hershey, PA 17033, USA
Tel: 717-533-8845 x100 • Fax: 717-533-8661
E-Mail: cust@igi-global.com • www.igi-global.com

Editorial Advisory Board

Table of Contents

Preface xvi

Chapter 1
Enzymes and Dairy Products: Focus on Functional Products 1
Gabriela Piccolo Maitan-Alfenas, Federal University of Mato Grosso, Brazil
Sabrina Neves Casarotti, Federal University of Mato Grosso, Brazil

Chapter 2
Research Tools and Methods for the Analysis of Microbiota in Dairy Products 23
Sylvia Klaubauf, Dutch DNA Biotech, The Netherlands
Frank J. J. Segers, Westerdijk Fungal Biodiversity Institute, The Netherlands

Chapter 3
Functional Starter Cultures for Fermented Dairy Products 54
Sarang Dilip Pophaly, College of Dairy Science and Food Technology, India
Manorama Chauhan, College of Dairy Science and Food Technology, India
Vaibhao Lule, College of Dairy Technology, India
Poonam Sarang, National Dairy Research Institute, India
Jitesh Tarak, College of Dairy Science and Food Technology, India
Kiran Thakur, Hefei University of Technology, China
Sudhir Kumar Tomar, National Dairy Research Institute, India

Chapter 4
Biopreservatives for Improved Shelf-Life and Safety of Dairy Products: Biopreservatives for Dairy Products 69
Tejinder Pal Singh, College of Dairy Science and Food Technology, India
Sarang Dilip Pophaly, College of Dairy Science and Food Technology, India
Ruby Siwach, College of Dairy Science and Food Technology, India

Chapter 5
Aerobic Respiration in Lactic Acid Bacteria: Current and Future Applications in Dairy Starter Culture ... 87
Sarang Dilip Pophaly, College of Dairy Science and Food Technology, India
Manorama Chauhan, College of Dairy Science and Food Technology, India
Jitesh Tarak, College of Dairy Science and Food Technology, India
Shekhar Banala Bashetty, College of Dairy Science and Food Technology, India
Tejinder Pal Singh, College of Dairy Science and Technology, India
Sudhir Kumar Tomar, National Dairy Research Institute, India

Chapter 6
The Effects of Probiotic Cultures in Functional Foods: Technological Aspects of Probiotics ... 101
Gulden Basyigit Kilic, Mehmet Akif Ersoy University, Turkey

Chapter 7
Genomics Technologies for Enhanced Understanding of Robustness of LAB Starter Cultures ... 122
Annereinou R. Dijkstra, NIZO Food Research BV, The Netherlands
Peter A. Bron, NIZO Food Research BV, The Netherlands

Chapter 8
Probiotic Microorganisms and Encapsulation Method Approaches ... 132
Seydi Yıkmış, Namık Kemal University, Turkey
Harun Aksu, Istanbul University, Turkey
Mehmet Alpaslan, Namık Kemal University, Turkey
Osman Şimşek, Namık Kemal University, Turkey

Chapter 9
Transglutaminase Applications in Dairy Technology ... 152
H. Ceren Akal, Ankara University, Turkey
Celalettin Koçak, Ankara University, Turkey
H. Barbaros Özer, Ankara University, Turkey

Chapter 10
Role of Microbial Cultures and Enzymes During Cheese Production and Ripening ... 182
Şebnem Öztürkoğlu Budak, Ankara University, Turkey
Celalettin Koçak, Ankara University, Turkey
Peter A. Bron, NIZO Food Research BV, The Netherlands
Ronald P. de Vries, Westerdijk Fungal Biodiversity Institute, The Netherlands & Utrecht University, The Netherlands

Chapter 11
Microbial Non-Coagulant Enzymes Used in Cheese Making ... 204
Ekaterini Moschopoulou, Agricultural University of Athens, Greece

Chapter 12
Microbial Production of Recombinant Rennet: Recent Developments 222
Zumrut Begum Ogel, Konya Food and Agriculture University, Turkey

Chapter 13
Adjunct Cultures in Cheese Technology 234
Ayşe Gürsoy, Ankara University, Turkey
Nazlı Türkmen, Ankara University, Turkey

Chapter 14
Filamentous Fungi in Cheese Production 257
Banu Metin, Istanbul Sabahattin Zaim University, Turkey

Chapter 15
Potential Probiotic Microorganisms in Kefir 276
H. Ceren Akal, Ankara University, Turkey
Şebnem Öztürkoğlu Budak, Ankara University, Turkey
Atila Yetisemiyen, Ankara University, Turkey

Chapter 16
The Effects of Probiotic Cultures on Quality Characteristics of Ice Cream 297
Nihat Akın, Selçuk University, Turkey
Hale İnci Öztürk, Selçuk University, Turkey

Compilation of References 316

About the Contributors 404

Index 411

Detailed Table of Contents

Preface xvi

Chapter 1
Enzymes and Dairy Products: Focus on Functional Products 1
Gabriela Piccolo Maitan-Alfenas, Federal University of Mato Grosso, Brazil
Sabrina Neves Casarotti, Federal University of Mato Grosso, Brazil

The use of enzymes in food production, including dairy products, is below expected due the high costs associated with their production and purification. Microbial enzymes have great potential for industrial applications since they can be produced via large-scale fermentation and they are easily expressed by cloning in well-known cultivated microorganisms. The combination of different procedures such as over-expression techniques and the use of low costs induction sources has resulted in the production of enzymes to be used in high added-value dairy products. The addition of glucose oxidase to probiotic yogurts has been indicated as an alternative to the maintenance of probiotic functionality. Bile salt hydrolase contributes to prevention of hypercholesterolemia which is interesting to produce new functional dairy products. This chapter discusses enzyme sources and their relevance in dairy products, the production of enzymes using cloning and super-expression techniques, as well as enzymes related to functional dairy products.

Chapter 2
Research Tools and Methods for the Analysis of Microbiota in Dairy Products 23
Sylvia Klaubauf, Dutch DNA Biotech, The Netherlands
Frank J. J. Segers, Westerdijk Fungal Biodiversity Institute, The Netherlands

Microorganisms play important roles in dairy products. They can be a source of spoilage, or they promote health or cause diseases. In dairy fermentations, microorganisms are of great importance due to their function as starter cultures and during the production process of dairy products. In order to better understand and improve the process, it is essential to identify the species that are involved and to monitor the development of microbial communities. There are several different approaches for the detection and analysis of the microbiota. The methods can be culture dependent and, for example, make use of morphological and physiological characters or DNA sequencing. Culture-independent methods include direct PCR and qPCR, but also electrophoresis-based profiling techniques as well as metagenomics. Descriptions of relevant methods are provided and their applications are discussed in this chapter.

Chapter 3
Functional Starter Cultures for Fermented Dairy Products 54
Sarang Dilip Pophaly, College of Dairy Science and Food Technology, India
Manorama Chauhan, College of Dairy Science and Food Technology, India
Vaibhao Lule, College of Dairy Technology, India
Poonam Sarang, National Dairy Research Institute, India
Jitesh Tarak, College of Dairy Science and Food Technology, India
Kiran Thakur, Hefei University of Technology, China
Sudhir Kumar Tomar, National Dairy Research Institute, India

Fermented dairy products are known for their high nutritional and therapeutic value and are also having excellent sensory characteristics, which make them popular throughout the world. Most of the characteristics of fermented products can be attributed to the activity of their starter culture. The starter microorganisms, in the course of their metabolism, elaborate various compounds and enzymes, which impart desirable properties to the product. Lactic acid bacteria (LAB) are the most common starter bacterial group used extensively in fermentation of dairy products. Lactic acid bacteria synthesize different compounds belonging to classes of organic acids, gases, bacteriocins, flavoring agents, biothickners, nutrients, etc. Molecules within each class vary for each species and even within strains. The functionalities from such novel starter cultures include improved nutritional value, sensory characteristics, and therapeutic benefits. These starter cultures are thus important for development of functional fermented and probiotic products. This chapter explores functional starter cultures for fermented dairy products.

Chapter 4
Biopreservatives for Improved Shelf-Life and Safety of Dairy Products: Biopreservatives for Dairy Products 69
Tejinder Pal Singh, College of Dairy Science and Food Technology, India
Sarang Dilip Pophaly, College of Dairy Science and Food Technology, India
Ruby Siwach, College of Dairy Science and Food Technology, India

Globally, there is an increasing demand for minimally processed, easily prepared, and ready-to-eat fresh food, globalization of food trade, and distribution from centralized processing which pose major challenges for food safety and quality because perishable food may get contaminated with undesirable microorganisms. Food spoilage adversely affects the economy and also erodes the consumer's confidence. On other hand, food-borne illness leads to loss of earnings and productivity, unemployment and litigation, and weakens trade and tourism. Another challenge for the food producers is to produce less stable foods by processes that confer less harm to the detrimental microflora. A challenge for food producers is to develop products with a sufficiently longer shelf-life and at a competitive price. This brings them to the most promising approach to this end, the so-called biopreservation. This chapter provides a scientific background, functionality, as well as food applications and further commercial aspects of biopreservatives derived from microbial sources.

Chapter 5
Aerobic Respiration in Lactic Acid Bacteria: Current and Future Applications in Dairy Starter Culture............ 87
Sarang Dilip Pophaly, College of Dairy Science and Food Technology, India
Manorama Chauhan, College of Dairy Science and Food Technology, India
Jitesh Tarak, College of Dairy Science and Food Technology, India
Shekhar Banala Bashetty, College of Dairy Science and Food Technology, India
Tejinder Pal Singh, College of Dairy Science and Technology, India
Sudhir Kumar Tomar, National Dairy Research Institute, India

Lactic acid bacteria (LAB) are used as food-grade microorganisms for production of a variety of fermented milk products. They are also the most common probiotic organisms used for making functional foods. Lactic acid bacteria are well known for their fermentative metabolism wherein they convert simple carbohydrates to organic acids and other end products. Fermentation helps the bacteria to generate ATP required for various cellular activities via substrate level phosphorylation reaction. Fermentation results in incomplete oxidation of substrate and hence is an inefficient process with a low ATP yield. However, some LAB are genetically capable of activating an auxiliary respiratory metabolism in which a quinol oxidase serves as the final electron acceptor and high ATP production is achieved due to oxidative phosphorylation. The respiratory process is associated with high biomass production of LAB and more robust bacterial cells, which are essentially required for manufacture of high viability starter culture. This chapter explores LAB's current and future applications in dairy starter cultures.

Chapter 6
The Effects of Probiotic Cultures in Functional Foods: Technological Aspects of Probiotics........... 101
Gulden Basyigit Kilic, Mehmet Akif Ersoy University, Turkey

Functional foods are an important part of an overall healthy lifestyle that includes a balanced diet and physical activity. The consumption of probiotic foods has many benefits. Dairy products that contain probiotic bacteria are those that are produced with various fermentation methods, especially lactic acid fermentation, by using starter cultures and those that have various textures and aromas. Fermented dairy probiotic products are popular due to their differences in taste and their favourable physiological effects. However, recent upsurge in interest of consumers towards dairy alternatives has opened up new research areas for developing non-dairy probiotic products. Different substrates such as cereals, fruit juices, vegetables can be used utilized for delivering these beneficial microorganisms. This chapter provides an insight on the recent research/developments about selection criteria of bacteria as probiotics and in the field of technological properties of probiotics.

Chapter 7
Genomics Technologies for Enhanced Understanding of Robustness of LAB Starter Cultures........ 122
Annereinou R. Dijkstra, NIZO Food Research BV, The Netherlands
Peter A. Bron, NIZO Food Research BV, The Netherlands

During production and application, starter culture strains encounter several stresses. To ensure an adequate contribution to the fermentation process, starter culture strains should remain viable, justifying the increasing industrial efforts to unravel robustness characteristics of LAB starter culture strains. Nowadays, large numbers of genome sequences are publicly available, which enables the employment of several genomics technologies to increase our understanding of robustness. This aids the improvement of currently applied LAB starter culture strains and supports the industrial application of novel strains with specific desirable traits but currently inadequate robustness characteristics. This chapter explores genomics technologies for enhanced understanding of robustness of LAB starter cultures.

Chapter 8
Probiotic Microorganisms and Encapsulation Method Approaches 132
Seydi Yıkmış, Namık Kemal University, Turkey
Harun Aksu, Istanbul University, Turkey
Mehmet Alpaslan, Namık Kemal University, Turkey
Osman Şimşek, Namık Kemal University, Turkey

Nowadays, interest in probiotics, which are useful and necessary for healthy life, is rapidly increasing, and studies on the beneficial effects of probiotics on human health continue intensely. Every year, increasing efforts to prevent cancer, which has been anticipated, has increased the interest in probiotics and therefore synbiotics. Encapsulation methods are one of the most important protection methods currently used to ensure the viability of probiotics and their effectiveness. Especially milk and dairy products are used for many purposes such as increasing the shelf life, increasing the nutritional value, providing digestibility, shortening the ripening period, improving taste and aroma substances. The use of the microencapsulation technique alone can improve probiotic vitality. Combining microencapsulation with various food processing technologies is thought to help improve the vitality of probiotics in production and storage. In this chapter, probiotic microorganisms and encapsulation applications are explored.

Chapter 9
Transglutaminase Applications in Dairy Technology 152
H. Ceren Akal, Ankara University, Turkey
Celalettin Koçak, Ankara University, Turkey
H. Barbaros Özer, Ankara University, Turkey

Consumers' expectations from a dairy product have changed dramatically during the last two decades. People are now more eager to purchase more nutritious dairy foods with improved sensory characteristics. Dairy industry has made many efforts to meet such expectations and numerious production strategies and alternatives have been developed over the years including non-thermal processing, membrane applications, enzymatic modifications of milk components, and so on. Among these novel approaches, transglutaminase (TG)-mediated modifications of milk proteins have become fairly popular and such modifications in dairy proteins offer many advantages to the dairy industry. Since late 1980s, a great number of researches have been done on TG applications in milk and dairy products. Especially, milk proteins-based edible films and gels from milk treated with TG have found many application fields at industrial level. This chapter reviews the characteristics of microbial-origin TG as well as its mode of action and recent developments in TG applications in dairy technology.

Chapter 10
Role of Microbial Cultures and Enzymes During Cheese Production and Ripening 182
Şebnem Öztürkoğlu Budak, Ankara University, Turkey
Celalettin Koçak, Ankara University, Turkey
Peter A. Bron, NIZO Food Research BV, The Netherlands
Ronald P. de Vries, Westerdijk Fungal Biodiversity Institute, The Netherlands & Utrecht University, The Netherlands

Many different kinds of cultures, enzymes, and methods are used during the production and ripening of a variety of cheese types. In this chapter, the importance, types, and applications of microbial cultures during cheese production are discussed. Moreover, an overview of the important role of enzymatic systems, either derived from these cultures or directly added to the milk fermentation, is presented. The main biochemical events including glycolysis, lipolysis, and proteolysis during cheese ripening are explained, focusing on their end products, which contribute to the development of the overall aroma of cheese.

Chapter 11
Microbial Non-Coagulant Enzymes Used in Cheese Making 204
Ekaterini Moschopoulou, Agricultural University of Athens, Greece

In this chapter, the use of microbial non-coagulant proteases, microbial lipases, and microbial transglutaminase in the cheese making procedure is discussed. Microbial proteases and lipases have been used for over 30 years to accelerate cheese ripening and consequently to enhance the cheese flavor development by increasing proteolysis and lipolysis level in a shorter time. They are commercially produced by bacteria and fungi species. Transglutaminase is a relative new enzyme, which catalyzes the cross-linking of peptide bonds and helps to improve the cheese texture and to increase the cheese yield. Today, cheeses from almost all cheese categories are produced using these enzymes.

Chapter 12
Microbial Production of Recombinant Rennet: Recent Developments 222
Zumrut Begum Ogel, Konya Food and Agriculture University, Turkey

Rennet, traditionally obtained from calves, is non-vegeterian and unethical due to the slaughter of unweaned animals. Chymosin is highly specific to the Phe105-Met106 bond of κ-casein and has low proteolytic activity. Microbial aspartic proteases can partly replace chymosin. However, recombinant DNA technology has allowed chymosin itself to be produced by bacteria, yeast, and molds. Not only rennet from calf, but from animals like goat kid, lamb, buffalo, camel, and others can be used in cheesemaking. Chymosins of these animals can be cloned and successfully expressed in microorganisms and can be employed in the production of novel as well as traditional cheese products from the milk of camel, goat, and even horse and donkey. This chapter outlines the recombinant DNA techniques applied over the past few years to improve the microbial production of recombinant rennet, from animals and plants.

Chapter 13
Adjunct Cultures in Cheese Technology.. 234
Ayşe Gürsoy, Ankara University, Turkey
Nazlı Türkmen, Ankara University, Turkey

Cheese ripening involves highly complex biochemical events. Coagulant enzymes as well as the utilized starters play an important role in these events. Two types of starters are used: primary and secondary. The main role of the primary culture, which consists of lactic acid bacteria, is to carry out lactic production during fermentation. They contribute to proteolysis and limited flavor formation with the enzymes they possess. Secondary or adjunct cultures are used to develop the texture and to accelerate the ripening. During the selection of this type of culture, enzyme profiles (i.e., proteolytic and lipolytic activities and their autolyse levels) in cheese are the primary factors to be taken into consideration. Apart from these, the other factors are their positive effects on health, availability, and economy. Adjunct cultures include yeast, molds, and bacteria. Some of the heterofermentative lactobacilli species, in particular weakened strains, are used as adjunct cultures in order to accelerate the ripening and shorten the ripening time in fat-reduced and low-fat cheeses. This chapter explores adjunct cultures in cheese technology.

Chapter 14
Filamentous Fungi in Cheese Production 257
Banu Metin, Istanbul Sabahattin Zaim University, Turkey

Filamentous fungi play important roles in the production of a variety of cheeses. The most famous are the blue cheeses, such as Roquefort or Gorgonzola, in which *Penicillium roqueforti* is the principal mold, and the moldy soft cheeses, such as Camembert or Brie, in which production involves *Penicillium camemberti.* There are also other filamentous fungi associated with certain types of cheeses, such as *Mucor* spp., *Trichothecium roseum*, and *Fusarium domesticum* in Saint Nectaire, *Sporendonema casei* in Cantal, Salers and Rodez cheeses, *Scopulariopsis* species in various French and Austrian cheeses, and *Mucor mucedo* and *Mucor racemosus* in the traditional Norwegian cheese Gamalost. These fungi are either inoculated on the cheese as a starter culture or stand out in mixed cultures during spontaneous fermentation. This chapter reviews the filamentous fungi used to produce different kinds of cheeses in terms of taxonomy, physiology, ecology, and mycotoxins, and the microbiological or biochemical effects of these fungi on cheese production.

Chapter 15
Potential Probiotic Microorganisms in Kefir 276
H. Ceren Akal, Ankara University, Turkey
Şebnem Öztürkoğlu Budak, Ankara University, Turkey
Atila Yetisemiyen, Ankara University, Turkey

Probiotic microorganisms are defined as living microorganisms that provide health benefits on the host when administered in adequate amounts. The benefits include improvement of microbial balance immune system and oral health, provision of cholesterol-lowering effect, and antimicrobial activity against a wide variety of bacteria and some fungi. Kefir microbiota contains active living microorganisms. Many researches were carried out that potential probiotic bacteria such as *Lactobacillus acidophilus, Lactobacillus plantarum, Lactobacillus kefir, Lactobacillus kefiranofaciens, Leuconostoc mesenteroides*, or yeasts like microorganisms such as *Saccharomyces cerevisiae, Kluyveromyces lactis,* and *Kluyveromyces marxianus* were isolated from kefir grains. This chapter presents the data both on the probiotic bacteria isolated from kefir grains or kefir and the probiotic properties of kefir produced with these microorganisms.

Chapter 16
The Effects of Probiotic Cultures on Quality Characteristics of Ice Cream 297
Nihat Akın, Selçuk University, Turkey
Hale İnci Öztürk, Selçuk University, Turkey

The manufacturing design and probiotic strain that are used in ice cream production affect sensorial, nutritional, and physicochemical properties of ice cream. Therefore, it is possible to produce better quality products or to develop existing methods by considering the effects of probiotic cultures on these properties of ice cream. The International Dairy Federation has recommended that probiotic products contain at least 10^7 CFU/g of viable probiotic strain before consumption, so that the therapeutic value of a probiotic product can be associated with the viability of these bacteria. The nutritional value of probiotic ice cream is related not only to the dietary components it contains, but also to the maintaining of the viability and the activity of the probiotic bacteria. This chapter provides information on the viability of probiotic strains during ice cream processing and storage. Furthermore, the effects of probiotic cultures on sensory and physicochemical properties of ice cream are also discussed.

Compilation of References 316

About the Contributors 404

Index 411

Preface

Microbial food cultures are living bacteria, yeasts or moulds that are used in food production. These cultures carry out the fermentation process in foodstuffs which helps to improve sensory and textural properties of the products. More than 260 different species of microbial food culture are identified and described for their beneficial use in fermented food products globally. Microbial enzymes are usually produced from microorganims during the fermentation processes and provide beneficial effects during dairy applications. These are mainly, milk-clotting enzymes or rennets, recombinant fungal and bacterial rennets for cheese manufacture, and fungal lactases for the manufacture of some milk products with reduced content of lactose. They are also used in cheese ripening, such as proteinases for accelerated ripening period for good flavour and textural development, proteases to reduce allergic properties of cow milk products for infants, and lipases for the development of lipolytic flavours in speciality cheeses. Microbial enzymes are preferred to chemically obtained enzymes for both their health effects and also because of their economical point of view. As an example, utilization of pure microbial strain as an adjunct in cheese production could a more feasible alternative than usage of a purified enzyme, due to the high cost of enzyme isolation and purification.

This book contains 16 chapters which were written by authors who are expert in their respective fields. Nine chapters out of 16 include knowledge about starter cultures and enzymes used in dairy technology, while five chapters deal with microbial cultures and enzymes used in cheese technology and the other two chapters address to probiotic organisms used in ice cream and kefir technology.

The relationship between dairy products and enzymes are outlined in Chapter 1. Chapter 2 includes novel methods and research tools which could be applied in determination of microbiota of dairy products. Chapter 3 deals with starter cultures, which bring functional properties to fermented dairy products, by virtue of different metabolites they produce. Biopreservatives, which are used for safety and prolonged shelf life of dairy products, are discussed in Chapter 4. Respiration mechanisms of metabolites produced by lactic acid bacteria which have an important role in production of dairy products are depicted in Chapter 5, while effects of probiotic cultures on functional foods are explained in detail, in Chapter 6. Genomic technology are discussed in Chapter 7 in order to bring a better understanding about robustness of lactic acid bacteria which is the most important factor for them to keep their vitality. Probiotic bacteria which have important benefits for human health and of which a new benefit is discovered day by day, have an undeniable importance for dairy industry. Different methods are used for bring probiotic characteristics in a food, or enhancing its probiotic features of a food material. Application of encapsulation method on probiotic bacteria, which is the most widely used one among mentioned methods is explained in Chapter 8. Detailed information is provided in Chapter 9 on transglutaminase enzyme practises in dairy technology. Cheese is the most remarkable product among dairy products in terms of culture and enzyme activities. Differences which could be applied during production and ripening are able to bring about major changes in cheese characteristics. These differences enable the production of different types of cheeses and development of quality characteristics of cheese. The cultures and enzymes which have a major importance in cheese technology are deal with Chapters 10 to 14. Changes in cheese production and ripening which stem from cultures and enzymes are depicted in Chapter 10. Information about microbial non-coagulant enzymes, which are used in cheese production are provided in Chapter 11. Rennet has a great importance in enzyme technology, which forms a basis for cheese production. Recombinant rennet technology has been particularly emphasized in recent years. As such, microbial production of recombinant rennet is dealt with in Chapter 12. Adjunct cultures and filamentous fungi were analysed in terms of cheese technology in Chapter 13 and 14, respectively. Potential probiotic microorganisms contained in kefir is explained in Chapter 15 and effects of probiotic culture usage on product quality characteristics in ice cream are discussed in Chapter 16.

This compiled book includes a lot of up-to-date information about cultures and enzymes which are used or could be potentially used in dairy products, from different perspectives. Not only scientists but also culture or enzyme producers in dairy industry can also benefit from this information. This book is thought to be beneficial for both academic research and dairy sector by virtue of inclusion of results of recent studies and newly discovered data in a systematic way.

Chapter 1
Enzymes and Dairy Products:
Focus on Functional Products

Gabriela Piccolo Maitan-Alfenas
Federal University of Mato Grosso, Brazil

Sabrina Neves Casarotti
Federal University of Mato Grosso, Brazil

ABSTRACT

The use of enzymes in food production, including dairy products, is below expected due the high costs associated with their production and purification. Microbial enzymes have great potential for industrial applications since they can be produced via large-scale fermentation and they are easily expressed by cloning in well-known cultivated microorganisms. The combination of different procedures such as over-expression techniques and the use of low costs induction sources has resulted in the production of enzymes to be used in high added-value dairy products. The addition of glucose oxidase to probiotic yogurts has been indicated as an alternative to the maintenance of probiotic functionality. Bile salt hydrolase contributes to prevention of hypercholesterolemia which is interesting to produce new functional dairy products. This chapter discusses enzyme sources and their relevance in dairy products, the production of enzymes using cloning and super-expression techniques, as well as enzymes related to functional dairy products.

INTRODUCTION

Enzymes are usually proteins that occur in all living organisms with biological catalytic activity, and they are used in several industrial applications, including food processing. As a catalyst, an enzyme works to accelerate biological and biochemical reactions necessary to support life by lowering the activation energy involved in the reaction (Olempska-Beer et al., 2006). These macromolecules appeared for the first time more than 5,000 years ago with the storage of milk in an animal's stomach that contained enzymes called rennet, which turn milk into cheese (Novozymes, 2017).

Enzymes have several applications in the food industry because of their high specificity, effectiveness in small quantities and catalytic activity under mild temperature and pH conditions. Their high

DOI: 10.4018/978-1-5225-5363-2.ch001

specificity minimizes unwanted byproducts formation, since they catalyze the conversion of only a single substrate or a small group of closely related compounds or a specific bond. Furthermore, because they are active in small amounts, low usage levels make enzymes economical and practical for commercial application. Another advantage is related to their capacity to actuate under mild pH and temperature conditions, which reduce energy costs (James and Simpson, 1996). However, the use of enzymes in the food industry, including dairy products, is still below expected levels, mainly due the high costs involved in their production and purification, which will be further discussed.

Enzymes can be produced by animals, plants and microorganisms, which mean that they originate from natural sources and therefore are desired for food process. Microbial sources have great potential for industrial applications since they can be produced by fermentation on a large scale. Furthermore, they are easily expressed by cloning in well-known cultivated microorganisms (Gandhi, 1997). The use of recombinant DNA technology has also made it possible to manufacture novel enzymes suitable for specific food processing conditions and to increase enzyme production, reducing processing costs. Nevertheless, there are also immobilized enzyme systems which are used to overcome inhibition problems and to lower the costs of enzymes, since the purification step can be omitted.

The application of enzymes in dairy products is well established, where proteases, lipases, esterases, lactase and catalase are the most common enzymes found in these products. According to Afroz et al. (2015), rennet, a mixture of chymosin (or rennin) and pepsin produced by animal and microbial sources, are used for milk coagulation in the first stage of cheese production. Proteases are used to accelerate cheese ripening, to modify functional properties and also to promote the alteration of milk proteins to reduce the allergenic properties of cow's milk products for babies. Lipases are used for cheese ripening and contribute to the development of lipolytic flavors. Lactase or β-galactosidase (from microorganisms) is used to hydrolyze lactose to glucose and galactose so the product is more digestible by lactose intolerant individuals, and also to improve the solubility and sweetness of dairy products. Peptidases are used to break down bitter peptides that appear in ripened cheese and thus assist in providing the traditional cheese flavor.

It is certain that increased concern regarding energy scarcity and required protection of the environment may lead to deeper knowledge on enzymatic processes, including those for food processing. Therefore, studies on enzymology are currently being conducted to make improvements to dairy products. The aim is to find enzymes with technological advantages that do not noticeably alter the sensory characteristics of the dairy products. Another tendency is to use enzymes with functional properties that permit the production of high added-value dairy products, such as glucose oxidase and bile salt hydrolase (BSH).

BACKGROUND

Production of Enzymes

Microorganisms play an essential role in the production of enzymes for food processing and different strategies are used for the prospection of novel and/or more efficient enzymes, especially those destined for dairy products. In 1874, the scientist Christian Hansen extracted rennin (chymosin) from calves' stomachs for use in cheese manufacturing and this corresponds to the beginning of the industrial production of enzymes for application in food products (Nielsen et al., 1994). Nowadays, chymosin is widely produced by microorganisms that contain the bovine pro-chymosin gene introduced via the recombinant deoxyribonucleic acid (rDNA) technique (Olempska-Beer et al., 2006).

For production of an enzyme of interest the microorganism is first selected and cultivated in a medium formulated to induce high enzyme production yields. Industrial fermentations are normally carried out under liquid conditions, and to a lesser extent on solid media. Enzyme production is usually developed in batch fermentation and in fewer applications via a continuous process. The last step is related to recovery and purification of the enzyme. The enzyme of interest may be secreted into the culture medium (extracellular enzyme) or may remain inside the cells (intracellular enzyme), where in the latter a procedure for cell disruption is necessary to liberate the enzyme.

Enzymes used in food production are sold as enzyme preparations, which contain the enzyme of interest, and substances that facilitate its incorporation into food, such as preservatives and stabilizers which are suitable and allowed to be used in foods. Other metabolites from the producer microorganism and/or residues of raw materials used in the culture media and constituents that appeared during manipulation and purification of the enzyme can also be found in these enzymatic preparations. In all situations, these materials must present appropriate purities consistent with good manufacturing practices, i.e., they must be of food grade and comply with the regulatory policy standards where the enzyme is applied (Olempska-Beer et al., 2006; Pariza and Johnson, 2001).

According to Gaynor (2006), enzyme preparations must have the generally recognized as safe (GRAS) status, which is valid for native and recombinant enzymes. The GRAS affirmation is obtained based on a history of safe use in foods before 1958 or on scientific techniques, and it is given by the FDA (Food and Drug Administration) or determined independently by qualified experts. Therefore, FDA regulations in Title 21 of the Code of Federal Regulations (21 CFR) do not include all GRAS substances and the specific uses described in the GRAS regulations may not be comprehensive for the listed ingredients. Within European Union, enzymes should have Qualified Presumption of Safety (QPS) status given by EFSA (European Food Safety Authority).

For prospecting new microbial enzymes, some alternatives are available. The first consists of bioprospecting microorganisms in specific environmental niches with posterior investigation of their ability to hydrolyze substrates of interest. Then, a screening of the best candidates that possess interesting enzymes is performed to make the final decision. Another strategy is application of the metagenomic technique, which is extensively utilized for the genetic composition analysis of microorganism mixtures using probes or group-specific primers seeking new enzymes. A major problem is that this tool is unable to detect the metabolic potential of the microorganisms since the isolation and cultivation steps are not performed. A third alternative is comparative genomics of different microorganisms which is also commonly used for the screening of several candidates of interest in a short period of time. This strategy permits the genome analysis of a determined microorganism for evaluation of its proteome (Maitan-Alfenas et al., 2015).

To ensure lower enzyme production costs there are few alternatives, such as the use of low costs substrates for fermentation and enzyme immobilization. The utilization of chemically synthesized media for growing enzyme-producing microorganisms is costly and large scale production remains economically challenging for many types of enzymes. Therefore, the use of cheaper and renewable substrates from various sectors such as the agricultural (sugarcane bagasse, molasses, corn straw and corn cobs, wheat straw, plant oils, oil wastes, starchy substances, lactic whey), distillery wastes, animal fat and oil industries have been reported by different researchers (Banat et al., 2014).

The immobilization procedure dates back to 1960 and consists of confining the enzyme in a support to permit that it be reused in multiple operations. When the enzyme is immobilized, there is a reduction in the need for large enzyme quantities since they can be reused, thus reducing costs. Moreover, this technique increases the enzyme stability since free enzymes usually have poor stability in varying pH and high temperatures, and are difficult to recover and reuse. Other advantages are that there are no residual enzymes left in the product and less additional processing, such as the purification steps, which are often expensive and increase labor costs. When the enzyme is immobilized, the substrate can be continuously added and the product continuously removed, while the enzyme is maintained for many cycles until it becomes inactive (James and Simpson, 1996).

It is known that free enzymes are more active than immobilized enzymes since the support used for immobilization may interact with the enzyme. Even with some loss of activity, immobilization typically provides increased stability and reusability when subject to variable temperatures, pH and other factors since the interaction between the enzyme and the support can stabilize the enzyme configuration and reduce leakage (Cao et al., 2012). Furthermore, immobilization has proven to be a very useful tool to decrease inhibition caused by substrates, reaction products or components of the bulk medium via different mechanisms (Mateo et al., 2007).

In food processing, special attention is required when the enzyme is immobilized, mainly with respect to toxicology problems. The material used for immobilization needs to be innocuous to consumers. Solid supports are normally used for enzyme immobilization, including polysaccharides, inorganic supports, fibrous proteins, synthetic polymers, hydrogels and hollow fibers, and the enzymes can be constrained or held to these supports by adsorption, covalent attachment, crosslinking, entrapment, microencapsulation or by other methods (Olson and Korus, 1976).

To give an example of the importance of immobilized enzymes in dairy products, Table 1 shows different immobilized β-galactosidases, which are used for hydrolysis of lactose to glucose and galactose.

Another possibility is immobilization of the whole cells instead of isolated enzymes. In this case, the enzyme environment is maintained and enzyme degradation is reduced. Sophisticated transformations can also be performed, mainly those requiring several enzymes and cofactor regeneration. However, in this situation the microorganisms need to be considered GRAS to be applied in food processing, and the diffusion of substrate and products through the cell wall is more complex which may result in undesirable reactions, causing specificity losses (James and Simpson, 1996).

Table 1. Immobilization of microbial β-galactosidases in different matrices

Enzyme Source	Immobilization Matrix	Reference
Aspergillus niger	Porous ceramic monolith	Papayannukos et al. (1993)
Kluyveromyces lactis	Thiolsulfinate/thiolsulfonate	Ovsejevi et al. (1998)
Aspergillus oryzae	Chitosan and nylon supports	Portaccio et al. (1998)
Thermus sp. T2 (recombinant)	Sepabeads-epoxy supports	Pessela et al. (2003)
Geobacillus stearothermophilus (recombinant)	Chitosan with Tris(hydroxymethyl)phosphine and glutaraldehyde	Chen et al. (2009)
Aspergillus oryzae	Cellulose acetate-polymethylmethacrylate membrane	Ansari et al. (2014)
Aspergillus oryzae	Agarose	Guerrero et al. (2017)

The production and utilization of new or more efficient enzymes for food processing is certainly beneficial for both food manufacturers and consumers. Therefore, the search for low costs enzymes is continuous so as to produce better products that are economically accessible.

Recombinant Enzymes: Cloning and Over-Expression

Recombinant DNA techniques play an important role with regards to new developments in the food enzymology field. It may always be used to improve enzymatic production, in homologous and heterologous systems, and several methods have been developed to increase recombinant protein production in microorganisms. Moreover, enzymes produced via biotechnology are identical to those found in nature and they do not produce by-products or off flavors in foods.

Homologous expression is related to over-expression of a gene in a system from where it originates, while heterologous expression is the expression of a gene in either a different species or cell type. Heterologous expression in yeast and fungi has many advantages over bacteria, including the production of soluble and correctly folded recombinant proteins that have undergone post-translational modifications, the capacity of growth at high cell densities and the ability to secrete reasonable protein quantities (Maitan-Alfenas et al., 2016).

Recombinant enzymes have a long history of use in foods. Most recombinant enzymes currently used in food are derived from well-characterized and easy cultivated microorganisms. Modern PCR (Polymerase Chain Reaction) techniques are used to introduce the specific DNA into the host organism to the sequence encoding the desired enzyme. Moreover, enzyme properties can be adapted to specific use conditions using modern genetic techniques. Site-specific or random mutagenesis can be used to introduce changes in the amino acid sequence of the enzyme. Site-specific mutagenesis is preferred when the tridimensional structure of the enzyme is known and the relationships between structure and enzyme properties have been elucidated. In recent years, a powerful approach for improving enzyme properties known as molecular or directed evolution has been developed. The FDA recently reviewed several GRAS notices on enzymes improved by using either one or a combination of modern genetic techniques (Olempska-Beer et al., 2006).

In a recombinant expression system, a gene that codes for an enzyme of interest is transferred to another microorganism (Figure 1). This is performed through an expression vector, which is a DNA plasmid that carries the expression cassette that will be inserted in the host after being constructed with the gene of interest. The expression cassette essentially contains a promoter, the gene encoding the desired enzyme, and a terminator. The promoter and the terminator are regulatory sequences that control the transcription of the gene of interest. In summary, the gene of interest is cut from the DNA of the producing microorganism and it is glued into an expression vector for expression in a new microorganism (identical or not to the one which originated the DNA, characterizing homologous and heterologous expression, respectively). Using sophisticated recombinant techniques, an expression plasmid contains the gene that codes for the enzyme of interest under the control of suitable expression signals and it is transferred into the host strain (cloning) that will enable high-yield production of the enzyme of interest (over-expression).

The expression vectors usually have a marker gene to select the growth of the transformants. In some situations, the vector which does not present the marker gene needs to be co-transformed with another vector, with the selectable marker gene. This is important for selection of the best transformation candidates. The transformation consists of production of the enzyme of interest by the host and there are

Figure 1. Summary of a recombinant expression system

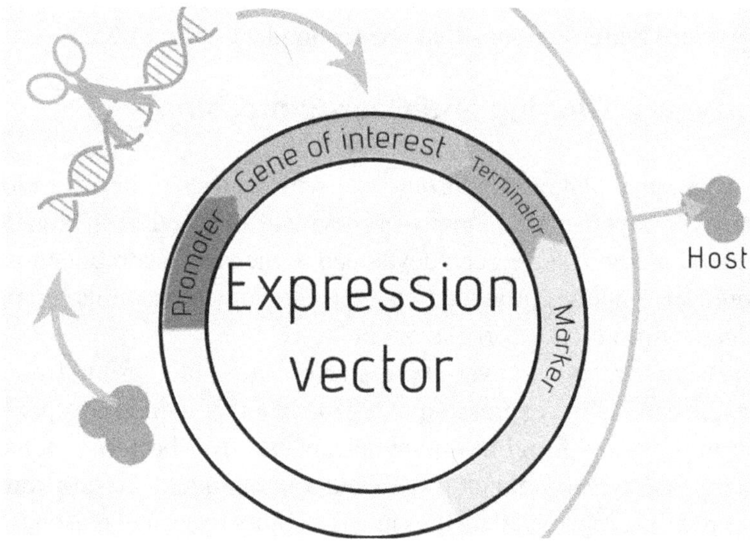

different techniques for transformation of microbial hosts. Because the gene that codes for the enzyme of interest is normally under control of an efficient promoter, there is an increase in the production of the enzyme, referred to as over-expression.

For the construction of a successful production strain, before the production of recombinant enzymes there is an extensive research and development phase. This involves the following stages: development of the host strain, construction of the expression vector, transformation of the host strain, identification of the best recombinant strain, additional improvements and characterization of the production strain (Olempska-Beer et al., 2006).

The host strains for food-processing enzymes are normally derived from a relatively small number of bacterial and fungal species such as *B. subtilis*, *B. licheniformis*, *A. niger* or *A. oryzae*. These microorganisms must be nonpathogenic and non-toxigenic, and are normally used as safe sources of native enzymes and present efficient growth under industrial conditions, meaning that they are well-adapted to large-scale industrial fermentations. They are also easily submitted to genetic manipulations and secrete large amounts of enzymes into the fermentation media, which facilitates purification steps. The wild-type strains of the hosts normally produce many extracellular enzymes that may be removed from the final enzyme preparation to avoid the catalysis of undesirable reaction. Another challenge is related to extracellular proteases, since they can degrade heterologous enzymes. To increase enzyme yields, researches have built protease-deficient host strains. Another concern regards the development of sporulation-deficient mutant strains, which further facilitates the enzymatic production process (Olempska-Beer et al., 2006).

Recombinant strains can be further improved using classical mutagenesis. For instance, fungal expression vectors, which are more complex, can integrate into the host genome at different loci and have several copies. Therefore, the transformation technique will yield transformants with different levels of the enzyme of interest. These transformants are subsequently grown under different conditions and a satisfactory transformant is identified when subjected to mutagenesis using either a chemical mutagen or radiation (UV or ionization). Subsequently, the population of mutants is screened with regards to enzyme yield to identify the best performer (Novozymes, 2017).

Recombinant β-galactosidases from different sources, expressed in *Escherichia coli* as the expression host, are shown on Table 2.

Several recombinant enzymes are used in dairy products and advances in molecular biology has enhanced enzyme yields. The cloning of genes that code enzymes of interest and their over-expression in host cells allows for obtaining food products with higher quality. Furthermore, mutagenesis techniques may produce enzymes with high performance and well-adapted to food processing conditions such as varying temperature and pH. It is very important to focus on the innocuity of the enzyme since it will directly affect the consumers.

Purification Steps

Because an enzyme is produced by a microorganism, it must be purified for application in food processing if it is not added in an immobilized form. For purifying an enzyme many properties are investigated including size, affinity and molecular charge, and the steps are normally tested by trial and error.

The most common procedures used for enzyme purification are chromatography, precipitation, ultrafiltration, centrifugation and dialysis. Although these conventional techniques are very useful, they normally require considerable effort and time, and the results may be incomplete with low yield (Cuatrecasas et al., 1968). Furthermore, they also have some limitations with respect to expensive instrumentation and trained operators (Cao et al., 2012). These factors contribute to increased costs of the process, especially when utilizing equipment such as HPLC (High Performance Liquid Chromatography) or FPLC (Fast Protein Liquid Chromatography) to reduce analysis time.

Chromatography, normally performed in columns, includes many principles of separation such as gel filtration, ion exchange, size exclusion, hydrophobic interaction and affinity biding. These principles can be combined for total purification of the enzyme of interest. Normally, a column is filled with a resin that interacts with the enzyme or non-enzyme components and separates the enzyme from the remaining media constituents to increase the purity.

To give an example of how columns work, if considering affinity chromatography the column is filled with a polymer or gel chemically linked to a specific competitive inhibitor of the enzyme. When the enzyme solution passes through this column, all proteins without affinity for the bound inhibitor pass directly, while the enzyme recognizes the inhibitor and is retarded in relation to its affinity constant. For

Table 2. Recombinant microbial β-galactosidases developed for dairy products

Enzyme Source	Expression Vector	Reference
Bifidobacterium infantis HL96	pET24(+) (Novagen)	Hung and Lee (2002)
Pseudoalteromonas sp. 22b	pUC18 (Stratagene, USA)	Cieśliński et al. (2005)
Sulfolobus solfataricus	pET-24a	Park et al. (2008)
Geobacillus stearothermophilus KVE39	pHWG509	Placier et al. (2009)
Paracoccus sp. 32d	pBAD/Myc-His A	Wierzbicka-Woś et al. (2011)
Pyrococcus furiosus	pET24a_(+)	Dong et al. (2014)
Paenibacillus barengoltzii	pMD19	Liu et al. (2017)

elution of the bounded enzyme, parameters such as salt concentration or pH are changed, or a competitive inhibitor is added and the enzyme is recovered (Cuatrecasas et al., 1968).

When recombinant enzymes are purified the process is usually easier, less expensive and less time consuming since affinity tags are inserted on the enzyme of interest. These are highly efficient tools for protein purification and allow for purification without knowledge on the biochemical properties of the enzyme. Therefore, the use of affinity tags is very common in several fields, including food processing. However, in some cases the presence of the affinity tag in the recombinant enzyme is unwanted or may represent a disadvantage, and some approaches have been developed to remove the affinity tag from the recombinant protein. Currently, more than 60% of the recombinant enzymes produced include a polyhistidine tag (his-tag). The big advantage of using affinity tags is that purification results in high yields and there is a decrease in the number of unit operations needed, resulting in reduced time and cost for purification (Arnau et al., 2006).

In all cases, if the enzyme is not immobilized the purification step is necessary for application of the enzyme in dairy products. Consequently, this is an essential procedure and researchers have concentrated efforts on improving enzyme purification techniques for producing high quality foods.

Production and Application of Bile Salt Hydrolase (BSH) and Glucose Oxidase (GOX) as Novel Functional Enzymes for the Dairy Industry

In recent decades consumers have shown more interest in products with beneficial health properties and the demand for such products has increased on a global scale. This demand has grown due to awareness of the relationship between diet and well-being, and the socio-economic alterations such as longer life expectancy, the rise of health care costs, the impact of chronic diseases on public health, and the universal desire for a better quality of life (Bimbo et al., 2017).

Sales of health-enhancing foods totaled approximately $168 billion in 2013 and are expected to exceed $300 billion by 2020 (Research and Markets, 2014). Food industries are attracted by the high profits derived from this expansion, and for this reason have been investing in the development of innovative functional products. Functional foods can be defined as "foods and food components that provide health benefits beyond basic nutrition" (Serafini et al., 2012).

A series of strategies to develop new and innovative functional foods have been undertaken, including the addition of probiotics, mainly *Lactobacillus* and *Bifidobacterium,* and use of a variety of ingredients including prebiotics, essential fatty acids, whey protein powder, fruits, cereals, and food by-products from different sources (okara, fruits, etc.). In this sense, the addition of enzymes or enzyme-producer microorganisms to obtain foods with greater health appeal is a possibility that can also be explored.

Although enzymes have been widely used in the field of dairy technology for many years, new studies are important to assess the feasibility of supplementing enzymes which can contribute to meet consumer demands for healthy and innovative foods. Therefore, the addition of such enzymes to dairy products may contribute to the growing market of functional foods. A list of some enzymes produced by microorganisms and their potential health functions in human body is shown on Table 3. This section focuses on the bile salt hydrolase (BSH) and glucose oxidase enzymes with regards to their characterization, importance, production and potential applications in functional dairy products.

Table 3. Enzymes produced by microorganisms and their functional properties

Enzymes	Main Function	Reference
Superoxide dismutase, glutathione, and catalase	Reduce the levels of reactive oxygen species, decrease certain metabolites and enzymes supposed to be involved in synthesis or activation of carcinogens, genotoxins and tumor promoters such as β-glucuronidase, β-glucosidase, nitrate reductase and ammonia	Kumar et al., 2010
Aldehyde dehydrogenase	Metabolizes acetaldehyde to acetate, resulting in reduced levels of acetaldehyde, a substance that can be toxic to human body	Konkit et al., 2016
Cell envelope-associated proteinases	Promote the first step of casein breakdown releasing bioactive peptides during milk fermentation	Brown et al., 2017

Bile Salt Hydrolase (BSH): Characterization and Beneficial Properties for Consumer Health

Bile has several functions, including emulsification of dietary fat, lipid absorption and cholesterol homeostasis. Bile also presents antimicrobial effects due to its ability to bind to bacterial membranes, causing damages. It contains cholesterol, phospholipids, water, biliverdin pigment and mainly bile salts, which correspond to 50% of the organic elements of bile (Patel et al., 2010; Bi et al., 2013).

Bile salts are amphipathic molecules synthesized in the liver from cholesterol in a process controlled by multiple enzymes. They are formed by bile acids conjugated with glycine and taurine, and act as natural ionic detergents. After conjugation, bile salts are stored in the gallbladder before secretion into the duodenum via the common bile duct. Bile acids conjugated with other amino acids, for example leucine or lysine, are rapidly hydrolyzed by pancreatic carboxypeptidases (Joyce et al., 2014).

The conjugation of bile acids enhances its solubility and decreases its passive absorption in the biliary tract and small intestine. As a result, there is a high intraluminal bile concentration which supports the excretion of lipids such as cholesterol and it is crucial for facilitating fat digestion and absorption (Begley et al., 2005). Although the majority of bile salts are reabsorbed in the intestine and rapidly sent back to hepatocytes, reconjugated and resecreted into bile (enterohepatic circulation) after digestion, a smaller portion (5%) is modified by microbial enzymes (Chand et al., 2017).

Besides being chemically modified by bacterial enzymes in the intestine, bile salts have an effect on the gut microbiota composition. They are involved in the connection between microbiota and the host, and as a result bile salts may influence the host metabolic pathways and the microbial community structure and diversity (Joyce et al., 2014; Staley et al., 2017).

The bile salt hydrolase enzyme (E.C.3.5.1.24) plays a significant role in the bile salts metabolism. Bile salt hydrolase enzyme (BSH) is produced by several microorganisms that inhabit the gastrointestinal tract of mammals (Bi et al., 2013). BSH belongs to the colloylglycine hydrolase family of enzymes and is classified as an N-terminal nucleophilic hydrolase (Ntn) with an N-terminal cysteine residue (Patel et al., 2010). BSHs are characterized as intracellular enzymes that show insensitivity to oxygen, have an optimum pH usually between pH 5 and 6 and activity dependent on biomass production (Begley et al., 2005).

In the terminal illeum, this enzyme catalyzes the hydrolysis of bile acids conjugated with glycine or taurine in amino acid residues and unconjugated bile acids, which are then amenable to further bacterial modification to yield secondary bile acids. Some studies have shown that BSH are capable of cleaving the amide bond of a wide variety of bile salts conjugates, including the six main human conjugated bile salts - glycocholic acid (GCA), glycodeoxycholic acid (GDCA), glycochenodeoxycholic acid (GCDCA), taurocholic acid (TCA), taurodeoxycholic acid (TDCA) and taurochenodeoxycholic acid (TCDCA), releasing glycine or taurine and free bile acids (Chand et al., 2017).

After cleavage of bile salts conjugates, deconjugated bile acids are less soluble in an aqueous environment and therefore excreted via the feces. This results in reduced absorption in the intestinal lumen and stimulates the formation of new bile salts to compensate for those lost in the feces. New bile salts are generated from dietary cholesterol which maintains the homoeostasis of bile salts (Bi et al., 2013). In this sense, a part of the cholesterol in the body is used to replace bile salts which were lost due to their deconjugation in the small intestine and this may prevent hypercholesterolaemia. This is an important action performed by the BSH enzyme, since hypercholesterolaemia is considered a major risk factor for the development of coronary heart disease (Liong and Shah, 2005).

Although there are a number of drugs available today for the treatment of high cholesterol, these products are expensive and are known to have severe side effects. For this reason, the possibility of using the BSH enzyme to lower serum cholesterol levels in hypercholesterolemic patients and prevent hypercholesterolemia in normal people has arisen as an interesting alternative to conventional drugs (Chae et al., 2013).

Moreover, a recent study recognized BSH as a mechanism through which the microbiota modulates the host lipid metabolism and demonstrated that this mechanism corresponds to a viable alternative for the development of intervention approaches for the management of obesity and metabolic syndrome (Joyce et al., 2014). The effect of some BSH-producing strains on cholesterol reduction using *in vitro* and *in vivo* tests is presented in Table 4.

In addition to the cholesterol lowering activity of BSH discussed above, the bile salt deconjugation process has further beneficial effects on host health. This process forms amino acids which act as carbon, nitrogen, and energy sources. For this reason, BSH-positive strains could confer a nutritional advantage (Patel et al., 2010). Also, strains that produce the BSH enzyme show greater resistance against bile found in the gastrointestinal tract, allowing for effective colonization of bacteria in the gut and increased competitiveness against pathogenic microbiota (Jones et al., 2008).

In light of the beneficial effects of BSH containing bacteria, it is clear that BSH production by a bacterial strain is desirable. Production of the BSH enzyme is considered an important criterion for the selection of bacteria with probiotic potential. In fact, the Guidelines for the Evaluation of Probiotics in Food published by the Food Agriculture Organization (FAO, 2002) include the bile salt hydrolase activity among the main enzymes currently used in *in vitro* tests for the study of probiotic strains. Several studies have isolated BSH-positive strains as show in Table 5.

The BSH loci show different architecture in various bacteria and some *Lactobacillus* species have been observed to exhibit more than one BSH isoform (Chand et al., 2017). On the other hand, some species do not have the genes responsible to produce this enzyme and for this reason some molecular techniques can be used to generate BSH-positive strains as a therapeutic probiotic bacterium.

Some studies have also shown that is possible to clone and overexpress the BSH genes from *Lactobacillus* and *Bifidobacterium* species (Chae et al., 2013; Kaya et al., 2017). Cho et al. (2015) cloned the bile salt hydrolase gene (*bsh*) from *Lactobacillus plantarum* and expressed it in *L. citreum* using the pCB4170 vector. They found that *L. citreum* transformants successfully transcribed the *bsh* gene, and expressed the BSH enzyme. The recombinant strain grew in up to 0.3% (w/v) GDCA, conditions unsuitable for the host strain. In *in vitro* digestion conditions of 10 mM bile salt, the transformant was over 67.6% viable, whereas only 0.8% of the host strain survived. The *Lactobacillus plantarum cbh* gene,

Table 4. Effects of BSH-producer strains on cholesterol levels using in vitro and in vivo assays

Strains	Research Purpose	Conditions of Assay	Main Results	Reference
L. johnsonii BFE 1061, *L. johnsonii* BFE 1059 and *L. reuteri* BFE 1058	To evaluate the effect of three BSH-positive isolates on minipigs.	*In vivo*, using minipigs, on a high cholesterol diet for 17 weeks.	Reduction of serum cholesterol levels after three weeks of probiotic feeding, associated with an increase in the moisture content of the feces and *Lactobacillus* cell numbers.	Du Toit et al. (1998)
L. buchneri ATCC 4005	To assess the effect of oral administration of the immobilized BSH enzyme on serum cholesterol, triglyceride, high density lipoprotein levels of rats.	*In vivo*, using Wistar rats, induced with hypercholesteremia by triton X-100.	Serum cholesterol and triglycerides were reduced by 50% and 15%, respectively, in the group fed with the immobilized enzyme at the lower dose, whereas administration of the higher dose of the immobilized enzyme resulted in reduction of serum cholesterol by 58% and triglycerides by 45%, respectively.	Sridevi et al. (2009)
Lactobacillus acidophilus ATCC 314, *L. acidophilus* FTCC 0291, *L. bulgaricus* FTCC 0411, *L. bulgaricus* FTDC 1311 and *L. casei* ATCC 393	To evaluate cholesterol removing mechanisms by different strains of lactobacilli	*In vitro*, using conditions that mimic the human gastrointestinal tract	The concentration of cholesterol removal ranged from 4.92 to 62.42 µg/mL.	Lye et al. (2010)
L. reuteri NCIMB 30242	To determine the cholesterol-lowering efficacy and mechanism of action of bile salt hydrolase-active *Lactobacillus reuteri* NCIMB 30242 capsules in hypercholesterolemic adults	*In vivo*, using 127 subjects who consumed *L. reuteri* NCIMB 30242 capsules or placebo capsules over a 9-week intervention period	*Lactobacillus* reduced LDL-C by 11.64%, total cholesterol by 9.14%, non-HDL-C by 11.30% and apoB-100 by 8.41% relative to the placebo. Triglycerides and HDL-C were unchanged.	Jones et al. (2012)
B. breve MB 113, *and B. animalis* subsp. *lactis* MB 2409	To evaluate the effects of the probiotic mixture on lipemic values of hypercholesterolemic rats	*In vivo*, using male Wistar rats consuming a hypercholesterolemic diet for 30 days	Reduction of total cholesterol and LDL-C, whereas it did not affect the HDL-C and HDL-C/LDL-C ratio.	Bordoni et al. (2013)
L. plantarum GD2 and LA3, *B. breve* A19 and A26, *B. longum* BASO9 and BASO15	To examine cholesterol removal abilities via coprecipitation with deconjugated bile of lactobacilli and bifidobacteria strains isolated from the feces of breast-fed infants.	*In vitro*, using sterile medium containing 0, 0.2, 0.4 and 0.6% oxgall supplemented with 100 µg/ml filter-sterilized cholesterol.	The highest cholesterol precipitation (81%) was found in a 0.2% oxgall-containing medium; a decrease in cholesterol-removing ability of the isolates was observed at oxgall concentrations exceeding 0.4%.	Öner et al. (2014)
Lab4 consortium of probiotics (Lab4, composed of *L. acidophilus* CUL21 and CUL60, *B. bifidum* CUL20 and *B. animalis* subsp. *lactis* CUL34) plus *L. plantarum* CUL66	To investigate the effect of the probiotic mixture on plasma total cholesterol levels in mice	*In vivo*, using male C57BL/6 J mice fed a high fat diet, 2-weeks supplementation with probiotic mixture	The probiotic consortium is effective in promoting cholesterol-lowering during short-term feeding.	Michael et al. (2017)
L. plantarum Lp3	To assess potential of the strain to be used as cholesterol-reducing probiotics in rats	*In vivo*, using male Sprague-Dawley rats fed a high cholesterol diet for 7 weeks	The strain could be a potential probiotic to treat hyperlipidemia since it promoted notable declines in serum and liver cholesterol and triglycerides levels.	Ding et al. (2017)
L. plantarum LP96	To evaluate the effect of the *Lactobacillus* strain on the high-cholesterol diet-fed rats	*In vivo*, using male Kunming rats fed a high cholesterol diet for 5 weeks	The levels of serum triglyceride, total cholesterol, LDL-C, and liver total cholesterol and triglyceride were reduced significantly in the groups that received the *Lactobacillus* strain.	Liu et al. (2017)

HDL-C: high-density lipoprotein cholesterol; LDL-C: low-density lipoprotein cholesterol.

encoding the conjugated BSH, was cloned and successfully expressed in the heterologous host *L. casei* LK1 with the aid of pSMA23-derived vectors (Sudhamani et al., 2008).

Xiong et al. (2017) cloned and sequenced 4 *bsh* genes from the genome of *L. plantarum* AR113 in *L. casei.* The results indicated that heterologous expression of BSH in lactic acid bacteria could significantly enhance host activity/resistance to bile salts during industrial processes and gastrointestinal transit, and also improved BSH activity of *L. casei* for various food-industry applications, mainly of fermented dairy products.

Table 5. BSH-positive strains isolated from different sources

Sources	Strains	Bile Salts Used	Main Results	Reference
Human feces and traditional fermented products	5 *L. plantarum* strains and 2 *L. johnsonii* strains	TDCA (0.5% w/v)	All strains exhibited bile salt hydrolase activity.	Pinto et al. (2006)
Swine feces and intestines	15 Lactic Acid Bacteria (LAB) strains	TDCA (0.5% w/v)	75% of the strains from feces and 89% from intestines were BSH-positive.	Guo et al. (2010)
Buffalo and human milk, human feces, rat feces, silage, infant intestine, yogurt and culture collections	102 *Lactobacillus* spp. strains	TDCA (0.5% w/v)	A total of twenty-two *Lactobacillus* strains were able to hydrolyse TDCA.	Kumar et al. (2012)
Naturally fermented olives	71 LAB strains, including *Leuconostoc* spp and *Lactobacillus* spp	TDCA(0.5% w/v)	Of the seventy-one strains, twelve exhibited partial bile salt hydrolase activity	Argyri et al. (2013)
Vegetables and traditional Indian fermented food products	8 LAB strains, including *Lactobacillus* spp, *Weisella* spp and *Pediococcus*	TDCA (0.5% w/v)	A total of four strains were found to be BSH-positive, three of the isolates showed poor BSH activity while one of the isolates showed a BSH negative character.	Patel et al. (2014)
Traditional fermented food *idli*	10 strains, including *Leuconostoc* spp and *Bacillus* spp	TDCA (0.5% w/v)	Screening of BSH activity showed a maximum zone of 13 mm for *Bacillus* spp strains.	Saravanan et al. (2015)
Fruits and fermented foods	8 *L. plantarum* strains	Oxgall bile salts (mixture of bile salts)	Seven strains showed BSH-activity.	Shekh et al. (2016)
Goat milk	2 *L. brevis* and 2 *L. paracasei*	TDCA and GDCA (0.5% w/v)	All four demonstrated the ability to hydrolyze both sodium GDCA and sodium TDCA.	Ferrari et al. (2016)
Camel milk	9 LAB strains	GCA, TCA, and mixture of bile salts (GCA, GCDCA, TCA, TCDCA, TDCA)	All LAB strains showed the ability to hydrolyze all three investigated bile salts	Abushelaibi et al. (2017)
Digestive tracts of Indonesian native chickens	26 LAB strains	TDCA (0.5% w/v)	All twenty-six isolates showed precipitation zones with different diameters; the largest diameter values were isolated from chicken intestine and the smallest diameter values were isolated from chicken colon.	Julendra et al. (2017)

Glucose Oxidase: Characterization and Effects on Probiotic Dairy Products

Glucose oxidase (β-D-glucose oxygen 1-oxidoreductase, EC 1.1.3.4) is classified as an oxidoreductase enzyme (flavoprotein) which catalyzes the oxidation of β-D-glucose to gluconic acid and hydrogen peroxide (H_2O_2) (Kovačević et al., 2014). The H_2O_2 resultant from the GOX reaction has antimicrobial activity and has been shown to be active against Gram-negative and Gram-positive bacteria (Derakshan et al., 2017).

A broad variety of fungi, mostly of the genera *Aspergillus* and *Penicillium*, produce GOX due to their outstanding capacities to use a range of carbon sources and to accumulate a large quantity of natural GOX under stressed conditions. The mechanisms of GOX accumulation by different fungi are not completely known (Dubey et al., 2017). Industrial production of GOX is complicated by the fact that wild fungal strains have some issues including low yield, production of other enzymes as by-products which makes the purification process difficult and high sensitivity to metal ions present in raw materials (Derakshan et al., 2017).

These extensive applications of GOX by industry have resulted in an increased demand for more research to overcome these obstacles so as to enhance production, characterization, and stability to allow for long term application (Meng et al., 2014). Studies dealing with the improvement of production and properties of GOX still receive a great deal of attention and among the innovative strategies explored with this purpose are immobilization, mutagenesis and recombination, and enzymatic engineering techniques (rational design, rational redesign, and directed evolution). A recent study extensively reviewed and detailed these approaches (Dubey et al., 2017).

GOX is an enzyme of great importance and many applications in the industry, including the chemical, pharmaceutical, medical diagnostics, textile, food, beverage, and biotechnology industries (Abbasi et al., 2016). GOX has also been used by the dairy industry, mainly in the production of fermented milk and yogurts.

These are popular dairy products and they are recognized for their good nutritional value and positive health appeal. This is one of the reasons responsible for the increase in yogurt sales throughout the world, and the possibility of adding probiotic microorganisms to yogurt has contributed to further grow the global market. Probiotics are defined as live microorganisms that are able to colonize the gastrointestinal (GI) tract, and when administered in adequate amounts, confer a health benefit to the host (FAO/WHO, 2002).

The inclusion of probiotic strains to foods and the preservation of their numbers above the minimum required to promote consumer health throughout the storage period characterize a continuous challenge to the dairy industry, and both intrinsic and extrinsic factors associated with processing of probiotic food must be taken in consideration. Probiotic counts in fermented milk products can be impaired by different factors, including the ingredients used for product manufacture (Coman et al., 2013), heat treatment and incubation temperatures during processing (Mortazavian et al. 2006), the presence of fruit pulps (Bedani et al., 2014), the interactions between starter cultures and probiotic microorganisms (Casarotti et al., 2014) and others.

Exposure to oxygen during storage of the product should be as low as possible to avoid oxidative stress in order to prevent the death of probiotic strains and the loss or reduction of functionality of the product, because most probiotic bacteria have anaerobic or microaerophilic metabolisms, with some alternatives (Mattila-Sandholm et al., 2002; Cruz et al., 2012b). Among the options that have been proposed to overcome the problem of oxygen toxicity in yogurt and fermented milk is the addition of GOX (Cruz et al., 2010).

Considering that oxygen degradation by GOX generates $H_2O_{2,}$ an antimicrobial substance, it is fundamental to add catalase along with this enzyme in the case of probiotic dairy product, which will help to maintain microorganisms viability. It is also crucial to highlight that even though GOX is a potent antioxidant, some studies have shown that it may cause adverse effect on the aromatic compounds amount and on the sensorial characteristics of dairy products (Cruz et al., 2013; Pereira et al., 2016), as discussed in the text below. The generation of off flavor may be associated with the presence of other antioxidants compounds in yogurt matrix, and its interaction with the lactic acid and probiotic microorganisms. Nevertheless, such effect does not need to be considered as an obstacle for GOX use, since the inclusion of a commercial fruit pulp can minimize this issue, as these products contain sucrose, which gives a sweet taste.

Cruz et al. (2012a) analyzed the effect of glucose oxidase on some quality parameters of probiotic stirred yogurt during 15 days of refrigerated storage. They reported an increase in the viable population (6.9-8.7 log CFU/g) of *Bifidobacteria longum,* lower levels of oxygen and pH, and greater proteolytic activity the probiotic yogurts containing different levels of glucose oxidase in comparison with the control treatment. These findings indicate that GOX supported a superior metabolic activity of probiotic microorganisms in yogurts with the enzyme. In addition, acetaldehyde and diacetyl, the typical yogurt aroma compounds, were found to be produced in adequate amounts, along with the formation of moderate amounts of organic acids, particularly lactic acid and acetic acid in yogurts with GOX (Cruz et al., 2012b).

Despite the benefits to probiotic viability, the addition of GOX to dairy products may reduce its sensorial acceptance. A study evaluated the consumer perception of two probiotic yogurts with different concentrations of GOX and four commercial strawberry-flavored probiotic yogurts using four different consumer profiling techniques. The presence of GOX changed the sensory perception of the yogurts which may be explained by the high acidity, caused by the increased production of organic acids by the probiotic bacteria, and by more intense bitter taste, supported by higher proteolysis and consequent release of bitter hydrophobic peptides. The improvement of yogurt supplemented with GOX from a sensory standpoint requires the decrease of acid and bitter tastes, the reduction of the presence of grains as well as the improvement of viscosity of the product (Cruz et al., 2013).

It is important to highlight that supplementation of yogurts and fermented milk with excessive amounts of GOX may be ineffective because there would be a lack of glucose available in the matrix, as revealed by Cruz et al. (2012b). The authors concluded that intermediate amounts of the enzyme resulted in improved results throughout the shelf life, i.e., lower values of dissolved oxygen and moderate post acidification and proteolysis, which represent a noteworthy outcome on the economic aspects of production.

Moreover, adding GOX can impact the texture of other types of dairy products. The addition of GOX to petit suisse affected its reological properties, decreasing the gel strength and leading to less viscous and consistent products. It is likely that GOX interferes with formation of the protein network, which may be broken down due to harsh shearing conditions in the antioxidant incorporation stage during the petit suisse processing (Pereira et al., 2016).

In summary, GOX has a potential as a feasible alternative to lessen the oxygen dissolved in probiotic dairy products and consequently maintain the viability of probiotic microorganisms during storage. However, some questions need to be better understood and it is also essential to optimize the formulations before use of this enzyme in commercial dairy products.

CONCLUSION

Enzymes play a key role in the food industry, since they allow for the production of foods with appropriate characteristics and increase the diversity of products obtained via industrial processes. The dairy industry has experienced great success in exploiting enzymes and will continue to explore them to generate profits and increase global sales.

FUTURE PROSPECTS

Over the last decades, significant improvements have been made in the field of enzyme production and many innovations have been developed in order to optimize this process. However, considering its economic importance, it is still necessary to find new and/or more efficient enzymes, as well as to make advances in technologies available for their production to reduce costs. Additionally, studies must address the use of enzymes to obtain foods with health-enhancing properties.

ACKNOWLEDGMENT

The authors would like to thank Fundação de Amparo à Pesquisa do Estado de Mato Grosso (FAPEMAT/ Edital Universal 042/2016) for the support and Fernanda Kleber Curtarelli for the kindness, motivation and speed for doing the Figure 1 of this chapter.

REFERENCES

Abbasi, M., Amiri, R., Bordbar, A.-K., Ranjbakhsh, E., & Khosropour, A. R. (2016). Improvement of the stability and activity of immobilized glucose oxidase on modified iron oxide magnetic nanoparticles. *Applied Surface Science*, *364*, 752–757. doi:10.1016/j.apsusc.2015.12.120

Abushelaibi, A., Al-Mahadin, S., El-Tarabily, K., Shah, N. P., & Ayyash, M. (2017). Characterization of potential probiotic lactic acid bacteria isolated from camel milk. *Lebensmittel-Wissenschaft + Technologie*, *79*, 316–325. doi:10.1016/j.lwt.2017.01.041

Afroz, Q. M., Khan, K. A., & Ahmed, P. & S, U. (2015). Enzymes used in dairy industries. *International Journal of Applied Research*, *1*(10), 523–527.

Ansari, S. A., Satar, R., Zaidi, S. K., & Ahmad, A. (2014). Immobilization of *Aspergillus oryzae* β-Galactosidase on cellulose acetate-polymethylmethacrylate membrane and its application in hydrolysis of lactose from milk and whey. *International Scholarly Research Notices*, *163987*, 1–6. PMID:27350979

Argyri, A. A., Zoumpopoulou, G., Karatzas, K. A., Tsakalidou, E., Nychas, G. J., Panagou, E. Z., & Tassou, C. C. (2013). Selection of potential probiotic lactic acid bacteria from fermented olives by *in vitro* tests. *Food Microbiology*, *33*(2), 282–291. doi:10.1016/j.fm.2012.10.005 PMID:23200662

Arnau, J., Lauritzen, C., Petersen, G. E., & Pedersen, J. (2006). Current strategies for the use of affinity tags and tag removal for the purification of recombinant proteins. *Protein Expression and Purification*, *48*(1), 1–13. doi:10.1016/j.pep.2005.12.002 PMID:16427311

Banat, I. M., Satpute, S. K., Cameotra, S. S., Patil, R., & Nyayanit, N. V. (2014). Cost effective technologies and renewable substrates for biosurfactants' production. *Frontiers in Microbiology*, *5*(697), 1–18. PMID:25566213

Bedani, R., Vieira, A. D. S., Rossi, E. A., & Saad, S. M. I. (2014). Tropical fruit pulps decreased probiotic survival to in vitro gastrointestinal stress in synbiotic soy yoghurt with okara during storage. *Lebensmittel-Wissenschaft + Technologie*, *55*(2), 436–443. doi:10.1016/j.lwt.2013.10.015

Begley, M., Gahan, C. G., & Hill, C. (2005). The interaction between bacteria and bile. *FEMS Microbiology Reviews*, *29*(4), 625–651. doi:10.1016/j.femsre.2004.09.003 PMID:16102595

Bhushan, B., Tomar, S. K., & Chauhan, A. (2017). Techno-functional differentiation of two vitamin B12 producing *Lactobacillus plantarum* strains: An elucidation for diverse future use. *Applied Microbiology and Biotechnology*, *101*(2), 697–709. doi:10.100700253-016-7903-z PMID:27752783

Bi, J., Fang, F., Lu, S., Du, G., & Chen, J. (2013). New insight into the catalytic properties of bile salt hydrolase. *Journal of Molecular Catalysis. B, Enzymatic*, *96*, 46–51. doi:10.1016/j.molcatb.2013.06.010

Bimbo, F., Bonanno, A., Nocella, G., Viscecchia, R., Nardone, G., De Devitiis, B., & Carlucci, D. (2017). Consumers' acceptance and preferences for nutrition-modified and functional dairy products: A systematic review. *Appetite*, *113*, 141–154. doi:10.1016/j.appet.2017.02.031 PMID:28235616

Bordoni, A., Amaretti, A., Leonardi, A., Boschetti, E., Danesi, F., Matteuzzi, D., ... Rossi, M. (2013). Cholesterol-lowering probiotics: *In vitro* selection and *in vivo* testing of bifidobacteria. *Applied Microbiology and Biotechnology*, *97*(18), 8273–8281. doi:10.100700253-013-5088-2 PMID:23872958

Brown, L., Pingitore, E. V., Mozzi, F., Saavedra, L., Villegas, J. M., & Hebert, E. M. (2017). Lactic Acid Bacteria as Cell Factories for the Generation of Bioactive Peptides. *Protein and Peptide Letters*, *24*(2), 146–155. doi:10.2174/0929866524666161123111333 PMID:27890004

Cao, M., Lia, Z., Wanga, J., Gea, W., Yuea, T., Lib, R., ... Yuc, W. W. (2012). Food related applications of magnetic iron oxide nanoparticles: Enzyme immobilization, protein purification, and food analysis. *Trends in Food Science & Technology*, *27*(1), 47–56. doi:10.1016/j.tifs.2012.04.003

Casarotti, S. N., Monteiro, D. A., Moretti, M. M. S., & Penna, A. L. B. (2014). Influence of the combination of probiotic cultures during fermentation and storage of fermented milk. *Food Research International*, *59*, 67–75. doi:10.1016/j.foodres.2014.01.068

Chae, J. P., Valeriano, V. D., Kim, G. B., & Kang, D. K. (2013). Molecular cloning, characterization and comparison of bile salt hydrolases from *Lactobacillus johnsonii* PF01. *Journal of Applied Microbiology*, *114*(1), 121–133. doi:10.1111/jam.12027 PMID:23035872

Chand, D., Avinash, V. S., Yadav, Y., Pundle, A. V., Suresh, C. G., & Ramasamy, S. (2017). Molecular features of bile salt hydrolases and relevance in human health. *Biochimica et Biophysica Acta*, *1861*(1), 2981–2991. doi:10.1016/j.bbagen.2016.09.024 PMID:27681686

Chen, W., Chen, H., Xia, Y., Yang, J., Zhao, J., Tian, F., ... Zhang, H. (2009). Immobilization of recombinant thermostable β-galactosidase from *Bacillus stearothermophilus* for lactose hydrolysis in milk. *Journal of Dairy Science*, *92*(2), 491–498. doi:10.3168/jds.2008-1618 PMID:19164659

Cho, S. K., Lee, S. J., Shin, S. Y., Moon, J. S., Li, L., Joo, W., ... Han, N. S. (2015). Development of Bile salt-resistant *Leuconostoc citreum* by expression of bile salt hydrolase gene. *Journal of Microbiology and Biotechnology*, *25*(12), 2100–2105. doi:10.4014/jmb.1505.05072 PMID:26282688

Cieśliński, H., Kur, J., Białkowska, A., Baran, I., Makowski, K., & Turkiewicz, M. (2005). Cloning, expression, and purification of a recombinant cold-adapted β-galactosidase from antarctic bacterium *Pseudoalteromonas sp.* 22b. *Protein Expression and Purification*, *39*(1), 27–34. doi:10.1016/j.pep.2004.09.002 PMID:15596357

Coman, M. M., Verdenelli, M. C., Cecchini, C., Silvi, S., Vasile, A., Bahrim, G. E., ... Cresci, A. (2013). Effect of buckwheat flour and oat bran on growth and cell viability of the probiotic strains *Lactobacillus rhamnosus* IMC 501®, *Lactobacillus paracasei* IMC 502® and their combination SYNBIO®, in synbiotic fermented milk. *International Journal of Food Microbiology*, *167*(2), 261–268. doi:10.1016/j.ijfoodmicro.2013.09.015 PMID:24140807

Cruz, A. G., Cadena, R. S., Castro, W. F., Esmerino, E. A., Rodrigues, J. B., Gaze, L., ... Bolini, H. M. A. (2013). Consumer perception of probiotic yogurt: Performance of check all that apply (CATA), projective mapping, sorting and intensity scale. *Food Research International*, *54*(1), 601–610. doi:10.1016/j.foodres.2013.07.056

Cruz, A. G., Castro, W. F., Faria, J. A., Lollo, P. C., Amaya-Farfan, J., Freitas, M. Q., ... Godoy, H. T. (2012b). Probiotic yogurts manufactured with increased glucose oxidase levels: Postacidification, proteolytic patterns, survival of probiotic microorganisms, production of organic acid and aroma compounds. *Journal of Dairy Science*, *95*(5), 2261–2269. doi:10.3168/jds.2011-4582 PMID:22541455

Cruz, A. G., Castro, W. F., Faria, J. A. F., Bogusz, S. Jr, Granato, D., Celeguini, R. M. S., ... Godoy, H. T. (2012a). Glucose oxidase: A potential option to decrease the oxidative stress in stirred probiotic yogurt. *Lebensmittel-Wissenschaft + Technologie*, *47*(2), 512–515. doi:10.1016/j.lwt.2012.01.037

Cruz, A. G., Faria, J. A. F., Walter, E. H. M., Andrade, R. R., Cavalcanti, R. N., Oliveira, C. A. F., & Granato, D. (2010). Processing optimization of probiotic yogurt containing glucose oxidase using response surface methodology. *Journal of Dairy Science*, *93*(11), 5059–5068. doi:10.3168/jds.2010-3336 PMID:20965320

Cuatrecasas, P., Wilchek, M., & Anfinsen, C. B. (1968). Selective enzyme purification by affinity chromatography. *Proceedings of the National Academy of Sciences of the United States of America*, *61*(2), 636–643. doi:10.1073/pnas.61.2.636 PMID:4971842

Derakshan, K. F., Darvishi, F., Dezfulian, M., & Madzak, C. (2017). Expression and characterization of glucose oxidase from *Aspergillus niger* in *Yarrowia lipolytica*. *Molecular Biotechnology*, *59*(8), 307–314. doi:10.100712033-017-0017-8 PMID:28631058

Ding, W., Shi, C., Chen, M., Zhou, J., Long, R., & Guo, X. (2017). Screening for lactic acid bacteria in traditional fermented Tibetan yak milk and evaluating their probiotic and cholesterol-lowering potentials in rats fed a high-cholesterol diet. *Journal of Functional Foods*, *32*, 324–332. doi:10.1016/j.jff.2017.03.021

Dong, Q., Yan, X., Zheng, M., & Yang, Z. (2014). Characterization of an extremely thermostable but cold-adaptive β-galactosidase from the hyperthermophilic archaeon *Pyrococcus furiosus* for use as a recombinant aggregation for batch lactose degradation at high temperature. *Journal of Bioscience and Bioengineering*, *117*(6), 706–710. doi:10.1016/j.jbiosc.2013.12.002 PMID:24462527

Du Toit, M., Franz, C. M., Dicks, L. M., Schillinger, U., Haberer, P., Warlies, B., ... Holzapfel, W. H. (1998). Characterisation and selection of probiotic lactobacilli for a preliminary minipig feeding trial and their effect on serum cholesterol levels, faeces pH and faeces moisture content. *International Journal of Food Microbiology*, *40*(1-2), 93–104. doi:10.1016/S0168-1605(98)00024-5 PMID:9600615

Dubey, M. K., Zehra, A., Aamir, M., Meena, M., Ahirwal, L., Singh, S., ... Bajpai, V. K. (2017). Improvement strategies, cost effective production, and potential applications of fungal glucose oxidase (GOD): Current updates. *Frontiers in Microbiology*, *8*, 1032. doi:10.3389/fmicb.2017.01032 PMID:28659876

FAO/WHO. (2002). *Guidelines for the evaluation of probiotics in food: Report of a joint FAO/WHO working group on drafting guidelines for the evaluation of probiotics in food*. London, Canada: FAO/WHO.

Ferrari, I. S., de Souza, J. V., Ramos, C. L., da Costa, M. M., Schwan, R. F., & Dias, F. S. (2016). Selection of autochthonous lactic acid bacteria from goat dairies and their addition to evaluate the inhibition of *Salmonella* typhi in artisanal cheese. *Food Microbiology*, *60*, 29–38. doi:10.1016/j.fm.2016.06.014 PMID:27554143

Gandhi, N. N. (1997). Applications of Lipase. *Journal of the American Oil Chemists' Society*, *74*(6), 621–634. doi:10.100711746-997-0194-x

Gaynor, P. (2006). How U.S. FDA's GRAS notification program works. *Food Safety Magazine*, *11*, 16–19.

Guerrero, C., Vera, C., Serna, N., & Ilanes, A. (2017). Immobilization of *Aspergillus oryzae* β-galactosidase in an agarose matrix functionalized by four different methods and application to the synthesis of lactulose. *Bioresource Technology*, *232*, 53–63. doi:10.1016/j.biortech.2017.02.003 PMID:28214445

Guo, X. H., Kim, J. M., Nam, H. M., Park, S. Y., & Kim, J. M. (2010). Screening lactic acid bacteria from swine origins for multistrain probiotics based on *in vitro* functional properties. *Anaerobe*, *16*(4), 321–326. doi:10.1016/j.anaerobe.2010.03.006 PMID:20304081

Hung, M. N., & Lee, B. H. (2002). Purification and characterization of a recombinant beta-galactosidase with transgalactosylation activity from *Bifidobacterium infantis* HL96. *Applied Microbiology and Biotechnology*, *58*(4), 439–445. doi:10.100700253-001-0911-6 PMID:11954789

James, J., Simpson, B. K., & Marshall, M. R. (1996). Application of enzymes in food processing. *Critical Reviews in Food Science and Nutrition*, *36*(5), 437–463. doi:10.1080/10408399609527735 PMID:8725673

Jones, B. V., Begley, M., Hill, C., Gahan, C. G., & Marchesi, J. R. (2008). Functional and comparative metagenomic analysis of bile salt hydrolase activity in the human gut microbiome. *Proceedings of the National Academy of Sciences of the United States of America*, *105*(36), 13580–13585. doi:10.1073/pnas.0804437105 PMID:18757757

Jones, M. L., Martoni, C. J., & Prakash, S. (2012). Cholesterol lowering and inhibition of sterol absorption by *Lactobacillus reuteri* NCIMB 30242: A randomized controlled trial. *European Journal of Clinical Nutrition*, *66*(11), 1234–1241. doi:10.1038/ejcn.2012.126 PMID:22990854

Joyce, S. A., MacSharry, J., Casey, P. G., Kinsella, M., Murphy, E. F., Shanahan, F., ... Gahan, C. G. (2014). Regulation of host weight gain and lipid metabolism by bacterial bile acid modification in the gut. *Proceedings of the National Academy of Sciences of the United States of America*, *111*(20), 7421–7426. doi:10.1073/pnas.1323599111 PMID:24799697

Julendra, H., Suryani, A. E., Istiqomah, L., Damayanti, E., Anwar, M., & Fitriani, N. (2017). Isolation of Lactic Acid Bacteria with cholesterol-lowering activity from digestive tracts of Indonesian native chickens. *Media Peternakan*, *40*(1), 35–41. doi:10.5398/medpet.2017.40.1.35

Kaya, Y., Kök, M. Ş., & Öztürk, M. (2017). Molecular cloning, expression and characterization of bile salt hydrolase from *Lactobacillus rhamnosus* E9 strain. *Food Biotechnology*, *31*(2), 128–140. doi:10.1080/08905436.2017.1303778

Konkit, M., Choi, W. J., & Kim, W. (2016). Aldehyde dehydrogenase activity in Lactococcus chungangensis: Application in cream cheese to reduce aldehyde in alcohol metabolism. *Journal of Dairy Science*, *99*(3), 1–7. doi:10.3168/jds.2015-10549 PMID:26723119

Kovačević, G., Blažić, M., Draganić, B., Ostafe, R., Gavrović-Jankulović, M., Fischer, R., & Prodanović, R. (2014). Cloning, heterologous expression, purification and characterization of M12 mutant of *Aspergillus niger* glucose oxidase in yeast *Pichia pastoris* KM71H. *Molecular Biotechnology*, *56*(4), 305–311. doi:10.100712033-013-9709-x PMID:24122283

Kumar, M., Kumar, A., Nagpal, R., Mohania, D., Behare, P., Verma, V., ... Yadav, H. (2010). Cancer-preventing attributes of probiotics: An update. *International Journal of Food Sciences and Nutrition*, *61*(5), 473–496. doi:10.3109/09637480903455971 PMID:20187714

Kumar, R., Grover, S., & Batish, V. K. (2012). Bile Salt Hydrolase (Bsh) activity screening of lactobacilli*: In vitro* selection of indigenous *Lactobacillus* strains with potential bile salt hydrolysing and cholesterol-lowering ability. *Probiotics and Antimicrobial Proteins*, *4*(3), 162–172. doi:10.100712602-012-9101-3 PMID:26782042

Liong, M. T., & Shah, N. P. (2005). Bile salt deconjugation and BSH activity of five bifidobacterial strains and their cholesterol co-precipitating properties. *Food Research International*, *38*(2), 135–142. doi:10.1016/j.foodres.2004.08.003

Liu, Y., Chen, Z., Jiang, Z., Yan, Q., & Yang, S. (2017). Biochemical characterization of a novel β-galactosidase from *Paenibacillus barengoltzii* suitable for lactose hydrolysis and galactooligosaccharides synthesis. *International Journal of Biological Macromolecules*, *104*, 1055–1063. doi:10.1016/j.ijbiomac.2017.06.073 PMID:28652150

Liu, Y., Zhao, F., Liu, J., Wang, H., Han, X., Zhang, Y., & Yang, Z. (2017). Selection of cholesterol-lowering Lactic Acid Bacteria and its effects on rats fed with high-cholesterol diet. *Current Microbiology*, *74*(5), 623–631. doi:10.100700284-017-1230-1 PMID:28286906

Lye, H.-S., Rahmat-Ali, G. R., & Liong, M.-T. (2010). Mechanisms of cholesterol removal by lactobacilli under conditions that mimic the human gastrointestinal tract. *International Dairy Journal*, *20*(3), 169–175. doi:10.1016/j.idairyj.2009.10.003

Maitan-Alfenas, G. P., Oliveira, M. B., Nagem, R. A., de Vries, R. P., & Guimarães, V. M. (2016). Characterization and biotechnological application of recombinant xylanases from *Aspergillus nidulans*. *International Journal of Biological Macromolecules*, *91*, 60–67. doi:10.1016/j.ijbiomac.2016.05.065 PMID:27235731

Maitan-Alfenas, G. P., Visser, E. M., & Guimarães, V. M. (2015). Enzymatic hydrolysis of lignocellulosic biomass: Converting food waste in valuable products. *Current Opinion in Food Science*, *1*, 44–49. doi:10.1016/j.cofs.2014.10.001

Mateo, C., Palomo, J. M., Fernandez-Lorente, G., Guisan, J. M., & Fernandez-Lafuente, R. (2007). Improvement of enzyme activity, stability and selectivity via immobilization techniques. *Enzyme and Microbial Technology*, *40*(6), 1451–1463. doi:10.1016/j.enzmictec.2007.01.018

Mattila-Sandholm, T., Myllärinen, P., Crittenden, R., Mogensen, G., Fondén, R., & Saarela, M. (2002). Technological challenges for future probiotic foods. *International Dairy Journal, 12*(2-3), 173–182. doi:10.1016/S0958-6946(01)00099-1

Meng, Y., Zhao, M., Yang, M., Zhang, Q., Hao, J., & Meng, Y. (2014). Production and characterization of recombinant glucose oxidase from *Aspergillus niger* expressed in *Pichia pastoris. Letters in Applied Microbiology*, *58*(4), 393–400. doi:10.1111/lam.12202 PMID:24283586

Michael, D. R., Davies, T. S., Moss, J. W. E., Calvente, D. L., Ramji, D. P., Marchesi, J. R., ... Hughes, T. R. (2017). The anti-cholesterolaemic effect of a consortium of probiotics: An acute study in C57BL/6J mice. *Scientific Reports*, *7*(1), 2883. doi:10.103841598-017-02889-5 PMID:28588193

Mortazavian, A. M., Ehsani, M. R., Mousavi, S. M., Reinheimer, J. A., Emamdjomeh, Z., Sohrabvandi, S., & Rezaei, K. (2006). Preliminary investigation of the combined effect of heat treatment and incubation temperature on the viability of the probiotic micro-organisms in freshly made yogurt. *International Journal of Dairy Technology*, *59*(1), 8–11. doi:10.1111/j.1471-0307.2006.00216.x

Nielsen, P. H., Malmos, H., Damhus, T., Diderichsen, B., Nielsen, H. K., Simonsen, M., ... Nielsen, T. K. (1994). Enzyme applications (industrial). In K. Othmer (Ed.), *Encyclopedia of Chemical Technology* (pp. 567–620). New York: John Wiley & Sons, Inc.

Novozymes. (2017). *What are enzymes?* Retrieved July 2nd, 2017, from http://www.novozymes.com/en/about-us/our-business/industrial-biotechnology/basic-technologies/recombinant-expression/for-the-experts-

Olempska-Beer, Z. S., Merker, R. I., Ditto, M. D., & DiNovi, M. J. (2006). Food-processing enzymes from recombinant microorganisms - a review. *Regulatory Toxicology and Pharmacology*, *45*(2), 144–158. doi:10.1016/j.yrtph.2006.05.001 PMID:16769167

Olson, A. C., & Korus, R. A. (1976). Immobilized enzymes. In R. L. Ory & A. J. St. Angelo (Eds.), Enzymes in Food and Beverage Processing (pp. 100-131). Washington, DC: American Chemical Society.

Öner, O., Aslim, B., & Aydas, S. B. (2014). Mechanisms of cholesterol-lowering effects of lactobacilli and bifidobacteria strains as potential probiotics with their bsh gene analysis. *Journal of Molecular Microbiology and Biotechnology*, *24*(1), 12–18. doi:10.1159/000354316 PMID:24158048

Ovsejevi, K., Grazu, V., & Batista-Viera, F. (1998). Beta-Galactosidase from *Kluyveromyces lactis* immobilized on to thiolsulfinate/thiolsulfonate supports for lactose hydrolysis in milk and dairy by-products. *Biotechnology Techniques*, *12*(2), 143–148. doi:10.1023/A:1008892518473

Papayannakos, N., Markas, G., & Kekos, D. (1993). Studies on modeling and simulation of lactose hydrolysis by free and immobilized β-galactosidase from *Aspergillus niger. Chemical Engineering Journal*, *52*(1), B1–B12. doi:10.1016/0300-9467(93)80044-O

Pariza, M. W., & Johnson, E. A. (2001). Evaluating the safety of microbial enzyme preparations used in food processing: Update for a new century. *Regulatory Toxicology and Pharmacology*, *33*(2), 173–186. doi:10.1006/rtph.2001.1466 PMID:11350200

Park, H. Y., Kim, H. J., Lee, J. K., Kim, D., & Oh, D. K. (2008). Galactooligosaccharide production by a thermostable β-galactosidase from *Sulfolobus solfataricus*. *World Journal of Microbiology & Biotechnology*, *24*(8), 1553–1558. doi:10.100711274-007-9642-x

Patel, A., Prajapati, J. B., Holst, O., & Ljungh, A. (2014). Determining probiotic potential of exopolysaccharide producing lactic acid bacteria isolated from vegetables and traditional Indian fermented food products. *Food Bioscience*, *5*, 27–33. doi:10.1016/j.fbio.2013.10.002

Patel, A. K., Singhania, R. R., Pandey, A., & Chincholkar, S. B. (2010). Probiotic bile salt hydrolase: Current developments and perspectives. *Applied Biochemistry and Biotechnology*, *162*(1), 166–180. doi:10.100712010-009-8738-1 PMID:19669939

Pereira, E. P., Cavalcanti, R. N., Esmerino, E. A., Silva, R., Guerreiro, L. R., Cunha, R. L., ... Cruz, A. G. (2016). Effect of incorporation of antioxidants on the chemical, rheological, and sensory properties of probiotic petit suisse cheese. *Journal of Dairy Science*, *99*(3), 1762–1772. doi:10.3168/jds.2015-9701 PMID:26805976

Pessela, B. C. C., Mateo, C., Fuentes, M., Vian, A., García, J. L., Carrascosa, A. V., ... Fernández-Lafuente, R. (2003). The immobilization of a thermophilic β-galactosidase on Sepabeads supports decreases product inhibition complete hydrolysis of lactose in dairy products. *Enzyme and Microbial Technology*, *33*(2-3), 199–205. doi:10.1016/S0141-0229(03)00120-0

Pinto, M. G., Franz, C. M. A. P., Schillinger, U., & Holzapfel, W. H. (2006). *Lactobacillus* spp. with *in vitro* probiotic properties from human faeces and traditional fermented products. *International Journal of Food Microbiology*, *109*(3), 205–214. doi:10.1016/j.ijfoodmicro.2006.01.029 PMID:16503361

Placier, G., Watzlawick, H., Rabiller, C., & Mattes, R. (2009). Evolved β-galactosidases from *Geobacillus stearothermophilus* with improved transgalactosylation yield for galacto-oligosaccharides production. *Applied and Environmental Microbiology*, *7*(19), 6312–6321. doi:10.1128/AEM.00714-09 PMID:19666723

Portaccio, M., Stellato, S., Rossi, S., Bencivenga, U., Mohy Eldin, M. S., Gaeta, F. S., & Mita, D. G. (1998). Galactose competitive inhibition of β-galactosidase (*Aspergillus oryzae*) immobilized on chitosan and nylon supports. *Enzyme and Microbial Technology*, *23*(1-2), 101–106. doi:10.1016/S0141-0229(98)00018-0

Research and Markets. (2014). *Global functional food and nutraceuticals market (2014-2020) - by type (foods, beverages, supplements); benefits (health and wellness, disease prevention, fitness, beauty); origin & ingredient.* Author.

Saravanan, C., Gopu, V., & Shetty, P. H. (2015). Diversity and functional characterization of microflora isolated from traditional fermented food idli. *Journal of Food Science and Technology*, *52*(11), 7425–7432. doi:10.100713197-015-1791-6

Serafini, M., Stanzione, A., & Foddai, S. (2012). Functional foods: Traditional use and European legislation. *International Journal of Food Sciences and Nutrition*, *63*(sup1), 7–9. doi:10.3109/09637486.2011.637488 PMID:22117621

Shekh, S. L., Dave, J. M., & Vyas, B. R. M. (2016). Characterization of *Lactobacillus plantarum* strains for functionality, safety and γ-amino butyric acid production. *Lebensmittel-Wissenschaft + Technologie*, *74*, 234–241. doi:10.1016/j.lwt.2016.07.052

Sridevi, N., Vishwe, P., & Prabhune, A. (2009). Hypocholesteremic effect of bile salt hydrolase from *Lactobacillus buchneri* ATCC 4005. *Food Research International*, *42*(4), 516–520. doi:10.1016/j.foodres.2009.02.016

Staley, C., Weingarden, A. R., Khoruts, A., & Sadowsky, M. J. (2017). Interaction of gut microbiota with bile acid metabolism and its influence on disease states. *Applied Microbiology and Biotechnology*, *101*(1), 47–64. doi:10.100700253-016-8006-6 PMID:27888332

Sudhamani, M., Ismaiel, E., Geis, A., Batish, V., & Heller, K. J. (2008). Characterisation of pSMA23, a 3.5 kbp plasmid of *Lactobacillus casei*, and application for heterologous expression in *Lactobacillus*. *Plasmid*, *59*(1), 11–19. doi:10.1016/j.plasmid.2007.09.001 PMID:17961648

Wierzbicka-Woś, A., Cieśliński, H., Wanarska, M., Kozłowska-Tylingo, K., Hildebrandt, P., & Kur, J. (2011). A novel cold-active β-d-galactosidase from the *Paracoccus sp.* 32d - gene cloning, purification and characterization. *Microbial Cell Factories*, *49*(12), 2122–2133. PMID:22166118

Xiong, Z. Q., Wang, Q. H., Kong, L. H., Song, X., Wang, G. Q., Xia, Y. J., ... Ai, L. Z. (2017). Short communication: Improving the activity of bile salt hydrolases in *Lactobacillus casei* based on *in silico* molecular docking and heterologous expression. *Journal of Dairy Science*, *100*(2), 975–980. doi:10.3168/jds.2016-11720 PMID:28012620

Chapter 2
Research Tools and Methods for the Analysis of Microbiota in Dairy Products

Sylvia Klaubauf
Dutch DNA Biotech, The Netherlands

Frank J. J. Segers
Westerdijk Fungal Biodiversity Institute, The Netherlands

ABSTRACT

Microorganisms play important roles in dairy products. They can be a source of spoilage, or they promote health or cause diseases. In dairy fermentations, microorganisms are of great importance due to their function as starter cultures and during the production process of dairy products. In order to better understand and improve the process, it is essential to identify the species that are involved and to monitor the development of microbial communities. There are several different approaches for the detection and analysis of the microbiota. The methods can be culture dependent and, for example, make use of morphological and physiological characters or DNA sequencing. Culture-independent methods include direct PCR and qPCR, but also electrophoresis-based profiling techniques as well as metagenomics. Descriptions of relevant methods are provided and their applications are discussed in this chapter.

INTRODUCTION

Raw milk naturally contains microorganisms. Microorganisms play a number of important roles in dairy products, for example in dairy fermentations, but they can also cause spoilage, promote health or cause disease.

In the process of dairy product production, microorganisms can be intentionally or unintentionally added and play an important role in the formation of the final product. In order to identify starter cultures or to monitor the development of microbial communities during the production process, different methods for the detection and analysis of the microbiota can be used.

DOI: 10.4018/978-1-5225-5363-2.ch002

Knowledge on the microbial content and development is not only important for the purpose of improving processes, but also to assure the quality of the products, to validate health benefits and to assess food spoilage.

This chapter presents tools and methods that are used for the detection and analysis of microbiota in dairy products. The principles of different methods using culture-dependent as well as culture-independent approaches are presented. The numerous different tools to characterise microorganisms at various levels using classical microbiological techniques as well as advanced molecular methods are elaborated and compared. The main advantages and disadvantages associated with these techniques are discussed and example protocols are provided.

BACKGROUND

There is an array of methods available to investigate the presence of microorganisms in dairy products. These methods require either culturing or are culture independent. Some methods involve expensive apparatus, while others can be performed by using simple equipment. Many methods require, besides knowledge of microbiology, knowledge of molecular biology and biochemistry.

Besides the most commonly used methods of classical microbiology and culture independent techniques, there are also methods to detect metabolic products and toxins produced by microorganisms. Certain situations require that several methods have to be applied in parallel in order to gain the best possible result.

The choice of method depends on the purpose of the investigation: whether for quality assurance purposes or an examination of spoilage. Which type of information is required for a certain study implies the set of methods to choose from. Is it, for example, relevant to determine the number of viable cells in a sample or is the identification of a particular spoilage species of interest?

Traditional culturing methods are still commonly used for the detection of yeasts and fungi. These methods are more time consuming than alternative culture-independent methods, especially for fungi, which grow more slowly than bacteria. It is therefore difficult to develop rapid methods based on culturing. Other methods may provide faster results, but the results are difficult and often inappropriate to correlate with traditional plate counts (Hocking, 2007). Different methods highlight different aspects and may not necessarily provide overlapping but rather complementary results.

SAMPLING AND SAMPLE PREPARATION

Any analysis starts with taking a sample of the product. Sampling is a crucial step and there are several considerations that should be taken into account.

Samples should be taken aseptically and stored in sterile containers, preferably analysed immediately, or kept under appropriate storage conditions until further treatment. Employing good hygienic practices is important in order to avoid contamination and transfer from gloves and utensils.

If food spoilage is the reason of sampling, control (non-spoiled or normal) samples should be analysed at the same time and the results compared to those obtained from spoiled products.

For certified institutions, international standards for sampling and procedures for analysis can be followed. For example, ISO 707:2008 or ISO 6887-5:2010 contain specific guidelines for sampling and preparation of milk and milk products.

Preparation of Test Samples

In order to extract microorganisms from solid food matrices such as hard cheese and from dry starters, an extraction procedure should be applied. In this way, cells are separated from the particles. It is important to assess the extraction efficiency of the procedure. Separation methods based on an initial crude homogenisation in a blender and/or a stomacher can be applied. For hard cheeses, an ultra-turrax or a stomacher is recommended (Coeuret, Dubernet, Bernardeau, Gueguen, & Vernoux, 2003). Softer products can be homogenized and diluted in buffer solution resulting in a sample suspension. If required, dilutions are prepared for further analysis.

For sampling only, the surface of a product without taking a part of the product, a solid culture medium preparation can be used, which is pressed directly on the surface of the sample, thereby getting directly inoculated with the microorganisms present on the surface. Also, a reference measurement from the surrounding air is recommended in order to avoid misinterpretation of results due to the presence of air-borne spores (Samson, Houbraken, Thrane, Frisvad, & Andersen, 2010).

Another method is to take a swab sample, whereby a sterile cotton tip is used to swab the surface of the sample and then inoculate a nutrient agar plate (Samson et al., 2010).

METHODS BASED ON CULTURING

Classical Microbiology Procedures

Classical microbiological methods rely on living microorganisms that can be cultured on microbiological culture media containing substrates. For bacteria as well as for fungi and yeasts, viability is typically assessed by plate counting on a suitable growth medium. Individual cells, spores or viable mycelial fragments will propagate and form colonies, which can be recorded as colony forming units (CFU).

Culturing methods to enumerate microbial populations, so called microbial counts, provide a picture of the number of microorganisms present in a sample (see Appendix protocol 1). Enumeration of fungi by dilution plating is usually done by spread-plating, which means, that the dilution is spread on the surface of the agar rather than by pour plating. The reason for this is that in general fungi are adapted to growing on surfaces rather than growing submerged. As a result, fungal colonies emerge faster and can develop their characteristic appearance, which can help with initial identification. Bacterial counts can be performed by the pour plate method (Oyeleke, 2009; Uzeh, Ohenhen, & Rojugbokan, 2006) (see Appendix protocol 2). During this method, the sample is mixed with molten agar medium and poured in a culture dish to solidify. A larger sample volume (e.g. 1 ml) can be used than with spread-plating, but the organism needs to be able to briefly withstand the temperature of liquid agar (around 40-45°C).

Dilution plating is suitable for liquid and soft products (Hocking, 2007). Results from dilution plating can be expressed as colony forming units per gram or ml.

An alternative and more rapid method for yeast and mould enumeration is the hydrophobic grid membrane filter technique (Tournas, 2009). The filtration unit consists of a square filter and a hydrophobic grid with 1600 compartments in which organisms are isolated during filtration. The membrane is then transferred to an agar plate and incubated. The grid facilitates counting of the colonies manually and electronically.

For isolation of specific microorganisms, selective media can be used. Selective media favour the growth of particular microorganisms by suppressing the growth of others or by containing specific nutrients. Several types of agar media are commercially available for isolation of certain groups of bacteria, for example, MRS agar for Lactobacilli, M17 for Lactococci, Violet Red Bile Agar for Enterobacteriaceae or Baird-Parker agar for Staphylococci (Ozturkoglu Budak, 2016) (see Appendix protocols 1, 2).

The limitation of culture-dependent methods lies in the fact that only those species will be detected, which are able to grow under the chosen conditions. These conditions can be the type of culture media, or physical parameters, such as temperature or oxygen availability. Considering the vast number of culture media and culturing conditions (pH, temperature), culture-dependent methods can become very laborious and time consuming if performed thoroughly. Additionally, microorganisms might be unevenly distributed in the product. Underestimation of the true microbial counts can result from oxidative killing of anaerobic microorganisms such as Bifidobacterium during plating or if bacteria occur in chains and/or clumps (Auty et al., 2001). Another point of consideration is the handling of samples, as during sampling and sample storage, certain microorganisms might be lost or killed.

Culturing methods not only record CFUs and thereby giving a quantitative result, they also afford the opportunity of a qualitative identification of species or groups of microorganisms. This offers a great advantage, as strains isolated from certain samples, can be purified and kept for further in-depth studies, for example activity assays (Ozturkoglu Budak, 2016). Microorganisms isolated from cheese or milk cream samples have also been effectively applied in screening experiments for lipase producing fungi (Griebeler et al., 2011).

Strain Isolation for Identification

After microorganisms are detected and enumerated, the next step towards identification is normally the preparation of pure cultures. Repeated subculturing onto fresh media is done by streak-inoculation. For fungi, the media used for streak-inoculation should support the formation of species specific differences in the culture appearance. This will also help in detecting contaminations (Samson et al., 2010). See Appendix protocols 3, 4, 5, 6 for protocols on how to isolate and identify strains.

Cultures can be maintained on agar slants for further identification. For future studies and long-term preservation, it is generally advised to store cultures as frozen glycerol stocks. Certain species, for example non-sporulating fungi may require alternative preservation methods for long-term storage.

Identification Media and Methods for Fungi

Although it is possible to identify some fungi directly from an enumeration plate, fungi need to be subcultured by streak-inoculation to ensure purity. Inoculation on specific identification media can help in confirming the identity of a fungus. For the identification up to species level, several characteristics, such as colony growth rates, colony morphology and microscopic morphology of reproduction structures are evaluated (Hocking, 2007). Standard media, such as Czapek Yeast Extract Agar (CYA), or malt extract agar (MEA) are particularly suitable for Penicillium and Aspergillus species (Hocking, 2007). To identify fungi that are able to grow under dry conditions (xerotolerant fungi), Dichloran 18% Glycerol agar

(DG18) and Malt Yeast 50% Glucose agar (MY50G) are used. Xerotolerant fungi are able to grow at low water activity (a_w) and for identification of fungi on food products with $a_w < 0.90$ (e.g. hard cheeses like parmesan) DG18 is recommended, while MY50G is recommended for identification of fungi on food products with $a_w < 0.70$ (Samson et al., 2010). For procedures on cultivation and identification of fungi see Appendix protocols 6, 7.

For identification of yeasts there are physiological and morphological standard methods (Appendix protocol 8).

Identification Media and Methods for Bacteria

Lactic acid bacteria are the major component of starters used in dairy fermentations. There are numerous tools available for their characterisation and identification. Phenotypic tests include the fundamental Gram staining (see Appendix protocol 9), which is based on the difference in cell walls of two major groups of bacteria.

Another morphological feature is the shape of the cell. Many bacteria are rod-shaped, others are spherical, and some are spirally curved (Walstra, Wouters, & Geurts, 2006). Spherical or roughly spherical cells are called cocci. They can exist as individual cells or in characteristic arrays, when cells adhere to each other after cell division, e.g. diplococci. Long chains of cocci are characteristic for *Streptococcus, Enterococcus* and *Lactococcus*. Repeated divisions in random planes result in irregular grapelike clumps, as in the case of *Staphylococcus*. Rod-shaped bacteria are often called bacilli. The shape of the end of the rod often varies between species, being flat, rounded, cigar-shaped or bifurcated. Other shapes are curved rods, spirals or helices, pear-shaped or flat boxes (Prescott, Harley, & Klein, 2001).

Flagella confer motility. They are threadlike appendages extending outward from the plasma membrane and cell wall. Flagella can be arranged in patterns which differ between species and therefore can be used as a taxonomic criterion. Examples of milk associated bacteria and their characteristics are listed in Table 1.

While differences purely based on morphology may be limited, physiological properties, on the contrary, are extremely variable. Identification criteria based on growth are, for example, the ability

Table 1. Some genera of bacteria possibly associated with milk (adapted from (Walstra et al., 2006))

Genus	Morphology	Motility	Gram Reaction	Oxygen Requirement
Micrococcus	Coccus	–	+	Aerobic
Staphylococcus	Clump of cocci	–	+	Aerobic
Lactococcus	Diplococcus	–	+	Aerotolerant
Streptococcus	Chain of cocci	–	+	Aerotolerant
Lactobacillus	Rod, chain	–	+	Aerotolerant
Bacillus	Rod, spores	–	+	Aerobic
Clostridium	Rod, spores	–	+	Anaerobic
Escherichia	Rod	+	–	Facultative aerobic
Salmonella	Rod	+	–	Facultative aerobic
Pseudomonas	Rod	+	–	Aerobic

to use different substrates, oxygen requirement, temperature or salt sensitivity and heterofermentation ability. Such tests have been successfully applied to discriminate between most of the common genera of starter bacteria, such as *Lactococcus, Leuconostoc, Enterococcus, Streptococcus,* or *Lactobacillus* (Cogan et al., 1997).

Covering a wide range of genera and species, lactic acid bacteria share common features, such as Gram-positive, catalase-negative, growth under microaerophilic to strictly anaerobic conditions and lactic acid production. *Lactobacillus* is one of the most important genera of lactic acid bacteria. Classical phenotypic tests have been performed for Lactobacillus strains from dairy products and allowed the characterisation on genus, species or strain level (Coeuret et al., 2003).

Classical Phenotypic Tests

For identification of bacteria, classical phenotypic tests based on physiological characteristics, can be performed. Additional to already mentioned features such as respiratory type, motility, or growth temperature and salt sensitivity, a number of biochemical characteristics can be analysed. Traits, such as homo/hetero-fermentation, production of lactic acid isomers, metabolism of carbohydrate substrates, coagulation of milk and presence of particular enzymes (e.g. arginine dihydrolase, antibiotic susceptibilities) can be assayed (Coeuret et al., 2003).

There are also ready-to-inoculate identification kits available. API® strips are standardised, easy-to-use test systems and allow rapid identification of pure cultures based on databases. However, the reliability of these tests has been questioned. The exact identification of closely related species can be problematic. Furthermore, variability may be observed within a single strain. Such variations may result from the loss or gain of plasmids which often encode proteins involved in carbohydrate fermentation (Coeuret et al., 2003).

MOLECULAR METHODS

Fingerprinting Techniques

Analogous to human fingerprints, which are detailed, durable and nearly unique to identify an individual, molecular markers, such as DNA, can be used for the identification of microorganisms. For example, a set of proteins produced by a bacterium or a set of genes present in the genome of an organism can be seen as a profile specific for this particular strain or species. Even though it might be technically possible to identify every single protein or gene of a set, this is often not necessary as these molecules can be visualised as patterns. And if these patterns differ between strains or species, such patterns or profiles can be used for identification. The methods are then called fingerprinting techniques or profiling methods.

A method that is commonly used to visualise protein or DNA patterns, is gel electrophoresis. This is a method for the separation and analysis of macromolecules based on their size and charge. After insertion in a gel, the macromolecules move through the pores of the gel (sieving matrix) when an electric field is applied. The result is a banding pattern of DNA or proteins sorted by their size, a results plot from an electrophoresis is also called electropherogram.

Protein Fingerprinting

Protein fingerprinting is based on the assumption that a bacterial strain always produces the same set of proteins if grown under standardised conditions. Total soluble cytoplasmic proteins are analysed by protein gel electrophoresis and the patterns obtained allow rapid, computer-based digital analysis for species identification (Coeuret et al., 2003). A combination of phenotypic characterization and whole-cell protein electrophoresis has been applied for the classification of non-starter lactic acid bacteria to species level (Voulgari et al., 2010).

Multilocus enzyme electrophoresis is the analysis of electrophoretic mobilities of certain intracellular enzymes. Proteins are separated based on their net charge. Mutations at the gene locus that cause amino acid substitutions with an effect on the net charge of the enzyme will result in a different mobility on the gel (Stanley & Wilson, 2003). Such analysis can discriminate between bacteria to the species level and beyond (Coeuret et al., 2003).

As for classical tests, the identification of isolates to species level by protein fingerprinting also harbours difficulties. Considerable variations in biochemical attributes can occur between strains of the same species. And some species may not be readily distinguishable in terms of phenotypic features (Coeuret et al., 2003).

DNA-Based Techniques

In the past, identification of microorganisms relied on phenotypic and physiological characters. Identification solely based on morphology is often experienced as being difficult. The analysis of metabolite profiles can also be used for species identification. Besides that, molecular DNA-based methods are nowadays routinely applied for species identification.

Various techniques, such as DNA sequencing, pulsed field gel electrophoresis, ribotyping and PCR based fingerprinting techniques have been performed for species recognition and differentiation. For identification purposes, however, the most frequently used method is DNA sequencing (Samson et al., 2010).

The first step in the procedure of any DNA-based method is the extraction of DNA from the microorganism. There are various DNA extraction protocols and many commercial kits available. A number of general steps have to be followed: First, cultures are grown in liquid or on solid media. Cells are then broken up by mechanical, enzymatic or chemical lysis. Next, DNA is separated from cells, proteins are removed and genomic DNA is precipitated, washed and dissolved in a suitable buffer. An example protocol for DNA extraction from cheese is provided in Appendix protocol 10.

Several methods continue with a step to amplify the piece of DNA of interest. This is performed by PCR.

Polymerase Chain Reaction (PCR)

Many techniques used in molecular biology rely on PCR. In the course of a PCR reaction a segment of DNA is specifically targeted and amplified resulting in thousands to millions of copies. It involves a thermostable DNA polymerase for DNA replication, primers of 20 to 30 base pairs long which bind specifically to the target region, and a thermal cycling system for repeated heating and cooling allowing different temperature-dependent reactions.

Important factors for a successful PCR amplification are the quality and quantity of the template DNA, optimal Mg^{2+} concentration, reduction of secondary structures in the DNA template, and an optimized primer set.

DNA Sequencing for Identification

The sequence of a DNA strand is defined by the order of nucleotides. There are different methods for the determination of a DNA sequence. A commonly used method is the chain-termination method developed by (Sanger, Nicklen, & Coulson, 1977). The sequencing reaction involves a DNA polymerase, a single primer, normal nucleotides and modified di-deoxynucleotides (ddNTPs). When a ddNTP is incorporated in the DNA strand, its elongation is terminated resulting in DNA fragments of different lengths. For dye-terminator sequencing labelled ddNTPs are used. The DNA fragments are separated by high resolution capillary electrophoresis and the fluorescent labels detected.

A sequence read arising from one sequencing reaction can be up to 900 bp long. A cleaned PCR product is a typical template for a sequencing reaction.

Not every gene, locus or region in the DNA is suitable for identification purposes. Therefore, specific loci, which have been proven useful phylogenetic markers, should be considered for the PCR and sequencing reaction. Genes encoding ribosomal RNAs (rRNAs) and their spacer regions are well established taxonomic markers. Comprehensive public databases can be used for comparative sequence-based identification.

DNA sequencing of specific loci is often the method of choice for identification, as it is reproducible in different laboratories, unambiguous and portable. However, interpretation of the results can be difficult and wrongly identified strains in public databases can be misleading and hamper the analysis.

For a description of the general procedure see Appendix protocol 11.

Ribosomal RNA Genes as Phylogenetic Markers

Ribosomes are the translation machinery of the cell. They are composed of protein and a special type of RNA called rRNA. The genes coding for rRNAs are arranged in the rRNA operon, a cluster of rRNA genes in the genome. In eukaryotes, the coding region for rRNA consists of 18S, 5.8S and 28S rRNA genes, which are separated by internal transcribed spacers ITS1 and ITS2. In bacteria, the rRNA genes 16S and 23S rRNA are separated by one internal transcribed spacer and 5.8S rRNA is part of the 23S rRNA molecule.

Ribosomes are present in all living systems and the high degree of functional constancy and different functional selective pressure results in both evolutionary conserved regions as well as highly variable regions within the same rRNA gene. Therefore, rRNA genes are commonly used for phylogenetic classification and identification of organisms. Islands of highly conserved sequence regions within rRNA genes allow the construction of nearly "universal" primers for amplification by PCR (Figure 1).

Figure 1. Eukaryotic rRNA gene repeat unit. Sections can be amplified by PCR through the use of primers indicated by arrows (modified from (Klaubauf, 2008))

The genes encoding 16S rRNA is commonly used for bacteria, while the internal transcribed spacer (ITS) region finds application for fungi. The microbiota in traditional Turkish Divle Cave cheese was identified by this molecular approach (Ozturkoglu Budak, Figge, Houbraken, & de Vries, 2016). The DNA sequences of Lactobacillus 16S rRNA gene provides an accurate basis for identification at genus level (Coeuret et al., 2003).

There are many more examples of studies using rRNA genes for identification of microorganisms isolated from dairy products (e.g. (Delbès, Ali-Mandjee, & Montel, 2007; El-Sharoud, Belloch, Peris, & Querol, 2009; Martín-Platero, Valdivia, Maqueda, Martín-Sánchez, & Martínez-Bueno, 2008).

Hybridisation Techniques

The property of nucleic acids to hybridise (specifically bind) to complementary sequences of another nucleic acid is used for many molecular biology techniques. The similarity between pools of DNA sequences reflect the genetic relatedness of these DNA pools derived from different strains (Rossello-Mora & Amann, 2001). More closely related organisms share more sequence similarity. When using DNA-DNA hybridisation, the melting profile of the hybridized DNA determines the degree of genetic similarity between organisms. The higher the temperature required to melt DNA hybrids, the closer related the organisms (Auch, von Jan, Klenk, & Göker, 2010).

Furthermore, hybridisation methods can be used to target specific areas in the genome or RNAs. Labelled probes are used to target the sequence of interest. Probes are single-stranded nucleic acid fragments that specifically bind to complementary sequences of the target nucleic acid. This method can be applied on a colony directly, or after DNA/RNA extraction. In particular probes that target rRNA sequences can be used for the identification of bacteria (Jaziri et al., 2014).

Strain Differentiation by Fingerprinting Techniques

As mentioned above, rRNA genes are widely used as phylogenetic markers. They are present in all organisms, but the exact sequence is unique to each species, serving as a genetic fingerprint.

Ribotyping

A hybridisation technique for bacterial identification based on the information from rRNA phylogenetics is ribotyping (Rodtong & Tannock, 1993). Genomic DNA is digested with specific restriction enzymes. Different restriction enzymes recognise different sequences, usually of four to eight nucleotides. When a restriction enzymes is applied, the DNA molecules are cut at this specific sequence, resulting in fragments of different lengths. After separation by gel electrophoresis, the DNA fragments are transferred to a membrane and hybridised with a labelled 16S or 23S rRNA probe. Thereby, only the fragments containing rRNA genes are visualised. The pattern is indicative for the organism which can be identified by comparison with reference organisms.

Polymerase Chain Reaction-Restriction Fragment Length Polymorphism (PCR-RFLP)

Another technique which makes use of the marker gene for rRNA, but does not involve hybridisation, is Restriction Fragment Length Polymorphism (RFLP). RFLP is a technique by which organisms may be differentiated by analysing patterns formed from cleavage of their DNA by restriction enzymes.

In contrast to ribotyping which uses whole genomic DNA, for RFLP, a genomic region of interest is PCR amplified first and the PCR product is subjected to restriction digestion. The fragments of different lengths are resolved on a gel and the resulting pattern is analysed (Figure 2).

The RFLP technique on the spacer regions of rRNA genes, was applied on yeasts isolated from traditional Egyptian dairy products. The resulting RFLP patterns were compared with patterns of a reference yeast database and could be assigned to known species (El-Sharoud et al., 2009). The authors further showed that Kariesh cheese was found to be the most diverse in its yeast floras compared with other examined dairy products. Its acidic pH and lower salt content of this cheese may favour the growth and survival of yeasts (El-Sharoud et al., 2009).

PCR-RFLP is a rapid profiling technique and can be used as a screening tool for subsequent deeper analyses of a selected number of strains or for identification if compared to a reference database of known species (see Appendix protocol 12).

Pulsed-Field Gel Electrophoresis (PFGE)

Pulsed field gel electrophoresis is a special form of gel electrophoresis for the separation of large DNA molecules. DNA moves in a gel matrix by applying an electric field that periodically changes direction. By alternating directions of the electric field even very large DNA molecules can move.

The method is used for strain typing. For this, chromosomal DNA is digested with restriction enzymes of low cutting frequency resulting in large fragments. These fragments are separated by PFGE. It is a highly reproducible method and the patterns formed are easily interpreted. However, the number of samples that can be analysed at one time is very limited (Coeuret et al., 2003).

Figure 2. RFLP-profiles of fungal rRNA genes and internal transcribed spacer regions (modified from (Klaubauf, 2008))

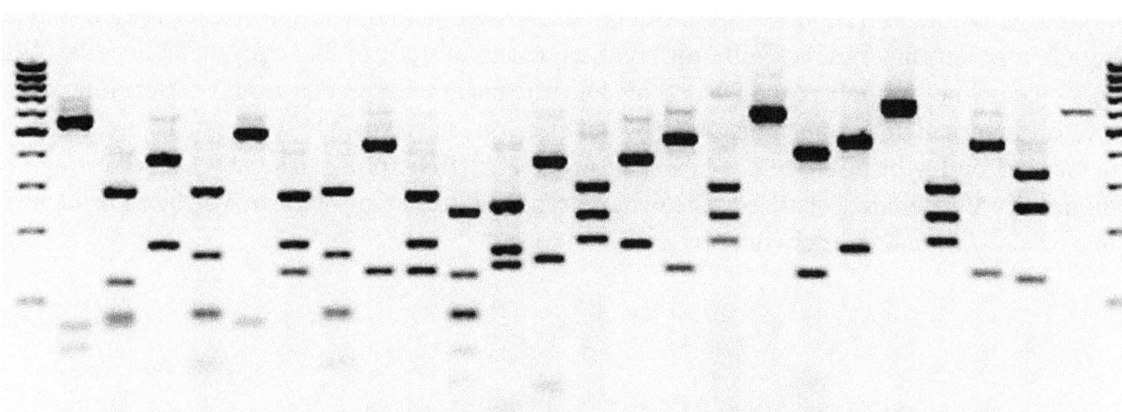

Amplified Fragment Length Polymorphism (AFLP)

By combining PCR and restriction enzyme analysis, amplified fragment length polymorphism (AFLP) can be used to identify species and even strains, for example of *Lactococcus lactis* (Kutahya et al., 2011). The AFLP method uses two restriction enzymes, usually resulting in DNA fragments with two different types of sticky ends. Templates for PCR are created by ligating adaptors to these sticky ends. Primers complementing the adaptor sequence are used to multiply all successfully created DNA fragments with two ligated adaptors at both ends. After multiplication, these fragments can be visualised using electrophoretic separation on a gel matrix. The resulting formation of bands can be used for identification of groups, species or even strains of microorganisms (Vos et al., 1995).

Isolate Identification by Modern Techniques

Some of the methods described above are fairly old, but can still be valuable for certain circumstances, for example in developing countries (Awad et al., 2015).

With the development of next generation sequencing and the rapid progress in bioinformatics, whole-genome sequencing (WGS) of bacterial isolates has become a standard practice (Awad et al., 2015; Gill, 2017; Land et al., 2015). While WGS provides unprecedented discrimination between isolates, standards for analysis and interpretation of WGS data for identification are still lacking (Gill, 2017).

MALDI-TOF MS

Matrix-assisted laser desorption/ionization (MALDI) time of flight (TOF) mass spectrometry (MS) has recently become a relatively fast method to determine species of bacteria (Patel, 2015), fungi (Normand et al., 2013) and yeasts (Quintilla et al., 2018). It is mostly used for clinical purposes, but has also found its way to the food industry (Pavlovic, Huber, Konrad, & Busch, 2013). The apparatus needed is costly and acquiring a useful database requires a lot of knowledge and time, but it is easy in use and fast once a good database exists (Singhal, Kumar, Kanaujia, & Virdi, 2015). The difficulty with using this method is that new species are not easily recognized and more research and other methods are needed when a test results in unknown species.

Nevertheless, MALDI-TOF MS represents a robust and reliable tool for the identification of microorganisms with the potential of reducing the turnaround time for a final identification and the number of isolates that require DNA sequencing (Rodríguez-Sánchez et al., 2016).

CULTURE-INDEPENDENT METHODS

Methods, such as viability tests, identification and isolation of novel starter cultures or in-depth physiological analysis of microorganisms require a living culture isolated from a sample. Possessing all the advantages as discussed above, the main disadvantage is the time and labour-intensive setup. Moreover, culturing approaches are selective and might not give a direct inspection of the microbial populations present in a sample. Tools which do not require culturing can overcome these limitations and open new possibilities for the assessment of microbial populations in dairy products.

Examples of culture-independent methods for the analysis of microbial viability in dairy products are the ATP-Bioluminescence technique (Cunha2014) or the direct microscopic differentiation of live and dead bacteria (Auty et al., 2001). For this method, cells are stained with fluorescent viability indicators and can be counted by confocal laser scanning microscopy. It has been used for in situ viability assessment of bacteria in probiotic dairy products. The study revealed a discrepancy between conventional plate counting and direct microscopic enumeration. It is assumed that clumping of cells leads to an underestimation of bacterial numbers, when culturing techniques are applied. These results indicate the value of the microscopic approach for rapid viability testing for probiotic milk and fermented milk (Auty et al., 2001). While the approach from Auty and co-workers could not distinguish bifidobacteria from other species, the use of fluorescent markers for the specific detection of certain species could be a solution (2001).

Another method that makes use of fluorescence probes is flow cytometry. A flow cytometer detects cells or particles in a stream of fluid. It is used for cell counting and sorting and can detect fluorescent markers. In food industry and especially for the analysis of milk, flow cytometry is a valuable tool for the detection of specific pathogens and for bacterial quality control (Gunasekera, Attfield, & Veal, 2000; McClelland & Pinder, 1994). As milk showed high background, procedures to increase sensitivity have been developed, for example, by applying a sample clearing procedure (Bunthof & Abee, 2002).

Apart from being rapid and sensitive, this technique can measure various physiological characteristics of individual cells. It has been applied on lactic acid bacteria for viability assessment and to differentiate between culturable, metabolically active, but non-culturable and permeable cells. The study of Bunthof & Abee (2002) showed that a significant proportion of intact and active but not cultivatable cells are present in dairy products. The active population of starter cells not able to be cultured, might well contribute to fermentation. Likewise, for probiotic products, those populations might contribute to many of the proposed health effects.

Further applications of flow cytometry are the specific detection of bacterial species with antibodies or fluorescently labelled rRNA probes. The analysis of physiological characteristics, such as membrane integrity, enzyme activities, membrane potential, intracellular pH, respiration, and antibiotic susceptibility is also possible.

Molecular Methods Based on DNA

The use of molecular methods has been a rapidly developing field. DNA-based techniques can be used to investigate population diversity and dynamics. Fingerprinting techniques, such as DGGE, are valuable tools which do not rely on culturing and can be used for high throughput applications. Another implementation of molecular methods is the culture-independent detection of particular species in a sample by specifically designed probes or primers.

In general, molecular methods involve a genomic DNA extraction step, followed by the analysis of the genetic material by sequencing techniques or profiling methods. At first, total genomic DNA from, for example dairy samples, is extracted. This is achieved by a cell lysing step on the basis of mechanical or enzymatic procedures, and thereby releasing cell content including nucleic acids. A step for protein degradation may also be incorporated in the procedure. DNA is further purified, inhibiting substances removed, and dissolved in buffers or water, often including an RNase treatment. Lipid-rich materials such as cheese may require additional steps for fat removal, as its presence can impair bacterial recovery and lysis, thereby strongly affecting DNA yields (Devirgiliis, Zinno, Stirpe, Barile, & Perozzi, 2014).

DNA extracted directly from a sample is a mixture of DNA molecules originating from theoretically all organisms present in the sample. Extracted total DNA may also contain DNA freely present in the sample matrix or in membrane compromised (dead) cells. In order to target the viable community, samples can be treated with propidium monoazide (PMA) prior to DNA extraction, which allows for selective amplification of DNA from intact cells (Nocker, Sossa-Fernandez, Burr, & Camper, 2007). PMA has been applied frequently in recent years for the selective analysis of viable cells, for example during Gouda cheese manufacturing and ripening (Erkus et al., 2016). Furthermore, PMA has been used for the detection of pathogens in milk and milk products in order to eliminate false-positive results (Yu et al., 2017).

Microbial Community Profiling Methods

These techniques, often PCR-based, give a picture of the microbial populations present in a complex matrix. Advantages of PCR-based methods are their simplicity, speed, cost-effectiveness and reliability (Settanni & Corsetti, 2007).

For cheese, the most commonly used community profiling techniques are terminal-RFLP (T-RFLP) and length heterogeneity-PCR (LH-PCR) analysis (reviewed by (Neviani, Bottari, Lazzi, & Gatti, 2013; Pogačić, Kelava, Zamberlin, Dolenčić-Špehar, & Samaržija, 2010).

Terminal-RFLP (T-RFLP)

T-RFLP is a molecular technique for profiling of microbial communities and is based on the detection of fluorescently labelled PCR fragments. Firstly, total DNA serves as template for a PCR reaction targeting a specific genomic region, usually a single gene, such as the rRNA gene. The PCR product consists of a mixture of PCR amplified variants of the target gene. PCR amplicons are labelled by incorporating a fluorescent dye on either one or both PCR primers. Similar to RFLP, PCR products are digested with restriction enzymes. The digested labelled fragments are subjected to capillary electrophoresis and detected with a laser. The result is a profile of terminal restriction fragments (TRFs) (Figure 3) which can be compared to TRF databases, provided as web-based tools or downloadable databases (Kent, Smith, Benson, Triplett, & Triplett, 2003; Rösch & Bothe, 2005). However, the data may be difficult to interpret.

Length Heterogeneity-PCR (LH-PCR)

Length heterogeneity-PCR (LH-PCR) analysis distinguishes different organisms based on the natural length variation of 16S rRNA genes. Variable regions of 16S rRNA genes are PCR amplified using fluorescently labelled universal primers. PCR products are separated by capillary electrophoresis and detected on an automated gene sequencer.

LH-PCR presents an easy, rapid and reproducible method with the limitation of incomplete reference databases.

LH-PCR results showed a variable microbial composition among whey starters for Parmigiano Reggiano opposing earlier findings using culture based approaches (B. Bottari, Santarelli, Neviani, & Gatti, 2010).

Alternative community profiling methods which do not rely on an automated gene sequencer, but on conventional gel electrophoresis are described below.

Figure 3. Peak profiles of terminal restriction fragments visualising microbial community in a series of complex samples

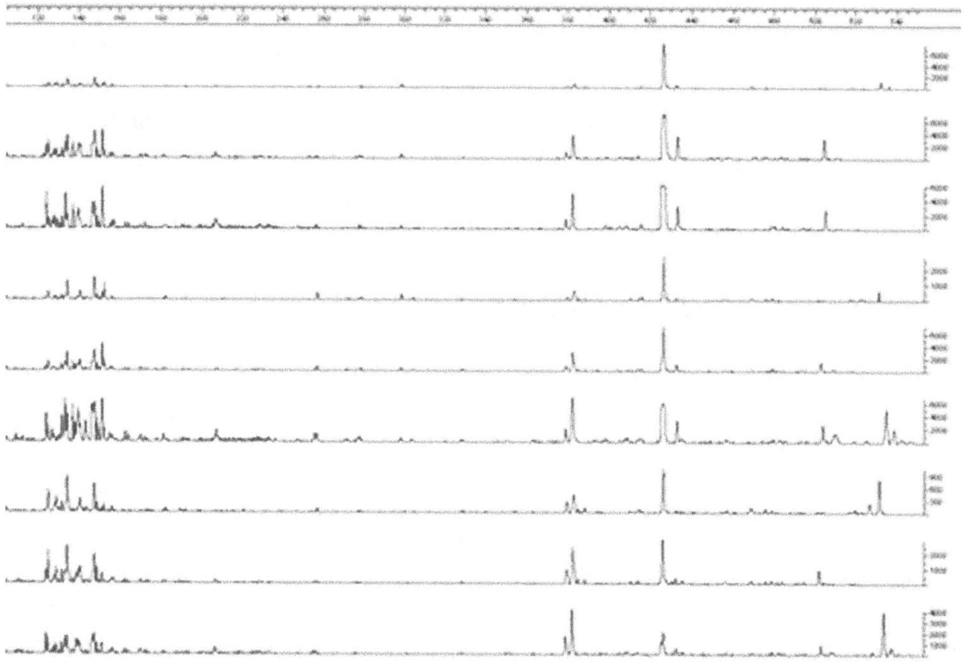

Denaturing Gradient Gel Electrophoresis (DGGE)

Denaturing gradient gel electrophoresis (DGGE) is an electrophoretic method that can separate DNA fragments of similar lengths. DNA molecules of the same length but different basepair sequence can be separated in a polyacrylamide gel with a denaturing gradient generated by urea and formamide. Fragments passing the gel are partially denaturing depending on the bases present in the fragment. Electrophoresis is performed at a constant temperature between 55 and 60°C.

Temporal Temperature Gel Electrophoresis (TTGE)

Temporal temperature gel electrophoresis (TTGE) is a technique similar to DGGE that can separate DNA fragments based on lower electrophoretic mobility of the partially melted double-stranded DNA molecules. Instead of using chemical denaturing agents as in DGGE, temperature is used to denature the DNA fragments. The DNA fragments are translocated over a gel matrix with a linear temperature gradient. The fragments will melt at specific melting temperatures (Tm) depending on the domains of base pairs present in the DNA molecules. This melting will lead to denaturing of the molecules and the transformation of a helical to a partially melted structure will stop migration on the gel.

DGGE as a culture independent method on 16s rRNA fragments was used to check the bacterial cultures in probiotics and has been proven to be more accurate than the culture dependent method (Temmerman, Scheirlinck, Huys, & Swings, 2003). DGGE is a useful method to study the development and changes in the microbial structure on artisanal cheeses as was shown on cheese from Sicily (Randazzo,

Torriani, Akkermans, De Vos, & Vaughan, 2002). This method and TTGE on 16s rRNA were used on several dairy products to follow development of bacterial population during food production and ripening of cheese (El-Baradei, Delacroix-Buchet, & Ogier, 2007; Jany & Barbier, 2008; Lafarge et al., 2004; Le Bourhis et al., 2005; Licitra et al., 2007; Ogier, Son, Gruss, Tailliez, & Delacroix-Buchet, 2002). Furthermore, the PCR-DGGE approach has been used to identify microbial species occurring in natural whey cultures used as starter for the production of water buffalo Mozzarella cheese (Ercolini, 2004).

Direct PCR for Detection of Certain Species

Targeted PCR reactions directly applied on the DNA extracted from a complex sample allow the detection of specific organisms, such as contaminants or pathogens. By the use of specific primers, the microorganism of choice can be targeted. This method has been used for decades and experienced many new developments and improved variations have been published.

It has been shown already early on, that PCR-based detection is more sensitive than conventional culture methods (Wegmuller, Luthy, & Candrian, 1993).

The design of specific primers is facilitated by the accumulation of sequence data over the years.

While, rRNA genes are still often the molecular target for species-specific detection, alternative loci have been shown to work as well or offer other benefits. For instance, other loci may show more sequence variability compared to 16S rRNA for better discrimination of closely related Bacteria. In addition, single copy genes may be more suitable as target for quantification purposes (Benedetta Bottari, Agrimonti, Gatti, Neviani, & Marmiroli, 2013).

Another improvement is the development of multiplex PCR assays. Simultaneous amplification of more than one locus with several specific primer sets enables the detection of multiple microorganisms in a single reaction. Multiplex PCR is a rapid method for the identification of selected species and represents the fastest culture-independent approach for strain-specific detection in complex matrices reviewed by (Settanni & Corsetti, 2007).

Quantitative PCR (qPCR)

Unlike conventional PCR, which gives the result at the end of the reaction, a real-time PCR or quantitative PCR (qPCR) monitors the amplification of a targeted DNA molecule during the reaction (in real-time). The detection is based on fluorescent dyes, which can be non-specific dyes that intercalate with any double-stranded DNA or labelled sequence-specific DNA probes. The possibility for quantification makes it an interesting tool if detection alone is not sufficient. For example, a qPCR assay has been developed for quantitative detection of tyramine-producing bacteria in dairy products. The data showed a good relationship between the number of detected tyramine-producing bacteria and tyramine concentration in cheese samples. The qPCR method could be used as a preventive tool for identifying starters, raw material and cheeses contaminated with tyramine-producing bacteria (Ladero, Martínez, Martín, Fernández, & Alvarez, 2010).

Furthermore, a rapid screening method for the simultaneous detection of several thermophilic lactic acid bacteria starters for hard cooked cheeses has been developed. This approach made use of a multiplex real time PCR, targeting the gene phenylalanine synthase with species-specific primers. Specific detection was achieved by the analysis of melting curves which were distinct for each of the different amplicons. Although not used for quantification, this method provides a fast and sensitive detection of the targeted species (Benedetta Bottari et al., 2013).

Additional to applying qPCR on genomic DNA templates, it is often used on reverse transcribed RNA (cDNA) templates. In this way, it can be used to specifically identify metabolically active and viable microorganisms. A real-time reverse transcription PCR assay has been developed for the detection and quantification of viable yeasts and moulds contaminating yogurts. The assay was based on the detection of actin mRNA as an indicator of cell viability (Bleve, Rizzotti, Dellaglio, & Torriani, 2003).

In food microbiology, qPCR finds several applications, such as quantification of pathogens, detection of beneficial microbial populations or the study of growth dynamics and metabolic activities for example during cheese ripening (Postollec, Falentin, Pavan, Combrisson, & Sohier, 2011).

Clone Libraries for Species Identification

For the analysis of microbial community composition in a matrix, there are different possibilities to achieve detailed sequence data. A very common method is amplicon-based sequencing. As for other methods described above, total DNA is extracted from a matrix and universal primers are used for PCR amplification of rRNA genes. Unlike fingerprinting methods, which are based on patterns, sequencing methods result in the actual nucleotide sequences of individual microorganisms present in the matrix. Before high throughput sequencing became available and affordable, classical cloning procedures have been used to separate the individual DNA molecules present in the amplicon mix resulting in a clone library. For this, DNA molecules are ligated into plasmids which, after transformed into Escherichia coli, can be multiplied, purified and the inserted fragment can be sequenced by the Sanger method. In this way, long high-quality reads can be obtained for taxonomic identification, while being rather expensive on the downside. It can also be used for quantification purposes. The number of clones analysed determines the depth of the analysis. If only a few clones are sequenced, the result will be the dominating species in the community. For example, dominant species from curd and cheese samples have been identified by the construction of a 16S rRNA clone library (Randazzo et al., 2002). Rare species will only be detected if the sequencing effort is proportionally large. The number of clones resulting in the same species can give an indication on the frequency of this species in the original sample. However, as this method includes a PCR step beforehand, results can be biased as even universal primers may exhibit different amplification efficiencies for different taxonomic groups. On the other hand, by deliberately selecting primers with a certain specificity, it is possible to focus on certain taxonomic groups.

Metagenomics and Modern Sequencing Technologies

An approach in which the total genomic content from a matrix is sequenced including taxonomic relevant genes as well as functional genes, refers to the term metagenomics (National Research Council (US) Committee on Metagenomics: Challenges and & Functional Applications, 2007).

During the last 10 years, high throughput sequencing technologies (so called next generation sequencing, NGS) have made considerable progress and have been applied extensively for studying microbial community diversity. For example, rRNA genes in fermented milk have been targeted and sequenced using the 454-pyrosequencing technology (Liu et al., 2015). The 454 system (Roche) was the first commercially available NGS platform and is based on pyrosequencing (Margulies et al., 2005). Pyrosequencing works on the principle of "sequencing by synthesis", detecting the incorporated nucleotides by the release of pyrophosphate (Ronaghi, Karamohamed, Pettersson, Uhlén, & Nyrén, 1996). Other NGS technologies include the SOLiD technology based on multiplex polony sequencing (Shendure, 2005), or the Illumina (Solexa) platform which involves clonal amplification of DNA on a surface and uses the reversible terminator sequencing chemistry (Turcatti, Romieu, Fedurco, & Tairi, 2008).

The major advantage of NGS is that the large quantity of data produced by the parallelized sequencing process has made (meta-)genome sequencing affordable. A recognized limitation of NGS technologies is that they rely on a PCR amplification step, which introduces a bias in read distribution (Cao, Fanning, Proos, Jordan, & Srikumar, 2017).

Third generation sequencing technologies are currently under active development and were designed for single molecule sequencing. Direct sequencing of single DNA molecules reduces error rates by avoiding amplification associated bias or synchronization problems (Cao et al., 2017).

NGS technologies have been widely applied to the study of food microbial consortia. In recent years, several studies on dairy and fermented milks used high throughput sequencing for the characterization of the microbial composition (reviewed by De Filippis et al, 2017).

NGS technologies enable detailed analysis of changes in microbial structure over time, along with changes in the gene expression patterns related to different fermentation stages (Cao et al., 2017; Lessard, Viel, Boyle, St-Gelais, & Labrie, 2014).

A broad variety of cheeses were studied through amplicon-based sequencing of rRNA gene regions, which allowed monitoring of curd fermentation or cheese ripening (De Filippis, Parente, & Ercolini, 2017). Amplicon-based sequencing methods are key for studies on microbial community composition, but do not give information about a community's functional capabilities (Ferrocino & Cocolin, 2017).

Only the combination of different approaches will provide a comprehensive view on how microbial ecosystems function (Ferrocino & Cocolin, 2017). Such information is especially interesting for cheese maturation processes. Dugat-Bony and co-workers have recently shown how data from metagenomic, metatranscriptomic and biochemical analyses can provide a better view on the metabolic activities of the different community members and their possible interactions (Dugat-Bony et al., 2015).

Fluorescence in Situ Hybridization (FISH)

Unlike the before mentioned molecular methods, fluorescence in situ hybridization (FISH) does not require a DNA extraction step, which makes it a powerful and rapid tool for identifying and quantifying microbial populations. For this non-PCR-based molecular technique, rRNA-targeted probes are used to identify single cells directly in a sample. FISH has been used, for example, for the detection of microorganisms in cheese (Ercolini, Hill, & Dodd, 2003) or to identify and quantify populations in starter cultures within 1 day (Olsen, Brockmann, & Molin, 2007).

FUTURE RESEARCH DIRECTIONS

Despite the disadvantages of culture-dependent methods (e.g. time- and labour-intensive, growth-related biases), an isolated strain presents a valuable piece for further research and applications. Continuous efforts on characterisation and classification of novel strains as well as an updated classification system are indispensable for correct species identification. Regarding identification of cultures, over the years a trend of moving from phenotypical to genotypical identification can be observed. Molecular tools, with their higher sensitivity and accuracy, facilitate species recognition. A fundamental reassessment of traditional systematics released by molecular phylogenetic studies led to new classifications.

Molecular, taxonomic and physiological data obtained from pure cultures are crucial for the validation of culture-independent methods.

The available methods have been developed for specific research needs. There is still room for adaptations and new research questions may require adjustments in a certain direction which will lead to improved approaches.

Before choosing a specific method, it is important to carefully identify the research question and purpose of the analysis. With the number of available methods, picking the right tool for optimal results, might be a difficult task. Several other factors may play a role as well. For example, is rapid identification of highest priority or the accurate detection of a pathogen? In the end, the choice of method depends on a balance of speed, accuracy and finances.

For the identification of isolates, we recommend rRNA gene/ITS sequencing for initial species determination. With the current efforts of establishing a comprehensive database for MALDI-TOF and the advantage of being fast, robust and reliable, this method has good potential for the future.

Another benefit of cultures is the analysis of their genomic content. With the progress of new sequencing technologies, WGS is becoming a standard method.

A better understanding of how enzymes are involved in processes, such as cheese ripening, can be achieved by combining biochemical data with genomics.

As for culture-independent techniques, flow cytometry and omics technologies display interesting prospects for the future. For quality control analyses, method validation is important and official international standards need to be developed. The application of omics methods on milk and dairy products will increase our knowledge on the identity of the microbiota, their metabolic activities and their interactions.

REFERENCES

Altschul, S. F., Gish, W., Miller, W., Myers, E. W., & Lipman, D. J. (1990). Basic local alignment search tool. *Journal of Molecular Biology, 215*(3), 403–410. doi:10.1016/S0022-2836(05)80360-2 PMID:2231712

Apajalahti, J. H., Särkilahti, L. K., Mäki, B. R., Heikkinen, J. P., Nurminen, P. H., & Holben, W. E. (1998). Effective recovery of bacterial DNA and percent-guanine-plus-cytosine-based analysis of community structure in the gastrointestinal tract of broiler chickens. *Applied and Environmental Microbiology, 64*(10), 4084–4088. PMID:9758849

Auch, A. F., von Jan, M., Klenk, H.-P., & Göker, M. (2010). Digital DNA-DNA hybridization for microbial species delineation by means of genome-to-genome sequence comparison. *Standards in Genomic Sciences, 2*(1), 117–134. doi:10.4056igs.531120 PMID:21304684

Auty, M. A. E., Gardiner, G. E., McBrearty, S. J., O' Sullivan, E. O., Mulvihill, D. M., Collins, J. K., ... Ross, R. P. (2001). Direct In Situ Viability Assessment of Bacteria in Probiotic Dairy Products Using Viability Staining in Conjunction with Confocal Scanning Laser Microscopy. *Applied and Environmental Microbiology, 67*(1), 420–425. doi:10.1128/AEM.67.1.420-425.2001 PMID:11133474

Awad, M., Ouda, O., El-Refy, A., El-Feky, F. A., Mosa, K. A., & Helmy, M. (2015). FN-identify: Novel restriction enzymes-based method for bacterial identification in absence of genome sequencing. *Advances in Bioinformatics*, *2015*, 1–14. doi:10.1155/2015/303605 PMID:26880910

Bleve, G., Rizzotti, L., Dellaglio, F., & Torriani, S. (2003). Development of reverse transcription (RT)-PCR and real-time RT-PCR assays for rapid detection and quantification of viable yeasts and molds contaminating yogurts and pasteurized food products. *Applied and Environmental Microbiology*, *69*(7), 4116–4122. doi:10.1128/AEM.69.7.4116-4122.2003 PMID:12839789

Bottari, B., Agrimonti, C., Gatti, M., Neviani, E., & Marmiroli, N. (2013). Development of a multiplex real time PCR to detect thermophilic lactic acid bacteria in natural whey starters. *International Journal of Food Microbiology*, *160*(3), 290–297. doi:10.1016/j.ijfoodmicro.2012.10.011 PMID:23290237

Bottari, B., Santarelli, M., Neviani, E., & Gatti, M. (2010). Natural whey starter for Parmigiano Reggiano: Culture-independent approach. *Journal of Applied Microbiology*, *108*(5), 1676–1684. doi:10.1111/j.1365-2672.2009.04564.x PMID:19849773

Bunthof, C. J., & Abee, T. (2002). *Development of a Flow Cytometric Method To Analyze Subpopulations of Bacteria in Probiotic Products and Dairy Starters*. Academic Press.

Cao, Y., Fanning, S., Proos, S., Jordan, K., & Srikumar, S. (2017). A review on the applications of next generation sequencing technologies as applied to food-related microbiome studies. *Frontiers in Microbiology*, *8*, 1–16. doi:10.3389/fmicb.2017.01829 PMID:29033905

Coeuret, V., Dubernet, S., Bernardeau, M., Gueguen, M., & Vernoux, J. P. (2003). Isolation, characterisation and identification of lactobacilli focusing mainly on cheeses and other dairy products. *Le Lait*, *83*(4), 269–306. doi:10.1051/lait:2003019

Cogan, T. M., Barbosa, M., Beuvier, E., Bianchi-Salvadori, B., Cocconcelli, P. S., Fernandes, I., ... Rodriguez, E. (1997). Characterization of the lactic acid bacteria in artisanal dairy products. *The Journal of Dairy Research*, *64*(3), 409–421. doi:10.1017/S0022029997002185

De Filippis, F., Parente, E., & Ercolini, D. (2017). Metagenomics insights into food fermentations. *Microbial Biotechnology*, *10*(1), 91–102. doi:10.1111/1751-7915.12421 PMID:27709807

Delavenne, E., Mounier, J., Asmani, K., Jany, J. L., Barbier, G., & Le Blay, G. (2011). Fungal diversity in cow, goat and ewe milk. *International Journal of Food Microbiology*, *151*(2), 247–251. doi:10.1016/j.ijfoodmicro.2011.08.029 PMID:21944758

Delbès, C., Ali-Mandjee, L., & Montel, M.-C. (2007). Monitoring bacterial communities in raw milk and cheese by culture-dependent and -independent 16S rRNA gene-based analyses. *Applied and Environmental Microbiology*, *73*(6), 1882–1891. doi:10.1128/AEM.01716-06 PMID:17259356

Devirgiliis, C., Zinno, P., Stirpe, M., Barile, S., & Perozzi, G. (2014). Functional Screening of Antibiotic Resistance Genes from a Representative Metagenomic Library of Food Fermenting Microbiota. *BioMed Research International*. PMID:25243126

Dugat-Bony, E., Straub, C., Teissandier, A., Onésime, D., Loux, V., Monnet, C., ... Bonnarme, P. (2015). Overview of a surface-ripened cheese community functioning by meta-omics analyses. *PLoS One*, *10*(4), 1–25. doi:10.1371/journal.pone.0124360 PMID:25867897

El-Baradei, G., Delacroix-Buchet, A., & Ogier, J. C. (2007). Biodiversity of bacterial ecosystems in traditional Egyptian Domiati cheese. *Applied and Environmental Microbiology*, *73*(4), 1248–1255. doi:10.1128/AEM.01667-06 PMID:17189434

El-Sharoud, W. M., Belloch, C., Peris, D., & Querol, A. (2009). Molecular identification of yeasts associated with traditional egyptian dairy products. *Journal of Food Science*, *74*(7), M341–M346. doi:10.1111/j.1750-3841.2009.01258.x PMID:19895478

Ercolini, D. (2004). PCR-DGGE fingerprinting: Novel strategies for detection of microbes in food. *Journal of Microbiological Methods*, *56*(3), 297–314. doi:10.1016/j.mimet.2003.11.006 PMID:14967221

Ercolini, D., Hill, P. J., & Dodd, C. E. R. (2003). Development of a fluorescence in situ hybridization method for cheese using a 16S rRNA probe. *Journal of Microbiological Methods*, *52*(2), 267–271. doi:10.1016/S0167-7012(02)00162-8 PMID:12459248

Erkus, O., de Jager, V. C. L., Geene, R. T. C. M., van Alen-Boerrigter, I., Hazelwood, L., van Hijum, S. A. F. T., ... Smid, E. J. (2016). Use of propidium monoazide for selective profiling of viable microbial cells during Gouda cheese ripening. *International Journal of Food Microbiology*, *228*, 1–9. doi:10.1016/j.ijfoodmicro.2016.03.027 PMID:27077825

Ferrocino, I., & Cocolin, L. (2017). Current perspectives in food-based studies exploiting multi-omics approaches. *Current Opinion in Food Science*, *13*(February), 10–15. doi:10.1016/j.cofs.2017.01.002

Gill, A. (2017). The importance of bacterial culture to food microbiology in the age of genomics. *Frontiers in Microbiology*, *8*, 1–6. doi:10.3389/fmicb.2017.00777 PMID:28507541

Griebeler, N., Polloni, A. E., Remonatto, D., Arbter, F., Vardanega, R., Cechet, J. L., ... Ninow, J. L. (2011). Isolation and screening of lipase-producing fungi with hydrolytic activity. *Food and Bioprocess Technology*, *4*(4), 578–586. doi:10.100711947-008-0176-5

Gunasekera, T. S., Attfield, P. V., & Veal, D. A. (2000). A flow cytometry method for rapid detection and enumeration of fungal spores in the atmosphere. *Applied and Environmental Microbiology*, *66*(3), 1228–1232. doi:10.1128/AEM.66.3.1228-1232.2000 PMID:10698799

Hocking, A. D. (2007). Current methods for the detection of molds and yeasts. In *Food Mycology 2007: Emerging mold problems and spoilage in food and beverages. Proceedings of Food Mycology, 2007*, 4–8.

Jany, J., & Barbier, G. (2008). Culture-independent communities in cheese methods for identifying microbial communities in cheese. *Food Microbiology*, *25*(7), 839–848. doi:10.1016/j.fm.2008.06.003 PMID:18721671

Jaziri, F., Parisot, N., Abid, A., Denonfoux, J., Ribière, C., & Gasc, C. (2014). … Peyret, P. (2014). Phy-lOPDb: A 16S rRNA oligonucleotide probe database for prokaryotic identification. *Database*, (1): 1–7.

Kent, A. D., Smith, D. J., Benson, B. J., Triplett, W., & Triplett, E. W. (2003). Web-Based phylogenetic assignment tool for analysis of terminal restriction fragment length polymorphism profiles of microbial communities. *Applied and Environmental Microbiology*, *69*(11), 6798. doi:10.1128/AEM.69.11.6768-6776.2003 PMID:14602639

Klaubauf, S. (2008). *The role of fungal nitrogen cycling in soil systems: assessment of fungal diversity and development of a sequence set of fungal nitrate reductases*. Universität Wien.

Klaubauf, S., Inselsbacher, E., Zechmeister-Boltenstern, S., Wanek, W., Gottsberger, R., Strauss, J., & Gorfer, M. (2010). Molecular diversity of fungal communities in agricultural soils from Lower Austria. *Fungal Diversity*, 44. PMID:23794962

Kutahya, O. E., Starrenburg, M. J. C., Rademaker, J. L. W., Klaassen, C. H. W., van Hylckama Vlieg, J. E. T., Smid, E. J., & Kleerebezem, M. (2011). High-Resolution Amplified Fragment Length Polymorphism Typing of Lactococcus lactis Strains Enables Identification of Genetic Markers for Subspecies-Related Phenotypes. *Applied and Environmental Microbiology*, *77*(15), 5192–5198. doi:10.1128/AEM.00518-11 PMID:21666014

Ladero, V., Martínez, N., Martín, M. C., Fernández, M., & Alvarez, M. A. (2010). qPCR for quantitative detection of tyramine-producing bacteria in dairy products. *Food Research International*, *43*(1), 289–295. doi:10.1016/j.foodres.2009.10.007

Lafarge, V., Ogier, J., Girard, V., Maladen, V., Leveau, J., Gruss, A., & Delacroix-Buchet, A. (2004). Raw cow milk bacterial population shifts attributable to refrigeration raw cow milk bacterial population shifts attributable to refrigeration. *Applied and Environmental Microbiology*, *70*(9), 5644–5650. doi:10.1128/AEM.70.9.5644-5650.2004 PMID:15345453

Land, M., Hauser, L., Jun, S.-R., Nookaew, I., Leuze, M. R., Ahn, T.-H., ... Ussery, D. W. (2015). Insights from 20 years of bacterial genome sequencing. *Functional & Integrative Genomics*, *15*(2), 141–161. doi:10.100710142-015-0433-4 PMID:25722247

Le Bourhis, A.-G., Saunier, K., Dore, J., Carlier, J.-P., Chamba, J.-F., Popoff, M.-R., & Tholozan, J.-L. (2005). Development and validation of PCR primers to assess the diversity of Clostridium spp. in cheese by temporal temperature gradient gel electrophoresis. *Applied and Environmental Microbiology*, *71*(1), 29–38. doi:10.1128/AEM.71.1.29-38.2005 PMID:15640166

Ledenbach, L. H., & Marshall, R. T. (2009). Microbiological Spoilage of Dairy Products. In W. H. Sperber & M. P. Doyle (Eds.), *Compendium of the Microbiological Spoilage of Foods and Beverages, Food Microbiology and Food Safety. Springer Science+Business Media, LLC*. doi:10.1007/978-1-4419-0826-1_2

Lessard, M.-H., Viel, C., Boyle, B., St-Gelais, D., & Labrie, S. (2014). Metatranscriptome analysis of fungal strains Penicillium camemberti and Geotrichum candidum reveal cheese matrix breakdown and potential development of sensory properties of ripened Camembert-type cheese. *BMC Genomics*, *15*(1), 235. doi:10.1186/1471-2164-15-235 PMID:24670012

Licitra, G., Ogier, J. C., Parayre, S., Pediliggieri, C., Carnemolla, T. M., Falentin, H., ... Lortal, S. (2007). Variability of bacterial biofilms of the "tina" wood vats used in the Ragusano cheese-making process. *Applied and Environmental Microbiology*, *73*(21), 6980–6987. doi:10.1128/AEM.00835-07 PMID:17720831

Liu, W., Zheng, Y., Kwok, L.-Y., Sun, Z., Zhang, J., Guo, Z., ... Zhang, H. (2015). High-throughput sequencing for the detection of the bacterial and fungal diversity in Mongolian naturally fermented cow's milk in Russia. *BMC Microbiology*, *15*(1), 45. doi:10.118612866-015-0385-9 PMID:25887414

Margulies, M., Egholm, M., Altman, W. E., Attiya, S., Bader, J. S., Bemben, L. A., ... Rothberg, J. M. (2005). Genome Sequencing in Open Microfabricated High Density Picoliter Reactors. *Nature*, *437*(7057), 376–380. doi:10.1038/nature03959 PMID:16056220

Martín-Platero, A. M., Valdivia, E., Maqueda, M., Martín-Sánchez, I., & Martínez-Bueno, M. (2008). Polyphasic approach to bacterial dynamics during the ripening of Spanish farmhouse cheese, using culture-dependent and -independent methods. *Applied and Environmental Microbiology*, *74*(18), 5662–5673. doi:10.1128/AEM.00418-08 PMID:18658288

McClelland, R. G., & Pinder, A. C. (1994). Detection of Salmonella typhimurium in dairy products with flow cytometry and monoclonal antibodies. *Applied and Environmental Microbiology*, *60*(12), 4255–4262. PMID:7811064

Moreira, S. R., Schwan, R. F., De Carvalho, E. P., & Wheals, A. E. (2001). Isolation and identification of yeasts and filamentous fungi from yoghurts in Brazil. *Brazilian Journal of Microbiology*, *32*(2), 117–122. doi:10.1590/S1517-83822001000200009

National Research Council (US) Committee on Metagenomics Challenges and Functional Applications. (2007). *The New Science of Metagenomics. The New Science of Metagenomics: Revealing the Secrets of Our Microbial Planet*. National Academies Press (US). Retrieved from http://www.ncbi.nlm.nih.gov/pubmed/21678629

Neviani, E., Bottari, B., Lazzi, C., & Gatti, M. (2013). New developments in the study of the microbiota of raw-milk, long-ripened cheeses by molecular methods: The case of Grana Padano and Parmigiano Reggiano. *Frontiers in Microbiology*, *4*(FEB), 1–14. PMID:23450500

Nocker, A., Sossa-Fernandez, P., Burr, M. D., & Camper, A. K. (2007). Use of propidium monoazide for live/dead distinction in microbial ecology. *Applied and Environmental Microbiology*, *73*(16), 5111–5117. doi:10.1128/AEM.02987-06 PMID:17586667

Normand, A.-C., Cassagne, C., Ranque, S., L'ollivier, C., Fourquet, P., Roesems, S., ... Piarroux, R. (2013). Assessment of various parameters to improve MALDI-TOF MS reference spectra libraries constructed for the routine identification of filamentous fungi. *BMC Microbiology*, *13*(1), 76. doi:10.1186/1471-2180-13-76 PMID:23565856

Ogier, J.-C., Son, O., Gruss, A., Tailliez, P., & Delacroix-Buchet, A. (2002). Identification of the bacterial microflora in dairy products by temporal temperature gradient gel electrophoresis. *Applied and Environmental Microbiology*, *68*(8), 3691–3701. doi:10.1128/AEM.68.8.3691-3701.2002 PMID:12147461

Olsen, K. N., Brockmann, E., & Molin, S. (2007). Quantification of Leuconostoc populations in mixed dairy starter cultures using fluorescence in situ hybridization. *Journal of Applied Microbiology, 103*(4), 855–863. doi:10.1111/j.1365-2672.2007.03298.x PMID:17897187

Oyeleke, S. B. (2009). Microbial assessment of some commercially prepared yoghurt retailed in Minna, Niger State. *African Journal of Microbiological Research, 3*(5), 245–248.

Ozturkoglu Budak, S. (2016). *Dynamics of complex microbiota and enzymes in Divle Cave cheese and their biochemical consequences*. Utrecht University.

Ozturkoglu Budak, S., Figge, M. J., Houbraken, J., & de Vries, R. P. (2016). The diversity and evolution of microbiota in traditional Turkish Divle Cave cheese during ripening. *International Dairy Journal, 58*, 50–53. doi:10.1016/j.idairyj.2015.09.011

Patel, R. (2015). MALDI-TOF MS for the diagnosis of infectious diseases. *Clinical Chemistry, 61*(1), 100–111. doi:10.1373/clinchem.2014.221770 PMID:25278500

Pavlovic, M., Huber, I., Konrad, R., & Busch, U. (2013). Application of MALDI-TOF MS for the Identification of Food Borne Bacteria. *The Open Microbiology Journal, 7*(1), 135–141. doi:10.2174/1874285801307010135 PMID:24358065

Peterson, S. W. (2008). Phylogenetic analysis of Aspergillus species using DNA sequences from four loci. *Mycologia, 100*(2), 205–226. doi:10.1080/15572536.2008.11832477 PMID:18595197

Pogačić, T., Kelava, N., Zamberlin, Š., Dolenčić-Špehar, I., & Samaržija, D. (2010). Methods for culture-independent identification of lactic acid bacteria in dairy products. *Food Technology and Biotechnology, 48*(1), 3–10.

Postollec, F., Falentin, H., Pavan, S., Combrisson, J., & Sohier, D. (2011). Recent advances in quantitative PCR (qPCR) applications in food microbiology. *Food Microbiology, 28*(5), 848–861. doi:10.1016/j.fm.2011.02.008 PMID:21569926

Prescott, L. M., Harley, J. P., & Klein, D. A. (2001). Microbiology (5th ed.). McGraw-Hill Higher Education.

Quintilla, R., Kolecka, A., Casaregola, S., Daniel, H. M., Houbraken, J., Kostrzewa, M., ... Groenewald, M. (2018). MALDI-TOF MS as a tool to identify foodborne yeasts and yeast-like fungi. *International Journal of Food Microbiology, 266*, 109–118. doi:10.1016/j.ijfoodmicro.2017.11.016 PMID:29202340

Randazzo, C. L., Torriani, S., Akkermans, A. D. L., De Vos, W. M., & Vaughan, E. E. (2002). Diversity, dynamics, and activity of bacterial communities during production of an artisanal sicilian cheese as evaluated by 16S rRNA analysis. *Applied and Environmental Microbiology, 68*(4), 1882–1892. doi:10.1128/AEM.68.4.1882-1892.2002 PMID:11916708

Rodríguez-Sánchez, B., Alcalá, L., Marín, M., Ruiz, A., Alonso, E., & Bouza, E. (2016). Evaluation of MALDI-TOF MS (Matrix-Assisted Laser Desorption-Ionization Time-of-Flight Mass Spectrometry) for routine identification of anaerobic bacteria. *Anaerobe, 42*, 101–107. doi:10.1016/j.anaerobe.2016.09.009 PMID:27702604

Rodtong, S., & Tannock, G. W. (1993). Differentiation of Lactobacillus strains by ribotyping. *Applied and Environmental Microbiology*, *59*(10), 3480–3484. PMID:7504432

Ronaghi, M., Karamohamed, S., Pettersson, B., Uhlén, M., & Nyrén, P. (1996). Real-Time DNA Sequencing Using Detection of Pyrophosphate Release. *Analytical Biochemistry*, *242*(1), 84–89. doi:10.1006/abio.1996.0432 PMID:8923969

Rösch, C., & Bothe, H. (2005). Improved assessment of denitrifying, N2 -fixing, and total-community bacteria by terminal restrictionfragment length polymorphism analysis using multiple restriction enzymes. *Applied and Environmental Microbiology*, *71*(4), 2026–2035. doi:10.1128/AEM.71.4.2026-2035.2005 PMID:15812035

Rossello-Mora, R., & Amann, R. (2001). The species concept for procaryotes. *FEMS Microbiology Ecology*, *25*(1), 39–67. doi:10.1111/j.1574-6976.2001.tb00571.x PMID:11152940

Samson, R. A., Houbraken, J., Thrane, U., Frisvad, J. C., & Andersen, B. (2010). *Food and indoor fungi - CBS laboratory manual series*. Utrecht, The Netherlands: CBS-KNAW Fungal Biodiversity Centre.

Sanger, F., Nicklen, S., & Coulson, A. R. (1977). DNA sequencing with chain-terminating inhibitors. *Proceedings of the National Academy of Sciences of the United States of America*, *74*(12), 5463–5467. doi:10.1073/pnas.74.12.5463 PMID:271968

Settanni, L., & Corsetti, A. (2007). The use of multiplex PCR to detect and differentiate food- and beverage-associated microorganisms: A review. *Journal of Microbiological Methods*, *69*(1), 1–22. doi:10.1016/j.mimet.2006.12.008 PMID:17280731

Shendure, J. (2005). Accurate Multiplex Polony Sequencing of an Evolved Bacterial Genome. *Science*, *309*(5741), 1728–1732. doi:10.1126cience.1117389 PMID:16081699

Singhal, N., Kumar, M., Kanaujia, P. K., & Virdi, J. S. (2015). MALDI-TOF mass spectrometry: An emerging technology for microbial identification and diagnosis. *Frontiers in Microbiology*, *6*(AUG), 1–16. PMID:26300860

Stanley, T., & Wilson, I. G. (2003). Multilocus enzyme electrophoresis: A practical guide. *Molecular Biotechnology*, *24*(2), 203–220. doi:10.1385/MB:24:2:203 PMID:12746557

Tappler, P., Twrdik, F., Reinthaler, F. F., Haas, D., Galler, H., Gorfer, M. … Samson, R. A. (2013). Beurteilung, Messmethoden, Identifizierung. In Pilze in Innenräumen und am Arbeitsplatz (pp. 195–422). Vienna: Springer Vienna.

Temmerman, R., Scheirlinck, I., Huys, G., & Swings, J. (2003). Culture-independent analysis of probiotic products by denaturing gradient gel electrophoresis culture-independent analysis of probiotic products by denaturing gradient gel electrophoresis. *Applied and Environmental Microbiology*, *69*(1), 220–226. doi:10.1128/AEM.69.1.220-226.2003 PMID:12513998

Tournas, V. H. (2009). Evaluation of the Hydrophobic Grid Membrane Filter for the Enumeration of Moulds and Yeasts in Naturally- Contaminated Foods. *Microbiology Insights*, *2*, 31–37. doi:10.4137/MBI.S2906

Turcatti, G., Romieu, A., Fedurco, M., & Tairi, A. P. (2008). A new class of cleavable fluorescent nucleotides: Synthesis and optimization as reversible terminators for DNA sequencing by synthesis. *Nucleic Acids Research*, *36*(4), e25. doi:10.1093/nar/gkn021 PMID:18263613

Uzeh, R. E., Ohenhen, R. E., & Rojugbokan, A. K. (2006). Microbiological and nutritional qualities of dairy products : Nono and Wara. *Nature and Science*, *4*(3), 37–40.

Vos, P., Hogers, R., Bleeker, M., Reijans, M., van de Lee, T., Hornes, M., ... Kuiper, M. (1995). AFLP: A new technique for DNA fingerprinting. *Nucleic Acids Research*, *23*(21), 4407–4414. doi:10.1093/nar/23.21.4407 PMID:7501463

Voulgari, K., Hatzikamari, M., Delepoglou, A., Georgakopoulos, P., Litopoulou-Tzanetaki, E., & Tzanetakis, N. (2010). Antifungal activity of non-starter lactic acid bacteria isolates from dairy products. *Food Control*, *21*(2), 136–142. doi:10.1016/j.foodcont.2009.04.007

Walstra, P., Wouters, J. T. M., & Geurts, T. J. (2006). *Dairy science and technology (Second Edi)*. Boca Raton, FL: CRC Press, Taylor & Francis Group.

Wegmuller, B., Luthy, J., & Candrian, U. (1993). Direct polymerase chain reaction detection of Campylobacter jejuni and Campylobacter coli in raw milk and dairy products. *Applied and Environmental Microbiology*, *59*(7), 2161–2165. PMID:8357250

Yu, S., Yan, L., Wu, X., Li, F., Wang, D., & Xu, H. (2017). Multiplex PCR coupled with propidium monoazide for the detection of viable Cronobacter sakazakii, Bacillus cereus, and Salmonella spp. in milk and milk products. *Journal of Dairy Science*, *100*(10), 7874–7882. doi:10.3168/jds.2017-13110 PMID:28755939

KEY TERMS AND DEFINITIONS

DNA: Desoxyribonucleic acid. Genetic material in form of nucleic acid.

Gel-Electrophoresis: A basic method to separate molecules in a gel matrix by applying an electric field.

Locus: Positon on the chromosome, can also refer to a gene.

Microbial Community Profiling: A method to analyse the microbial diversity in a sample usually applicable for high-throughput.

PCR: Polymerase chain reaction. A basic method in biology to multiply a specific segment of DNA.

Phylogenetic Marker: An easily identifiable piece of DNA suitable for phylogenetic reconstructions and the identification of populations, species, or strains.

Restriction Enzyme: An endonuclease enzyme that cuts a DNA molecule at a particular place by recognizing a specific sequence.

Ribosomal RNA Operon: Genomic DNA containing a cluster of rRNA genes.

APPENDIX

Protocols and Procedures

The procedures described here should give an indication on how certain protocols may look like. However, each method should be set up for its individual purpose and parameters have to be adjusted accordingly.

1. Microbial Counts Using Spread Plate Method

- Sampling and sample preparation:
  - Place raw milk samples in sterile bottles and keep at 4 °C
  - Homogenise sample (e.g. 10 g in 90 ml) under sterile conditions
  - Start the analysis within 3 h of sampling
- Dilute an appropriate volume of the sample with e.g. distilled water, phosphate buffered saline, 0.1% peptone water (pH 7), quarter strength Ringers' solution (Oxoid), or another appropriate diluent
- Prepare serial dilutions in steps of a factor of 10 (decimal dilutions)
- Spread small volumes of the dilutions on appropriate media, let dry and incubate
  - At 25°C for 3 to 5 days for fungal and yeast enumeration
  - At 30°C for 6 days for lactic acid bacteria (anaerobic incubation)
  - At 30°C for 3 days for mesophilic aerobic bacteria
- Count the number of colonies formed on the plates (CFU = colony forming unit) and calculate the number of viable bacteria or fungal cells per sample unit

Media for fungal/yeast enumeration:

- **YEGC:** Yeast extract glucose chloramphenicol agar
- **SMA:** Standard methods agar (SMA)

Media for bacteria enumeration:

- **MRS:** Man-Rogosa-Sharpe agar for Lactobacilli
- **PCA:** Plate count agar for total aerobic bacteria
- **M17:** For Lactococci
- **Baird-Parker agar** supplemented with egg yolk tellurite solution for *Staphylococci*

(Delavenne et al., 2011; El-Sharoud et al., 2009; Ledenbach & Marshall, 2009; Ozturkoglu Budak, 2016)

2. Microbial Counts Using Pour Plate Method

- Prepare decimal dilutions (as for spread plate method) up to a 10^{-8} dilution
- Mix one ml of each 10^{-7} and 10^{-8} dilutions with molten nutrient agar cooled to approximately 40-45 °C in Petri dishes

- Let the agar solidify and incubate
    - e.g. at 37°C for 24 h (coliform counts)

Media for coliform counts:

- MacConkey medium
- Violet Red Bile Agar

(Oyeleke, 2009; Ozturkoglu Budak, 2016)

3. Isolation of Lactobacilli

- Resuspend duplicate samples in peptone or CRM resuspension medium
- If required, dilute with resuspension medium*
- Incubate for 30 min at 23°C or 37°C before plating
- Plate on MRS and incubate at 37°C for 2 to 3 days

* For cheeses, dilution in Trisodium citrate (2% w/v) is generally recommended, and peptone salt or phosphate buffer salt is generally used for dairy products such as yoghurts and fermented milks (Coeuret et al., 2003).

- **CRM:** Calicchia resuspension medium

Lactobacilli are generally isolated on rich media such as MRS, which is routinely used for the isolation and counting of lactobacilli from most (fermented) food products (Coeuret et al., 2003).

4. Isolation of Thermophilic Bacterial Strains (Lactobacilli)

- Prepare decimal dilutions (as for spread plate method)
- Spread small volumes of the dilutions on milk (100 g/l) agar containing 0.2 g/l bromcresol purple
- Incubate under anaerobic conditions at 42°C

(Cogan et al., 1997)

5. Selective Media for the Isolation of Bacterial Strains

For selective enumeration of bifidobacteria from fermented milk and cheddar cheese:

- Tryptone-phytone-yeast extract agar containing NPNL selective solution (20 mg/l neomycin sulfate, 40 mg/l paromomycin sulfate, 3 mg/l nalidixic acid, 600 mg/l lithium chloride)
- Incubate plates under anaerobic conditions at 37°C for 3 days

(Auty et al., 2001)

6. Cultivation and Identification of Filamentous Fungi

- Prepare decimal dilutions
- Spread small volumes of the dilutions on agar plates
- Incubate at 25°C for several days
- If required, induce sporulation by visible or UV light (depending on fungal species)
- Purify culture for identification
- Preparation of sample for microscopic slide

Standard culture media:

- **DG18:** Dichloran Glycerol Agar
- **MEA:** Malt extract agar
- **PDA:** Potato dextrose agar

(Tappler et al., 2013)

7. Morphological Identification of Fungi Using Agar Slice on Glass Slide

- Inoculate fungi on a slice of agar laid on a sterile glass slide and cover by a sterile coverslip
- Place the slide in a Petri dish
- Incubate for 5 days at 25 °C
- Withdraw the coverslip with the adhered hyphae and spores and stain with cotton blue dye
- Identify the fungi genus based on the macroscopic morphology of the colonies and on the study of fructification structures of the strains

(Griebeler et al., 2011)

8. Identification of Yeasts to Species Level by Physiological and Morphological Standard Methods

- Test assimilation of glucose, galactose, L-sorbose, sucrose, maltose, cellobiose, trehalose, lactose, melibiose, raffinose, melezitose, inulin, soluble amides, D-xylose, L- and D-arabinose, D-ribose, L-rhamnose, D-glucosamine, N-acetyl-D-glucosamine, methanol, ethanol, glycerol, erythritol, ribitol, galactitol (dulcitol), D-mannitol, D-sorbitol, α-methyl-D-glucoside, salicin, D-gluconic acid, DL-lactic acid, sodium succinate, sodium citrate, inositol, hexadecane, α-ketoglutaric acid, xylitol, L-arabinitol, propane 1.2 diol, butane 2.3 diol, lysine, ethylamine, potassium nitrate, cadaverine, creatine and glucosamine.
- Test fermentation of glucose, sucrose, maltose, galactose, raffinose, lactose, trehalose, melezitose, cellobiose and inulin.
- Test starch formation.
- Test growth on presence of cycloheximide (1%, 0.1% and 0.01%).
- Test growth on media containing 50% of glucose.

- Test growth at different temperatures (40°C, 37°C, 34°C, 25°C, 19°C, 17°C).
- Test growth on vitamin-free medium.
- Test growth on 10% NaCl with 5% glucose.
- Test for the production of a true mycelium or pseudo-mycelium.
- Test for vegetative reproduction by budding or splitting.
- Test for production of ascospores or basidiospores.
- Test for growth on 5% malt extract.

(Moreira, Schwan, De Carvalho, & Wheals, 2001)

9. Identification of Bacterial Cultures Based on Classical Methods

- **Test Gram Reaction:** Bacteria are first stained with crystal violet and next treated with iodine to promote dye retention. The cells are decolorized by washing with ethanol or acetone. Gram-positive bacteria retain the purple dye, while gram-negative become colourless. This step generates the differential aspect of the Gram stain. This difference is due to the cell wall component peptidoglycan, which in Gram-negative bacteria is very thin, not as highly cross-linked and has larger pores leading to dye removal during the alcohol wash. Finally, cells are counterstained by a red stain (most commonly safranin) so that Gram-negative cells can be visualized. Microscopic examination allows the differentiation of dark purple Gram-positive bacteria and pink to red Gram-negative bacteria (Prescott et al., 2001).
- Test catalase using 3 g H_2O_2/L.
- Examine shape by phase contrast microscopy of overnight cultures.
- Test for heterofermentation or homofermentation of sugars.

(Cogan et al., 1997)

10. DNA Extraction From Lipid-Rich Materials Such as Cheese

For efficient bacterial recovery and lysis, it is important to remove viscous polysaccharides and other soluble compounds interfering with these procedures. A four-cycle dilution and washing process is presented as suggested by (Apajalahti et al., 1998). Fat removal and DNA extraction efficiencies can be improved by introducing serial washes of dairy homogenates in Na-citrate buffer (Devirgiliis et al., 2014).

Bacterial recovery:

- Suspend 1g samples in 30 ml of wash buffer.
- Shake for 10 min on a reciprocating horizontal platform shaker at moderate speed at room temperature.
- Subject suspension to centrifugation at 30,000 x g for 15 min (to collect the bacterial fraction).
- Suspend pellet and wash three more times in 30 ml of fresh wash buffer.
- Take samples of the suspended bacteria at each step for direct microscopic enumeration.

Bacterial lysis and DNA recovery:

- Resuspend bacterial pellets in 3 ml of TE buffer (10 mM Tris [pH 8], 1 mM EDTA).
- Subject suspensions to five freeze-thaw cycles of incubation at -70°C for 60 min followed by 40°C for 15 min.
- Add 0.7 ml Lysozyme (200-mg/ml stock) to each sample and incubate at 37°C for 3 h.
- Add 0.2 ml of sodium dodecyl sulfate (10% [wt/vol]) and 20 ul of proteinase K solution (20 mg/ml in TE buffer) and incubate the mixture at 37°C for an additional hour.
- Add 0.72 ml of 5 M NaCl, 0.6 ml of 10% CTAB (hexadecyltrimethyl ammonium bromide) in 0.7 M NaCl, and 1 g of 1,000-um-diameter glass beads.
- Incubate at 65°C for 20 min with vortexing for 30 s after every 5 min.
- Take samples of the suspended bacterial fraction before and after lysis to assess lysis efficiency by direct microscopic enumeration of bacteria.
- Extract cell lysate with an equal volume of chloroform-isoamyl alcohol (24:1).
- Perform centrifugation at 6,000 x g for 10 min at room temperature to separate the aqueous and organic phases.
- Transfer of the aqueous phase to a clean test tube.
- Precipitate DNA by addition of 0.6 volume of 100% isopropanol and incubate for 1 h at room temperature.
- Collect DNA by centrifugation at 10,000 x g for 15 min at room temperature.
- Wash DNA pellet briefly with 70% ethanol, vacuum dry and dissolve in 2 ml of TE buffer.

(Apajalahti et al., 1998)

11. Identification of a Pure Culture Based on DNA Sequence

- DNA extraction from microbial cells (DNA extraction kits can be commercially obtained)
- **Amplification by PCR:** A typical temperature profile used in a thermal cycler:
    - Initial denaturation at 95°C for 5 min
    - 35 cycles of
        - Denaturation at 94°C for 30 sec
        - Annealing at primer depending temperature for 40 sec
        - Elongation at 72°C for 60-90 sec
    - Final elongation step at 72°C for 10 min
    - Hold at 10°C

An overview of commonly used loci and primers for fungi identification can be found in (Samson et al., 2010).

- Electrophoresis for analysis of PCR products
- PCR product purification (PCR cleanup kits are commercially available)
- Sequencing reaction by dye-terminator sequencing (Sanger et al., 1977)

One PCR fragment can be sequenced from more positions resulting in several reads that can be assembled to a single continuous read by software programs.

- Comparison of DNA sequence with database

Public databases such as GenBank (www.ncbi.nlm.nih.gov/genbank) can be used. Special search tools, such as BLAST (Basic Local Alignment Search Tool, (Altschul, Gish, Miller, Myers, & Lipman, 1990) can find similar sequences to the query. If the query sequence is 100% identical to one or more deposited sequences, it indicates that it is the same species or closely related. One has to be aware that there can be misidentified sequences in the database. It is therefore recommended to compare with reference sequences of well recognised strain collections, such as CBS or ATCC or obtain original paper to check the used taxonomy. If there is less than 100% sequence identity, the result should be interpreted manually. Different taxonomic groups and different loci may require different cut-offs.

For fungal identification, ITS or rRNA genes may not provide sufficient information for species identification. Other genomic regions have been shown to provide better resolution. For example, the genera *Aspergillus* and *Penicillium* can be identified by using beta-tubulin (Peterson, 2008).

(Samson et al., 2010)

12. PCR-RFLP

- PCR amplification of region of interest (usually rRNA genes)
- Perform reaction with restriction endonuclease (e. g. BsuRI) as indicated by the manufacturer, e.g. for 2h at 37 °C
- Separate fragments on a 3% high resolution agarose gel
- For identification, compare patterns with reference strains or database
- For sequence-based identification, select representative patterns for sequencing

(Klaubauf et al., 2010)

Chapter 3
Functional Starter Cultures for Fermented Dairy Products

Sarang Dilip Pophaly
College of Dairy Science and Food Technology, India

Manorama Chauhan
College of Dairy Science and Food Technology, India

Vaibhao Lule
College of Dairy Technology, India

Poonam Sarang
National Dairy Research Institute, India

Jitesh Tarak
College of Dairy Science and Food Technology, India

Kiran Thakur
Hefei University of Technology, China

Sudhir Kumar Tomar
National Dairy Research Institute, India

ABSTRACT

Fermented dairy products are known for their high nutritional and therapeutic value and are also having excellent sensory characteristics, which make them popular throughout the world. Most of the characteristics of fermented products can be attributed to the activity of their starter culture. The starter microorganisms, in the course of their metabolism, elaborate various compounds and enzymes, which impart desirable properties to the product. Lactic acid bacteria (LAB) are the most common starter bacterial group used extensively in fermentation of dairy products. Lactic acid bacteria synthesize different compounds belonging to classes of organic acids, gases, bacteriocins, flavoring agents, biothickners, nutrients, etc. Molecules within each class vary for each species and even within strains. The functionalities from such novel starter cultures include improved nutritional value, sensory characteristics, and therapeutic benefits. These starter cultures are thus important for development of functional fermented and probiotic products. This chapter explores functional starter cultures for fermented dairy products.

DOI: 10.4018/978-1-5225-5363-2.ch003

INTRODUCTION

Fermented foods are an important part of tradition as well as nutrition across all human civilizations. The long traditions of manufacturing these fermented products over a geo-temporal space have closely imbibed the cultures associated with such products with human health and well-being. Fermented products are known for their rich nutritional as well as therapeutic benefits. The health promoting aspects of these products can either be assigned to the culture consortia involved in the fermentation or the biochemical alterations in the substrate matrix brought about by the fermentation or both. The microbial cultures associated with fermentation have a wide biodiversity and impart specific functional characteristic to the product. These attributes can be broadly classified into technological and therapeutic functionalities. Growth of fermenting bacteria, mostly lactic acid bacteria (LAB) have a distinct effect on the food functionality depending upon the species or strain of the culture and the composition of food matrix. Besides this basic feature, the technological and health promoting activity of the cultures vary widely.

Manufacturing of fermented milk products like yogurt, cultured buttermilk, cheese etc. require fermentation of milk by suitable microorganisms. These starter organisms result into breakdown of carbohydrates and other nutritional macromolecules present in milk to lactic acid and other constituents. The production of metabolic end products of bacterial metabolism results in specific physico-chemical changes in milk, which are determined by combination of factors like type of starter bacteria, growth-processing conditions and composition of milk. One of the key activities of starter cultures is to produce organic acid especially lactic acid, formic acid and acetic acid. The acid production causes secondary effects such as drop in pH of the medium inducing a protein mediated coagulation of milk. The lactic acid imparts a characteristic acid flavor and results in preservation of milk solids. Flavor production by starter or adjunct cultures is important for development of the delicate flavor and aroma of fermented milk products. The major flavoring compounds include diacetyl, acetoin and acetaldehyde. The biosynthesis of flavor compounds is a species and/or strain specific phenomenon. The milk in its natural form is very prone to microbial spoilage and has a very limited shelf-life. Production of fermented milk products is a unique and natural way to preserve the milk solids for a longer period. This preservative action is effected through organic acids, antimicrobial and antifungal compounds synthesized by the starter cultures. The starter culture used for manufacture of fermented products, are carefully selected based on their biochemical and metabolic characteristics and their suitability to impart a specific functionality or characteristic to a particular product (Table 1). Traditionally, such fermented products have been the habitat for these cultures. In today's times, an industrial practice is to isolate the cultures from their niches, identify, characterize, and prepare freeze-dried starter cultures to use them for commercial scale operations.

FUNCTIONALITY OF STARTER CULTURES

Functional starter cultures are starters that possess at least one inherent functional property. They can offer several organoleptic, technological, nutritional or health advantages and have great potential in food fermentation industry. They can be applied for food preservation, improve food safety, texture and flavor, accelerate the maturation process of cheese and give functionalities to the end product by producing nutraceuticals and reducing toxic or anti-nutritional factors (Table 2). The use of phage-resistant starter

Table 1. Starter cultures used in different fermented dairy products

S. No.	Fermented Dairy Product	Starter Culture	General Functionality
1	Yogurt	*Streptococcus thermophilus* & *Lactobacillus delbruckii* subsp. *bulgaricus*	Symbiotic interaction of cultures;
2	Acidophilus Milk	*Lactobacillus acidophilus*	Therapeutic value
3	Bulgarian Butter Milk	*Lactobacillus delbruckii* subsp. *bulgaricus*	Sharp acid taste and thick consistency
4	Swiss and Italian style cheeses	*Streptococcus thermophilus, Lactobacillus helveticus, Lactobacillus delbruckii* subsp. *bulgaricus*	Facilitating rennet activity, Ripening of cheese
5	Hard Cheese Varieties without eyes	*Lactococcus lactis* subsp. *lactis* *Lactococcus lactis* subsp. *cremoris*	Facilitating rennet activity, Ripening of cheese
6	Cheeses with eyes	*Lactococcus lactis* subsp. *lactis* *Lactococcus lactis* subsp. *cremoris, Leuconostoc mesenteroides* subsp. *cremoris*	Creating conditions for growth of Propionibacteria.
7	Sour Cream, Ripened Cream	*Lactococcus lactis* subsp. *lactis* *Lactococcus lactis* subsp. *cremoris, Leuconostoc mesenteroides* subsp. *cremoris*	Mild acidic taste and delicate diacteyl flavor
8	Kefir	*Lactobacillus kefiri, Lactobacillus kefiranofaciens, Lactobacillus brevis*	Production of "Kefiran" and kefir grains
9	Dahi	*Lactococcus lactis* subsp. *lactis* *Lactococcus lactis* subsp. *cremoris,* *Leuconostoc mesenteroides* subsp. *cremoris*	Mild acidic taste & flavor
	Koumiss	*Lactobacillus delbruckii* subsp. *bulgaricus, Lactobacillus acidophilus,* Lactose fermenting yeast (*Saccharomyces lactis, Torula koumiss*)	Alcohol & acid production
	Leben	*Streptococcus thermophilus* & *Lactobacillus delbruckii* subsp. *bulgaricus,* Lactose fermenting yeast	Sharp acid taste
	Villi	*Lactococcus lactis* subsp. *lactis* *Lactococcus lactis* subsp. *cremoris, Lactococcus lactis* subsp. *lactis* biovar diacteylactis, *Kluveromyces marxianus*	EPS production/ Slime formation

cultures also offers a solution for phage contamination in dairy industry. The desired functional starters may be obtained by genetic engineering or as wild-type organisms. They provide better control during fermentation and new healthy food for consumers.

Technological Functionalities

Texture Improvements

Certain LAB species and strains are capable of producing polymeric sugar compounds in their surrounding medium called as exopolysachharides (EPS). EPS are long chained polysaccharides consisting of branched units of sugar or sugar derivatives. These sugars mainly include glucose, galactose and rhamnose (Welman & Maddox, 2003). Two forms of EPSs are produced by LAB, capsular and unattached. Capsular EPSs form capsules around the cell wall and are not secreted into the medium. Unattached EPSs are secreted outside the cell wall of the bacteria. *In-situ* exo-polysaccharide production by LAB is widely being used for improvement of texture and rheology of fermented milk products. EPS produced

Table 2. Functionality of starter culture

S. No.	Functionality	Mediating Molecule(s) or Cellular Property	Culture	Current or Potential Application	Reference
1	EPS production	Dextran, Mutan, Alternan, Levan, Inulin etc.	Leuconostoc spp. Lactococcus spp. Streptococcus	Improvement of body & texture characteristics of fermented foods Manufacture of low-fat range of products with improved consistency	(Petry *et al.*, 2000; Ruas-Madiedo *et al.*, 2002; Lule *et al.*, 2016)
2	Flavor production	Diacetyl/Acetoin	*Lactococcus lactis* subsp. *lactis* biovar. *diacteylactis* *Leuconostoc sp.*.	*In-situ* flavoring of fermented dairy products Industrial production of food-grade flavors	(Beshkova *et al.*, 1998; Daly *et al.*, 1998; Leroy & De Vuyst, 2004; Patra *et al.*, 2009; Patra *et al.*, 2011)
3	Galactose metabolism	Galactose utilization gene	*S. thermophilus*	Development of low browning pizza cheese Low galactose yogurt with extended shelf life	(Anbukkarasi *et al.*, 2014)
4	Polyol production	Mannitol	Leuconsotoc	Development of low calorific foods	(Wisselink *et al.*, 2002; Patra *et al.*, 2009)
5	Antimicrobial production	Bacteriocins (nisin), lactic acid, alcohol	*Lactococcus lactis* subsp. *lactis*	Biopreservation of foods	(Hugas, 1998; Suomalainen & Mäyrä-Makinen, 1999; Gálvez *et al.*, 2014)
6	Phage resistance	Phage resistance genes	*Lactococcus lactis* subsp. *lactis*	Phage resistance starters for product stability and consistency	(Moineau, 1999; Coffey & Ross, 2002)
7	Vitamin Production	Vitamin B12 Folate Riboflavin	*Propionibacterium freundenrechii* *S. thermophilus* *Lb. plantarum*	Nutritional enrichment of foods	(Hugenholtz & Smid, 2002; Iyer *et al.*, 2010; LeBlanc *et al.*, 2013; LeBlanc *et al.*, 2015)
8	Amino acid synthesis and catabolism	Amino acids	Lactic acid bacteria	Nutritional enrichment of foods Flavor enhancement in cheese	(Yvon & Rijnen, 2001)
9	Antioxidant production	Antioxidant peptides, Glutathione, thioredoxin reductase	*S. thermophilus* *Lb. plantarum* *Lb. rhamnosus*	Increase in therapeutic value of foods	(Coda *et al.*, 2012; Pophaly *et al.*, 2012)
10	Metal Binding	Binding, transformation of Selenium Zinc, Lead, Arsenic etc.	LAB	Reduced toxicity of metals & their biotransformation to organic form	(Bhakta *et al.*, 2012; Mrvčić *et al.*, 2012; Pophaly *et al.*, 2014)
11	Detoxification of Antinutritional factors/ Toxins		LAB	Reduced toxicity of ANF, mycotoxins	(Peltonen *et al.*, 2001; Gote *et al.*, 2004; Refstie *et al.*, 2005; Roger *et al.*, 2015)
12	Probiotic health benefits (Alleviation of Gastrointestinal disorders)	Various therapeutic biomolecules and enzymes	Probiotic cultures	Development of functional probiotic food with specific health benefits	(Oh *et al.*, 2002; Yang *et al.*, 2008; Bekar *et al.*, 2011; Tabbers *et al.*, 2011)

by such GRAS (Generally Recognized as Safe) status cultures act as natural biothickeners and contribute towards improved consistency of products. The EPS production is particularly useful trait for developing low fat fermented products (Ruas-Madiedo *et al.*, 2002). The reduction in fat percent has a detrimental effect on texture which is partly compensated by exopolysaccharides. The EPS gives the product a firm consistency and desired mouthfeel. EPS-producing strains increase the water-holding capacity (WHC) and viscosity of the fermented products (Ruas-Madiedo *et al.*, 2002).

In situ production of natural texture improving sugar polymers (exopolysaccharides) can be produced by LAB for the manufacturing of yoghurts (Vuyst & Degeest, 1999), fat-free *Dahi* (Behare *et al.*, 2009) and low-fat cheeses (Broadbent *et al.*, 2001) and low-fat Mozzarella (Petersen *et al.*, 2000).The EPS production by starter bacteria could be enhanced by optimizing different factors such as oxygen content, pH, temperature, choice of carbohydrate and medium (Petry *et al.*, 2000; Lule *et al.*, 2016).

Improved Aroma and Flavors

Lactic fermentation causes acidification of the food matrix which results in a peculiar acidic taste of these products. Three main pathways are associated with flavour production in fermented products: glycolysis (conversion of lactose); lipolysis (conversion of fat) and proteolysis (conversion of caseins). Homofermentative LAB convert the available energy source (sugar) almost completely into lactic acid via pyruvate to produce energy and to equilibrate the redox balance. Pyruvate can lead to the generation of many other metabolites such as acetate, ethanol, diacetyl, and acetaldehyde. LAB produce volatile substances that contribute to the typical flavour of certain fermented products, such as sourdough (determined by the lactate/acetate ratio), kefir and Koumiss (ethanol), butter and buttermilk (diacetyl), and yoghurt (acetaldehyde) (Leroy & De Vuyst, 2004). The contribution of LAB to lipolysis is relatively little, but proteolysis is the key biochemical pathway for flavour formation. Degradation of caseins by the activities of rennet enzymes and the cell-envelope proteinase and peptidases yields small peptides and free amino acids, the latter of which can be further converted to various alcohols, aldehydes, acids, esters and sulphur compounds for specific flavour development. Some lactococci and Leuconostoc spp. are also able to ferment a number of non-carbohydrates. The most important of these is citrate whose metabolism usually results in the formation of the end-products diacetyl, acetoin, butanediol and acetaldehyde.

The final aroma and taste of the product is a blend of such complex flavor molecules produced by bacterial metabolism. Fermented foods have a distinct sensory profile because of microbial action on food macromolecules results in production of an array of flavor compounds. Diacetyl in particular is a significant flavour and aroma compound in products such as buttermilk, lactic butter, cottage cheese, etc. The diacetyl is synthesized by strains of *Lactococcus lactis* subsp. *lactis* biovar. *diacteylactis* and *Leuconostoc* spp. Extensive biochemical and genetic information describing the synthetic pathway of this compound has enabled researchers to improve the yield of diacetyl in dairy applications. For example, understanding that production of this metabolite is favoured by low pH and aerobic conditions has been used to improve production processes of butter and margarine (Daly *et al.*, 1998). Acetaldehyde is the major flavoring compound in yogurt which is produced by *Streptococcus thermophilus* by the action of threonine aldolase activity (Beshkova *et al.*, 1998). Some LAB are also capable of producing polyols like mannitol which has low calorific value and imparts sweet taste to products (Patra *et al.*, 2009; Patra *et al.*, 2011). The cheese flavor biochemistry is more complex and is dependent on activity of starter bacteria as well as NSLAB (Non Starter Lactic Acid Bacteria)during the ripening process. The proteolysis and hydrolysis reactions by cultures give a blend of flavor compounds specific to the particular cheese variety (Steele *et al.*, 2013). Wild strain starter cultures and NSLAB play an important role in flavour chemistry because they have a high biosynthetic capacity and produce important aromatic compounds (Fox *et al.*, 1998).The addition of NSLAB as adjunct cultures for cheese manufacturing increases the level of free amino acids, peptides, and free fatty acids, leading to flavour intensity and accelerated cheese ripening.

Biopreservation

Fermented milk products have been seen as a natural form of milk solid preservation by employing starter bacteria. This type of preservation is a generic effect observed due to factors such as acid or alcohol production, depletion of essential nutrients and available sugars, lowering of pH and redox potential. However, a stronger biopreserative effect has been observed with cultures with ability to produce bacteriocins. The application of bacteriocins from LAB as biopreservatives is particularly significant in view of the strong market demand for foods that have been minimally processed and that contain less artificial additives. The use of starter cultures that produce bacteriocins during fermentations, therefore, is of major interest to the industry. In addition to preventing the growth of undesirable microorganisms they have the added bonus of not requiring a food additive label.

Biopreservation refers to extension of shelf-life of foods using natural or biological agents such as bacteria or their elaborated antimicrobial compounds (Leroy & De Vuyst, 2004). The functionality of biopreservation in starter cultures has been derived from the evolutionary and ecological needs of the bacteria to kill other microorganisms to establish dominance. This property is now widely considered in screening of cultures with ability to destroy spoilage microflora in dairy and food products and extend their shelflife. The use of such bioprotective cultures in product fermentation enables to eliminate or reduce the addition of chemical preservatives, which have harmful health implications to human body. The preservative action of such cultures may be due to production of one or more antimicrobial compounds such bacteriocins, organic acids etc. The biopreservative cultures can either be added as primary starters, adjunct starters or in form of prefermented slurry to the product. Also, these cultures can be used to manufacture the active molecule involved in preservative action.

Bacteriocins are ribosomally synthesized antimicrobial proteins, usually showing bactericidal activity against genetically related species (Gálvez *et al.*, 2014). Lactic acid bacteria can elaborate several bacteriocins, which could be potentially used for biopreservation of fermented foods. Lactic acid bacteria have several features which make them attractive for food preservation (Gálvez *et al.*, 2014). Firstly, LAB have a long history of safe use in form of fermented foods and hence the bacteria themselves and their cell products have been awarded GRAS status. Secondly, LAB bacteriocins are non-toxic towards eukaryotic cells. Thirdly, due to the proteinaceous nature of bacteriocins, they are expected to be rendered inactive by digestive proteases and hence should not exert antimicrobial activity against gut microflora (Gálvez *et al.*, 2014). Nisin is one of the most widely used biopreservative in dairy and food industry. Nisin is synthesized by some strains of *Lactococcus lactis* subsp. *lactis* biovar diacetylactis. Use of nisin producing strains has been used to extend shelf-life of different products (Hugas, 1998; Suomalainen & Mäyrä-Makinen, 1999). LAB starter strains are able to produce their bacteriocins in food matrices and consequently display documented for fermented sausage (Albano *et al.*, 2009), fermented vegetables and olives (Ruiz-Barba *et al.*, 1994) and dairy products (Benkerroum *et al.*, 2002).

Phage Resistance

Bacteriophage or phage are viruses that attack and can destroy bacterial cells. Phage particles infect the bacterial cultures by penetrating the cells and highjack the bacterial metabolism towards making new phage particles. In doing so, they destroy the bacterial cell by causing its lysis. Thus phage is a major cause of "Starter failure" in fermented dairy. The inhibitory effect of phages on dairy fermentations has incurred heavy losses to the dairy industry (Moineau, 1999). Understanding the biology of phage infection has been elaborated extensively in *L. lactis* and to some extent in *S. thermophilus* (Coffey & Ross, 2002). Phage control measures in dairy plant include starter rotation, use of mixed culture, proper disposal of whey etc. Many bacterial strains have evolved phage resistant mechanisms to protect themselves from the viral attack. Phage resistance may be caused by natural resistance mechanisms (restriction and modification enzymes), prevention of intracellular phage development through phage adsorption and abortive phage infection, or by intracellular defense strategies. Strains that have acquired natural mechanisms of phage resistance, e.g. through in vivo recombination (conjugation) or *in vitro* self-cloning, are currently applied on a large scale in the dairy industry (Leroy & De Vuyst, 2004). Isolation of natural phage resistant strains with rapid acidification ability are employed for production of commercial starter cultures.

Health Promoting Functionalities

Nutritional Enrichment

Fermentation of different food substrates by lactic acid bacteria contributes towards improvement of nutritional status of the final product. These nutritional benefits include increased protein content as a result of large bacterial biomass accumulated during fermentation and other nutritional compounds produced in course of bacterial metabolism (Leroy & De Vuyst, 2004). These additional nutrients also compensates for the loss of nutrients as a result of processing treatments like heating, drying etc. The improved protein content is particularly important in fermented fruits and vegetables which have little protein content (Steinkraus, 1994). The biomass of bacterial cultures, which predominantly consists of proteins, serves as an important source of nitrogen and protein for the consumers. Besides increase in net protein content, the bacterial starters are also responsible for extensive proteolysis of native food proteins generating peptides and free amino acids. Some of these peptides have specific physiological role in host like antihypertensive, antimicrobial, antioxidant and anticarcinogenic activities (Pihlanto & Korhonen, 2014). Some LAB are also capable of producing glutathione (Pophaly *et al.*, 2012; Pophaly *et al.*, 2017) which acts as a potent antioxidant and may contribute towards nutritive value of fermented foods.

Lactic acid bacteria and Bifidobacteria which are commonly used in fermented and probiotic foods are able to *de-novo* synthesize many vitamins and thus capable of delivering these high value metabolites for human consumption (Hugenholtz & Smid, 2002; LeBlanc *et al.*, 2013). Fermentation with these selected cultures may be used for bio-enrichment of vitamins in such foods (Steinkraus, 1997). Certain species and strains of lactic acid bacteria are capable of synthesis of many B-group vitamins including riboflavin, folate and vitamin B_{12} (LeBlanc *et al.*, 2015) and also vitamin K_2, niacin, biotin (LeBlanc *et al.*, 2013). Numerous researchers have reported that LAB, such as the industrial starter bacteria *L. lactis* and *S. thermophilus,* have the ability to synthesize folate. *S. thermophilus* has a strain-specific ability for folate production and has been reported to produce higher quantities of folate compared with other LAB; the majority of this is excreted into milk (Iyer *et al.*, 2010). In general, Lactobacillus strains do not produce folate with the exception of *Lactobacillus plantarum* where as *S. thermophilus* are the best known folate producers. Propionibacteria used in Swiss cheese can produce several compounds such as vitamins, conjugated linoleic acid, trehalose, propionic acid etc. that contributes towards enrichment of nutritive value of fermented products (Poonam *et al.*, 2012).

Degradation of Toxic and Undesirable Compounds

Many fermenting bacteria are capable of utilizing and/or degrading toxic and undesirable molecules present in natural food systems. The degradation of such antinutritional compounds renders the product safer for human consumption. *Lactobacillus brevis* and *Lactobacillus fermentum* were shown to reduce the phytate and tannin in Kutukutu (corn paste) (Roger *et al.*, 2015). Soymilk and other soya bean derived products are particularly known for their antinutritional factors (ANF) like phytate, tannin, trypsin inhibitors and some oligosaccharides. Lactic fermentation is able to reduce the levels and activity of many ANFs and flatulence factors in soya-derived products (Gote *et al.*, 2004; Refstie *et al.*, 2005). Many LAB are capable of binding aflatoxin AFB_1 with potential to control the mycotoxin induced toxicity in food system (Peltonen *et al.*, 2001; Dalié *et al.*, 2010). Lactic acid bacteria are also known for their metal binding ability which helps to reduce metal induced toxicity caused by Cadmium, Lead, Arsenic, Nickle and Copper (Bhakta *et al.*, 2012; Mrvčić *et al.*, 2012). Selenium toxicity is also reduced as the toxic inorganic form of the mineral is converted to organic or elemental selenium (Pophaly *et al.*, 2014).

Alleviation of Lactose Maldigestion Symptoms

A considerable percentage of world population exhibits low activity of β-galactosidase in small intestine making them intolerant towards lactose rich milk. Intestinal microorganisms ferment the undigested lactose and causes bloating, flatulence, abdominal pain, loose stools and nausea. Fermented milk products are well known for their ability to alleviate symptoms of lactose intolerance (Solomons, 2002). Most lactic cultures exhibit β-galactosidase (lactase) activity and degrade the lactose to constituent monosaccharaides. Thus partial hydrolysis reduces the amount of effective lactose in the product. Moreover, probiotic cultures also continue this activity in the intestinal environment and helps to digest the lactose (Leroy & De Vuyst, 2004).

Probiotic Cultures

A probiotic has been defined as a live microbial feed supplement that beneficially affects the host by improving its intestinal microbial balance. To be effective, probiotic bacteria must be capable of being prepared in a viable manner on an industrial scale and must remain viable during storage and use. They must also be able to survive the intestinal ecosystem of the host, to then ultimately be capable of exerting their beneficial effects. However, there are many factors, which affect cell viability, such as oxygen, moisture, heat and acid.

Functional fermented foods are attractive options for delivery of probiotic organisms in human system. These organisms have the capability to ferment milk or meat products and also impart health benefits to the host. A number of genera of bacteria (and yeast) are used as probiotics starter cultures, including Lactobacillus, Leuconostoc, Pediococcus, Bifidobacterium, and Enterococcus. Members of the genera Lactobacillus and Bifidobacterium have a long and safe history in the manufacture of dairy products and are also found as a part of gastrointestinal microflora. Probiotic bacteria with desirable properties and well-documented clinical effects include *Lb. johnsonii* La1, *Lb.* rhamnosus GG (ATCC 53103), *Lb. casei* Shirota, *Lb. acidophilus* NCFB 1478, *B. animalis* Bb12 and *L. reuteri.*

A number of health benefits are claimed in favour of products containing probiotic organisms including antimicrobial activity and gastrointestinal infections, improvement in lactose metabolism, antimutagenic properties, anticarcinogenic properties, reduction in serum cholesterol, anti-diarrhoeal properties, immune system stimulation, improvement in inflammatory bowel disease and suppression of *Helicobacter pylori* infection (Shah, 2007).

Some of the health benefits are well established, while other benefits have shown promising results in animal models. However, additional studies are required in humans to substantiate these claims. Health benefits imparted by probiotic bacteria are strain specific, and not species- or genus-specific. It is important to note that no strain will provide all proposed benefits, not even strains of the same species, and not all strains of the same species will be effective against defined health conditions (Shah, 2007).

The primary interest in application of probiotics has been in the GI infections and diseases. Gut microbiota deviations have been associated with enhanced risk of specific diseases; therefore, modulation of an unbalanced indigenous microbiota forms the rationale of probiotic therapy (Turnbaugh *et al.*, 2006). Probiotic therapy is also gaining importance owing to the rapid emergence of antibiotic resistant pathogens and adverse consequences of antibiotics on gut microbiota. Probiotics have also been reported to have antimutagenic and anticancer activities. Several mechanisms have been proposed as to how lactic acid bacteria may inhibit colon cancer, which includes enhancing the host's immune response, altering the metabolic activity of the intestinal microbial communities, binding and degrading carcinogens, producing antimutagenic compounds, and altering the physiochemical conditions in the colon. Oral administration of LAB has been shown to effectively reduce DNA damage, induced by chemical carcinogens, in gastric and colonic mucosa in rats (Nagpal *et al.*, 2012).

Gastrointestinal Disorders

Gastrointestinal disorders are a range of diseases involving the entire GI tract but mainly affecting the stomach and the intestines. Stomach disorders mainly include gastritis, peptic ulcers and stomach cancer. Fermented products in form of traditional knowledge have long been used to relive in GI disorders. Recent scientific evidence in support of their beneficial role have emerged and established these products as therapeutic foods. *Helicobacter pylori* is the causative agent in chronic gastritis and idiopathic peptic ulcer. Consumption of yogurt (Oh *et al.*, 2002) and kefir (Bekar *et al.*, 2011) fermented with specific starter microorganisms either helps to control *H. pylori* infection or compliments the other therapies for its eradication.

Intestine being the interface between the outer tract and inner organs, is the hotspot of metabolic and immune activities. The presence of a large number of microbial species in the intestine makes the gut ecology more complex. Intestinal disorders are largely correlated with disturbances in normal gut microbiota. Common intestinal disorders include constipation, enterocolitis, Inflammatory bowel diseases (IBD), Irritable bowel syndrome (IBS) etc. Fermented dairy products like yogurt and dahi are well known for their laxative effect. These products helps to decrease the GI transit time and improve bowel movement alleviating symptoms of constipation (Yang *et al.*, 2008; Tabbers *et al.*, 2011).

Production of Bioactive Peptides

Milk proteins are not only a source of nutrients for maintenance of proper body functions but are also considered as a source of bioactive peptides, which can promote health and prevent diseases. BAPs are released from food proteins by endogenous proteolysis during gastric digestion or by exogenous hydrolysis using physical (heat), chemical (acid/alkali) or plant/animal/microbial proteases during food processing. The size of active sequences may vary from 2 to 20 amino acid (AA) residues and generally have molecular weight <6000 Da. BAP from different sources are found to possess common structural properties with hydrophobic AA residues in addition to lysine, proline or arginine (Anusha & Bindhu, 2016).

The fermentation process by lactic acid bacteria starter cultures, used in the dairy industry, involves the hydrolysis of milk proteins (particularly casein) into peptides and AAs which are used as nitrogen sources necessary for their growth. Several lactic acid bacteria are widely used and their role can be divided into starter and non-starter cultures. The proteolytic system of lactic acid bacteria consists of a cell wall-bound protease and other intracellular proteases . Different starter cultures and fermentation conditions affect the composition and nature of released BAPs. Studies have exhibited that LAB proteases especially from *Lactococcus* sp. and *Lactobacillus* sp. could hydrolyse ≥40% of peptide bonds of α_1 and β caseins resulting in the release of various oligopeptides that are further hydrolysed by complex peptidases (Anusha & Bindhu, 2016). Milk borne bioactive peptides have been found to exhibit various physiological activities such as antihypertensive, opioid, immunomodulatory antimicrobial, antioxidative, antithrombotic, and cytomodulatory. Such bioactive peptides may find use in the treatment of diarrhea, hypertension, thrombosis, dental carries, oxidative stress, mineral malabsorption, and immunodeficiency (Haque *et al.*, 2008). Fermentation of milk proteins using the proteolytic systems of lactic acid bacteria (LAB) is an attractive approach for generation of functional foods enriched in bioactive peptides given the low cost and positive nutritional image associated with fermented milk drinks and yoghurt (Hayes *et al.*, 2007).

CONCLUSION

Traditionally role of starter culture was considered only for carrying out fermentation and acid production. But, in recent years starter cultures with added or novel functions are gaining attention. Functional starter cultures are starters that possess at least one inherent functional property which may impart a technological or a health benefit or both. Functional starter cultures offer several organoleptic, technological, nutritional or health advantages and have great potential in food fermentation industry. Functionalities like phage resistance, EPS production and bacteriocin production have been widely used for technological benefits. Health promoting attributes of functional starters have largely gained attention in last two decades after the emergence of probiotics and nutraceuticals. Whole genome sequencing of large number of LAB strains have now empowered the microbiologists with genomic knowledge on the functional feature of particular strain. Evolutionary engineering and other strain development methodologies could be used for improvement of functionalities of these cultures for their better utilization in industry.

REFERENCES

Albano, H., Pinho, C., Leite, D., Barbosa, J., Silva, J., Carneiro, L., ... Teixeira, P. (2009). Evaluation of a bacteriocin-producing strain of *Pediococcus acidilactici* as a biopreservative for "Alheira", a fermented meat sausage. *Food Control, 20*(8), 764–770. doi:10.1016/j.foodcont.2008.09.021

Anbukkarasi, K., UmaMaheswari, T., Hemalatha, T., Nanda, D. K., Singh, P., & Singh, R. (2014). Preparation of low galactose yogurt using cultures of Gal+ *Streptococcus thermophilus* in combination with *Lactobacillus delbrueckii* ssp. *bulgaricus. Journal of Food Science and Technology, 51*(9), 2183–2189. doi:10.100713197-014-1262-5 PMID:25190881

Anusha, R., & Bindhu, O. S. (2016). *Bioactive Peptides from Milk. In Milk Proteins-From Structure to Biological Properties and Health Aspects.* InTech.

Behare, P., Singh, R., & Singh, R. P. (2009). Exopolysaccharide-producing mesophilic lactic cultures for preparation of fat-free Dahi–an Indian fermented milk. *The Journal of Dairy Research, 76*(1), 90–97. doi:10.1017/S0022029908003865 PMID:19121242

Bekar, O., Yilmaz, Y., & Gulten, M. (2011). Kefir improves the efficacy and tolerability of triple therapy in eradicating *Helicobacter pylori*. *Journal of Medicinal Food*, *14*(4), 344–347. doi:10.1089/jmf.2010.0099 PMID:21186984

Benkerroum, N., Ghouati, Y., Ghalfi, H., Elmejdoub, T., Roblain, D., Jacques, P., & Thonart, P. (2002). Biocontrol of *Listeria monocytogenes* in a model cultured milk (lben) by in situ bacteriocin production from *Lactococcus lactis* ssp. *lactis*. *International Journal of Dairy Technology*, *55*(3), 145–151. doi:10.1046/j.1471-0307.2002.00053.x

Beshkova, D., Simova, E., Frengova, G., & Simov, Z. (1998). Production of flavour compounds by yogurt starter cultures. *Journal of Industrial Microbiology & Biotechnology*, *20*(3-4), 180–186. doi:10.1038j.jim.2900504

Bhakta, J. N., Ohnishi, K., Munekage, Y., Iwasaki, K., & Wei, M. Q. (2012). Characterization of lactic acid bacteria-based probiotics as potential heavy metal sorbents. *Journal of Applied Microbiology*, *112*(6), 1193–1206. doi:10.1111/j.1365-2672.2012.05284.x PMID:22404232

Broadbent, J. R., McMahon, D. J., Oberg, C. J., & Welker, D. L. (2001). Use of exopolysaccharide-producing cultures to improve the functionality of low fat cheese. *International Dairy Journal*, *11*(4), 433–439. doi:10.1016/S0958-6946(01)00084-X

Coda, R., Rizzello, C. G., Pinto, D., & Gobbetti, M. (2012). Selected lactic acid bacteria synthesize antioxidant peptides during sourdough fermentation of cereal flours. *Applied and Environmental Microbiology*, *78*(4), 1087–1096. doi:10.1128/AEM.06837-11 PMID:22156436

Coffey, A., & Ross, R. P. (2002). Bacteriophage-resistance systems in dairy starter strains: molecular analysis to application. In R. J. Siezen, J. Kok, T. Abee, & G. Schasfsma (Eds.), *Lactic Acid Bacteria: Genetics, Metabolism and Applications* (pp. 303–321). Springer. doi:10.1007/978-94-017-2029-8_20

Dalié, D. K. D., Deschamps, A. M., & Richard-Forget, F. (2010). Lactic acid bacteria–Potential for control of mould growth and mycotoxins: A review. *Food Control*, *21*(4), 370–380. doi:10.1016/j.foodcont.2009.07.011

Daly, C., Fitzgerald, G. F., O'Connor, L., & Davis, R. (1998). Technological and health benefits of dairy starter cultures. *International Dairy Journal*, *8*(3), 195–205. doi:10.1016/S0958-6946(98)00042-9

Fox, P. F., McSweeney, P. L. H., & Lynch, C. M. (1998). Significance of non-starter lactic acid bacteria in chedder cheese. *Australian Journal of Dairy Technology*, *53*(2), 83.

Gálvez, A., López, R. L., Pulido, R. P., & Burgos, M. J. G. (2014). *Application of Lactic Acid Bacteria and their Bacteriocins for Food Biopreservation. In Food Biopreservation* (pp. 15–22). Springer.

Gote, M., Umalkar, H., Khan, I., & Khire, J. (2004). Thermostable α-galactosidase from Bacillus stearothermophilus (NCIM 5146) and its application in the removal of flatulence causing factors from soymilk. *Process Biochemistry*, *39*(11), 1723–1729. doi:10.1016/j.procbio.2003.07.008

Haque, E., Chand, R., & Kapila, S. (2008). Biofunctional properties of bioactive peptides of milk origin. *Food Reviews International*, *25*(1), 28–43. doi:10.1080/87559120802458198

Hayes, M., Ross, R. P., Fitzgerald, G. F., & Stanton, C. (2007). Putting microbes to work: dairy fermentation, cell factories and bioactive peptides. Part I: overview. *Biotechnology Journal*, *2*(4), 426–434. doi:10.1002/biot.200600246 PMID:17407210

Hugas, M. (1998). Bacteriocinogenic lactic acid bacteria for the biopreservation of meat and meat products. *Meat Science*, *49*, S139–S150. doi:10.1016/S0309-1740(98)90044-4 PMID:22060706

Hugenholtz, J., & Smid, E. J. (2002). Nutraceutical production with food-grade microorganisms. *Current Opinion in Biotechnology*, *13*(5), 497–507. doi:10.1016/S0958-1669(02)00367-1 PMID:12459344

Iyer, R., Tomar, S. K., Maheswari, T. U., & Singh, R. (2010). *Streptococcus thermophilus* strains: Multifunctional lactic acid bacteria. *International Dairy Journal*, *20*(3), 133–141. doi:10.1016/j.idairyj.2009.10.005

LeBlanc, J. G., Laiño, J. E., del Valle, M. J., de Giori, G. S., Sesma, F. & Taranto, M. P. (2015). B-Group Vitamins Production by Probiotic Lactic Acid Bacteria. *Biotechnology of Lactic Acid Bacteria: Novel Applications,* 279.

LeBlanc, J. G., Milani, C., de Giori, G. S., Sesma, F., van Sinderen, D., & Ventura, M. (2013). Bacteria as vitamin suppliers to their host: A gut microbiota perspective. *Current Opinion in Biotechnology*, *24*(2), 160–168. doi:10.1016/j.copbio.2012.08.005 PMID:22940212

Leroy, F., & De Vuyst, L. (2004). Lactic acid bacteria as functional starter cultures for the food fermentation industry. *Trends in Food Science & Technology*, *15*(2), 67–78. doi:10.1016/j.tifs.2003.09.004

Lule, V. K., Singh, R., Pophaly, S. D., & Tomar, S. K. (2016). Production and structural characterisation of dextran from an indigenous strain of *Leuconostoc mesenteroides* BA08 in Whey. *International Journal of Dairy Technology*, *69*(4), 520–531. doi:10.1111/1471-0307.12271

Moineau, S. (1999). *Applications of phage resistance in lactic acid bacteria. In Lactic Acid Bacteria: Genetics, Metabolism and Applications* (pp. 377–382). Springer. doi:10.1007/978-94-017-2027-4_23

Mrvčić, J., Stanzer, D., Šolić, E., & Stehlik-Tomas, V. (2012). Interaction of lactic acid bacteria with metal ions: Opportunities for improving food safety and quality. *World Journal of Microbiology & Biotechnology*, *28*(9), 2771–2782. doi:10.100711274-012-1094-2 PMID:22806724

Nagpal, R., Kumar, A., Kumar, M., Behare, P. V., Jain, S., & Yadav, H. (2012). Probiotics, their health benefits and applications for developing healthier foods: A review. *FEMS Microbiology Letters*, *334*(1), 1–15. doi:10.1111/j.1574-6968.2012.02593.x PMID:22568660

Oh, Y., Osato, M. S., Han, X., Bennett, G., & Hong, W. K. (2002). Folk yoghurt kills *Helicobacter pylori. Journal of Applied Microbiology*, *93*(6), 1083–1088. doi:10.1046/j.1365-2672.2002.01779.x PMID:12452966

Patra, F., Tomar, S. K., & Arora, S. (2009). Technological and functional applications of low-calorie sweeteners from lactic acid bacteria. *Journal of Food Science*, *74*(1), R16–R23. doi:10.1111/j.1750-3841.2008.01005.x PMID:19200114

Patra, F., Tomar, S. K., Rajput, Y. S., & Singh, R. (2011). Characterization of mannitol producing strains of Leuconostoc species. *World Journal of Microbiology & Biotechnology*, *27*(4), 933–939. doi:10.100711274-010-0536-y

Peltonen, K., El-Nezami, H., Haskard, C., Ahokas, J., & Salminen, S. (2001). Aflatoxin B1 binding by dairy strains of lactic acid bacteria and bifidobacteria. *Journal of Dairy Science*, *84*(10), 2152–2156. doi:10.3168/jds.S0022-0302(01)74660-7 PMID:11699445

Petersen, B. L., Dave, R. I., McMahon, D. J., Oberg, C. J., & Broadbent, J. R. (2000). Influence of Capsular and Ropy Exopolysaccharide-Producing *Streptococcus thermophilus* on Mozzarella Cheese and Cheese Whey. *Journal of Dairy Science*, *83*(9), 1952–1956. doi:10.3168/jds.S0022-0302(00)75071-5 PMID:11003223

Petry, S., Furlan, S., Crepeau, M.-J., Cerning, J., & Desmazeaud, M. (2000). Factors affecting exocellular polysaccharide production by *Lactobacillus delbrueckii* subsp. *bulgaricus* grown in a chemically defined medium. *Applied and Environmental Microbiology*, *66*(8), 3427–3431. doi:10.1128/AEM.66.8.3427-3431.2000 PMID:10919802

Pihlanto, A., & Korhonen, H. (2014). Bioactive peptides from fermented foods and health promotion. In W. Holzapfel (Ed.), *Advances in Fermented Foods and Beverages: Improving Quality, Technologies and Health Benefits* (p. 39). Elsevier.

Poonam, S., Pophaly, S. D., Tomar, S. K., De, S., & Singh, R. (2012). Multifaceted attributes of dairy propionibacteria: A review. *World Journal of Microbiology & Biotechnology*, *28*(11), 3081–3095. doi:10.100711274-012-1117-z PMID:22806746

Pophaly, S. D., Pophaly, S. D., Kapila, S., Nanda, D. K., Tomar, S. K., & Singh, R. (2017). Glutathione biosynthesis and activity of dependent enzymes in food grade lactic acid bacteria harboring multidomain bifunctional fusion gene (gshF). *Journal of Applied Microbiology*, *123*(1), 194–203. doi:10.1111/jam.13471 PMID:28403558

Pophaly, S. D., Singh, P., Kumar, H., Tomar, S. K., & Singh, R. (2014). Selenium enrichment of lactic acid bacteria and bifidobacteria: A functional food perspective. *Trends in Food Science & Technology*, *39*(2), 135–145. doi:10.1016/j.tifs.2014.07.006

Pophaly, S. D., Singh, R., Pophaly, S. D., Kaushik, J. K., & Tomar, S. K. (2012). Current status and emerging role of glutathione in food grade lactic acid bacteria. *Microbial Cell Factories*, *11*(1), 114. doi:10.1186/1475-2859-11-114 PMID:22920585

Refstie, S., Sahlström, S., Bråthen, E., Baeverfjord, G., & Krogedal, P. (2005). Lactic acid fermentation eliminates indigestible carbohydrates and antinutritional factors in soybean meal for Atlantic salmon (Salmo salar). *Aquaculture (Amsterdam, Netherlands)*, *246*(1), 331–345. doi:10.1016/j.aquaculture.2005.01.001

Roger, T., Ngoumé Léopold, T., & Carl Moses Funtong, M. (2015). Nutritional Properties and Antinutritional Factors of Corn Paste (Kutukutu) Fermented by Different Strains of Lactic Acid Bacteria. *International Journal of Food Sciences*. PMID:26904660

Ruas-Madiedo, P., Hugenholtz, J., & Zoon, P. (2002). An overview of the functionality of exopolysaccharides produced by lactic acid bacteria. *International Dairy Journal*, *12*(2), 163–171. doi:10.1016/S0958-6946(01)00160-1

Ruiz-Barba, J. L., Cathcart, D. P., Warner, P. J., & Jimenez-Diaz, R. (1994). Use of Lactobacillus plantarum LPCO10, a bacteriocin producer, as a starter culture in Spanish-style green olive fermentations. *Applied and Environmental Microbiology*, *60*(6), 2059. PMID:16349291

Shah, N. P. (2007). Functional cultures and health benefits. *International Dairy Journal*, *17*(11), 1262–1277. doi:10.1016/j.idairyj.2007.01.014

Solomons, N. W. (2002). Fermentation, fermented foods and lactose intolerance. *European Journal of Clinical Nutrition*, *56*(S4), S50–S55. doi:10.1038j.ejcn.1601663 PMID:12556948

Steele, J., Broadbent, J., & Kok, J. (2013). Perspectives on the contribution of lactic acid bacteria to cheese flavor development. *Current Opinion in Biotechnology*, *24*(2), 135–141. doi:10.1016/j.copbio.2012.12.001 PMID:23279928

Steinkraus, K. H. (1994). Nutritional significance of fermented foods. *Food Research International*, *27*(3), 259–267. doi:10.1016/0963-9969(94)90094-9

Steinkraus, K. H. (1997). Bio-enrichment: production of vitamins in fermented foods. In B. J. B. Wood (Ed.), *Microbiology of Fermented Foods* (pp. 603–621). Springer.

Suomalainen, T. H., & Mäyrä-Makinen, A. M. (1999). Propionic acid bacteria as protective cultures in fermented milks and breads. *Le Lait*, *79*(1), 165–174. doi:10.1051/lait:1999113

Tabbers, M. M., Chmielewska, A., Roseboom, M. G., Crastes, N., Perrin, C., Reitsma, J. B., ... Benninga, M. A. (2011). Fermented milk containing *Bifidobacterium lactis* DN-173 010 in childhood constipation: A randomized, double-blind, controlled trial. *Pediatrics*, *127*(6), e1392–e1399. doi:10.1542/peds.2010-2590 PMID:21606153

Turnbaugh, P. J., Ley, R. E., Mahowald, M. A., Magrini, V., Mardis, E. R., & Gordon, J. I. (2006). An obesity-associated gut microbiome with increased capacity for energy harvest. *Nature*, *444*(7122), 1027–1131. doi:10.1038/nature05414 PMID:17183312

Vuyst, L., & Degeest, B. (1999). Heteropolysaccharides from lactic acid bacteria. *FEMS Microbiology Reviews*, *23*(2), 153–177. doi:10.1111/j.1574-6976.1999.tb00395.x PMID:10234843

Welman, A. D., & Maddox, I. S. (2003). Exopolysaccharides from lactic acid bacteria: Perspectives and challenges. *Trends in Biotechnology*, *21*(6), 269–274. doi:10.1016/S0167-7799(03)00107-0 PMID:12788547

Wisselink, H. W., Weusthuis, R. A., Eggink, G., Hugenholtz, J., & Grobben, G. J. (2002). Mannitol production by lactic acid bacteria: A review. *International Dairy Journal*, *12*(2), 151–161. doi:10.1016/S0958-6946(01)00153-4

Yang, Y.-X., He, M., Hu, G., Wei, J., Pages, P., Yang, X.-H., & Bourdu-Naturel, S. (2008). Effect of a fermented milk containing *Bifidobacterium lactis* DN-173010 on Chinese constipated women. *World Journal of Gastroenterology*, *14*(40), 6237. doi:10.3748/wjg.14.6237 PMID:18985817

Yvon, M., & Rijnen, L. (2001). Cheese flavour formation by amino acid catabolism. *International Dairy Journal*, *11*(4), 185–201. doi:10.1016/S0958-6946(01)00049-8

KEY TERMS AND DEFINITIONS

Aflatoxins: A Class of fungal toxins produced mainly by *Aspergillus flavus* and *Aspergillus parasiticus*.

Biopreservation: Biopreservation refers to extension of shelf-life of foods using natural or biological agents such as bacteria or their elaborated antimicrobial compounds.

Fermentation: A metabolic process involves anaerobic breakdown of carbohydrates with an organic molecule acting as the final electron acceptor.

Phage: Viruses that infect a bacteria.

Polyols: A polyol is an alcohol containing multiple hydroxyl groups.

Probiotics: Live microbial cultures which when administered in adequate amount have a specific health benefit on the host.

Starter Culture: Starter culture refers to the intentionally added carefully selected microorganisms to a medium for carrying out desired fermentation.

Starter Functionality: A specific trait of starter microorganisms with regard to a technological or therapeutic characteristic which is imparted to the fermented product after fermentation.

Chapter 4
Biopreservatives for Improved Shelf-Life and Safety of Dairy Products:
Biopreservatives for Dairy Products

Tejinder Pal Singh
College of Dairy Science and Food Technology, India

Sarang Dilip Pophaly
College of Dairy Science and Food Technology, India

Ruby Siwach
College of Dairy Science and Food Technology, India

ABSTRACT

Globally, there is an increasing demand for minimally processed, easily prepared, and ready-to-eat fresh food, globalization of food trade, and distribution from centralized processing which pose major challenges for food safety and quality because perishable food may get contaminated with undesirable microorganisms. Food spoilage adversely affects the economy and also erodes the consumer's confidence. On other hand, food-borne illness leads to loss of earnings and productivity, unemployment and litigation, and weakens trade and tourism. Another challenge for the food producers is to produce less stable foods by processes that confer less harm to the detrimental microflora. A challenge for food producers is to develop products with a sufficiently longer shelf-life and at a competitive price. This brings them to the most promising approach to this end, the so-called biopreservation. This chapter provides a scientific background, functionality, as well as food applications and further commercial aspects of biopreservatives derived from microbial sources.

INTRODUCTION

The concept of biopreservation is not new to us, as we humans have continuously been using this biotechnological approach for preserving food, for many years, without even knowing the underlying mechanisms. In present world, the changing food trends, lifestyle and increasing consumer's awareness has brought up the challenge to food producers to meet totally contradictory trends and demands (Table 1).

DOI: 10.4018/978-1-5225-5363-2.ch004

Table 1. Highlighting consumer's preferences, changing food trends and demands, and problems associated with them

Preferences	Trends and Demands	Problems Associated
Health trends	Consumers concern for health has raised the demand for food products with reduced levels of salt, sugar and fat.	This also confers an increase in water activity, which provides a much suitable environment for microorganisms.
Taste preferences	In many products, trends are towards a milder (i.e. less acidic) taste.	This results in a higher pH that again is less adverse for microorganisms.
Perception of "natural"	Demand for milder or minimally processed foods (More natural/fresh food). Demand for "preservative-free" products.	Less inactivation of unwanted microorganisms.
Convenience trends ("practically homemade")		Two main risks associated with this trend—namely, more extensive processing, which results in more steps in which contamination with detrimental microorganisms can occur, and the need for proper handling by the consumer (e.g. sufficient heating), which may be neglected.
Durability and open shelf-life	Market access and economically viable logistics require a long shelf-life.	Furthermore, a sufficient open shelf-life is required to ensure customer loyalty.
Ethical issues		Concerns such as corporate social responsibility, carbon dioxide (CO_2) footprint, and fair-trade and organic products put restrictions on which solutions a food producer can employ.

Overall, it has been seen that these trends and demands lead to food formulations that provide conditions much more favourable for microbial growth; milder processing results in minimal reduction, more processing steps increase the risk of contamination, more difficult to maintain longer shelf-life, and pressure to minimize the waste. Moreover, the trendsetters and consumers are not in favour to use conventional preservatives. But the fact is that we cannot meet the demands of growing population and maintain present society living standards, and certainly cannot reduce the global food waste problems, with food that is not preserved.

Normally, many consumers associate the term 'Preservatives' with harmful, modern chemicals in foodstuffs. But, as a brief look back into the past will show that there are possibilities to preserve the food by more natural means (such as, fermentation). Despite a number of misgivings, preservatives have nowadays become an indispensable part of the food we eat; due to increasing demand from consumers for greater choice, ease and convenience of foods, and high food safety standards. Thus, there is a strong market need for natural food preservation methods that can ensure both food safety (i.e. reduce the number and/or outgrowth of pathogenic microorganisms) and longer shelf-life (i.e. delayed development of the spoilage microflora). One of the few possible solutions is biopreservation based on the concept of using food-grade microorganisms (mainly, Lactic acid bacteria; LAB) as so-called cell factories (Figure 1).

Generally Regarded As Safe (GRAS)/Food-grade microorganisms can be exploited to a great extent for restricting microbial growth as promising means to preserve food. These mechanisms naturally exist to balance the complex microbial ecosystems. Thereby, the naturally occurring fittest microorganisms can be exploited to design preservation methodologies that ensure food safety and shelf life while keeping up the desired quality of the food product.

The biopreservation principles from food-grade microorganisms can be categorized according to the antimicrobial compound (e.g. bacteriocin, other metabolites, bacteriophages, enzymes) as well as product format (purified antimicrobial, fermentate, protective culture), as outlined in Figure 1 (Elsser-

Figure 1. Main categories of biopreservatives that can be produced by using lactic acid bacteria and other suitable microorganisms as "cell factories"

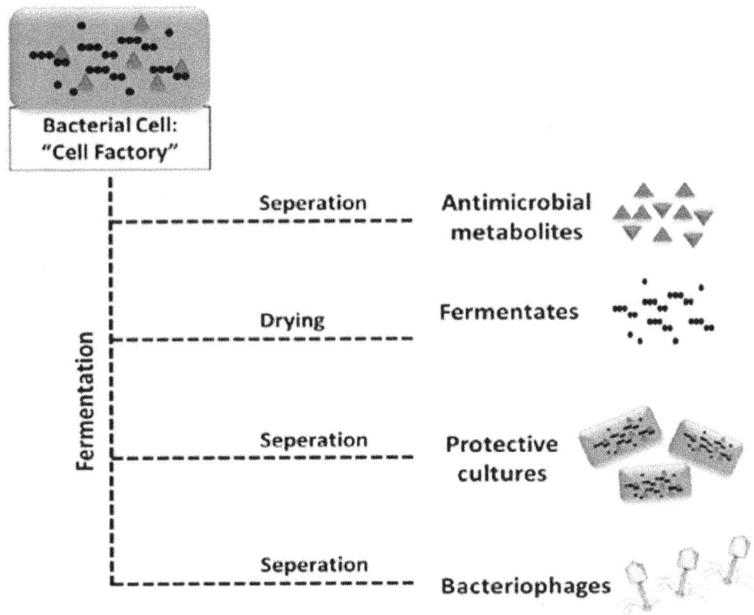

Gravesen & Elsser-Gravesen, 2014). Knowing the fact that not all antimicrobial metabolites from food-grade microorganisms have yet been discovered, and even the underlying mechanisms of several well-known antimicrobials are not well understood. Therefore, clear understanding and discoveries of new antimicrobial metabolites will further expand the options for natural food biopreservation systems. Roller (2003) suggested that one should not search for "the silver bullet" for biopreservation systems, as antimicrobial compounds in nature rarely function in isolation. Hence, targeted intelligent strategies based on multifactorial systems are the most likely to succeed for protecting food against detrimental microorganisms.

This chapter aims to provide a scientific background, current knowledge on biopreservatives, functionality, as well as covering solutions that are actually being used industrially today, and points out perceivable directions for future solution development.

BACTERIOCINS

The empirical utilization of microorganisms as well as their metabolites for food security and protection has been a typical practice ever (Rose et al., 2003). Among the wide range of antibacterial metabolites produced by microorganisms, the bacteriocins, particularly those synthesized by lactic acid bacteria (LAB), have pulled in the best consideration as promising means for food safety & biopreservation. Several studies have exclusively been done focusing the pivotal roles that this group of bacteria plays in food and health, structure, biosynthesis, genetics, and food application of LAB bacteriocins (Deegan et al. 2006; Franz et al. 2007; G´alvez et al. 2007).

Ribosomally synthesized antimicrobial peptides or proteins, called Bacteriocins, are nontoxic to eukaryotic cells and thereby, are generally recognized as safe (GRAS) substances. The antimicrobial activity of bacteriocins against pathogenic and saprophytic bacteria has raised significant interest for their application as food preservative. Use of bacteriocins may help decrease the utilization of synthetic additives as well as the intensity of heat and other physical treatments, satisfying customers demand for fresh tasting, ready to eat, and lightly preserved foods. Lately, impressive efforts have been made for applications of different bacteriocins and bacteriocinogenic strains. Depending on the raw materials, processing conditions, distribution, and consumption, the different types of foods offer a great variety of scenarios where food poisoning, pathogenic or spoilage bacteria may proliferate. Therefore, the bacteriocins efficacy needs to be carefully tested in the food systems for which they are intended to be applied against the selected target bacteria (Kaur et al., 2014).

With changing trend, there is also a growing interest in traditional foods and in the adaptation of local production processes to an industrial scale without substantial loss of the original value. Several developing countries have already succeeded but in many others, there is a need to meet the nutritional requirements of the population and to provide a minimal framework of food safety. The need to avoid economic losses due to microbial spoilage of raw materials and food products, to ameliorate the incidence of food borne illnesses, and to meet the food requirements of the growing world population strengthen the relevance of preservation methods in the food industry. In this context, the preservation of foods by natural, biological methods may be a satisfactory approach to solve many of the current food-related issues.

Biopreservation employing bacteriocins is a novel approach aims to eliminate or control pathogens in food while the emergence of resistant variants of the organism remains the major concern and limits their application. Several systematic studies investigating bacteriocin resistance in bacterial pathogens concluded that improved control of the target microorganisms and inhibition of bacteriocin-resistant strains and species can be achieved by using a hurdle concept. Hurdle technology refers to the manipulation of multiple factors (intrinsic and extrinsic) intended to prevent microbial entry, survival and growth in food stuff. In this approach, multiple factors/preservation methods coordinate to provide greater protection than a single method alone, hence the food produced will be more safe and secure. While in some foods intrinsic properties, for example, high salt may provide adequate protection, the conscious addition of an extra hurdle(s) can ensure safety to a greater extent (Leistner, 2000).

Classification and Biosynthesis of Bacteriocins of LAB

There is a wide range of bacteriocins produced by different LAB (Table 2) which can be classified based on their primary structures, molecular weights, post-translational modifications and genetic characteristics (Klaenhammer et al, 1993; Güllüce et al, 2013). However, there is no universally accepted classification scheme for bacteriocins. Most LAB bacteriocins are small (3-6 kDa), cationic, heat-stable, amphiphilic, and membrane-permeabilizing peptides.

Bacteriocinogenic strains of LAB produces bacteriocins throughout the growth phase and production ceases at the end or sometimes before the end of the exponential phase (Parente et al., 1997; Lejeune et al., 1998). Initially, bacteriocins are produced/synthesized as pre-propeptide which are processed/modified before their efflux by dedicated transport machinery (Nes et al., 1996). Bacteriocin production is highly affected by factors like type and level of the carbon, nitrogen and phosphate sources, cations surfactants and inhibitors. For example, nisin can be produced by *Lactococcus lactis* IO-1 from media containing different carbohydrate sources (glucose, sucrose and xylose) but better yield was obtained with

Table 2. Few examples of bacteriocins produced by lactic acid bacteria (LAB)

Type	Characteristic	Example
Class I (Lantibiotics)	• Ribosomally synthesized peptides that undergo posttranslational modifications. • Molecular weight 2–5 kDa. • Contain lanthionine and-methyl lanthionine.	• Nisin A from *Lc. lactis* ssp. lactis ATCC114 • Nisin Z from *Lc. lactis* ssp. lactis NIZ022186 • Enterocin A & Enterocin B from *Ent. faecium* T136
Class II	• Heat stable peptides formed exclusively by unmodified amino acids. • Ribosomally synthesized as inactive pre peptides to get activated by posttranslational cleavage of the N-terminal leader peptide. • Molecular weight < 10 kDa.	• Lactococcin A from *Lc. lactis* ssp. lactis LMG2130 • Lactococcin B from *Lc. lactis* ssp. cremoris 9B4 • Lactococcin M from *Lc. lactis* ssp lactis • Pediocin PA-1 from *Ped. acidilactici* PAC1.0
ClassIII	• Nonlantibiotics—Large, Heat-labile Bacteriocins. • Molecular weight >30 kDa.	• Helveticin J from *Lb. helveticus* 481 • Lacticin A from *Lb. delbrukii* JCM1106 • Lacticin B from *Lb. delbrukii* JCM 1248
Class IV	Complex bacteriocins carrying lipid or carbohydrate moieties.	Sublancin from *Bacillus subtilis* 168
Class V	Circular bacteriocins.	• Enterocin RM6 from *Enterococcus faecalis* • Gassericin from *Lactobacillus gasseri* LA39 • Reutericin A from *Lactobacillus reuteri*

glucose compared to xylose (Matsuaki et al., 1996; Chinachoti et al., 1997a,b). Similarly, for production of Pediocin AcH, glucose followed by sucrose, xylose and galactose were the best carbon sources in an unbuffered medium (Biswas et al., 1991).

All bacteriocins are synthesized with an N terminal leader sequence whose function seems to prevent the bacteriocin from being biologically active while still inside the producer and provide the recognition signal for the transporter system (Holo et al., 1991; Muriana and Klaenhammer, 1991; Klaenhammer, 1993; Havarstein et al., 1994). Lantibiotics are well studied upto genetic level and number of genes found in genome includes: (a) *Lan A*, the structural gene, (b) *Lan I* (and in some cases *Lan E, Lan F* and *Lan G*), immunity genes, encodes for proteins that protect the producer from the produced lantibiotic, (c) Lan T gene encodes a membrane associated ABC transporter that transfers the lantibiotic across the membrane, (d) *lan P* gene encodes a serine proteinase which removes the leader sequence of the lantibiotic prepeptide, (e) two genes, *lan B* and *Lan C* (or in some cases only one gene, *Lan M*) encode enzymes involved in the formation of lanthionine and methyl lanthionine, and (f) lan k and lan R genes encodes for two component regulatory proteins that transmit an extracellular signal and thereby induce lantibiotic production.

Bacteriocins: Mode of Action

Bacteriocins recognizes and bind to the specific receptors located on the target microbial cell surface. Once they bind to the cell surface, they alter the functioning of the cytoplasmic membrane (affecting energy synthesis and permeability). The initial electrostatic attraction between the target cell membrane and the bacteriocin peptide is thought to be the driving force for subsequent events whether the case is of broad spectrum bacteriocin or a narrow spectrum bacteriocin. While many bacteriocins, including well studied nisin, have been shown to induce pore formation in sensitive microorganisms. Nisin forms pores that disrupt the proton motive force and the pH equilibrium causing leakage of ions and hydrolysis of ATP resulting in cell death (Figure 2). Lacticin 3147, Pep5, subtilin and epidermin are examples of other lantibiotics that also kill the target cell by pore formation (Schuller et al. 1989; Brotz et al. 1998). Later, it has been recognised that nisin also interferes with cell wall biosynthesis (Reisinger et al. 1980). This was claimed that the nisin has an ability to bind lipid II, a peptidoglycan precursor, thus inhibit cell wall biosynthesis. Such binding is also intrinsic to the ability of nisin to form pores. The possession of dual mechanisms of action renders nisin active at nM concentrations (Breukink et al. 1999). Some other bacteriocins can inhibit nucleic acid synthesis, interference with the protein synthesis and change cell translator mechanism.

Bacteriocins Applications in Food Safety

In recent years, several studies have indicated that application of bacteriocins in food can offer following benefits: (i) longer shelf-life of foods, (ii) ensure extra protection during temperature abuse conditions, (iii) lower pathogen transmission risk through the food chain, (iv) alleviate economic losses due to food spoilage, (v) less use of chemical preservatives, (vi) permit the application of less severe heat treatments without compromising food safety: better preservation of food nutrients and vitamins, as well as organoleptic properties of foods, (vii) permit the marketing of "novel" foods (less acidic, with a lower salt content, and with a higher water content), and (viii) serve to satisfy industrial and consumers demands from safety and natural aspects (Thomas et al. 2000). The applications of bacteriocins studied in dairy sector/dairy foods are given in Table 3.

Figure 2. Proposed model of cell killing by pore-forming bacteriocins (Jack et al. 1995; Garneau et al. 2002)

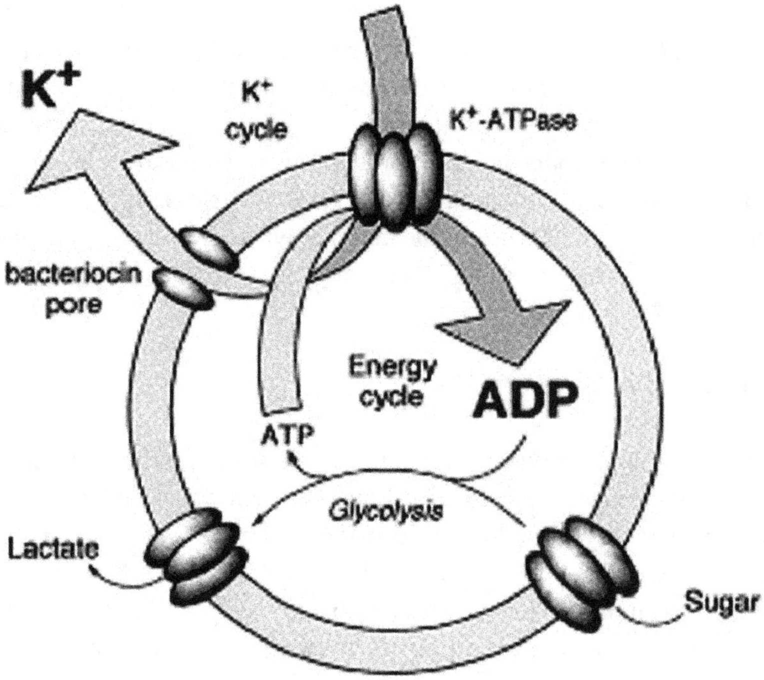

Table 3. Applications of bacteriocins studied in dairy sector/dairy foods

Bacteriocin	Producer Strain	Application	Reference
Nisin	*Lactococcus lactis*	prevent gas blowing in cheese caused by *C. tyrobutyricum*	De Vuyst and Vandamme 1994
Nisin	*Lactococcus lactis*	prevent proliferation of surviving endospore formers, mainly the gas-producing clostridia and *Clostridium botulinum*, as well as other post-process contaminating bacteria such as *L. monocytogenes* in processed cheeses and cheese products (e.g., block cheese, soft white cheeses, slices, spreads, sauces, dips)	Thomas and Delves-Broughton 2001
Enterocin AS-48	*Enterococcus faecalis*	added to a rice-based infant formula dissolved in whole milk completely inactivated *B. cereus* and prevented its growth for at least 15 days at 37°C	Grande et al. 2006
Pediocin 5	*P. acidilactici* UL5	Reported to have bactericidal effects against *L. monocytogenes* in milk	Huang et al. 1994
Partially purified Carnocin CP5	*C. piscicola* CP5	Reduced viable counts of *L. monocytogenes* in skim milk	Mathieu et al. 1994
Propionicin PLG-1	*Propionibacterium thoenii* P127	Kill or inhibit several psychrotrophic spoilage or pathogenic bacteria (*L. monocytogenes, P. fluorescens, Vibrio parahaemolyticus, Yersinia enterocolitica, and Corynebacterium sp.*)	Lyon et al. 1993
Jenseniin G	*Propionibacterium jensenii (thoenii)* P126	Control over acidification of yogurt and other fermented products because of its inhibitory activity to *Lactobacillus delbrueckii* ssp. bulgaricus	Weinbrenner et al. 1997
Variacin	*Kocuria varians*	Inhibited the proliferation of B. cereus in chilled dairy products, vanilla, and chocolate desserts in a concentration-dependent way	O'Mahony et al. 2001
Pediocin 34	*Pediococcus pentosaceus* 34	Enhanced the shelf-life of Paneer to about 75 days	Malik et al. 2005

Factors Influencing Bacteriocins Activity in Food System

Food may be commercially sterile, raw or fermented, being nutritionally rich may act as suitable matrix to harbour and support growth of wide variety of microorganisms. Even if commercially sterile foods get post-process contamination, the microorganisms may easily proliferate because of the lack of competitors. Under such conditions, the bacteriocins activity will greatly depend on a number of food-related factors (Table 4) which involve interaction with food components, precipitation, inactivation, or uneven distribution of bacteriocin molecules in the food matrix.

Bacteriocins and Hurdle Technology

The concept of hurdle technology began to apply in the food industry in a rational way after the observation that survival of microorganisms greatly decreased when they were confronted with multiple antimicrobial factors (Leistner 1978; Leistner and Gorris 1995; Leistner 2000). Over 60 potential hurdles have been described to improve food stability and/or quality (Leistner 2000). The application of bacteriocins as part of hurdle technology has received great attention in recent years (Chen and Hoover 2003; Rose et al. 2003; Deegan et al. 2006), since bacteriocins can be used purposely in combination with selected hurdles in order to increase microbial inactivation. Improved control of the target microorganisms and inhibition of bacteriocin-resistant strains and species can be obtained by using a hurdle concept.

Table 4. Factors affecting bacteriocin efficacy in foods

Food-Related Factors	Food Microbiota	Target Bacteria
Food processing conditions	Microbial load	Microbial load
Food storage temperature, Food pH, and bacteriocin unstability to pH changes	Microbial diversity	
Inactivation by food enzymes	Bacteriocin sensitivity	Bacteriocin sensitivity (Gram-type, genus, species, strains)
Interaction with food additives/ingredients	Microbial interactions in the food system	Physiological stage (growing, resting, starving orviable but non-culturable cells, stressed or sub lethally injured cells, endospores)
Low solubility and uneven distribution in the food matrix and Limited stability of bacteriocin during food shelf life		Protection by physico-che mical barriers (microcolonies, biofilms, slime)
Bacteriocin adsorption to food components		Development of resistance /adaptation

FERMENTATES: THE FOOD INGREDIENTS

Fermentates refers to ingredients produced by the fermentation of a variety of raw materials (typically milk, sugar, or plant derived materials such as corn or wheat starch) by food grade microorganisms, typically lactic acid bacteria or propionic acid bacteria (Elsser-Gravesen and Elsser-Gravesen, 2014). Such fermentation gives complex products that inherently do not have a well-defined composition but comprises of high proportion of antimicrobial metabolites, which may include organic acids (lactic, acetic or propionic acid), diacetyl, bacteriocins, and other secondary metabolites, depending on strain(s) used for the fermentation. Fermentates are marketed as a dry, cell-free powder which has very wide applications and thereby, their demand is increasing in food sector but has limited scientific background (Elsser-Gravesen and Elsser-Gravesen, 2014). Therefore, fermentates must be evaluated for their sensory suitability, safety and applicability by *in situ* testing (Crowley et al., 2013).

Currently, MicroGARD range (DuPont) and the DuraFresh range (Kerry) are commercially available fermentates for dairy applications. This also includes the former Alta and Perlac products from Quest, and many other products (namely spray-dried vinegar or fermented wheat flour products) that are claimed to be shelf-life enhancers. Besides wide applications of fermentates very limited scientific reports are available on their functionality in dairy foods. The MicroGARD products (produced by fermenting skimmed milk or dextrose with *Propionibacterium shermanii* or specific lactococci) and Alta (whey-based products prepared by *P. acidilactici*) were demonstrated to inhibit the psychotropic spoilage flora and thereby enhance the shelf-life of cottage cheese (al-Zoreky et al., 1991). Also, these products showed inhibition of *Pseudomonas*, *Salmonella*, *Yersinia*, and certain fungi. In contrast, no such significant effect of the MicroGARD and Alta products was seen against aerobic mesophilic counts, *Escherichia coli* or *Brochothrix thermosphacta* when tested in an acidified chicken meat model stored at 22°C (Lemay et al., 2002). Enhanced anti-listerial effect was reported in liquid cheese whey when nisin was used in combination with MicroGARD fermentate (von Staszewski and Jagus, 2008).

As already stated, there is limited scientific documentation for the fermentates in comparison to bacteriocins. Their wide industrial applications have raised the demands and United States leads the markets for fermentates. Such products are commercially available as "cultured milk" or "cultured sugar", based on the substrate used for producing the fermentate. A cultured dextrose version of MicroGARD has been evaluated for toxicity and no detrimental effects were observed (Buard et al., 2003). The MicroGARD products are used for a wide range of applications in dairy items including cottage cheese, yogurt, sour cream, dairy desserts. It has been estimated that approximately 30% of the US cheese production is made with MicroGARD (Sani et al., 2005). In 2011, FDA approved Durafresh (labeled as ''cultured grade A skim milk and skim milk powder'') use in cottage cheese to control *Listeria*.

Labelling is a major issue in the European Union (EU), as it would be required to label all active components in the fermentate. This presents two main concerns (i) not all active components are known, and (ii) most of the known ones have E-numbers. Therefore, the use of fermentates as natural preservatives is so far quite limited in the EU.

BIOPROTECTIVE CULTURES

Food preservation can be accomplished either by supplementing the food product with *ex-situ* produced bacteriocins (purified or partially purified), or by using fermentates as food ingredient (raw concentrate). The addition of partially purified or purified bacteriocin concentrates obtained by fermentation at industrial scale; require specific approval as preservatives from the legislative point of view. Till date, nisin is the only bacteriocin licensed as a food preservative. Legally, the raw concentrates/fermentates may also be regarded as food additives or ingredients, since their addition may affect the physico-chemical properties of the food (thickening, protein content etc). Also, the cell-derived antimicrobial metabolites (such as lactic acid, bacteriocins, reuterin etc.) present in these fermentates perform an additional bioprotectant function.

Considering both the processing cost as well as the legal issues related to addition of purified bacteriocins and/or fermentates, the use of bioprotective starter or adjunct cultures in fermented dairy foods is one promising solution. The fermentation of perishable food to enhance their shelf life is one of the oldest known uses of biotechnology. The value addition properties of the fermentation processes includes: (a) enhanced durability through production of antimicrobial metabolites (e.g. organic acids, bacteriocins, ethanol), often in conjunction with decreased water activity (drying and/or salting); (b) enhanced safety by reducing the level of either pathogenic microorganisms or their toxins; (c) enhanced nutritional value; and (d) enhanced organoleptic quality (Bourdichon et al., 2012). Also, there are some unwanted and unacceptable detrimental properties associated with food cultures that include virulence, toxicity, and antibiotic resistance. In the US, acceptable food microorganisms are granted the GRAS status, and in EU they are included in the Qualified Presumption of Safety list. An inventory list of currently used microbial food cultures, comprising 195 bacterial species and 69 fungal species, has recently been compiled (Bourdichon et al., 2012).

Microbial cultures used for fermentation can either be starter cultures (providing nutritional and organoleptic characteristics) or bioprotective cultures (providing durability and safety). All these properties are inherently linked for e.g. Formation of organic acids enhances durability and also contributes to the characteristic taste and texture of fermented product. All starter cultures are per se also protective cultures, but not all protective cultures are also starter cultures. Therefore, it is neither possible nor meaningful to differentiate between starter cultures and bioprotective cultures.

Research activities performed in past decades have aimed to develop cultures that can (i) enhance food safety either by direct killing or inhibiting the growth of pathogens or by suppressing toxin production or (ii) enhance shelf-life by controlling growth of spoilage microorganisms. Comprehensive reviews summarizing research findings including solutions for dairy products are available (Beshkova and Frengova, 2012; Grattepanche et al., 2008). Overall, most of the reports can be allocated to one of the following main categories:

1. **Use of Bacteriocinogenic LAB Cultures:** Mode of action relies on production of bacteriocins. In dairy industries, nisin-producing *L. lactis* strains have been reported to inhibit *L. monocytogenes* in several varities of cheeses, such as cottage cheese (Benkerroum and Sandine 1988) or Camembert (Maisnier- Patin et al. 1992). Also, enterocin producing *Enterococcus faecium* strains in smear of soft cheese controlled outgrowth of *L. monocytogenes* (Izquierdo et al., 2009).
2. **Use of Antifungal LAB and/or PAB to Delay Spoilage of Various Types of Food:** Mode of action relies on production of several antifungal metabolites. For e.g. *Lactobacillus rhamnosus* and *Propionibacterium freudenreichii* ssp. shermanii were used for inhibiting yeast in yogurt (Liptakova et al., 2006). *Lactobacillus harbinensis* was used as an antifungal culture in yogurt (Delavenne et al., 2013).
3. **Use of Non-Bacteriocinogenic LAB With Other Competitive Properties:** Mode of action relies on antimicrobial metabolites other then bateriocins. For e.g. commercial culture had a protective effect by depletion of oxygen, production of organic acids etc (Seibert, 2010).

Protective cultures are produced and marketed in the same way as starter cultures. The protective strains are produced in batch fermenters, subsequently concentrated by centrifugation, and finally formulated either as frozen pellets or freeze-dried powders. Protective cultures are now well established and recognized as an efficient tool to ensure the food safety and security. Few examples of the commercially available protective cultures for dairy products are given in Table 5.

BACTERIOPHAGES

Continuously increasing food-borne diseases caused by pathogens, such as *Salmonella, Campylobacter, Escherichia coli, Listeria* and others, shows that the current technologies employed to inactivate bacterial pathogens are not infallible (DuPont 2007). Contaminating bacteria can get access to food during milking, fermentation, processing, storage or packaging. In past decades, several strategies have been

Table 5. Examples of commercially available protective cultures

Protective Function	Microorganisms	Producer
Growth inhibition of *Listeria monocytogenes*	Lactic acid bacteria (e.g. *Lb. sakei,* *Lb. curvatus,* *Lb. plantarum)*	Chr. Hansen (Denmark) DuPont (USA)
Inhibition of mold and yeasts	*Lactobacillus* sp. *Lb. rhamnosus,* *Lb. paracasei,* *Propionobacterium* sps.	Chr. Hansen (Denmark) DuPont (USA)
Inhibition of *Clostridia tyrobutyricum*; prevention of late blowing	*Lactococcus lactis*	CSK (Netherlands)

Table 6. Advantages of using bacteriophages as biocontrol tools in food industry

History of safe use	• Ubiquitous in nature including food ecosystems • Natural commensals of humans and animals • Extensive clinical use in Eastern Europe
Highly active and specific	• No adverse effects on the intestinal microbiota • Innocuous to mammalian cells • Autoreplicative • Can be active against biofilms
Genetically amenable	
Versatile use along the food chain	• Phage therapy • Biosanitation • Biopreservation
Tools for detecting pathogens	
Source of potent antimicrobials	Endolysins and other peptidoglycan hydrolases

tried to minimize the microbial load of raw products, such as the use of antibiotics and physical treatments like steam, dry heat and UV light. But their applications are restricted due to the negative impact of antibiotics on human antimicrobial therapies, development of antibiotic resistant strains and physical treatments deteriorate the organoleptic properties. Also, some approaches used in the dairy industry to reduce contamination cannot be directly applied to dairy products.

Bacteriophages (Greek, meaning ''bacteria-eater'') are viruses that invade bacterial cells (host) and, in the case of lytic phages, disrupt bacterial metabolism and cause the host to lyse. Bacteriophages are host-specific that means the phage recognizes its specific host and can only propagate on a certain bacterial species. Their long history of safe use, relatively easy handling and highly specific antimicrobial activity make them suitable biocontrol tool to ensure food safety (Table 6).

In the dairy industry, starter cultures failure due to phage attack is a common problem. In contrast, phages which can attack pathogens can be useful in food processing. So to speak, ''the enemy of my enemy is my useful friend''. Applications of bacteriophages as natural antimicrobials in food industry to inhibit undesirable bacteria have become widely recognized in past years (Hagens and Loessner 2007; Strauch et al. 2007). They have been proposed as biocontrol tool to ensue food safety (Garcı´a et al., 2008).

Bacteriophages can be used for (i) biotherapy (to prevent or reduce colonization and infection in livestock), (ii) biosanitation and biocontrol (to decontaminate raw products and to disinfect equipment and contact surfaces) and (iii) biopreservation (to extend the shelf life of perishable food products). Bacteriophages should also be considered in hurdle technology in combination with different preservation methods (Leverentz et al. 2003; Martı´nez et al. 2008). Several examples of phage application in dairy industry are outlined below.

- *E. coli* O157:H7 remains a continuous public health threat because its ingestion even in small concentration (10–100 cells) may result in potent toxin exposure. Ruminants being the principal reservoir for this strain can release contaminated milk during milking. Raya et al. (2006) observed a 2-log-unit reduction in intestinal *E. coli* O157:H7 within 2 days after oral administration of phage CEV1 in sheep. However, phage KH1 did not show such effect. Sheng et al. (2006) observed significantly less cell count when a combination of phages KHI and SH1 were administered rectally to cattle and maintained at 10^6 PFU ml^{-1} in the drinking water.

- Modi et al. (2001) tested the activity of the *Salmonella* phage SJ2 in cheddar cheese manufacturing. In the presence of phages (MOI 10^4), *Salmonella* did not survive in the pasteurized cheeses after 89 days, whereas about 50 CFU ml^{-1} were still viable in raw milk cheeses.
- Kim et al. (2007a) addressed the issue of *Enterobacter sakazakii* in reconstituted infant milk formula. The authors observed that newly isolated phage at concentration of 10^9 PFU ml^{-1} was able to effectively suppress growth in prepared infant formula, both at 24 °C and 37°C (Kim et al. 2007b).
- Staphylococcal mastitis caused by *Staphylococcus aureus* is a major concern to the dairy industry and infected udder is the most important source of milk contamination by this pathogen. Staphylococcal food poisoning is due to the absorption of preformed staphylococcal enterotoxins (Le Loir et al. 2003). Gill et al. (2006a) evaluated the ability of the lytic *Staph. aureus* bacteriophage K to eliminate bovine *Staph. aureus* intramammary infection during lactation. The authors reported Phage K inactivation in raw milk, likely due to the adsorption of whey proteins to the cell surface that interfere with phage attachment (O'Flaherty et al. 2005; Gill et al. 2006b). However, Garcı´a et al. (2007) found that a cocktail of two lytic phages of dairy origin inhibited *Staph. aureus* in acid and enzymatic curd manufacturing processes.
- Recently, a commercial product named Listex P100 was approved by the FDA as a food biopreservative and granted as GRAS (Federal Register: August 18, 2006. Volume 71, Number 160, pp. 47729–47732). This product is based on the virulent phage P100 (Carlton et al. 2005) and, depending on dosage and treatment, a complete eradication of *L. monocytogenes* on surface-ripened red-smear soft cheese.

Bacteriophages may act as vectors that can transfer virulent and antibiotic-resistance traits, therefore, have raised safety concerns. Also, phages may not show the antimicrobial activity in food systems as observed in laboratory conditions. Several factors such as reduced diffusion rates that decrease the chance of host-phage collisions, the microbial load which might also act as a mechanical barrier by providing unspecific phage binding sites and other adverse factors such as temperature, pH and inhibitory compounds may limit their performance *in situ*.

CONCLUSION

The use of biopreservatives in the dairy industry can help to achieve the target of providing minimally processed, ready to eat, naturally preserved food. Although most efforts have been devoted to find out and understand different approaches to preserve the food products in more natural way but still many limitation to these approaches need to be overcome. Several researches on isolation and characterization of bacteriocinogenic LAB strains and their bacteriocins, their application in food safety preservation of dairy products have already been done. The careful selection of bacteriocinogenic strain with best performance for variety of food type and choice of bacteriocin treatments (either alone or in combination with other hurdles) may greatly help in development of suitable food safety systems. But the developing bacteriocins resistance among the pathogens is of major concern. Therefore, more study is needed to determine the distribution of bacteriocin-resistance phenomena among spoilage causing and pathogenic microorganisms. Also, knowledge of the characteristics of bacteriocin- resistant variants and the conditions that prevent their emergence will help in determining the optimal conditions for application of bacteriocins in foods and minimize the incidence of resistance. Fermentates have limitations as they can affect the organoleptic and physico-chemical properties of the product. No single bioprotective culture can be used for wide range of products i.e. why different bioprotective culture for different products need to be studied. Also, the maintainence of the bioprotective culture and its activity to meet the industrial demand is a challenge in itself. Bacteriophages can act as vector of undesired traits (virulence and antibiotic resistance genes) therefore, have raised safety concerns. Also, their performance *in situ* may not be as good as in laboratory conditions which may be attributed to several factors related to food matrices. Such limiting factors need to be considered while selecting the bacteriophage for particular food product. Lastly, it can be considered that there is no single magic bullet that can be used to preserve the food. With Hurdle technology or manipulation of several factors in combination with the biopreservatives, the aim of natural preservation can only be accomplished.

REFERENCES

al-Zoreky, N., Ayres, J. W., & Sandine, W. E. (1991). Antimicrobial activity of microgard against food spoilage and pathogenic microorganisms. *Journal of Dairy Science*, *74*(3), 758–763. doi:10.3168/jds.S0022-0302(91)78222-2 PMID:1906486

Benkerroum, N., & Sandine, W. E. (1988). Inhibitory action of nisin against *Listeria monocytogenes*. *Journal of Dairy Science*, *71*(12), 3237–3245. doi:10.3168/jds.S0022-0302(88)79929-4 PMID:3148644

Beshkova, D., & Frengova, G. (2012). Bacteriocins from lactic acid bacteria: Microorganisms of potential biotechnological importance for the dairy industry. *Engineering in Life Sciences*, *12*(4), 419–432. doi:10.1002/elsc.201100127

Biswas, S. R., Ray, P., Johnson, M. C., & Ray, B. (1991). Influence of growth conditions on the prod of a bacteriocin, pediocin AcH, by *Pediococcus acidilactic* H. *Applied and Environmental Microbiology*, *57*, 1265–1267. PMID:16348467

Bourdichon, F., Casaregola, S., Farrokh, C., Frisvad, J. C., Gerds, M. L., Hammes, W. P., ... Hansen, E. B. (2012). Food fermentations: Microorganisms with technological beneficial use. *International Journal of Food Microbiology*, *154*(3), 87–97. doi:10.1016/j.ijfoodmicro.2011.12.030 PMID:22257932

Breukink, E., Weidemann, I., van Kraaij, C., Kuipers, O. P., Sahl, H. G., & de Kruijff, B. (1999). Use of cell wall precursor lipid II by a poreforming peptide antibiotic. *Science*, *286*(5448), 2361–2364. doi:10.1126cience.286.5448.2361 PMID:10600751

Brotz, H., Bierbaum, G., Leopold, K., Reynolds, P. E., & Sahl, H. G. (1998). The lantibiotic mersacidin inhibits peptidoglycan synthesis by targeting lipid II. *Antimicrobial Agents and Chemotherapy*, *42*, 154–160. PMID:9449277

Buard, A., Carlton, B. D., Floch, F., & Simon, G. S. (2003). Subchronic toxicity, mutagenicity and allergenicity studies of a cultured dextrose food product. *Food and Chemical Toxicology*, *41*(5), 689–694. doi:10.1016/S0278-6915(03)00006-1 PMID:12659722

Carlton, R. M., Noordman, W. H., Biswas, B., de Meester, E. D., & Loessner, M. J. (2005). Bacteriophage P100 for control of *Listeria monocytogenes* in foods: Genome sequence, bioinformatic analyses, oral toxicity study, and application. *Regulatory Toxicology and Pharmacology, 43*(3), 301–312. doi:10.1016/j.yrtph.2005.08.005 PMID:16188359

Chen, H., & Hoover, D. G. (2003). Bacteriocins and their food applications. *Comprehensive Reviews in Food Science and Food Safety, 2*(3), 82–100. doi:10.1111/j.1541-4337.2003.tb00016.x

Chinachoti, N., Matsuaki, H., Sonomoto, K., & Ishikazi, A. (1997a). Utilization of xylose as an alternative carbon source for nisin Z production by *Lactococcus lactis* IO-1. *Fac Agric. Kyushu Univ, 42*, 171–181.

Chinachoti, N., Zaima, T., Masuaki, H., Sonomoto, K., & Ishikazi, A. (1997b). Relationship between nisin Z fermentaire prod and aeration condition using *Lactococcus lactis* I0-1. *Journal of the Faculty of Agriculture, Kyushu University, 43*, 437–448.

Crowley, S., Mahony, J., Morrissey, J. P., & van Sinderen, D. (2013). Transcriptomic and morphological profiling of *Aspergillus fumigatus* Af293 in response to antifungal activity produced by *Lactobacillus plantarum* 16. *Microbiology, 159*(Pt_10), 2014–2024. doi:10.1099/mic.0.068742-0 PMID:23876797

Deegan, L. H., Cotterm, P. D., Hill, C., & Ross, P. (2006). Bacteriocins: Biological tools for bio-preservation and shelf-life extension. *International Dairy Journal, 16*(9), 1058–1071. doi:10.1016/j.idairyj.2005.10.026

Delavenne, E., Ismail, R., Pawtowski, A., Mounier, J., Barbier, G., & Le Blay, G. (2013). Assessment of lactobacilli strains as yogurt bioprotective cultures. *Food Control, 30*(1), 206–213. doi:10.1016/j.foodcont.2012.06.043

DeVuyst, L., & Vandamme, E. (1994). *Bacteriocins of Lactic acid bacteria: microbiology, genetics and application*. London, UK: Chapman & Hill Ltd. doi:10.1007/978-1-4615-2668-1

DuPont, H. L. (2007). The growing threat of foodborne bacterial enteropathogens of animal origin. *Clinical Infectious Diseases, 45*(10), 1353–1361. doi:10.1086/522662 PMID:17968835

Elsser-Gravesen, D., & Elsser-Gravesen, A. (2014). Biopreservatives. *Advances in Biochemical Engineering/Biotechnology, 143*, 29–49. doi:10.1007/10_2013_234 PMID:24185748

Franz, C. M. A. P., van Belkum, M. J., Holzapfel, W. H., Abriouel, H., & G'alvez, A. (2007). Diversity of enterococcal bacteriocins and their grouping into a new classification scheme. *FEMS Microbiology Reviews, 31*(3), 293–310. doi:10.1111/j.1574-6976.2007.00064.x PMID:17298586

G'alvez, A., Abriouel, H., Lucas L'opez, R., & Ben Omar, N. (2007). Bacteriocin based strategies for food biopreservation. *International Journal of Food Microbiology, 120*(1-2), 51–70. doi:10.1016/j.ijfoodmicro.2007.06.001 PMID:17614151

Garcı'a, P., Madera, C., Martı'nez, B., & Rodrı'guez, A. (2007). Biocontrol of *Staphylococcus aureus* in curd manufacturing processes using bacteriophages. *International Dairy Journal, 17*(10), 1232–1239. doi:10.1016/j.idairyj.2007.03.014

Gill, J. J., Pacan, J. C., Carson, M. E., Leslie, K. E., Griffiths, M. W., & Sabour, P. M. (2006a). Efficacy and pharmacokinetics of bacteriophage therapy in treatment of subclinical *Staphylococcus aureus* mastitis in lactating dairy cattle. *Antimicrobial Agents and Chemotherapy*, *50*(9), 2912–2918. doi:10.1128/AAC.01630-05 PMID:16940081

Gill, J. J., Sabour, P. M., Leslie, K. E., & Griffiths, M. W. (2006b). Bovine whey proteins inhibit the interaction of *Staphylococcus aureus* and bacteriophage K. *Journal of Applied Microbiology*, *101*(2), 377–386. doi:10.1111/j.1365-2672.2006.02918.x PMID:16882145

Grande, M. J., Lucas, R., Abriouel, H., Valdivia, E., Ben Omar, N., Maqueda, M., ... G'alvez, A. (2006). Inhibition of toxicogenic *Bacillus cereus* in rice based foods by enterocin AS-48. *International Journal of Food Microbiology*, *106*(2), 185–194. doi:10.1016/j.ijfoodmicro.2005.08.003 PMID:16225949

Grattepanche, F., Miescher-Schwenninger, S., Meile, L., & Lacroix, C. (2008). Recent developments in cheese cultures with protective and probiotic functionalities. *Dairy Science & Technology*, *88*(4-5), 421–444. doi:10.1051/dst:2008013

Güllüce, M., Karaday, M., & Barış, Ö. (2013). Bacteriocins: promising natural antimicrobials. In *Microbial pathogens and strategies for combating them: science, technology and education* (pp. 1016–1027). Badajoz, Spain: FORMATEX.

Hagens, S., & Loessner, M. J. (2007). Application of bacteriophages for detection and control of foodborne pathogens. *Applied Microbiology and Biotechnology*, *76*, 513–519.

Håvarstein, H., Holo, H., & Nes, I. F. (1994). he leader peptide of colicin V shares consensus sequences with leader peptides that are common among peptide bacteriocins produced by gram-positive bacteria. *Microbiology*, *140*(9), 2383–2389. doi:10.1099/13500872-140-9-2383 PMID:7952189

Holo, H., Nilssen, O., & Nes, I. F. (1991). Lactococcin A, a new bacteriocin from *Lactococcus lactis* subsp. *cremoris*: Isolation and characterization of the protein and its gene. *Journal of Bacteriology*, *173*(12), 3879–3887. doi:10.1128/jb.173.12.3879-3887.1991 PMID:1904860

Huang, J., Lacroix, C., Daba, H., & Simard, R. E. (1994). Growth of *Listeria monocytogenes* in milk and its control by pediocin 5 produced by *Pediococcus acidilactici* UL5. *International Dairy Journal*, *4*(5), 429–443. doi:10.1016/0958-6946(94)90057-4

Izquierdo, E., Marchioni, E., Aoude-Werner, D., Hasselmann, C., & Ennahar, S. (2009). Smearing of soft cheese with *Enterococcus faecium* WHE 81, a multi-bacteriocin producer, against *Listeria monocytogenes*. *Food Microbiology*, *26*(1), 16–20. doi:10.1016/j.fm.2008.08.002 PMID:19028299

Kaur, M. R. K., & Singh, T. P. (2014). Bacteriocins and their potential applications. In Microbes in the service of mankind: tiny bugs with huge impact (pp. 309-345). JBC Press.

Kim, K., Jang, S. S., Kim, S. K., Park, J. H., Heu, S., & Ryu, S. (2007a). Prevalence and genetic diversity of *Enterobacter sakazakii* in ingredients of infant foods. *International Journal of Food Microbiology*, *122*(1-2), 196–203. doi:10.1016/j.ijfoodmicro.2007.11.072 PMID:18177966

Kim, K. P., Klumpp, J., & Loessner, M. J. (2007b). *Enterobacter sakazakii* bacteriophages can prevent bacterial growth in reconstituted infant formula. *International Journal of Food Microbiology, 115*(2), 195–203. doi:10.1016/j.ijfoodmicro.2006.10.029 PMID:17196280

Klaenhammer, T. R. (1993). Genetics of bacteriocins produced by lactic acid bacteria. *FEMS Microbiology Reviews, 12*(1-3), 39–86. doi:10.1111/j.1574-6976.1993.tb00012.x PMID:8398217

Le Loir, Y., Baron, F., & Gautier, M. (2003). *Staphylococcus aureus* and food poisoning. *Genetics and Molecular Research, 31*, 63–76. PMID:12917803

Leistner, L. (1978). Hurdle effect and energy saving. In W. K. Downey (Ed.), *Food Quality and Nutrition* (pp. 553–557). London, UK: Applied Science Publishers.

Leistner, L. (2000). Basic aspects of food preservation by hurdle technology. *International Journal of Food Microbiology, 55*(1-3), 181–186. doi:10.1016/S0168-1605(00)00161-6 PMID:10791741

Leistner, L., & Gorris, L. G. M. (1995). Food preservation by hurdle technology. *Trends in Food Science & Technology, 6*(2), 41–46. doi:10.1016/S0924-2244(00)88941-4

Lejeune, R., Callewaert, R., Crabbé, & De Vuyst. (1998). Modelling the growth and bacteriocin production by *Lactobacillus amylovorus* DCE 471 in batch cultivation. *Journal of Applied Microbiology, 84*(2), 159–168. doi:10.1046/j.1365-2672.1998.00266.x

Lemay, M. J., Choquette, J., Delaquis, P. J., Claude, G., Rodrigue, N., & Saucier, L. (2002). Antimicrobial effect of natural preservatives in a cooked and acidified chicken meat model. *International Journal of Food Microbiology, 78*(3), 217–226. doi:10.1016/S0168-1605(02)00014-4 PMID:12227640

Leverentz, B., Conway, W. S., Camp, M. J., Janisiewicz, W. J., Abuladze, T., Yang, M., ... Sulakvelidze, A. (2003). Biocontrol of *Listeria monocytogenes* on fresh-cut produce by treatment with lytic bacteriophages and a bacteriocin. *Applied and Environmental Microbiology, 69*(8), 4519–4526. doi:10.1128/AEM.69.8.4519-4526.2003 PMID:12902237

Liptakova, D., Valik, L., & Bajusova, B. (2006). Effect of protective culture on the growth of *Candida maltosa* YP1 in yoghurt. *Journal of Food and Nutrition Research, 45*, 147–151.

Lyon, W. J., Sethi, J. K., & Glatz, B. A. (1993). Inhibition of psychrotrophic organisms by propionicin PLG-1, a bacteriocin produced by *Propionibacterium thoenii*. *Journal of Dairy Science, 76*(6), 1506–1513. doi:10.3168/jds.S0022-0302(93)77482-2 PMID:8326023

Maisnier-Patin, S., Deschamps, N., Tatini, S. R., & Richard, J. (1992). Inhibition of *Listeria monocytogenes* in Camembert cheese made with a nisin-producing starter. *Le Lait, 72*(3), 249–263. doi:10.1051/lait:1992318

Malik, R. K., Rao, K. N., Bandhopadhyay, P., & Kumar, N. (2005). Bacteriocins: Natural and safe anti microbial peptides for food preservation. *Indian Food Industry, 24*(1), 69–70.

Martı'nez, B., Obeso, J. M., Rodrı'guez, A., & Garcı'a, P. (2008). Nisin-bacteriophage crossresistance in *Staphylococcus aureus*. *International Journal of Food Microbiology, 122*(3), 253–258. doi:10.1016/j.ijfoodmicro.2008.01.011 PMID:18281118

Mathieu, F., Michel, M., Lebrihi, A., & Lefebvre, G. (1994). Effect of the bacteriocin carnocin CP5 and of the producing strain *Carnobacterium piscicola* CP5 on the viability of Listeria monocytogenes ATCC 15313 in salt solution, broth and skimmed milk, at various incubation temperatures. *International Journal of Food Microbiology*, *22*(2-3), 155–172. doi:10.1016/0168-1605(94)90139-2 PMID:8074969

Matsuaki, H., Endo, N., Sonomoto, K., & Ishikazi, A. (1996). Lantibiotic nisin Z fermentative production by *Lactococcus lactis* IO-1: Relationship between production of the lantibiotic and lactate and cell growth. *Applied Microbiology and Biotechnology*, *45*(1-2), 36–40. doi:10.1007002530050645 PMID:8920177

Modi, R., Hirvi, Y., Hill, A., & Griffiths, M. W. (2001). Effect of phage on survival of *Salmonella enteritidis* during manufacture and storage of Cheddar cheese made from raw and pasteurized milk. *Journal of Food Protection*, *64*(7), 927–933. doi:10.4315/0362-028X-64.7.927 PMID:11456198

Muriana, P. M., & Klaenhamer, T. R. (1991). Purification and partial characterization of lactacin F, a bacteriocin produced by *Lactobacillus acidophilus* 11088. *Applied and Environmental Microbiology*, *57*, 114–121. PMID:1903624

Nes, I. F., Diep, D. B., Håvarstein, L. S., & Brurberg, M. B. (1996). Biosynthesis of bacteriocins in lactic acid bacteria. *Antonie van Leeuwenhoek*, *70*, 113–128. doi:10.1007/BF00395929 PMID:8879403

O'Flaherty, S., Coffey, A., Meaney, W. J., Fitzgerald, G. F., & Ross, R. P. (2005). Inhibition of bacteriophage K proliferation on *Staphylococcus aureus* in raw bovine milk. *Letters in Applied Microbiology*, *41*(3), 274–279. doi:10.1111/j.1472-765X.2005.01762.x PMID:16108920

O'Mahony, T., Rekhif, N., Cavadini, C., & Fitzgerald, G. F. (2001). The application of a fermented food ingredient containing 'variacin', a novel antimicrobial produced by *Kocuria varians*, to control the growth of *Bacillus cereus* in chilled dairy products. *Journal of Applied Microbiology*, *90*(1), 106–114. doi:10.1046/j.1365-2672.2001.01222.x PMID:11155129

Parente, E., Giglio, M. A., Riccardi, A., & Clementi, F. (1998). The combined effect of nisin, leucocin F10, pH, NaCl and EDTA on the survival of *Listeria monocytogenes* in broth. *International Journal of Food Microbiology*, *40*(1-2), 65–75. doi:10.1016/S0168-1605(98)00021-X PMID:9600612

Raya, R. R., Varey, P., Oot, R. A., Dyen, M. R., Callaway, T. R., Edrington, T. S., ... Brabban, A. D. (2006). Isolation and characterization of a new T-even bacteriophage, CEV1, and determination of its potential to reduce *Escherichia coli* O157:H7 levels in sheep. *Applied and Environmental Microbiology*, *72*(9), 6405–6410. doi:10.1128/AEM.03011-05 PMID:16957272

Reisinger, H., Seidel, H., Tschesche, P., & Hammes, W. P. (1980). The effect of nisin on murein synthesis. *Archives of Microbiology*, *127*(3), 187–193. doi:10.1007/BF00427192 PMID:6255884

Roller, S. (2003). Introduction. In S. Roller (Ed.), *Natural antimicrobials for the minimal processing of foods* (pp. 1–10). Cambridge, UK: Woodhead Publishing Ltd.

Rose, R. P., Sporns, P., Dodd, H. M., Gasson, M. J., Mellon, F. A., & McMullen, L. M. (2003). Involvement of dehydroalanine and dehydrobutyrine in the addition of glutathione to nisin. *Journal of Agricultural and Food Chemistry*, *51*(10), 3174–3178. doi:10.1021/jf026022h PMID:12720411

Sani, A. M., Ehsani, M. R., & Asadi, M. M. (2005). Effect of *Propionibacterium shermanii* metabolites on sensory properties and shelf life of UF-Feta cheese. *Nutrition & Food Science*, *35*(2), 88–94. doi:10.1108/00346650510585877

Schuller, F., Benz, R., & Sahl, H. G. (1989). The peptide antibiotic subtilin acts by formation of voltage-dependent multi-state pores in bacterial and artificial membranes. *European Journal of Biochemistry*, *182*(1), 181–186. doi:10.1111/j.1432-1033.1989.tb14815.x PMID:2471644

Seibert, T. M. (2010). Protective culture eliminates residual oxygen. *Die Fleischwirtschaft (Frankfurt)*, *90*, 59–61.

Sheng, H., Knecht, H. J., Kudva, I. T., & Hovde, C. J. (2006). Application of bacteriophages to control intestinal *Escherichia coli* O157:H7 levels in ruminants. *Applied and Environmental Microbiology*, *72*(8), 5359–5366. doi:10.1128/AEM.00099-06 PMID:16885287

Strauch, E., Hammerl, J. A., & Hertwig, S. (2007). Bacteriophages: New tools for safer food? *Journal für Verbraucherschutz und Lebensmittelsicherheit*, *2*(2), 138–143. doi:10.100700003-007-0188-5

Thomas, L. V., Clarkson, M. R., & Delves-Broughton, J. (2000). Nisin. In A. S. Naidu (Ed.), *Natural food antimicrobial systems* (pp. 463–524). Boca-Raton, FL: CRC Press.

Thomas, L. V., & Delves-Broughton, J. (2001). New advances in the application of the food preservative nisin. *Advances in Food Sciences*, *2*, 11–22.

von Staszewski, M., & Jagus, R. J. (2008). Natural antimicrobials: Effect of Microgard (TM) and nisin against *Listeria innocua* in liquid cheese whey. *International Dairy Journal*, *18*(3), 255–259. doi:10.1016/j.idairyj.2007.08.012

Weinbrenner, D. R., Barefoot, S. F., & Grinstead, D. A. (1997). Inhibition of yogurt starter cultures by Jenseniin G, a *Propionibacterium* bacteriocin. *Journal of Dairy Science*, *80*(7), 1246–1253. doi:10.3168/jds.S0022-0302(97)76053-3

Chapter 5
Aerobic Respiration in Lactic Acid Bacteria:
Current and Future Applications in Dairy Starter Culture

Sarang Dilip Pophaly
College of Dairy Science and Food Technology, India

Manorama Chauhan
College of Dairy Science and Food Technology, India

Jitesh Tarak
College of Dairy Science and Food Technology, India

Shekhar Banala Bashetty
College of Dairy Science and Food Technology, India

Tejinder Pal Singh
College of Dairy Science and Technology, India

Sudhir Kumar Tomar
National Dairy Research Institute, India

ABSTRACT

Lactic acid bacteria (LAB) are used as food-grade microorganisms for production of a variety of fermented milk products. They are also the most common probiotic organisms used for making functional foods. Lactic acid bacteria are well known for their fermentative metabolism wherein they convert simple carbohydrates to organic acids and other end products. Fermentation helps the bacteria to generate ATP required for various cellular activities via substrate level phosphorylation reaction. Fermentation results in incomplete oxidation of substrate and hence is an inefficient process with a low ATP yield. However, some LAB are genetically capable of activating an auxiliary respiratory metabolism in which a quinol oxidase serves as the final electron acceptor and high ATP production is achieved due to oxidative phosphorylation. The respiratory process is associated with high biomass production of LAB and more robust bacterial cells, which are essentially required for manufacture of high viability starter culture. This chapter explores LAB's current and future applications in dairy starter cultures.

DOI: 10.4018/978-1-5225-5363-2.ch005

INTRODUCTION

Lactic acid bacteria (LAB) consist of a group of Gram-positive bacteria that are united by certain morphological, metabolic and physiological characteristics. LAB are the workhorses of most food fermentation processes, and have been extensively studied for their metabolic and genetic properties. LAB are primarily used as starter cultures in manufacturing of diverse range of fermented products from milk, cereals etc. LAB are also used as probiotic bacteria for their physiological stress resistance and health promoting attributes (Ljungh & Wadstrom, 2006). LAB metabolism is fermentative in nature which means energy production in these bacteria is depended on the substrate level phosphorylation. Fermentation involves anaerobic breakdown of carbohydrates with an organic molecule acting as the final electron acceptor. Fermentative metabolism has peculiar disadvantages with regard to bacterial growth. First, since fermentation does not involve an electron transport system, it provides little energy due to partial breakdown of glucose. Secondly, fermentation results in high amount of lactic acid production in the medium which limits the growth of cells and do not allow to attain high biomass concentration. These hindrances limit the use of fermentation in manufacturing of commercial starter cultures and probiotics, which require high cell densities. These problems can be overcome by switching over from fermentative to respiratory metabolism (Pedersen *et al.*, 2012).

The use of alternative terminal electron acceptor such as cytochrome oxidase enzyme allows the mounting of a conditional respiration system in LAB. Respiration is a high energy yielding process in which organic or reduced inorganic compounds are oxidized by inorganic compounds. Aerobic respiration involves glycolysis, citric acid cycle (Kreb's Cycle) and an electron transport chain (ETC) in model organisms such as *E. coli*. However, LAB lack a functional Kreb's Cycle and ETC. Instead, the heme dependent cytochrome quinol oxidase can act as a terminal electron acceptor for supporting respiration. This conditional respiration system generates a high biomass and also improves stress resistance in LAB. These advantages can be used for enhanced technological and physiological performance of LAB during industrial processing conditions and *in-situ* settings.

Fermentation

In biological systems, energy is produced or conserved through reduction-oxidation (redox) reactions. Redox reactions involve transfer of electrons between molecules. Oxidation refers to the removal of electrons and 'reduction' refers to the addition of electrons. Fermentation is basically an energy production reaction-using breakdown of organic molecules such as glucose to pyruvic acid. In this process, glucose is actually oxidized (stripped of electrons) and pyruvate is produced. The actual fermentative step involves conversion of pyruvate to lactate.

In fermentation an organic molecule serves as the final electron acceptor. It does not involve an electron transport chain and hence results in low ATP yield. Glycolysis is the initial stage of fermentation. In glycolysis, a six carbon glucose molecule is partially broken down into two, three carbon molecules of pyruvate, 2NADH, 2 H+ and 2 ATP as a result of substrate level phosphorylation. If NADH is not oxidized back to NAD+, there will be no further catabolism. Thus NADH produced during glycolysis should be oxidized by transferring its electron to an electron acceptor. Therefore, a suitable terminal electron acceptor must be used. The NADH donates its electrons to the pyruvate molecule formed during glycolysis and regenerates NAD+. In this process the pyruvate molecule is in turn reduced and results in formation of lactic acid. This step is catalyzed by lactate dehydrogenase enzyme.

Aerobic Respiration

Aerobic respiration, involves diversion of pyruvate generated at the end of glycolysis to complete oxidation instead of formation of lactic acid. The transfer of pyruvate to Kreb's Cycle and Electron Transport Chain (ETC) in a generic respiration system yields high amount of ATP. The cellular respiration in model microorganisms is carried out in three phases

1. Glycolysis
2. Kreb's Cycle
3. Electron Transport Chain

During the first phase, a sugar molecule such as glucose is broken down by the complex enzymatic machinery, a process called glycolysis. The glycolysis is usually carried out via the Embden Meyerhof Parnas (EMP) pathway. The end product of glucose breakdown is the three-carbon molecule, pyruvate. The pyruvate produced at the end of glycolysis can be diverted towards production of lactic acid (in case of fermentation) or to Kreb's Cycle (for further breakdown) in case of aerobic respiration. In the presence of oxygen, the end product (pyruvate) of glycolysis enters the Kreb's Cycle (Citric Acid Cycle). In the Krebs cycle, pyruvate is broken down into carbon dioxide. In addition to generating ATP, the Krebs cycle and glycolysis generate hydrogen ions (H+), and electrons. The electrons and H+ are harvested in cellular respiration to generate energy on the electron transport chain. In electron transport chain, ATP is formed through a series of electron transfer reactions within the cytoplasmic membrane that drive the oxidative phosphorylation of ADP to ATP. Bacteria use various flavins, cytochrome, and non-heme iron components as well as multiple cytochrome oxidases for this process.

The bacterial aerobic respiratory chain requires three main membrane components, (a) An NADH dehydrogenase which acts as an electron donor. (b) A quinone then delivers electrons from the dehydrogenase to a final acceptor enzyme complex. In Gram-positive bacteria, this function is mediated by menaquinones. (c) The acceptor protein complex is a cytochrome oxidase whose activity is heme dependent. In addition to these essential features, the protons extruded by the respiratory chain can be utilized by the F0F1 ATP synthase to generate ATP (Lechardeur *et al.*, 2011).

AEROBIC RESPIRATION IN LACTIC ACID BACTERIA

Respiration system in LAB is more simplified in terms of enzymatic machinery but distinct in features from model respiratory organisms (Table 1). Lactic acid bacteria lack certain components of classical respiration system such as Kreb's Cycle, Electron transport chain and also the ability to synthesize heme and in some cases menaquinone (Pedersen *et al.*, 2012). Thus the respiration of LAB cells is conditional and depends heavily upon the supplemented media components. First, LAB require exogenous heme, as they lack enzymes for heme biosynthesis. Some LAB also lack *menFDHBEC*, the menaquinone biosynthesis genes, which means that two exogenous component are needed to activate respiration (Lechardeur *et al.*, 2011). The absence of these required cofactors means that respiration metabolism is strongly dependent on the media supplements. Second, respiring LAB all encode a single type of cytochrome oxidase, the quinol oxidase CydAB. This enzyme complex is known to work at low oxygen concentrations and may contribute to oxygen elimination from the bacterial environment (Pederson

Table 1. Components of aerobic respiration in lactic acid bacteria

S. No.	Activity/ Role	Component	LAB Specific Component
1	Pyruvate Production	Glycolysis	Present (EMP)
2	NADH production	Krebs' Cycle	Absent (generated by glycolysis)
3	Electron Transport	Electron Transport Chain	Absent
4	Electron Acceptor complex	Cytochrome oxidase	Heme dependent Cytochrome oxidase
5	Heme Biosynthesis	Heme for Cytochrome oxidase	Absent (hence require media supplementation)
6	Electron Shuttle	Quinone	Menaquinone (Absent in Lactobacilli/ present in lactococci)
7	Heme Transporter	Ferric uptake regulator (*fur*)	*fhuDBAR* operon (lactococci)

et. al, 2012). While other bacteria use the Krebs cycle to produce NADH, LAB require a sugar carbon source and glycolytic activity to generate NADH. Thus, in some way, fermentation metabolism is likely needed prior to, or during, LAB respiratory growth (Pedersen *et al.*, 2012).

The conditional aerobic respiration in LAB group has been proposed through an *in-silico* approach in certain species of the genus *Lactobacillus, Lactococcus, Streptococcus, Leuconostoc* (Pedersen *et al.*, 2012). The biochemical evidence of this system has been demonstrated in *Lactococcus lactis* (Duwat *et al.*, 2001; Gaudu *et al.*, 2002; Rezaïki *et al.*, 2004; Vido *et al.*, 2004), *Lactobacillus plantarum* (Brooijmans *et al.*, 2009a) and *Lactobacillus casei* group (Zotta *et al.*, 2014). Among the *Lb. casei* group *Lb. paracasei* was found to have least adaptation to aerobic growth as compared to *Lb. casei* and *Lb. rhamnosus* (Zotta *et al.*, 2014). Brooijmans *et. al.* (2009) also reported heme and menaquinone induced aerobic respiration in *Lb. plantarum, Lb. rhamnosus, Lb. brevis, Lb. paralimentarius, S. entericus* and *L. garviae.* A detailed list of different species and strains of LAB capable of aerobic respiration is given in Table 2. A complete heme biosynthesis pathway is absent in entire LAB group and hence its supplementation in medium is essential to activate respiration. The electron shuttle menaquinone can be synthesized by *Lactococcus, Leuconostoc, Enterococcus, Weissella* by the *menFDHBEC* gene set but this genetic assembly is lacking in *Lactobacillus*. Thus induction of aerobic respiration in lactobacilli is only possible through simultaneous supplementation of heme and menaquinone (Pedersen *et al.*, 2012).

The aerobic respiration in lactic acid bacteria is dependent upon factors such as aeration of the medium, heme concentration, menaquinone concentration, glucose levels etc. The aeration of the medium helps to dissolve oxygen into the medium, which can be utilized as the terminal electron acceptor and thus promotes the respiration capacity of cultures (Ianniello *et al.*, 2016). Aeration of LAB cultures is in general toxic to the growth but respiratory cultures show increase in tolerance of oxygen in presence of supplemented heme. The concentration of the heme in the growth medium has a direct bearing on performance of respiratory cultures as the activity of cytochrome oxidase is heme dependent. Free heme, however, is toxic at higher concentrations and the cells require to maintain a heme-homeostasis (Lechardeur *et al.*, 2011). Respiring LAB such as *L. lactis* lacks the enzymes required to degrade heme for iron recovery and thus rely on a heme efflux system (HrtRBA) which drives the excess heme out of the cell (Pedersen *et al.*, 2008). In an optimization experiment, Giridhar (2013) showed that heme concentration of 1µg/ml was most optimum for growth of *L. lactis* and a concentration of 3 and 5 µg/ml showed lowered growth parameters (Giridhar, 2013). Some workers although have reported higher

Table 2. Aerobic respiration in lactic acid bacteria

S. No.	Culture Name	Heme Conc.	Menaquinone Conc.	References
1	*L. lactis* MG1363	2 µg/ml	-	(Brooijmans *et al.*, 2007)
2	*L. lactis* MG1363	1.25 µg/ml	-	(Duwat *et al.*, 2001)
3	*L. lactis* subsp. *lactis* CHCC2862	8 µM	-	(Pedersen *et al.*, 2008)
4	*L. lactis* PJ4700 and PJ 4662	5 µg/ml	-	(Blank *et al.*, 2001)
5	*L. lactis* MG1363	10 µM	-	(Joubert *et al.*, 2014)
6	*L. lactis* subsp. *lactis* 3022	10 µM	-	(Kaneko *et al.*, 1990)
7	*L. lactis* MG1363 and BK 2337	5 µg/ml	-	(Koebmann *et al.*, 2008)
8	*L. lactis* MG1363	2.5 mg/L	-	(Zhao *et al.*, 2013)
9	*L. lactis subsp. lactis* ATCC 11454	1.25 µg/ml	-	(Nagayasu *et al.*, 2007)
10	*L. lactis* N8	3 µg/ml	-	(Kördikanlıoğlu *et al.*, 2015)
11	*L. lactis* MG1363	10 µg/ml	-	(Vido *et al.*, 2004)
12	*L. lactis* strains MG1363	1.25 µg/ml	-	(Rezaïki *et al.*, 2004)
13	Lactococcus spp.	1.25 µg/ml	-	(Jiao *et al.*, 2013)
14	*L. lactis* NZ9000	4 µg /ml	-	(Shi *et al.*, 2016)
15	*Lb. casei* N87	2.5 µg/ ml	1 µg /ml	(Zotta *et al.*, 2014)
16	*Lb. casei* N87, N811, N2014, CI4368	2.5 µg/ ml	1 µg /ml	(Ianniello *et al.*, 2015)
17	*Lb. casei*, *Lb. paracasei* and *Lb. rhamnosus*	2.5 µg /mL	1 µg /ml	(Zotta *et al.*, 2014)
18	*Lb. plantarum*, *Lb. paraplantarum* and *Lb. pentosus*	2.5 µg/ ml	1 µg /ml	(Guidone *et al.*, 2013)
19	*Lb. plantarum* WCFS1	2.5 µg/ ml	1 µg /ml	(Brooijmans *et al.*, 2009b)
20	*Lb. casei* N87	2.5 µg/ ml	1 µg /ml	(Reale *et al.*, 2016a)
21	*Lb. plantarum*, *Lb. rhamnosus*, *Lb. brevis*	2 µg/ml	20 ng/ml.	(Brooijmans *et al.*, 2009b)

heme concentration for biomass production by LAB (Table 2). Menaquinone concentration affects the respiratory capacity of lactobacilli, since the group lacks the synthesis assembly for the molecule. Oxidative stress resistance of respiring cultures is also affected by the manganese concentration in the medium (Watanabe *et al.*, 2012b). The leakage of electrons from the ETC may result in synthesis of ROS such as superoxide and hydroxyl radicals. *Lb. plantarum* accumulates high intracellular manganese concentration from the medium for scavenging ROS generated during aerobic respiration (Watanabe *et al.*, 2012b). The respiration efficiency was improved by diverting the metabolic flux from homolactic fermentation achieved by inhibition of lactate dehydrogenase (Arioli *et al.*, 2013). The aerobic respiration in LAB is linked to the fermentative metabolism wherein the former requires an initial glycolytic activity for respiration to progress. Thus, the initial substrate concentration also directly affects the respiration ability of cultures (Razvi, 2007).

ENHANCED STRESS RESISTANCE IN AEROBIC RESPIRATORY CELLS

Lactic acid bacteria face different stress conditions during processing which can be classified as technological stress and physiological stress. The use of LAB as starter culture is dependent on their inherent stress resistance and ability to maintain high viability under different conditions. Aerobic respiratory conditions have been shown to improve upon the inherent stress resistance of LAB to different challenges (Table 3). Oxidative conditions during industrial processing or in human intestinal transit inflict oxidative stress upon the cells. Lactic acid bacteria are known to be aerotolerant but their growth is impeded in aerobic conditions. Oxygen is a powerful oxidant in itself but incomplete reduction of oxygen generates reactive oxygen species (ROS) such as superoxide, hydrogen peroxide (H_2O_2) and hydroxyl radicals which can damage various cellular biomolecules and cause cell death (Yamamoto *et al.*, 2011). Lactic acid bacteria have evolved different defense systems to counter oxidative stress (Pophaly *et al.*, 2012; Pophaly *et al.*, 2017). In fermentative mode, the LAB are particularly sensitive to oxidative stress. On the other hand, the conditional respiration system is supported by oxygen enriched condition and thus serves a way to counter oxidative stress. Oxygen serves as a good electron acceptor and converts to H_2O by the action of cytochrome oxidase. Aeration of the medium to infuse oxygen was found beneficial for growth of *L. lactis* in presence of heme due to induction of aerobic respiration wherein oxygen was used as a final electron acceptor (Duwat *et al.*, 2001). Similar results were reported by Rezaiki *et. al.* (2004) who showed that oxygen is toxic during fermentative metabolism of *L. lactis* and causes severe protein and DNA damage. In contrast, the respiratory growth in presence of haem showed less cellular damage (Rezaïki *et al.*, 2004). High oxygen tolerance as a result of heme and menaquinone supplementation was also reported in the plantarum group including *Lb. plantarum, Lb. paraplantarum* and *Lb. pentosus* (Guidone *et al.*, 2013). The increased antioxidative activity of *Lb. plantarum* WCFS1 in heme supplemented medium was further improved by addition of manganese to the medium (Watanabe *et al.*, 2012b). Aerobic respiration in *Lb. plantarum* C17 doubled the tolerance to H_2O_2 compared to anaerobiosis (Zotta *et al.*, 2013). Similar results were observed for *Lb. casei* and *Lb. rhamnosus* (Zotta *et al.*, 2014). The probiotic cultures *Lb. rhamnosus* GG and *Lb. casei* Shirota were found to be inept for

Table 3. Enhanced stress resistance in LAB cultures undergoing aerobic respiration

S. No.	Type of Stress	Generic Phenotypic Change due to Respiratory Condition	Culture	Reference (s)
1	Oxidative Stress	No accumulation of H_2O_2 in respiratory culture, Oxygen-induced protein damage is reduced by respiration growth.	*L. lactis* MG1363	(Rezaïki *et al.*, 2004)
			Lb. plantarum, Lb. paraplantarum, Lb. pentosus	(Guidone *et al.*, 2013)
			Lb. casei N87 *Lb. casei* N811	(Ianniello *et al.*, 2015)
			Lb. plantarum WCFS1	(Watanabe *et al.*, 2012b)
2	Acid Stress	Diminished lactic acid production and a higher final medium pH	*L. lactis*	(Rezaïki *et al.*, 2004)
			Lb. casei N87 *Lb. casei* N811	(Ianniello *et al.*, 2015)
3	Cold Stress (Low Temperature and Freezing)	Greatly improved long-term storage survival at 4^{0}C and Freezing conditions	*L. lactis* strains MG1363	(Gaudu *et al.*, 2002)
			Lb. plantarum C17	(Zotta *et al.*, 2013)
4	Freeze Drying	High viability	*Lb. plantarum* C17	(Zotta *et al.*, 2013)

aerobic respiration but *Lb. casei* N87 and N811 showed signs of respiration and also oxygen tolerance (Ianniello *et al.*, 2015).

Acid stress in LAB is very often caused by their own acid generation during fermentative growth and also during gastrointestinal transit where the pH of stomach is extremely low. The acid stress causes rapid loss of viability and is a major challenge during starter culture production for LAB cells. The pyruvate obtained from degradation of glucose is converted to lactic acid by lactate dehydrogenase during fermentation. Inhibition of lactate dehydrogenase enzyme in fact, helps to increase the heme-dependent respiratory efficiency (Arioli *et al.*, 2013). The induction of aerobic respiration in LAB results in diminished lactic acid production and hence a higher final pH (Pedersen *et al.*, 2008). Aerated (non-heme) *L. lactis* cells showed a decline in cell numbers due to acid stress while heme supplemented cells maintained high viability (Rezaïki *et al.*, 2004). Similar results have been reported in *Lb. casei* N87 and N811 which exhibited aerobic respiration and resulted in high pH values as compared to non-heme conditions (Ianniello *et al.*, 2015). However, *Lb. plantarum* WCFS1 cells grown in aerobic respiratory conditions were reported to be sensitive towards acid stress (Watanabe *et al.*, 2012a). This sensitivity was attributed to a pH linked increase in radical formation and a decrease in membrane integrity in respiratory conditions.

Freezing and freeze-drying are important processing conditions during industrial starter manufacture. Respiratory *L. lactis* (Gaudu *et al.*, 2002) and *Lb. plantarum* (Zotta *et al.*, 2013) cells were reported to exhibit high viability at low temperature storage or freezing conditions.

APPLICATIONS OF AEROBIC RESPIRATION SYSTEM IN STARTER CULTURE IMPROVEMENT

Starter culture production at industrial scale requires high biomass of LAB cells, which should be robust towards different stress conditions. Respiratory pathway may confer several physiological (increased biomass and oxidative stress response, synthesis of antioxidant enzymes) and metabolic (increase of external pH, extra energy production, prevention of oxygen accumulation) advantages which can be used for development of competitive starter and probiotic cultures for use in food and health (Zotta *et al.*, 2014). Respiratory cultures could also be potentially employed for production of flavor compounds and various other biomolecules (Table 4).

Biomass Production

The most remarkable effect observable from aerobic respiration for LAB is an increased bacterial biomass yield as compared to fermentative metabolism (Duwat *et al.*, 2001; Gaudu *et al.*, 2002). This has been attributed to several interrelated factors. First, the high biomass is linked to the additional ATP production in respiratory system, which helps the respiring bacteria to contribute the cellular resources towards growth and reproduction. Second, the aerobic respiration system generates less acid in comparison to fermentation conditions, as NADH is regenerated to NAD+ by the respiratory NADH activity, which directly and negatively affects the enzymatic activity of lactate dehydrogenase (which produces lactic acid from pyruvate using NADH as cofactor) (Lechardeur *et al.*, 2011). The lowered acid production helps maintain a high pH and does not contribute towards acid stress in LAB. Thirdly, respiratory chain uses oxygen as terminal electron acceptor and convers it to H_2O, thus reducing the oxidative stress on

Table 4. Potential applications of aerobic respiration system

S. No.	Physiological Trait	Potential/Current Application	Reference
1	Increased Biomass	Starter Culture	(Duwat *et al.*, 2004; Pedersen *et al.*, 2005)
2	Low Acid Production	Starter Culture	(Duwat *et al.*, 2004)
3	Enhanced Stress Tolerance	Starter Culture	(Gaudu *et al.*, 2002; Zotta *et al.*, 2017a)
4	Flavour Production	Starter Culture	(Lan *et al.*, 2006)
5	Biomolecule Production	Nisin	(Kördikanlıoğlu *et al.*, 2015)
		Acetoin	(Lan *et al.*, 2006)
		Brazzein	(Berlec *et al.*, 2008)
		Folate	(Sybesma *et al.*, 2003)
6	Product Development (using aerobic starter cultures)	Cheddar cheese	(Pedersen *et al.*, 2005; Reale *et al.*, 2016b)
		Feta & Cottage cheese	(Pedersen *et al.*, 2005)
		Sourdough	(Reale *et al.*, 2016a; Zotta *et al.*, 2017a)
		Fromage Frais	(Pedersen *et al.*, 2005)

the bacteria (Cesselin *et al.*, 2010). Thus, the high ATP production coupled with abated stress conditions (acidic and oxidative stress) are together responsible for an enhanced biomass.

Duwat *et. al.* (2001) reported almost doubling of biomass of *Lactococcus lactis* as a result of heme mediated aerobic respiration. Similar results were observed by (Koebmann *et al.*, 2008). The increase in the increase in biomass could be potentially used for wide industrial purposes (Pedersen *et al.*, 2008). Starter culture production requires high biomass of LAB cells, which can be processed for making DVI starters. The lactococcal cultures produced with aerobic respiration system were extensively tested for making Cheddar, Feta & Cottage cheese and a fermented milk product Fromage Frais (Pedersen *et al.*, 2005). The starter cultures produced with respiratory and fermentative cultures showed no significant difference in terms of sensory profile and hence conditional respiratory system proved to be a novel starter technology for making high quality fermented products. The use of porphyrin compounds like heme for production of lactococcal DVI starter culture through induced aerobic respiration system has now been patented (Duwat *et al.*, 2004; Kringelum *et al.*, 2006). The conditional respiration system has, thus been successfully utilized for manufacture of commercial starter cultures by Chr. Hansen A/S (Horsholm, Denmark) (Pedersen *et al.*, 2005; Geppel *et al.*, 2016).

Flavor Production

Induction of aerobic respiration in LAB causes changes in central metabolism resulting in rerouting of pyruvate (end product of glycolysis) away from lactate. Respiration activity results in depletion of NADH causing a consequent decrease in NADH/NAD+ ratio. This ratio is determinant of the choice of metabolic pathway by the cell. The decrease in NADH/NAD+ ratio thus tends to reduce the activity of lactate dehydrogenase whose activity is dependent on the available NADH pool, driving the pyruvate away from conversion to lactate (Duwat *et al.*, 2001). This accumulated pyruvate is driven towards other end products such as acetoin or acetate through supporting pathways. The acetoin and its derivative diacetyl are major flavoring compounds in many food products. The increased acetoin and diacetyl production has been demonstrated in *L. lactis* cultures (Duwat *et al.*, 2001; Gaudu *et al.*, 2002). Recently, *Lactobacilli casei* N87 was also showed to produce acetoin and diacetyl under induced aerobic respiration conditions in chemically defined medium (Zotta *et al.*, 2017b).

Biomolecule Production

The LAB are used as a cellular factory for production of various food grade industrial biomolecules such as antimicrobial substances, sugar polymers, sweeteners, aromatic compounds, vitamins etc. (Hugenholtz, 2008). The biomass production and inherent stress resistance of LAB cells under aerobic respiratory conditions offer a unique metabolic platform for production of different biomolecules of industrial and therapeutic importance. Nisin, an important biopreservative is produced by some strains of *L. lactis.* Nisin production by respiration competent *L. lactis* N8 was increased 3.1 fold in heme induced respiration state as compared to the fermentative control (Kördikanlıoğlu *et al.*, 2015). Folate production in *L. lactis* was also reported to increase significantly as a result of heme supplementation (Sybesma *et al.*, 2003). *L. lactis* was used for increased production of menaquinone (essential component of aerobic respiration) by using a heme supplemented medium (Garault *et al.*, 2011). The production of menaquinone (Vitamin K_2) through respiratory system was proposed as a means for increasing the nutritive value of fermented products. Recombinant *L. lactis* carrying brazzein gene, however showed no increase in production of the sweetener when supplemented with heme in the medium (Berlec *et al.*, 2008).

Product Development

The altered functionalities and physiology of lactic acid bacteria as a result of conditional aerobic respiration could be potentially used for development of novel food products. Respirative lactobacilli with increased ROS scavenging enzymes and production of aroma compounds, have been used for improved sourdough production (Reale *et al.*, 2016a; Zotta *et al.*, 2017a) and Cheddar cheese (Pedersen *et al.*, 2005; Reale *et al.*, 2016b), Feta & Cottage cheese and a fermented milk product Fromage Frais (Pedersen *et al.*, 2005).

The use of *Lb. casei* N87 cultivated under respiratory conditions resulted in enhanced hydrolysis of wheat proteins in comparison to cells grown under anaerobic conditions in fermented dough (Reale *et al.*, 2016a), thus the respiratory grown cultures could be potentially used for texture improvement of bakery products. The respiratory culture also showed high production of metabolites with spoilage inhibiting properties and aroma compounds in sourdough. The increase in desirable aroma compounds such as diacetyl, acetoin was also demonstrated in Cheddar type cheeses produced with respiratory *Lb. casei* strains (Ianniello *et al.*, 2016). Aerobic respiration in LAB is thus a unique approach for designing novel fermented foods.

CONCLUSION

Lactic acid bacteria were earlier known to be obligatory fermentative in nature but a conditional aerobic respiration system is now reported in lactococci and some lactobacilli. The aerobic respiration is dependent on heme supplementation in lactococci and both heme and menaquinone in lactobacilli. Aerobic respiration is an energy rich metabolic pathway with high ATP production due to involvement of electron transport chain as compared to low ATP synthesis in substrate level phosphorylation during fermentation. The induction of aerobic respiration causes marked changes in metabolism and physiology of LAB. They are reported to produce more biomass and show remarkable resistance to oxidative and acidic stress conditions. Also, the lactic acid production during fermentation is replaced by synthesis of less acidic products such as acetoin during respiration. These alterations due to aerobic respiratory route of LAB have potential technological and therapeutic advantages. Respiratory system is helpful for development of robust starter cultures with high tolerance to stress conditions and also for improved biomolecule production from lactic acid bacteria.

REFERENCES

Arioli, S., Zambelli, D., Guglielmetti, S., De Noni, I., Pedersen, M. B., Pedersen, P. D., ... Mora, D. (2013). Increasing the heme-dependent respiratory efficiency of *Lactococcus lactis* by inhibition of lactate dehydrogenase. *Applied and Environmental Microbiology*, *79*(1), 376–380. doi:10.1128/AEM.02734-12 PMID:23064338

Berlec, A., Tompa, G., Slapar, N., Fonović, U. P., Rogelj, I., & Štrukelj, B. (2008). Optimization of fermentation conditions for the expression of sweet-tasting protein brazzein in *Lactococcus lactis. Letters in Applied Microbiology*, *46*(2), 227–231. doi:10.1111/j.1472-765X.2007.02297.x PMID:18215220

Blank, L. M., Koebmann, B. J., Michelsen, O., Nielsen, L. K., & Jensen, P. R. (2001). Hemin reconstitutes proton extrusion in an H+-ATPase-Negative Mutant of *Lactococcus lactis. Journal of Bacteriology*, *183*(22), 6707–6709. doi:10.1128/JB.183.22.6707-6709.2001 PMID:11673444

Brooijmans, R., De Vos, W. M., & Hugenholtz, J. (2009a). *Lactobacillus plantarum* WCFS1 electron transport chains. *Applied and Environmental Microbiology*, *75*(11), 3580–3585. doi:10.1128/AEM.00147-09 PMID:19346351

Brooijmans, R., Smit, B., Santos, F., Van Riel, J., de Vos, W. M., & Hugenholtz, J. (2009b). Heme and menaquinone induced electron transport in lactic acid bacteria. *Microbial Cell Factories*, *8*(1), 28. doi:10.1186/1475-2859-8-28 PMID:19480672

Brooijmans, R. J. W., Poolman, B., Schuurman-Wolters, G. K., De Vos, W. M., & Hugenholtz, J. (2007). Generation of a membrane potential by *Lactococcus lactis* through aerobic electron transport. *Journal of Bacteriology*, *189*(14), 5203–5209. doi:10.1128/JB.00361-07 PMID:17496098

Cesselin, B., Derré-Bobillot, A., Fernandez, A., Lamberet, G., Lechardeur, D., Yamamoto, Y., ... Gaudu, P. (2010). Respiration, a strategy to avoid oxidative stress in *Lactococcus lactis*, is regulated by the heme status. *Nihon Nyusankin Gakkaishi*, *21*(1), 10–15. doi:10.4109/jslab.21.10

Duwat, P., Bravard, A., Sourice, S., & Gruss, A. (2004). *Process for preparing starter cultures of lactic acid bacteria*. United States Patent No. US20050032196.

Duwat, P., Sourice, S., Cesselin, B., Lamberet, G., Vido, K., Gaudu, P., ... Gruss, A. (2001). Respiration capacity of the fermenting bacterium *Lactococcus lactis* and Its positive effects on growth and survival. *Journal of Bacteriology*, *183*(15), 4509–4516. doi:10.1128/JB.183.15.4509-4516.2001 PMID:11443085

Garault, P., Quere, G., Beal, C., Bomchil, N., Faurie, J.-M., Gobert, G., & Lipowski, G. (2011). *Method for obtaining variants of lactic acid bacteria usable for producing vitamin K2 and application to the preparation of food products*. United States Patent No. US7981657 B2.

Gaudu, P., Vido, K., Cesselin, B., Kulakauskas, S., Tremblay, J., Rezaïki, L., ... Gruss, A. (2002). *Respiration capacity and consequences in Lactococcus lactis. In Lactic Acid Bacteria: Genetics, Metabolism and Applications* (pp. 263–269). Springer. doi:10.1007/978-94-017-2029-8_16

Geppel, A., Kringelum, B. W., Hansen, K. F., Iversen, S. L., & Henriksen, C. M. (2016). *Porphyrin containing lactic acid bacterial cells and use therof.* United States Patent No. US 8486468 B2.

Giridhar, A. (2013). *Increasing the Lactococcus lactis biomass through aerobic growth* (Master's thesis). Massey University, Palmerston North, New Zealand.

Guidone, A., Ianniello, R. G., Ricciardi, A., Zotta, T., & Parente, E. (2013). Aerobic metabolism and oxidative stress tolerance in the *Lactobacillus plantarum* group. *World Journal of Microbiology & Biotechnology*, *29*(9), 1713–1722. doi:10.100711274-013-1334-0 PMID:23543191

Hugenholtz, J. (2008). The lactic acid bacterium as a cell factory for food ingredient production. *International Dairy Journal*, *18*(5), 466–475. doi:10.1016/j.idairyj.2007.11.015

Ianniello, R. G., Ricciardi, A., Parente, E., Tramutola, A., Reale, A., & Zotta, T. (2015). Aeration and supplementation with heme and menaquinone affect survival to stresses and antioxidant capability of *Lactobacillus casei* strains. *Lebensmittel-Wissenschaft + Technologie*, *60*(2), 817–824. doi:10.1016/j.lwt.2014.10.020

Ianniello, R. G., Zotta, T., Matera, A., Genovese, F., Parente, E., & Ricciardi, A. (2016). Investigation of factors affecting aerobic and respiratory growth in the oxygen-tolerant strain *Lactobacillus casei* N87. *PLoS One*, *11*(11), e0164065. doi:10.1371/journal.pone.0164065 PMID:27812097

Jiao, Y. H., Zhang, L. W., & Liu, F. (2013). *Screening of lactic acid bacteria strains with respiration ability in the present of heme.* Paper presented at the Advanced Materials Research. 10.4028/www.scientific.net/AMR.726-731.448

Joubert, L., Derré-Bobillot, A., Gaudu, P., Gruss, A., & Lechardeur, D. (2014). HrtBA and menaquinones control haem homeostasis in *Lactococcus lactis. Molecular Microbiology*, *93*(4), 823–833. doi:10.1111/mmi.12705 PMID:25040434

Kaneko, T., Takahashi, M., & Suzuki, H. (1990). Acetoin fermentation by citrate-positive *Lactococcus lactis* subsp. *lactis* 3022 grown aerobically in the presence of hemin or Cu2+. *Applied and Environmental Microbiology*, *56*(9), 2644–2649. PMID:16348274

Koebmann, B., Blank, L. M., Solem, C., Petranovic, D., Nielsen, L. K., & Jensen, P. R. (2008). Increased biomass yield of *Lactococcus lactis* during energetically limited growth and respiratory conditions. *Biotechnology and Applied Biochemistry*, *50*(1), 25–33. doi:10.1042/BA20070132 PMID:17824842

Kördikanlıoğlu, B., Şimşek, Ö., & Saris, P. E. J. (2015). Nisin production of *Lactococcus lactis* N8 with hemin-stimulated cell respiration in fed-batch fermentation system. *Biotechnology Progress*, *31*(3), 678–685. doi:10.1002/btpr.2075 PMID:25826783

Kringelum, B. W., Sorensen, N. M., Garrigues, C., Pedersen, M. B., & Gron, S. (2006). *Use of compounds involved in biosynthesis of nucleic acids to increase yield of bacterial cultures*. United States Patent No. US 20080171028 A1.

Lan, C. Q., Oddone, G., Mills, D. A., & Block, D. E. (2006). Kinetics of *Lactococcus lactis* growth and metabolite formation under aerobic and anaerobic conditions in the presence or absence of hemin. *Biotechnology and Bioengineering*, *95*(6), 1070–1080. doi:10.1002/bit.21070 PMID:16807924

Lechardeur, D., Cesselin, B., Fernandez, A., Lamberet, G., Garrigues, C., Pedersen, M., ... Gruss, A. (2011). Using heme as an energy boost for lactic acid bacteria. *Current Opinion in Biotechnology*, *22*(2), 143–149. doi:10.1016/j.copbio.2010.12.001 PMID:21211959

Ljungh, A., & Wadstrom, T. (2006). Lactic acid bacteria as probiotics. *Current Issues in Intestinal Microbiology*, *7*(2), 73–90. PMID:16875422

Nagayasu, M., Wardani, A. K., Nagahisa, K., Shimizu, H., & Shioya, S. (2007). Analysis of hemin effect on lactate reduction in *Lactococcus lactis. Journal of Bioscience and Bioengineering*, *103*(6), 529–534. doi:10.1263/jbb.103.529 PMID:17630124

Pedersen, M. B., Garrigues, C., Tuphile, K., Brun, C., Vido, K., Bennedsen, M., ... Gruss, A. (2008). Impact of aeration and heme-activated respiration on *Lactococcus lactis* gene expression: Identification of a heme-responsive operon. *Journal of Bacteriology*, *190*(14), 4903–4911. doi:10.1128/JB.00447-08 PMID:18487342

Pedersen, M. B., Gaudu, P., Lechardeur, D., Petit, M.-A., & Gruss, A. (2012). Aerobic respiration metabolism in lactic acid bacteria and uses in biotechnology. *Annual Review of Food Science and Technology*, *3*(1), 37–58. doi:10.1146/annurev-food-022811-101255 PMID:22385163

Pedersen, M. B., Iversen, S. L., Sørensen, K. I., & Johansen, E. (2005). The long and winding road from the research laboratory to industrial applications of lactic acid bacteria. *FEMS Microbiology Reviews*, *29*(3), 611–624. doi:10.1016/j.fmrre.2005.04.001 PMID:15935510

Pophaly, S. D., Pophaly, S. D., Kapila, S., Nanda, D. K., Tomar, S. K., & Singh, R. (2017). Glutathione biosynthesis and activity of dependent enzymes in food grade lactic acid bacteria harboring multidomain bifunctional fusion gene (gshF). *Journal of Applied Microbiology*, *123*(1), 194–203. doi:10.1111/jam.13471 PMID:28403558

Pophaly, S. D., Singh, R., Pophaly, S. D., Kaushik, J. K., & Tomar, S. K. (2012). Current status and emerging role of glutathione in food grade lactic acid bacteria. *Microbial Cell Factories*, *11*(1), 114. doi:10.1186/1475-2859-11-114 PMID:22920585

Razvi, A. (2007). *A study of the hemin-stimulated respirative growth kinetics of the lactis using batch cultivation* (Master's Thesis). University of Ottawa (Canada). Masters Abstracts International, Volume: 47-06, page: 3670. database.

Reale, A., Di Renzo, T., Zotta, T., Preziuso, M., Boscaino, F., Ianniello, R., ... Coppola, R. (2016a). Effect of respirative cultures of *Lactobacillus casei* on model sourdough fermentation. *Lebensmittel-Wissenschaft + Technologie*, *73*, 622–629. doi:10.1016/j.lwt.2016.06.065

Reale, A., Ianniello, R. G., Ciocia, F., Di Renzo, T., Boscaino, F., Ricciardi, A., ... McSweeney, P. L. H. (2016b). Effect of respirative and catalase-positive *Lactobacillus casei* adjuncts on the production and quality of Cheddar-type cheese. *International Dairy Journal*, *63*, 78–87. doi:10.1016/j.idairyj.2016.08.005

Rezaïki, L., Cesselin, B., Yamamoto, Y., Vido, K., Van West, E., Gaudu, P., & Gruss, A. (2004). Respiration metabolism reduces oxidative and acid stress to improve long-term survival of *Lactococcus lactis*. *Molecular Microbiology*, *53*(5), 1331–1342. doi:10.1111/j.1365-2958.2004.04217.x PMID:15387813

Shi, W., Li, Y., Gao, X., & Fu, R. (2016). Improvement of the respiration efficiency of *Lactococcus lactis* by decreasing the culture pH. *Biotechnology Letters*, *38*(3), 495–501. doi:10.100710529-015-1999-6 PMID:26585330

Sybesma, W., Starrenburg, M., Tijsseling, L., Hoefnagel, M. H. N., & Hugenholtz, J. (2003). Effects of cultivation conditions on folate production by lactic acid bacteria. *Applied and Environmental Microbiology*, *69*(8), 4542–4548. doi:10.1128/AEM.69.8.4542-4548.2003 PMID:12902240

Vido, K., le Bars, D., Mistou, M.-Y., Anglade, P., Gruss, A., & Gaudu, P. (2004). Proteome analyses of heme-dependent respiration in *Lactococcus lactis*: Involvement of the proteolytic system. *Journal of Bacteriology*, *186*(6), 1648–1657. doi:10.1128/JB.186.6.1648-1657.2004 PMID:14996795

Watanabe, M., van der Veen, S., & Abee, T. (2012a). Impact of respiration on resistance of *Lactobacillus plantarum* WCFS1 to acid stress. *Applied and Environmental Microbiology*, *78*(11), 4062–4064. doi:10.1128/AEM.00287-12 PMID:22447603

Watanabe, M., van der Veen, S., Nakajima, H., & Abee, T. (2012b). Effect of respiration and manganese on oxidative stress resistance of *Lactobacillus plantarum* WCFS1. *Microbiology*, *158*(1), 293–300. doi:10.1099/mic.0.051250-0 PMID:22016573

Yamamoto, Y., Gaudu, P. & Gruss, A. (2011). Oxidative stress and oxygen metabolism in lactic acid bacteria. *Lactic Acid Bacteria and Bifidobacteria: Current Progress in Advanced Research,* 91-102.

Zhao, R., Zheng, S., Duan, C., Liu, F., Yang, L., & Huo, G. (2013). NAD-dependent lactate dehydrogenase catalyses the first step in respiratory utilization of lactate by *Lactococcus lactis*. *FEBS Open Bio*, *3*(1), 379–386. doi:10.1016/j.fob.2013.08.005 PMID:24251099

Zotta, T., Guidone, A., Ianniello, R. G., Parente, E., & Ricciardi, A. (2013). Temperature and respiration affect the growth and stress resistance of *Lactobacillus plantarum* C17. *Journal of Applied Microbiology*, *115*(3), 848–858. doi:10.1111/jam.12285 PMID:23782242

Zotta, T., Parente, E., & Ricciardi, A. (2017a). Aerobic metabolism in the genus Lactobacillus: Impact on stress response and potential applications in the food industry. *Journal of Applied Microbiology*, *122*(4), 857–869. doi:10.1111/jam.13399 PMID:28063197

Zotta, T., Ricciardi, A., Ianniello, R. G., Parente, E., Reale, A., Rossi, F., ... Coppola, R. (2014). Assessment of aerobic and respiratory growth in the *Lactobacillus casei* group. *PLoS One*, *9*(6), e99189. doi:10.1371/journal.pone.0099189 PMID:24918811

Zotta, T., Tabanelli, G., Montanari, C., Ianniello, R. G., Parente, E., Gardini, F., & Ricciardi, A. (2017b). Tween 80 and respiratory growth affect metabolite production and membrane fatty acids in *Lactobacillus casei* N87. *Journal of Applied Microbiology*, *122*(3), 759–769. doi:10.1111/jam.13373 PMID:27981716

KEY TERMS AND DEFINITIONS

Cytochrome Oxidase: Cytochrome oxidase is the terminal enzyme in the respiratory electron chain. It serves as the electron acceptor protein complex for aerobic respiration.

Electron Transport Chain: An electron transport chain (ETC) is a series of complexes that transfers electrons from electron donors to electron acceptors via redox reactions.

Fermentation: A metabolic process involves anaerobic breakdown of carbohydrates with an organic molecule acting as the final electron acceptor.

Glycolysis: Glycolysis is the breakdown of a 6-carbon glucose molecule is partially broken down into two, three-carbon molecules of pyruvate, 2 NADH, $2H^+$ and 2ATP as a result of substrate level phosphorylation.

Heme: is an iron-containing compound of porphyrin class. Heme is required to be supplied externally for activation of cytochrome oxidase in LAB.

Respiration: Respiration is an energy producing process in which organic or reduced organic compounds are oxidized by inorganic compounds. In aerobic respiration oxygen acts as the terminal electron acceptor.

Starter Culture: Starter culture refers to the intentionally added carefully selected microorganisms to a medium for carrying out desired fermentation.

Chapter 6
The Effects of Probiotic Cultures in Functional Foods:
Technological Aspects of Probiotics

Gulden Basyigit Kilic
Mehmet Akif Ersoy University, Turkey

ABSTRACT

Functional foods are an important part of an overall healthy lifestyle that includes a balanced diet and physical activity. The consumption of probiotic foods has many benefits. Dairy products that contain probiotic bacteria are those that are produced with various fermentation methods, especially lactic acid fermentation, by using starter cultures and those that have various textures and aromas. Fermented dairy probiotic products are popular due to their differences in taste and their favourable physiological effects. However, recent upsurge in interest of consumers towards dairy alternatives has opened up new research areas for developing non-dairy probiotic products. Different substrates such as cereals, fruit juices, vegetables can be used utilized for delivering these beneficial microorganisms. This chapter provides an insight on the recent research/developments about selection criteria of bacteria as probiotics and in the field of technological properties of probiotics.

INTRODUCTION

For centuries, lactic acid bacteria (LAB) have been used for the preservation of food for the human consumption. LAB are a large group of fermentative, anaerobe facultative, aerotolerant microorganisms which are usually present in the gut of humans and other animals, raw vegetables, meat and meat products, and cereals. During the fermentation process, LAB also influences the sensory properties of a product, including the flavor development. Flavor compounds are formed by various processes, e.g. the conversions of lactose and citrate (glycolysis and pyruvate metabolism), fat (lipolysis), and proteins (proteolysis, peptidolysis, and amino acids catabolism) (Yoon et al., 2006). LAB quickly acidifies the raw material through the production of organic acids, mainly lactic acid. They also produce acetic acid, ethanol, aroma compounds, bacteriocins, exopolysaccharide (EPS)s, and several important enzymes

DOI: 10.4018/978-1-5225-5363-2.ch006

and they enhance shelf life and microbial safety, improve texture, and contribute to the pleasant sensory profile of the end product (Hati et al., 2013). Their numbers may vary with the animal species, the age of the host, or the location within the gut. In the food industry, lactic acid bacterial strains are widely employed either as starter cultures or as non-starter lactic acid bacteria (NSLAB). Furthermore, owing to their probiotic properties, several LAB strains are used as adjunctive cultures in foods and feed.

Probiotic which originated from a Greek term "probios" meaning "for life" as against "antibiotics" which means "against life" has become a significant concept in biomedical research. It refers live microbial food supplement which when administered in adequate amounts confer health benefit of consumers by maintaining or improving their intestinal microbial flora. The US Food and Drug Administration (FDA) uses other terms for live microbes for regulatory purposes; live microbes used in animal feeds are called "direct-fed microbials" (USDA, 2010), and when intended for use as human drugs, they are classified as "live biotherapeutics" (Vaillancourt, 2006). Probiotics are mainly members of genera *Lactobacillus* and *Bifidobacterium*, are normal residents of the complex ecosystem of the gastrointestinal (GI) tract of humans. Foods are carriers for the delivery of probiotic microorganisms to the human body. The growth and survival of probiotics during gastric transit is affected by the characteristics of the food carriers, like chemical composition and redox potential. Same probiotic strains could vary in functional and technological properties in the presence of different food ingredients or in different food environments (Ranadheera et al., 2010). In this chapter; selection of probiotic cultures, functionality of probiotics and culture production, effects of probiotics on quality parameters of fermented food products will be presented.

BACKGROUND

Probiotic bacteria, according to the definition adopted by the World Health Organization (WHO) in 2002, are live microorganisms, which when administered in adequate amounts confer a health benefit to the host (WHO/FAO, 2002a). The normal human digestive tract contains about 400 types of probiotic bacteria that reduce the growth of harmful bacteria and promote a healthy digestive system. The probiotic strain must be a normal inhabitant of the human intestinal tract and be able to survive acid in the stomach and bile in the small intestine. Probiotic strains should also persist in the GI tract to prevent their rapid removal by intestinal peristalsis. Colonization or at least temporary colonization is necessary for most probiotic organisms to exert their probiotic effects (Kailasapathy, 2013). The beneficial effects of food with added live microbes (probiotics) on human health, and in particular of milk products on children and other high-risk populations, are being increasingly promoted by health professionals. It has been reported that these probiotics can play an important role in immunological, digestive and respiratory functions and could have a significant effect in alleviating infectious disease in children (FAO/WHO, 2002b). Probiotic microorganisms, mainly those derived from the genera *Lactobacillus* and *Bifidobacterium* are closely related in terms of metabolism to lactic acid starter bacteria (Vinderola et al., 2011). *Lactobacillus* species from which probiotic strains have been isolated include *Lactobacilllus acidophilus, Lactobacilllus johnsonii, Lactobacilllus casei, Lactobacilllus rhamnosus, Lactobacilllus gasseri,* and *Lactobacilllus reuteri*. *Bifidobacterium* strains include *Bifidobacterium bifidum, Bifidobacterium longum,* and *Bifidobacterium infantis* (Heller, 2001). Other than these, some species of *Lactococcus, Enterococcus, Saccharomyces* and *Propionibacterium* are considered as probiotics due to their ability to promote health in the host (Zhang et al., 2010). These bacteria show symbiotic relationship with human.

They are present in the mucus membrane present on epithelial cells of the gut (Holzapfel & Schillinger, 2002) where they inhibit the growth and attachment of harmful bacteria by producing bactericidal chemicals against these bacteria. As advocated in 2008 (Reid, 2008), probiotic foods and supplements tested in human studies have since emerged that convey local gut and distant site effects on health. With the development of evidences regarding usefulness and safety of probiotics, these bacteria are replacing the traditional prophylactic and treatment regimes (Iqbal et al., 2014).

The quality control of probiotic cultures in foods traditionally has relied solely on tests to ensure that an adequate number of viable bacteria are present in the products throughout their shelf lives Viability is an important factor, but not the only criterion for quality assurance. To be effective, probiotic strains must retain the functional health characteristics for which they were originally selected (Tuomola et al., 2001). However, in the last years and, on the basis of different reports, there has been evidence of certain changes in functionality, i.e., in the qualitative and quantitative magnitude of the beneficial effect of a strain, without levels of viable cells being modified. Briefly, changes in the effect induced without changes in cell counts: same amount, different effects. In this context, the microbiological count would only partially reflect the functional capacity of the strain. Therefore, it appears necessary to know manage and control the microbiological variables that may alter the functionality of a strain in a particular food matrix, without having an impact on the number of viable cells. (Vinderola et al., 2011).

For selection of the probiotics; acid and bile tolerance, resistance to phenol, production of H_2O_2, adhesion to intestinal epithelial cell lines and the competitive exclusion of pathogens, antibiotic resistance patterns, phenotypic and genotypic stability, antimicrobial activity and biosafety of the strains should be tested. Another important issue for selection of the probiotics is their technological aspects. Technological properties of the cultures should be conducted under the food systems processes. Generally growth in milk, sensory properties, stability, phage resistance, resistance to stress conditions, viability in micro and macro processes are important quality assurance criteria for probiotics. Several studies were carried out to enhance the probiotic viability and functionality during storage. The fast growing characteristics of LAB and their metabolic activity have been the key in most applications including food production, agricultural industry, and probiotics. (Hayek & Ibrahim, 2013). Strains of LAB are also known as fast growing microorganisms that can explore different metabolic activities. Metabolic activities are associated with production of many beneficial compounds such as organic acids and antimicrobial compounds, unique enzymes that can breakdown complex organic compounds into simple functional compounds (Wright & Axelsson, 2011). Leite et al. (2015) isolated a total of 34 LAB from 4 different Brazilian kefir grains and identified and characterized among a group of 150 isolates, using the ability to tolerate acidic pH and resistance to bile salts as restrictive criteria for probiotic potential. Kefir grains showing good survival under normal GI conditions were evaluated in vitro for antimicrobial and antioxidative activities in this study. Among the analyzed strains, 3 *L. paracasei* strains were further selected. Among the 3 selected strains, *L. paracasei* MRS59 showed the highest number of in vitro probiosis properties; consequently, it was considered as the most appropriate kefir-derived candidate to be used as a probiotic.

FUNCTIONALITY OF PROBIOTICS AND CULTURE PRODUCTION

The increasing awareness about health, life expectancy, aspiration of people towards superior life quality, hectic lifestyles, increased consumption of fast foods, limited physical activities; increased incidence of self-medication; increased awareness from media and health authorities on nutrition and the connection between diet and health; scientific developments in nutrition research; and a competitive and crowded food market, exemplified by limited margins has resulted in the rise of functional foods markets (Roberfroid, 2007; Sir´o et al., 2008). Functional foods are the foods that affect functions of the body in a targeted manner so as to bring about positive effects on physiology and nutrition. In recent years, the commercial manufacture and marketing of functional foods particularly probiotic (acidophilus-bifidus) yogurts, has spread from the well-established Japanese niche market place into the lucrative and expanding European Union marketplace (Flynn et al., 2001). The launch of Activia yoghurt in Canada in 2004 literally transformed the awareness of probiotic concepts. Within one year, an estimated 75% of the 8 million people living in Quebec Province had consumed Activia. This incredible product uptake symbolized an awakening of the public to a concept that made sense healthwise, had a scientific rationale, and had not until then been available (Reid, 2015). To this regard, there is a growing need to identify new biofunctional strains, new strategies to assure survival of these cultures, and different sources from which to isolate strains (Awaisheh, 2012).

Probiotic bacteria should be able to survive during the food production process and passing through human digestive system in order to show their beneficial effects (Zago et al., 2011). Fermented and unfermented milks, cheese, fermented meat products, juices, smoothies, cereal, nutrition bars, chocolate and infant/toddler formula all are food vehicles for probiotic delivery (Wright & Axelsson, 2011). In addition to being sold as foods, probiotics are sold as dietary supplements, medical foods, and drugs. Often these products are composed of concentrated, dried microbes packaged into capsules, tablets, or sachets. This format is convenient for the delivery of large numbers of microbes that, if manufactured and stored properly, can be quite stable even at room temperature (CAST, 2007).

Identifying a single or a mixture of probiotic bacteria that inhibit the growth of spoilage and pathogenic bacteria is of growing interest for research to improve the shelf life and safety of the fermented food products. Therefore, it is important to search for probiotics for possible applications in fermented foods (Ertürkmen et al., 2016). Technological factors influencing the functionality of probiotics are; dose responce, strain characteristics, food matrix formulation, targeted strains, stable cultures, viability, microencapsulation to improve viability improved fermentation technology (Mattila-Sandholm & Saarela, 2002).

The first step in the production of a probiotic culture for its large scale implementation, the growth medium is one of critical parameter, which is likely to play a role upon survival subsequent to freeze-drying. In this sense, the variables involved during the growth of the strain (pH of the medium, type, and concentration of carbohydrates, occurrence of osmotic stress, growth phase at harvesting, cell physiological state, production of EPS, and altered fatty acid profile of the membrane etc.) may significantly affect both its resistance to biological barriers (gastric acidity and bile salts) and its capacity to interact with immune cells, thus conditioning its functionality (Carvalho et.al., 2004; Vinderola et al., 2011). Malek et al. (2010) screened seven media for batch cultivation of potential probiotic strain *Lactobacillus salivarius* WICC-BO8 isolated from human breast milk in bioreactor under controlled and uncontrolled pH conditions. Data showed that *L. salivarius* grew well in the medium was composed of (g L^{-1}): glucose, 20.0, yeast extract, 20 and meat peptone, 35. The maxiamum biomass of 5.71 g L^{-1} and 7.57 g L^{-1}, were obtained in shake flaks and controlled pH bioreactor, respectively.

On industrial settings, LAB may be preserved and distributed in liquid, spray dried, frozen or lyophilised forms. All these preparations are suitable for inoculation of bulk starter tanks (Broadbent & Lin, 1999). However, many leading starter culture manufacturers prepare generaly as a mixture of several strains of probiotics in highly concentrated form and most of them are prepared for direct vat (DVS) applications.

The highly concentrated DVS cultures are commonly used in yoghurt manufacture. Optimization of survival of LAB cultures during drying and subsequent storage for long periods is very important for both technological and economical purposes (Selmer-Olsen et al., 1999). The DVS cultures are supplied either as highly concentrated frozen cultures or freeze-dried cultures. Generally deep-frozen cultures contain $>10^{10}$ CFU/g, while frozen-dried cultures typically contain more than 10^{11} CFU/g (Obermann & Libudzisz, 1998). Microencapsulation has previously been reported as a technology to protect sensitive substances against the influences of adverse environments (Champagne & Fustier, 2007). The term ''microencapsulation'' designates a defined technology of packing solids, liquids or gases in small capsules, which can release their contents under specific circumstances. Microencapsulation technologies are hypothesized to be a promising prospect for introducing viable probiotic bacteria in foods because the encapsulation matrix can provide a physical barrier against harsh environmental conditions such as freezing and those encountered during gastric juice passage (Kailasapathy, 2002; Capela et al., 2006).

EFFECTS OF PROBIOTICS ON QUALITY PARAMETERS OF FERMENTED FOOD PRODUCTS

Probiotic Dairy Products

Consumer demand for natural nutrients and probiotic products, fermented dairy products have reached a different position and are considered to have an important impact on human health and nutrition. Fermented dairy products are popular due to their differences in taste and their favorable physiological effects. These products have favorable effects on human health such as, reducing lactose intolerance, prevention of diarrhea and constipation, increase in the effectiveness against *Helicobacter pylori* infection, preservation of oral health, partial prevention of cancer, cholesterol lowering, enhancement of mineral absorption. Along with their extensive effects on human health, they have the ability to form low molecular weight components such as conjugated linoleic acid (CLA), gamma aminobutyric acid and bacteriocin (Gobbetti et al., 2010; Divya et al., 2012; Kanmani et al., 2012). A typical example is lactic acid fermentation, which is widely used for the preparation of several fermented probiotic milk products, such as yoghurt, sour cream, various varieties of cheeses, ice cream, baby milk powder, dahi (curd), probiotic dairy drinks such as acidophilus milk, kefir, yakult, mil-mil, koumiss (Kailasapathy & Phillips, 2008; Cruz et al., 2009; Awaisheh, 2011). Especially fermented milks and fresh cheeses have been the food vehicles with the biggest technological and commercial success for incorporation of probiotic bacteria (Figueroa-González et al. 2011).

Dairy products are suggested to be a good matrix for transferring probiotic bacteria into the GI tract, as milk proteins is thought to function as buffering agents and thereby protect the cells during transit (Charteris et al., 1997). In the dairy industry, the main approach for improving the beneficial physiological properties of products has been the development of probiotic lines of traditional fermented products (Mattila-Sandholm & Saarela, 2000). Probiotic bacteria (mainly bifidobacteria and species of the *L. casei* and *Lactobacillus acidophilus* groups; Klein et al., 1998) and the starter bacteria used in the elaboration of fermented milks (*Streptococcus salivarius* subsp. *thermophilus* and *Lactobacillus delbrueckii* subsp. *bulgaricus*) or cheese (*Str. thermophilus*, *L. helveticus*, *Lactococcus lactis*, among other species), are microorganisms phylogenetically very closely related and with very similar nutritional requirements (Vinderola et al., 2011). An acidophilus milk product added with a prebiotic inulin was standardized using artificial neural network (Amiri et al. 2010). Supplementation of a probiotic-fermented soymilk with the fructo-oligosaccharide, inulin and pectin increased the angiotensin I-converting enzyme inhibitory activity and enhanced the in vitro antihypertensive effect (Yeo & Liong, 2010).

Cheese may be defined as the fresh or matured product made by coagulating or a combination of any of the following substances, namely milk, cream, skimmed milk, partly skimmed milk, concentrated milk, reconstituted dried milk and butter milk, and then partially draining the whey resulting from any such coagulation. Cheese substitutes and cheese analogs, made from either dairy and/or non-dairy ingredients, do not have standard identities but chemical and physical properties exist in the form of fat and moisture (CODEX STAN 283-1978; Mehta, 2014).

There are several types of soft, semi soft (semi hard), and hard probiotic cheese products have been developed and marketed in the last few years. Viability of probiotic bacteria during cheese processing and storage is the major challenge associated with the development of probiotic cheese. Probiotic bacteria should be technologically suitable for the incorporation into cheese products so that to retain both viability and functional efficacy during processing on a commercial scale and throughout consumption (Fortin et al., 2011). Furthermore, from a food processing perspective, it is desirable that such strains are suitable for large-scale industrial cheese production and withstand the processing conditions (Stanton et al., 2003).

It is important to produce a new interesting product within the functional food market. Daigle et al. (1999) produced Cheddar cheese from microfiltered milk standardized with cream and fermented with *Bifidobacterium infantis*. In this case, bifidobacteria showed good survival (> 3 x 10^6 CFU/g) on cheese packaged under vacuum and kept at 4°C for 84 days. Ong et al. (2006) added combinations of *L. acidophilus*, *L. casei* and *B. longum*; and *L. acidophilus, L. paracasei and Bifidobacterium lactis* to Cheddar cheese. In this case cheese was produce following a standard procedure, in which milk, after being standardized was tempered to 31°C before inoculation with cheese starter culture and probiotic bacteria. All probiotic adjuncts survived manufacturing process and maintained their viability until the end of the ripening process. Bzducha et al. (2007) studied to determine the influence of probiotic strains *L. acidophilus* La-5 and *L. casei* DN-114001 on the content of CLA in model ripening cheeses. Model cheeses with probiotic bacteria were subjected to ripening during 8 weeks at a temperature of 6°C. Fatty acids methyl esters including CLA (18:2 *cis*-9, *trans*-11) were analysed using gas chromatography coupled with mass spectrometry. The results showed that, the content of CLA in fresh cheeses was between 680 and 750 mg/100 g of fat. In the investigated model cheeses there were no essential changes in the CLA content during the eight weeks of ripening. Başyiğit Kılıç, G et al. (2009) investigated the technological and functional characteristics of Turkish Beyaz cheese which was produced with probiotic bacteria and also to determine the viability of the probiotic strains. The results indicate that *L. fermentum* (AB5-18 and AK4-120) and *L. plantarum* (AB16-65 and AC18-82) are suitable for development of an acceptable probiotic Turkish Beyaz cheese. Probiotic strains alone or in combination with commercial starter cultures can be adapted to industrial production of Turkish Beyaz cheese. Karlsson (2013) was evaluated the probiotic bacteria *Lactobacillus reuteri* DSM17938 as a starter culture in cheese production. Addition of casamino acids or MRS broth as supplements resulted in both higher levels of viable organisms and higher acidification rate. Moreover, the addition of an adjunct starter culture resulted in faster growth and metabolism. Results of this study suggest that strain DSM 17938 can be used as a starter in the production of fresh cheese with sufficient viable counts at the time of consumption. The possibility to use the bacteria when manufacturing a hard cheese is probably also good, as survival turned out to be high in this kind of cheese despite a low initial count in the milk inoculum.

Yogurt is one of the popular fermented milk products having different names and forms (Kurmann et al., 1992; Tamime & Robinson, 2007). It is a mixture of milk (whole, low-fat, or nonfat) and even cream fermented by a culture of lactic acid-producing bacteria, *Lactobacillus delbrueckii* subsp. *bulgaricus* (*L. bulgaricus*) and *Str. salivarius* subsp. *thermophilus* (*Str. thermophilus*) (Routray & Mishra, 2011). Recently new yogurt products, known as "Bio-Yogurt", have been manufactured by incorporating with different types of live probiotic strains; mainly species of *Lactobacillus* and *Bifidobacteria*; include *L. acidophilus; L. casei; L. gasseri; L. rhamnosus; L. reuteri; B. bifidum; B. animalis; B. infantis; and B. longum* in addition to the standard cultures, *S. thermophilus* and *L. bulgaricus*, into yogurt. Bio-Yogurt may give rise to claimed beneficial health effects. Different types of Bio-Yogurts are produced by far, including, plain, stirred, flavored, and fruits added Bio-Yogurts (Hekmat et al., 2009; Awaisheh, 2011; Awaisheh, 2012). Improving the health potential of yoghurt has become a popular field. Ng et al. (2010) aimed to determine the effects of yogurt starter cultures on the survival of five strains of *L. acidophilus*, with emphases on low pH and acid production. The researchers observed that, between the two starter culture species, *L. bulgaricus* exerted a greater detrimental effect on the survival of some *L. acidophilus* strains. They concluded that this situation maybe because of production of inhibitory metabolites such as H_2O_2. They also mentioned that, as with their intrinsic probiotic activities, the viabilities of *L. acidophilus* in yogurts showed marked strain-to-strain variations. Sarvari et al. (2014) investigated the viability of probiotic bacteria (*L. acidophilus* LA-5 and *B. lactis* BB-12) and yogurt bacteria (*Str. thermophilus* and *L. bulgaricus*) in yogurt during the fermentation, immediately after fermentation and during refrigerated storage. Researchers determined that, *Str. thermophilus* and *B. lactis* showed higher resistance than other bacteria when yogurt was stored at 4°C (21 days). Probiotic bacteria were able to maintain the recommended viable cell concentration (10^6 CFU/ml) until the end of the 21 day storage time. The survival loss of *B. lactis* was gradual and steady during the storage, and the organisms exhibited better stability during the storage compared to *L. acidophilus*. *S. thermophilus* showed the greatest viability among the starter bacteria. Acetaldehyde level significantly decreased in yogurt during the storage. It is concluded that the most remarkable changes in acetaldehyde level occurs at the end of the storage period. This may be due to the prolonged storage as a result of enzymatic reactions.

EPS improves the viscosity and texture of said matrix, and consequently reduces its susceptibility to syneresis (Low et al., 1998). EPS serves as prebiotics as a nondigestible food ingredients that stimulate the growth and activity of a limited numbers of bacteria in colon, thus EPS is used to improve the host health (Tu et. al., 2013). Therefore, EPS producing LAB are mostly preferred as starter cultures, especially in yoghurt manufacture.

Conjugated linoleic acid (CLA) is a mixture of linoleic acids that are geometric and positional isomers of linoleic acid with a conjugated double bond system. CLA has beneficial properties including antioxidant, anticarcinogen, antidiabetic, antiblood pressure, stimulating the body immune system and reducing cholesterol (Pariza et al., 2001). Khosravi-Darani et al. (2014) was assessed ability of *L. acidophilus* La5, *B.bifidum* and *Propionibacterium freudenerchii* for biotransformation of linoleic acid to conjugated linoleic acid. They determined that the best conditions to increase CLA in probiotic yogurt in present study were as follows: the amount of added whey powder 4% of the volume of milk inoculated, incubation temperature 35 °C, the amount of grape seed oil added 4% of inoculated milk, oil addition at pH 6, and time of incubation termination 27, inoculum age 36 h and size 0.8%, respectively. In another study, Khosravi-Darani et. al. (2016) studied with the strains of probiotic bacteria in yoghurt for their ability to produce CLA from free linoleic acid. The researchers investigated the effect of process variables on increasing of CLA in probiotic yoghurt. Two different set of microorganism consists of

L. acidophilus, *B. lactis* and traditional yoghurt starters and in second group *L. acidophilus*, *B. breve* was studied as probiotics besides starter culture. Results showed that in both sets, the highest amount of CLA was obtained by addition of 2% non-fat dry milk, the addition of safflower oil (1.4 ml/l milk) in pH=6.0, incubation temperature of 37°C, and termination of incubation at pH=4.5. In the most suitable condition, the amount of CLA in probiotic yoghurt containing *B. lactis* increased by 450% from an average of 0.04 mg/ml in non-treated yoghurt to 0.22 mg/ml fat in the probiotic yoghurt containing safflower oil. But in the case of *B. breve*, the content of produced CLA was increased 10 fold from 0.04 to 0.4 mg/ml. Anyway, the viability of *L. breve* was less than *L. acidophilus* by 2 log CFU/ml. So for further investigation encapsulation of *B. breve* is suggested to reach to a product with the high amount of CLA and enough viable cells. No significant difference between product and control was observed in the ranking sensory test.

Also, the incorporation of prebiotics (indigestible carbohydrates, such as fructooligosaccharides and inulin) (Liu et al., 2011) and neutraceuticals combination (isoflavones, phytosterols and omega-3-fatty acids) (Jiang & Fonden, 1998; Lin, 2000) in yoghurt formulations seemed to stimulate the viability and activity of probiotic bacteria. Başyiğit Kılıç & Akpınar (2013) determined the effects of adjunct probiotic *Lactobacillus plantarum* strains and oat based β-glucan in synbiotic functional non-fat yoghurt. The yoghurts were evaluated with respect to their microbiological, physicochemical properties, aroma compounds and organic acid profiles. The result of this study showed the potential for the production of synbiotic non-fat yoghurt involving probiotic *L. plantarum* strains and prebiotic ingredients. Yoghurts with β-glucan showed positive effects in terms of lactic acid content and pH reduction. However, at the beginning of the storage, the syneresis was higher in the 0.50, 1.0 and 1.5% β-glucan added groups. It was observed that β-glucan promotes growth and viability of LAB. However, some process modifications in β-glucan containing yoghurt production may be required to prevent quality defects such as syneresis. Further studies should be conducted to determine the effects of β-glucan on yoghurts produced with different milk types.

Ice cream is a frozen food made from dairy products, sugars, flavorings, and minor amounts of optional ingredients used to improve texture, enhance shipping and aging characteristics and make manufacturing more efficient (Mohan et al., 2014). During the last few decades, new type of the ice-cream products have been introduced to the markets, these products were developed by incorporating probiotic cultures into ice cream products. Ice cream can serve as an excellent environment to deliver probiotics without a loss in their beneficial effects. In this regard, ice cream has an advantage as a probiotic carrier over any other food. It can be successfully used commercially to deliver probiotic bacteria to children and other consumers from different age groups.

The effect of several probiotic strains on the sensory acceptance of ice cream was evaluated by Salem et al. (2005). Probiotic ice cream was manufactured by mixing fortified milk fermented with probiotic strains with an ice cream mix. They found that all the ice cream samples received a high score in the sensory evaluation. Ice cream containing *L. reuteri* was judged to be sourer and reached a higher score for "probiotic" flavor. Başyiğit et al. (2006) investigated the usage of probiotic cultures in ice cream manufacture. Ice cream with sucrose and ice cream with aspartame were prepared and each of these was divided into two subgroups: one with direct addition of the probiotic culture and one with milk fermented by the same probiotic culture. Results showed that the probiotic cultures remained unchanged in ice cream stored for up to 6 months regardless of the sweeteners used. Using probiotic cultures in ice cream mixes did not alter the characteristics of the product. Neither frozen conditions during the storage period nor the type of sweeteners used had any undesired effect on the survival of probiotic cultures. Corrales et al. (2007) evaluated the effect of the dynamic freezing operation on the viability of two different probiotic strains, *L. acidophilus* and *Bifidobacterium lactis,* during ice cream production. It was found that the reduction rate of both strains during this operation was not significant (P>0.05), but throughout the whole process of elaboration of the ice cream (dynamic freezing and then hardening at -30°C) there was a significant reduction on both populations. Homayouni et al. (2008) found that encapsulated *L. casei* (Lc-01) and *B. lactis* (Bb-12) bacteria in calcium alginate beads the probiotic survival raised at rate of 30% during the same period of storage at same temperature. In general, the results indicated that encapsulation can significantly increase the survival rate of probiotic bacteria in ice cream over an extended shelf-life. Salama & El-Sayed (2014) studied for determining the survivability of three probiotic strains *L. plantarum, L. casei* and *B. bifidum* in ice cream using microencapsulation by calcium alginate and whey protein concentrate. Also, the effect of different prebiotics (inulin, lactulose and Fructo-oligosaccharides) was studied by researchers. Six types of synbiotic ice cream containing free and microencapsulated *L. plantarum, L. casei* and *B. bifidum,* were manufactured using 2% fructo-oligosaccharides. The survival of all tested strains increased with different sugars even in the presence of FOS. Also, the viability of microencapsulated strains increased more than 1 log cycle compared with free cells. The survival of *L. plantarum, L. casei* and *B. bifidum* were monitored during the storage period of 90 days. The viable cell count of *L. plantarum, L. casei* and *B. bifidum* in the Free State in ice cream was 8.344, 8.413 and 8.230 log CFU/g at day one and the numbers were decreased to 7.841, 6.110 and 6.436 log CFU/ g after 90 days of storage respectively. Survival of the three encapsulated probiotic strains showed that the bacterial counts increased about two log cycles during the same period of storage and microencapsulated probiotic strains in ice cream had no significant effect on the physiological properties of ice cream and in the sensory properties.

Kefir is a naturally fermented milk product and is pro-duced using kefir grains or mother cultures prepared from kefir grains. It is produced not only with the traditional way but also with industrial processes and various types of milks can be used (Yerlikaya, 2004; Marsh et al., 2014).

Labaneh or labneh (concentrated yogurt) is a popular fermented milk product in the Middle East, which has a significant role in family nutrition (Abd El-Salam et al., 2011). Rocha et al. (2014) studied for developing a functional probiotic labneh using kefir as a fermenting agent and to perform a sensory analysis of the obtained product. Kefir was obtained by growing grains in pasteurized milk. Samples of skimmed and whole labneh were prepared from the inoculation of 5% kefir milk (skimmed/whole) at 28 °C for 24h, followed by cooling (12-18h) and whey drainage (12-24h), both at 4°C. Sensory analysis was performed. The labneh prepared showed good acceptance by the judges, and the whole labneh samples had the highest scores in the acceptance test. Further studies on the analysis of microbiological viability, nutritional composition, and determination of shelf life, also to improve acceptability of the low-fat version of the product, are needed.

Probiotic Meat Products

Preservation of meat products by fermentation has been used for hundreds of years. Starter cultures have a very important role in production of high quality meat products because of their effect on pH, the desired flavor development, as well as providing stability and safety. Therefore, it is important to determine which starter culture or combinations should be used to manufacture safe and high quality meat products for consumers (Ertürkmen et al., 2016). Using probiotic strains as starter cultures for fermented meat products requires that the probiotic strain be well adapted to the conditions found in fermented meat products. Probiotics in fermented meat products can be protected by fat and meat particles against human GI conditions. It is also believed that some of the meat and fat particles may be a source of energy for probiotic bacteria in human digestive tract (Tannock, 1999). Specifically, in order for the probiotic to become dominant in the final product, the starter culture needs to be well adapted to growing in meat material which has a naturally high numbers of background microbiota (Tsuda et al., 2012).

It is important to search for probiotics for possible applications in fermented meat products. Target products with probiotic bacteria are mainly dry sausages, which are processed by fermentation without heat treatment (Arihara, 2006). LAB and coagulase negative cocci, prevailingly species of the genus *Lactobacillus*, *Pediococcus* and *Staphylococcus* are the main bacteria in fermented sausages (Hutkins, 2006). *Lactobacilli* are of the most importance in meat fermentation because of their ability to provide, in the presence of fermentable sugars, rapid and effective acidification thus preserving the sausages from the development of spoilage and pathogenic bacteria. This is why they are often used as starters in dry fermented sausage production. Combining probiotic potential and technological performances of *Lactobacillus* strains would lead to interesting probiotic starters for use in novel dry fermented sausages. Demonstrated the feasibility of using probiotic *Lactobacillus* represented mainly by *Lactobacillus sakei*, *Lactobacillus curvatus* and *L. plantarum* and by coagulase-negative cocci (Ertürkmen et al., 2016). Erkkilä (2001) reported that probiotic and bioprotective *Lactobacillus rhamnosus* strains GG, LC-705 and E-97800 can produce high quality dry sausage with decreased risk for *L. monocytogenes* or *E. coli* O157:H7. Erkkilä et al., (2001) also reported that the use of potential probiotic cultures (*L. rhamnosus* GG, *L. rhamnosus* E-97800 and *L. plantarum* E-98098) had no adverse effect on sensory and technological properties of fermented sausages. Pennacchia et al., (2004) mentioned that usage of probiotics (*L. acidophilus* and *B. lactis*) in fermented sausage manufacture reduces lipid oxidation, total aerobic bacteria, LAB and *micrococcus/staphylococcus* counts in fermented sausage. Frece et al. (2014) have compared the effects of indigenous strains isolated from traditional sausages and the effects of commercial starters on the quality of industrial sausages. Results of this research have not only shown that indigenous strains have the ability to survive even during industrial production, but, compared with commercial strains, they have also displayed better results in the development of sensory properties, stability and microbiological safety of sausages. Tsuda et al. (2012) isolated *L. plantarum* JAB2001 strain from fermented sushi products and determined that this strain can survive passage through the GI tract. The researchers also observed that *L. plantarum* JAB2001 was capable of growth in meat patty and successfully suppressed the growth of *E. coli* and *S. aureus*. Therefore, this strain could be suitable for use as a starter culture of fermented meat products.

Conjugated linoleic acid (CLA) is an attractive bioactive compound for designing functional products. Such novel functional fermented meat products with probiotics and CLA would have a market. Since the effect of probiotic bacteria on the formation of CLA in media and fermented milk products has been demonstrated (Sieber et al., 2004), such effect would be also expected in fermented meat products. Özer et al. (2016) screened twenty-three probiotic *L. plantarum* strains *in vitro* to determine their ability to produce CLA. *L. plantarum* AA1-2 and *L. plantarum* AB20-961 were identified as potential strains for CLA production. Optimum conditions for these strains to produce high levels of CLA were determined by evaluating the amount of added hydrolyzed sunflower oil (HSO) and initial pH levels in a nutrient medium. The highest CLA production was obtained in medium with pH 6.0 and 2% HSO (P b 0.05). Researchers used these cultures as starter culture in sucuk fermentation. Five sucuk treatments included a control (no starter culture), two sucuk groups with *L. plantarum* AA1-2 at the initial pH of 5.8 or 6.0 and two sucuk groups with *L. plantarum* AB20-961 at the initial pH of 5.8 or 6.0. Results indicate that

L. plantarum AB20-961 produced higher amount of CLA in sucuk at initial pH of 5.8 and 6.0 levels during first 24 h of fermentation compared with other groups. CLA isomer concentration decreased in all sucuk groups during the rest of the fermentation period and remained quite stable during the storage. This study concluded that *L. plantarum* AB20-961 with probiotic properties can be used in sucuk manufacture to enhance CLA content at the early stages of fermentation without any adverse effects on quality characteristics of final product. In addition, the use of CLA producing *L. plantarum* AB20-961 with probiotic properties may provide benefits for the promotion of human health. On the other hand, *L. plantarum* AA1-2 was not able to produce CLA in sucuk fermentation. The manufacturing conditions such as temperature and pH were most probably the limiting factors for CLA production in sucuk, therefore further investigation is needed to determine the optimum manufacturing conditions to accomplish desired increase in the levels of CLA.

Probiotic Other Products

In recent years, there is an increasing consumer demand for nondairy-based probiotic food and probiotics are being incorporated into drinks or marketed as dietary supplements in the form of tablets and freeze-dried preparations (Wedajo, 2015). It is known that the allergy to dairy products affects negatively many people around the world. Traditions and economic reasons that limit the use of dairy products in developing countries, such as Japan, China, and some African countries, promote the idea of reducing milk components as vehicles for the probiotic agents or even replacing milk with other media, such as cereals, fruits, and vegetables. Lactose intolerance, cholesterol content, and allergenic milk proteins are the major drawbacks related to the intake of dairy products, which makes the development of new nondairy probiotic foods essential (Kumar et al., 2015). Technological and functional properties, besides the viability of the probiotics used in fermented products of non-dairy origin are extremely important to get a competitive advantage in the world market. However, cell viability is also an important attribute to cell functionality (Ouwehand & Salminen, 1998) and cell functionality is mainly influenced by the food matrix components (Ranadheera et al., 2010). Fruit juices, desserts, and cereal-based products featuring probiotics may be suitable media for delivering probiotics.

Consumers have a genuine interest in fruit juice based functional beverages prepared with probiotics because they offer varied taste profiles that are appealing to all age groups and also they are perceived as healthy and refreshing in contrast to dairy foods (do Espirito Santo et al. 2011). There are a wide variety of traditional non-dairy fermented beverages produced around the world. Much of them are non-alcoholic beverages manufactured with cereals as principal raw material. The non dairy probiotic beverages may be made from a variety of raw materials, such as cereals, millets, legumes, fruits and vegetables (Granato et al., 2010).

Fortified fruit juices represent a good alternative delivery vehicle for probiotic cultures. Fruit juices are consumed commonly therefore it is convenient to take full benefits of probiotics through their consumption. Therefore, it is necessary to choose such strains that can withstand highly acidic conditions and have acceptable viability and shelf life. Sheehan et al. (2007) have tested the viability of *Lactobacillus salivarius* ssp. *salivarius* UCC118 and UCC500, *Lactobacillus paracasei* ssp. *paracasei* NFBC43338, *L. rhamnosus* GG, *L. casei* DN-114 001 and *B. lactis* Bb-12 in varios fruit juice matrices like pineapple, orange, and cranberry juice. Survival was better in orange and pineapple juices as compared to cranberry juice. *L. casei* DN-114 001, *L. rhamnosus* GG and *L. paracasei* NFBC43338 were able to survive for about 12 weeks at concentration of $>10^7$ CFU/ml in pineapple juice and $>10^6$ CFU/ml in orange juice.

These strains were further tested for their ability to survive pasteurization conditions, but none was found suitable. So these cultures were recommended to be used as probiotics in orange or pineapple juice matrix, but they must be added before pasteurization.

Pimentel et al. (2015) reported the production of a probiotic beverage using clarified apple juice supplemented with oligofructose (as prebiotic and sugar substitute) and *L. paracasei* ssp. *paracasei*. The product has similar acceptability to the pure juice, however, more acidic, red and turbid. The shelf life of the probiotic beverage is up to 28 days under refrigeration. Pimentel et al. (2015) also proposed glass as the ideal packaging material for the new beverage since it had no effect on the characteristics of the product but proved capable to maintain the viability of *L. paracasei* ssp. *paracasei*.

Hardaliye is a lactic acid fermented beverage that is produced from crushed red grapes by adding crushed mustard seeds, sour cherry leaf and benzoic acid (Arici & Coskun, 2001). Başyiğit Kılıç et al., (2016) investigated the effects of probiotic *L. plantarum* AK4-11 and grape varieties on some properties of hardaliye. The results showed that grape variety did not have any effect on pH during fermentation period, but using red grapes resulted in higher pH 4.10 in hardaliye after 90 day storage. On the other hand using white grape resulted in higher brix values ranged from 12.90 to 14.00 at the end of the 14th day of fermentation. Using red grapes also increased the amount of phenolic compounds in hardaliye samples. Results of this study indicated that using *L. plantarum* AK4-11 and different grape varieties in hardaliye manufacture affected some quality parameters of hardaliye.

Raw materials in particular, cereals seem good substrates to develop new probiotic foods (Rivera-Espinoza & Gallardo-Navarro, 2010). Freire et al., (2017) aimed to develop a nondairy fermented beverage from a blend of cassava and rice based on Brazilian indigenous beverage cauimusing probiotic LAB and yeast. The indigenous strains *L. plantarum* CCMA 0743 and *Torulaspora delbrueckii* CCMA 0235, and the commercial probiotic, *L. acidophilus* LAC-04, were used as starter cultures in single and co-cultivations. The bacteria populations were around 8.0 log (CFU/mL) at the end of all fermentations as recommended for probiotic products. Higher residual starch contents were noted in the single LAB cultures (10.6% [w/w]) than in co-cultures (<6% [w/w]), showing that co-culture may help the digestibility. For all different assays (single and co-culture), lactic acid was the main organic acid detected (>1.6 g/L) and ethanol was lower than 0.5% (w/v) consisting in a non-alcoholic beverage. The assays containing yeast showed the highest antioxidant activity. Therefore, a nondairy fermented beverage was successfully obtained, and the co-culture of LAB and *T. delbrueckii* could increase the product's functional properties.

Probiotic vegetable preparations will have the potential to attract more consumers to functional products. Beganović et al. (2011) have successfully attempted to produce improved probiotic sauerkraut by incorporating *Leuconostoc mesenteroides* LMG 7954 and *L. plantarum* L4 probiotic cultures during controlled fermentation of cabbage. The probiotic count of the final product was $>10^6$ CFU/g. Sisto & Lavermicocca (2012) used human origin *L. paracasei* IMPC2.1 strain for the development of a new tableolive-based probiotic food. The researhers determined that *L. paracasei* IMPC 2.1 is a good example of a probiotic strain suitable for industrial fermentation of debittered table olives. In fact, it combines the positive characteristics of a probiotic culture with the efficacy of a starter culture which can control the fermentation process and protect table olives during storage even under adverse conditions of temperature and salt concentration, leading to a final low-salt-probiotic product with functional appeal and high organoleptic quality.

Kimchi, or gimchi (kimchee/ kim chee), is also prepared from vegetables and is popular in Korea. Kimchi is served as a side dish with meal and generally eaten raw. In addition to its active ingredients it also contains beneficial LAB which include different strains of *L. plantarum*, *Lactobacillus sakei* and a distinct species named *as Lactobacillus kimchii*. These LAB provide additional health benefits like help in digestion (Bansal et al., 2015). Lee et al. (2011) have isolated low acid producing microorganisms from kimchi and tested their probiotic potential and functional properties. They were successful in isolating potential probiotic strains which could be utilized to develop probiotic kimchi with consistent quality. Finally new studies should be carried out to produce several new probiotic beverages combining dairy and non-dairy substrates. This may result in new products with the accepted taste and flavor of dairy products and the numerous nutrients of non-dairy products such as fruit juices, vegetables, and cereals (Kandylis et al., 2016).

FUTURE REMARKS

Sales and marketing of probiotic dairy and non dairy products represent a huge growth potential all over the world. An important factor limiting the election of new probiotic cultures is linked to the industrial costs of detection, characterization, and clinical validation of new candidate LAB strains of probiotic interest. Stability of the probiotic strains is another important issue. The stability of probiotic strains could be improved by stable strain selection, mutation/ selection of strains against stress factors, applying protective materials or encapsulation in suitable matrices. For this purpose, new studies must be carried out to develop and optimize different sets of simple *in vitro* screening tests, new formulations and process technologies. Technological issues that can affect the survival of probiotic cultures throughout the production process and during storage should also be addressed while formulating new probiotic products. Functional properties are extremely important to get a competitive advantage in the world market. Hence, care should be taken while confirming the functional attributes of starters before incorporating in the product. The development of scientific knowledge on the microbial composition and characteristics of traditionally fermented foods, new probiotic products could be developed with suitable sensory and nutritional appeal, and beneficial properties, when regularly consumed. Developing of new traditional functional probiotic products, optimization of the process, determination of the growth and survival of the probiotic bacteria during processing and storage, composition, organoleptic properties and the nutritional value of the final product are the important issues.

CONCLUSION

Functional products are crucial to the human diet. Today, due to the increasing demand on safe and functional foods, consumption of new and enriched foods has shown growth to higher rates. With the developing scientific techniques today it is simpler to determine the microbial composition and characteristics of traditionally fermented foods. Formulation of traditional dairy and non dairy probiotic products is an important topic for researchers. Increasing the functional properties of traditional probiotic products by using new probiotic strains, adding prebiotics/CLA into the food composition, using encapsulation technology and different substrates are hot research areas. For this purpose, new studies must be carried out to optimize the production of the dairy and non dairy probiotic foods and enhance their functional properties to improve their usage in different food systems.

REFERENCES

Abd El-Salam, M. H., Hippen, A. R., El-Shafie, K., Assem, F. M., Abbas, H., Abd El-Aziz, M., ... El-Aassar, M. (2011). Preparation and properties of probiotic concentrated yoghurt (labneh) fortified with conjugated linoleic acid. *International Journal of Food Science & Technology*, *46*(10), 2103–2110. doi:10.1111/j.1365-2621.2011.02722.x

Amiri, Z. R., Khandelwal, P., & Aruna, B. R. (2010). Development of acidophilus milk via selected probiotics and prebiotics using artificial neural network. *Advances in Bioscience and Biotechnology*, *1*(03), 224–231. doi:10.4236/abb.2010.13031

Arici, M., & Coskun, F. (2001). Hardaliye: Fermented grape juice as a traditional Turkish beverage. *Food Microbiology*, *18*(4), 417–421. doi:10.1006/fmic.2001.0413

Arihara, K. (2006). Strategies for designing novel functional meat products. *Meat Science*, *74*(1), 219–229. doi:10.1016/j.meatsci.2006.04.028 PMID:22062731

Awaisheh, S. S. (2011). Development of probiotic soft cheese manufactured using goat's milk with the addition of thyme. *Milchwissenschaft. Milk Science International*, *66*(1), 51–54.

Awaisheh, S. S. (2012). Probiotic food products classes, types, and processing. In C. R. Everlon (Ed.), *Probiotics* (pp. 551–582). Croatia: Intech Publisher.

Awaisheh, S. S., Al-Dmoor, H. M., Omar, S. S., Hawari, A., & Al-Rwaily, M. M. (2012). Impact of Selected nutraceuticals on viability of probiotic strains in milk during refrigerated storage at 4°C for 15 days. *International Journal of Dairy Technology*, *65*(2), 268–273. doi:10.1111/j.1471-0307.2011.00817.x

Bansal, S., Mangal, M., Sharma, S. K., & Gupta, R. K. (2016). Non-dairy based probiotics: A healthy treat for intestine. *Critical Reviews in Food Science and Nutrition*, *56*(11), 1856–1867. doi:10.1080/10408398.2013.790780 PMID:25747894

Başyiğit, Kılıç, G., Ağdaş, K., Karahan, A.G., & Çakmakçı, M. L. (2016). Effect of *Lactobacillus plantarum* AK4-11 and different grape varieties on the properties of Hardaliye. *Journal of Agricultural Sciences*, *22*(4), 512–521.

Başyiğit, G., Kuleaşan, H., & Karahan, A. G. (2006). Viability of human derived probiotic lactobacilli in ice-cream produced with sucrose and aspartame. *Journal of Industrial Microbiology & Biotechnology*, *33*(9), 796–800. doi:10.100710295-006-0128-x PMID:16639576

Başyiğit Kılıç, G., & Akpınar, D. (2013). The effects of different levels of β-glucan on yoghurt manufactured with *Lactobacillus plantarum* strains as adjunct culture. *Journal of Food Agriculture and Environment*, *11*(1), 281–287.

Başyiğit Kılıç, G., Kuleaşan, H., Eralp, I., & Karahan, A. G. (2009). Manufacture of Turkish Beyaz cheese added with probiotic strains. *Lebensmittel-Wissenschaft + Technologie*, *42*(5), 1003–1008. doi:10.1016/j.lwt.2008.12.015

Beganović, J., Pavunc, A. L., Gjuračić, K., Spoljarec, M., Šušković, J., & Kos, B. (2011). Improved sauerkraut production with probiotic strain *Lactobacillus plantarum* L4 and *Leuconostoc mesenteroides* LMG 7954. *Journal of Food Science*, *76*(2), 124–129. doi:10.1111/j.1750-3841.2010.02030.x PMID:21535775

Broadbent, J. R., & Lin, C. (1999). Effect of heat shock or cold shock treatment on the resistance of Lactococcus lactis to freezing and lyophilization. *Cryobiology*, *39*(1), 88–102. doi:10.1006/cryo.1999.2190 PMID:10458904

Bzducha, A., Mieczysław, W., & Obiedziński, M. W. (2007). Influence of two probiotic *Lactobacillus* strains on CLA content in model ripening cheeses. *Polish Journal of Food And Nutrition Sciences, 57*(4A), 65-69.

Capela, P., Hay, T. K. C., & Shah, N. P. (2006). Effect of cryoprotectants, prebiotics and microencapsulation on survival of probiotic organisms in yoghurt and freeze-dried yoghurt. *Food Research International*, *39*(2), 203–211. doi:10.1016/j.foodres.2005.07.007

Carvalhoa, A. S., Silvaa, J., Hob, P., Teixeiraa, P., Malcataa, F. X., & Gibbsa, P. (2004). Relevant factors for the preparation of freeze-dried lactic acid bacteria. *International Dairy Journal*, *14*(10), 835–847. doi:10.1016/j.idairyj.2004.02.001

Champagne, C. P., & Fustier, P. (2007). Microencapsulation for the improved delivery of bioactive compounds into foods. *Current Opinion in Biotechnology*, *18*(2), 184–190. doi:10.1016/j.copbio.2007.03.001 PMID:17368017

Charteris, W. P., Kelly, P. M., Morelli, L., & Collins, J. K. (1998). Development and application of an in vitro methodology to determine the transit tolerance of potentially probiotic *Lactobacillus* and *Bifidobacterium* species in the upper human gastrointestinal tract. *Journal of Applied Microbiology*, *84*(5), 759–768. doi:10.1046/j.1365-2672.1998.00407.x PMID:9674129

Codex Stan., 283-1978. General standard for cheese.

Corrales, A., Henderson, M., & Morales, I. (2007). Survival of probiotic microorganisms *Lactobacillus acidophilus* and *Bifidobacterium lactis* in whipped ice cream. *Revista Chilena de Nutrición*, *34*(2), 157–163.

Council for Agricultural Science and Technology (CAST). (2007). *Probiotics: their potential to impact human health. Issue Paper 36*. Ames, Iowa: CAST.

Cruz, A. G., Antunes, A. E., Sousa, A. L. O. P., Faria, J. A. F., & Saad, S. M. I. (2009). Ice-cream as a probiotic food carrier. *Food Research International*, *42*(9), 1233–1239. doi:10.1016/j.foodres.2009.03.020

Daigle, A., Roy, D., Belanger, D., & Vuillemard, J. (1999). Production of probiotic cheese (Cheddar like cheese) using enriched cream fermented by *Bifidobacterium infantis*. *Journal of Dairy Science*, *82*(6), 1081–1091. doi:10.3168/jds.S0022-0302(99)75330-0 PMID:10386294

Divya, J. B., Varsha, K. K., Nampoothiri, K. M., Ismail, B., & Pandey, A. (2012). Probiotic fermented foods for health benefits. *Engineering in Life Sciences*, *12*(4), 377–390. doi:10.1002/elsc.201100179

do Espirito Santo, A. P., Perego, P., Converti, A., & Oliveira, M. N. (2011). Influence of food matrices on probiotic viability – a review focusing on the fruity bases. *Trends in Food Science & Technology*, *22*(7), 377–385. doi:10.1016/j.tifs.2011.04.008

El-Sayed, H. S., Salama, H. H., & El-Sayed, S. M. (2014). Production of synbiotic ice cream. *International Journal of Chemtech Research*, *15*(01), 138–147.

Erkkilä, S. (2001). *Bioprotective and probiotic meat starter cultures for the fermentatation of dry sausages* (Dissertation). Department of Food Technology, University of Helsinki, Finland.

Erkkilä, S., Suihko, M. L., Eerola, S., Petäjä, E., & Mattila-Sandholm, T. (2001). Dry sausage fermented by *Lactobacillus rhamnosus* strains. *International Journal of Food Microbiology*, *64*(1-2), 205–210. doi:10.1016/S0168-1605(00)00457-8 PMID:11252505

Ertürkmen, P., Başyiğit Kılıç, G., & Kılıç, B. (2016). Utilization of lactic acid bacteria and probiotics on meat products. *Journal of Hygienic Engineering and Design*, 78–82.

FAO/WHO. (2002b). *Report of a Joint FAO/WHO Working Group on Drafting Guidelines for the Evaluation of Probiotics in Food*. Author.

Figueroa-González, I., Quijano, G., Ramírez, G., & Cruz-Guerrero, A. (2011). Probiotics and prebiotics. Perspectives and challenges. *Journal of the Science of Food and Agriculture*, *91*(8), 1341–1348. doi:10.1002/jsfa.4367 PMID:21445871

Flynn, S., Fitzgerald, G., Daly, C., Kiely, B., O'Sullivan, G. C., Shanahan, F., & Collins, J. K. (2001). *In vitro* selection criteria for probiotic bacteria of human origin: Correlation with in vivo findings. *The American Journal of Clinical Nutrition*, *73*, 386–392. PMID:11157346

Fortin, M. H., Champagne, C. P., St-Gelais, D., Britten, M., Fustier, P., & Lacroix, M. (2011). Effect of time of inoculation, starter addition, oxygen level and salting on the viability of probiotic cultures during Cheddar cheese production. *International Dairy Journal*, *21*(2), 75–82. doi:10.1016/j.idairyj.2010.09.007

Frece, J., Kovačević, D., Kazazić, S., Mrvčić, J., Vahčić, N., Ježek, D., ... Markov, K. (2014). Comparison of sensory properties, shelf-life and microbiological safety of industrial sausages produced with autochthonous and commercial starter cultures. *Food Technology and Biotechnology*, *52*, 307–316.

Freire, A. L., Ramos, C. L., da Costa Souza, P. N., Cardoso, M. G. B., & Schwan, R. F. (2017). Nondairy beverage produced by controlled fermentation with potential probiotic starter cultures of lactic acid bacteria and yeast. *International Journal of Food Microbiology*, *248*, 39–46. doi:10.1016/j.ijfoodmicro.2017.02.011 PMID:28242421

Gobbetti, M., di Cagno, R., & de Angelis, M. (2010). Functional microorganisms for functional food quality. *Critical Reviews in Food Science and Nutrition*, *50*(8), 716–727. doi:10.1080/10408398.2010.499770 PMID:20830633

Granato, D., Branco, G. F., Nazzaro, F., Cruz, A. G., & Faria, A. F. (2010). Functional foods and nondairy probiotic food development: Trends, concepts, and products. *Comprehensive Reviews in Food Science and Food Safety*, *9*(3), 292–303. doi:10.1111/j.1541-4337.2010.00110.x

Hati, S., Mandal, S., & Prajapati, J. B. (2013). Novel starters for value added fermented dairy products. *Current Research in Nutrition and Food Science*, *1*(1), 83–91. doi:10.12944/CRNFSJ.1.1.09

Hayek, S. A., & Ibrahim, S. A. (2013). Current limitations and challenges with lactic acid bacteria: A review. *Food and Nutrition Sciences*, *4*(11), 73–87. doi:10.4236/fns.2013.411A010

Hekmat, S., Soltani, H., & Reid, G. (2009). Growth and survival of *Lactobacillus reuteri* RC-14 and *Lactobacillus rhamnosus* GR-1 in yogurt for use as a functional food. *Innovative Food Science & Emerging Technologies*, *10*(2), 293–296. doi:10.1016/j.ifset.2008.10.007

Heller, K. J. (2001). Probiotic bacteria in fermented foods: Product characteristics and starter organisms. *The American Journal of Clinical Nutrition*, *73*, 374–379. PMID:11157344

Holzapfel, W. H., & Schillinger, U. (2002). Introduction to pre and probiotics. *Food Research International*, *35*(2-3), 109–116. doi:10.1016/S0963-9969(01)00171-5

Homayouni, A., Ehsani, M. R., Azizi, A., Yarmand, M. S., & Razavi, S. H. (2008). Effect of microencapsulation and resistant starch on the probiotic survival and sensory properties of synbiotic ice cream. *Food Chemistry*, *111*(1), 50–55. doi:10.1016/j.foodchem.2008.03.036

Hutkins, R. W. (2006). Meat fermentation. In R. W. Hutkins (Ed.), *Microbiology and technology of fermented foods* (pp. 207–232). Blackwell Publishing. doi:10.1002/9780470277515.ch6

Iqbal, M. Z., Qadir, M. I., Hussain, T., Janbaz, K. H., Khan, Y. H., & Ahmad, B. (2014). Probiotics and their beneficial effects against various diseases. *Pakistan Journal of Pharmaceutical Sciences*, *27*(2), 405–415. PMID:24577933

Jiang, J., & Fonden, R. (1998). Production of CLA by dairy starter cultures. *Journal of Applied Microbiology*, *85*, 95–102. doi:10.1046/j.1365-2672.1998.00481.x PMID:9721660

Kailasapathy, K. (2002). Microencapsulation of probiotic bacteria: Technology and potential applications. *Current Issues in Molecular Biology*, *3*, 39–48. PMID:12400637

Kailasapathy, K. (2013). Commercial sources of probiotic strains and their validated and potential health benefits - a review. *International Journal of Fermented Foods*, *2*(1), 1–17.

Kailasapathy, K., & Phillips, H. M. (2008). Survival of *Lactobacillus acidophilus* and *Bifidobacterium animalis* ssp. *lactis* in stirred fruit yogurts. *Lebensmittel-Wissenschaft + Technologie*, *41*(7), 1317–1322. doi:10.1016/j.lwt.2007.08.009

Kandylis, P., Pissaridi, K., Bekatorou, A., Kanellaki, M., & Koutinas, A. A. (2016). Dairy and non-dairy probiotic beverages. *Current Opinion in Food Science*, *7*, 58–63. doi:10.1016/j.cofs.2015.11.012

Kanmani, P., Kumar, R. S., Yuvaraj, N., Paari, K. A., Pattukumar, V., & Arul, V. (2013). Probiotics and its functionally valuable products: A review. *Critical Reviews in Food Science and Nutrition*, *53*(6), 641–658. doi:10.1080/10408398.2011.553752 PMID:23627505

Karlsson, M. (2013). *Evaluation of Lactobacillus reuteri DSM17938 as starter in cheese production* (Unpublished master's thesis). Department of Microbiology, Institutionen för mikrobiologi, Uppsala, Sweden.

Khosravi-Darani, K., Karamadi, D., Nikoopour, H., & Reyhani, S. F. S. (2016). Comparison of conjugated linoleic acid production in *Bifidobacterium* sp. and *Lactobacillus acidophilus* probiotic yoghurt. *Journal of Chemical and Pharmaceutical Research*, *8*(8), 45–53.

Khosravi-Darani, K., Reihani, F. S., & Feili, R. (2014). Bioproduction of conjugated linoleic acid in yogurt by probiotic bacteria. *International Journal of Biotechnology for Wellness Industries*, *3*(2), 62–68. doi:10.6000/1927-3037.2014.03.02.4

Kumar, B. V., Vijayendra, S. V. N., & Reddy, O. V. S. (2015). Trends in dairy and non-dairy probiotic products - a review. *Journal of Food Science and Technology*, *52*(10), 6112–6124. doi:10.100713197-015-1795-2 PMID:26396359

Kurmann, J. A., Rasic, J. L., & Kroger, M. (1992). *Encyclopedia of fermented fresh milk products: an international inventory of fermented milk, cream, buttermilk, whey, and related products*. New York: Van Nostrand Reinhold.

Lee, H., Yoon, H., Ji, Y., Kim, H., Park, H., Lee, J., ... Holzapfel, W. (2011). Functional properties of *Lactobacillus* strains isolated from kimchi. *International Journal of Food Microbiology*, *145*(1), 155–161. doi:10.1016/j.ijfoodmicro.2010.12.003 PMID:21215484

Leite, A. M. O., Miguel, M. A. L., Peixoto, R. S., Ruas-Madiedo, P., Paschoalin, V. M. F., Mayo, B., & Delgado, S. (2015). Probiotic potential of selected lactic acid bacteria strains isolated from Brazilian kefir grains. *Journal of Dairy Science*, *98*(6), 3622–3632. doi:10.3168/jds.2014-9265 PMID:25841972

Lin, T. Y. (2000). Conjugated linoleic acid concentration as affected by lactic cultures and additives. *Food Chemistry*, *69*(1), 27–31. doi:10.1016/S0308-8146(99)00218-6

Liu, P., Shen, S., Ruan, H., Zhou, Q., Ma, L., & He, G. (2011). Production of conjugated linoleic acids by *Lactobacillus plantarum* strains isolated from naturally fermented Chinese pickles. *Journal of Zheijang University Science B - Biomedicine & Biotechnology, 12*(11), 923-930.

Low, D., Ahlgren, J. A., Horne, D., McMahon, D. J., Oberg, C. J., & Broadbent, J. R. (1998). Role of *Streptococcus thermophilus* MR- 1C capsular exopolysaccharide in cheese moisture retention. *Applied and Environmental Microbiology*, *64*(6), 2147–2151. PMID:9603827

Malek, R. A., Hamdan, S., El Enshasy, H. A., Othman, N. Z., Zainol, N. A., Sarmidi, M. R., & Aziz, R. A. (2010). Production of *Lactobacillus salivarius*, a new probiotic strain isolated from human breast milk, in semi-industrial scale and studies on its functional characterization. In A. Médez-Vilas (Ed.), *Current research, technology and education topics in applied microbiology and microbial biotechnology* (Vol. 2, pp. 1196–1204). Badajoz, Spain: Formatex Research Centre.

Marsh, A. J., Hill, C., Ross, R. P., & Cotter, P. D. (2014). Fermented beverages with health-promoting potential: Past and future perspectives. *Trends in Food Science & Technology*, *38*(2), 113–124. doi:10.1016/j.tifs.2014.05.002

Mattila-Sandholm, T., & Saarela, M. (2000). *Functional dairy products*. Boca Raton, FL: CRC Press.

Mohan, M. S., Hopkinson, J., & Harte, F. (2014). Milk and ice cream processing. Food Processing: Principles and Applications (2nd ed.). *John Wiley & Sons, Ltd.*

Ng, E. W., Yeung, M., & Tong, P. S. (2010). Effects of yogurt starter cultures on the survival of *Lactobacillus acidophilus. International Journal of Food Microbiology*, *145*(1), 169–175. doi:10.1016/j.ijfoodmicro.2010.12.006 PMID:21196060

Oberman, H., & Libudzisz, Z. (1998). Fermented milks. In B. J. B. Wood (Ed.), *Microbiology offermented foods* (pp. 308–350). London: Blackie Academic and Professional. doi:10.1007/978-1-4613-0309-1_11

Ong, L., Henriksson, A., & Shah, N. P. (2006). Development of probiotic Cheddar cheese containing *Lactobacillus acidophilus, Lb. casei, Lb. paracasei* and *Bifidobacterium* spp. and the influence of these bacteria on proteolytic patterns and production of organic acid. *International Dairy Journal, 16*(5), 446–456. doi:10.1016/j.idairyj.2005.05.008

Ouwehand, A. C., & Salminen, S. J. (1998). The health effects of cultured milk products with viable and non-viable bacteria. *International Dairy Journal, 8*(9), 749–758. doi:10.1016/S0958-6946(98)00114-9

Özer, C. O., Kılıç, B., & Başyiğit Kılıç, G. (2016). *In-vitro* microbial production of conjugated linoleic acid by probiotic *L. plantarum* strains: Utilization as a functional starter culture in sucuk fermentation. *Meat Science, 114*, 24–31. doi:10.1016/j.meatsci.2015.12.005 PMID:26720888

Pariza, M. W., Park, Y., & Cook, M. E. (2001). The biologically active isomers of conjugated linoleic acid. *Progress in Lipid Research, 40*(4), 283–298. doi:10.1016/S0163-7827(01)00008-X PMID:11412893

Pennacchia, C., Ercolini, D., Blaiotta, G., Pepe, O., Mauriello, G., & Villani, F. (2004). Selection of *Lactobacillus* strains from fermented sausages for their potential use as probiotics. *Meat Science, 67*(2), 309–317. doi:10.1016/j.meatsci.2003.11.003 PMID:22061328

Pimentel, T. C., Madrona, G. S., Garcia, S., & Prudencio, S. H. (2015). Probiotic viability, physicochemical characteristics and acceptability during refrigerated storage of clarified apple juice supplemented with *Lactobacillus paracasei* ssp. *paracasei* and oligofructose in different package type. *Lebensmittel-Wissenschaft + Technologie, 63*(1), 415–422. doi:10.1016/j.lwt.2015.03.009

Ranadheera, R. D. C. S., Baines, S. K., & Adams, M. C. (2010). Importance of food in probiotic efficacy. *Food Research International, 43*(1), 1–7. doi:10.1016/j.foodres.2009.09.009

Reid, G. (2008). Probiotics and prebiotics e progress and challenges. *International Dairy Journal, 18*(10-11), 969–975. doi:10.1016/j.idairyj.2007.11.025

Reid, G. (2015). The growth potential for dairy probiotics. *International Dairy Journal, 49*, 16–22. doi:10.1016/j.idairyj.2015.04.004

Rivera Espinoza, Y., & Gallardo Navarro, Y. (2010). Non-dairy probiotic products. *Food Microbiology, 27*(1), 1–11. doi:10.1016/j.fm.2008.06.008 PMID:19913684

Roberfroid, M. B. (2007). Concepts and strategy of functional food science: The European perspective. *The American Journal of Clinical Nutrition, 71*, 1660–1664. PMID:10837311

Rocha, D. M. U. P., Martins, J. F. L., Santos, T. S. S., & Moreira, A. V. B. (2014). Labneh with probiotic properties produced from kefir: Development and sensory evaluation. *Food Science and Technology (Campinas), 34*(4), 694–700. doi:10.1590/1678-457x.6394

Routray, W., & Mishra, H. N. (2011). Scientific and technical aspects of yogurt aroma and taste: A Review. *Comprehensive Reviews in Food Science and Food Safety, 10*(4), 208–220. doi:10.1111/j.1541-4337.2011.00151.x

Salem, M., Fathi, F., & Awad, R. (2005). Production of probiotic ice cream. *Polish Journal of Food and Nutrition Sciences*, *14*(3), 267–271.

Sarvari, F., Mortazavian, A. M., & Fazeli, M. R. (2014). Biochemical characteristics and viability of probiotic and yogurt bacteria in yogurt during the fermentation and refrigerated storage. *Applied Food Biotechnology*, *1*(1), 55–61.

Selmer-Olsen, E., Birkeland, S. E., & Sørhaug, T. (1999). Effect of protective solutes on leakage from and survival of immobilized *Lactobacillus* subjected to drying, storage and rehydration. *Journal of Applied Microbiology*, *87*(3), 429–437. doi:10.1046/j.1365-2672.1999.00839.x PMID:10540246

Sheehan, V. M., Ross, P., & Fitzgerald, G. F. (2007). Assessing the acid tolerance and the technological robustness of probiotic cultures for fortification in fruit juices. *Innovative Food Science & Emerging Technologies*, *8*(2), 279–284. doi:10.1016/j.ifset.2007.01.007

Sieber, R., Collomb, M., Aeschlimann, A., & Eyer, H. (2004). Impact of microbial cultures on conjugated linoleic acid in dairy products – a review. *International Dairy Journal*, *14*(1), 1–15. doi:10.1016/S0958-6946(03)00151-1

Sir'o, I., Kapolna, E., Kapolna, B., & Lugasi, A. (2008). Functional food. Product development, marketing and consumer acceptance-a review. *Appetite*, *51*(3), 456–467. doi:10.1016/j.appet.2008.05.060 PMID:18582508

Sisto, A., & Lavermicocca, P. (2012). Suitability of a probiotic *Lactobacillus paracasei* strain as a starter culture in olive fermentation and development of the innovative patented product "probiotic table olives". *Frontiers in Microbiology*, *3*(174), 1–5. PMID:22586426

Stanton, C., Desmond, C., Coakley, M., Collins, J. K., Fitzgerald, G., & Ross, R. P. (2003). Challenges facing development of probiotic-containing functional foods. In F. Er (Ed.), *Handbook of fermented functional foods* (pp. 50–79). Boca Raton, FL: CRC Press. doi:10.1201/9780203009727.ch2

Tamime, A. Y., & Robinson, R. K. (2007). Yoghurt science and technology (3rd ed.). Abington, UK: Woodhead Publishing Ltd. LLC.

Tannock, G. A. (1999). Fresh look at the intestinal microflora. In G. Tannock (Ed.), *Probiotics: A Critical Review* (p. 514). Norfolk, UK: Horizon Scientific Press.

Tsuda, H., Matsumoto, T., & Ishimi, Y. (2012). Selection of lactic acid bacteria as starter cultures for fermented meat products. *Food Science and Technology Research*, *18*(5), 713–721. doi:10.3136/fstr.18.713

Tu, N., Hoang Khue, H., & Ngoc, N. H. (2013). Exopolysaccharide in *Lactobacillus rhamnosus* Pn04 after co-culture with *Leuconostoc mesenteroides* Vtcc-B-643. *Journal of Applied Pharmaceutical Science, 3*(7), 14-17.

Tuomola, E., Crittenden, R., Playne, M., Isolauri, E., & Salminen, S. (2001). Quality assurance criteria for probiotic bacteria. *The American Journal of Clinical Nutrition*, *73*(2), 393–398. PMID:11157347

USDA. (2010). *Food and Drug Administration Compliance Policy Guide Sec. 527.300 Dairy Products - Microbial Contaminants and Alkaline Phosphatase Activity is now available*. Retrieved from Federal Register notice of availability: http://origin.www.gpo.gov/fdsys/pkg/FR-2010-12-12/pdf/2010-32232.pdf

Vaillancourt, J. (2006). *Regulating pre- and probiotics: a U.S. FDA perspective. In Ending the war metaphor: The changing agenda for unraveling the host-microbe relationship* (pp. 229–237). Washington, DC: National Academies Press.

Vinderola, G., Binetti, A., Burns, P., & Reinheimer, J. (2011). Cell viability and functionality of probiotic bacteria in dairy products. *Frontiers in Microbiology, 28*(70), 1–6. PMID:21833320

Wedajo, B. (2015). Lactic acid bacteria: Benefits, selection criteria and probiotic potential in fermented food. *Journal of Probiotics & Health, 3*(02), 129. doi:10.4172/2329-8901.1000129

WHO/FAO. (2002a). *Guidelines for the Evaluation of Probiotics in Food*. London: Author.

Wright, A. V., & Axelsson, L. (2011). Lactic acid bacteria: an introduction. In S. Lahtinen, A. C. Ouwehand, S. Salminen, & A. V. Wright (Eds.), *Lactic acid bacteria: microbiology and functional aspects*. Boca Raton, FL: CRC Press. doi:10.1201/b11503

Yeo, S. K., & Liong, M. T. (2010). Angiotensin I-converting enzyme inhibitory activity and bioconversion of isoflavones by probiotics in soymilk supplemented with prebiotics. *International Journal of Food Sciences and Nutrition, 61*(2), 161–181. doi:10.3109/09637480903348122 PMID:20085504

Yerlikaya, O. (2014). Starter cultures used in probiotic dairy product preparation and popular probiotic dairy drinks. *Food Science and Technology (Campinas), 34*(14), 221–229. doi:10.1590/fst.2014.0050

Yoon, K. Y., Woodams, E. E., & Hang, Y. D. (2006). Production of probiotic cabbage juice by lactic acid bacteria. *Bioresource Technology, 97*(12), 1427–1430. doi:10.1016/j.biortech.2005.06.018 PMID:16125381

Zago, M., Fornasari, M. E., Carminati, D., Burns, P., Suarez, V., Vinderola, G., ... Giraffa, G. (2011). Characterization and probiotic potential of *Lactobacillus plantarum* strains isolated from cheeses. *Food Microbiology, 28*(5), 1033–1040. doi:10.1016/j.fm.2011.02.009 PMID:21569949

Zhang, W., Xiao, S., Samaraweera, H., Lee, E. J., & Ahn, D. U. (2010). Improving functional value of meat products. *Meat Science, 86*(1), 15–31. doi:10.1016/j.meatsci.2010.04.018 PMID:20537806

KEY TERMS AND DEFINITIONS

Gastrointestinal Tract: The system of organs within multicellular animals that takes in food, digests it to extract energy and nutrients, and expels the remaining waste.

Prebiotics: Nondigestible substances that, when consumed, provide a beneficial physiological effect on the host by selectively stimulating the favorable growth or activity of a limited number of indigenous bacteria.

Probiotics: Live microorganisms that confer a health benefit on the host when administered in adequate amounts.

Starter Culture: *Starter cultures* are preparations of live microorganisms or their resting forms whose metabolic activity has desired effects in the fermentation substrate, the food.

Chapter 7
Genomics Technologies for Enhanced Understanding of Robustness of LAB Starter Cultures

Annereinou R. Dijkstra
NIZO Food Research BV, The Netherlands

Peter A. Bron
NIZO Food Research BV, The Netherlands

ABSTRACT

During production and application, starter culture strains encounter several stresses. To ensure an adequate contribution to the fermentation process, starter culture strains should remain viable, justifying the increasing industrial efforts to unravel robustness characteristics of LAB starter culture strains. Nowadays, large numbers of genome sequences are publicly available, which enables the employment of several genomics technologies to increase our understanding of robustness. This aids the improvement of currently applied LAB starter culture strains and supports the industrial application of novel strains with specific desirable traits but currently inadequate robustness characteristics. This chapter explores genomics technologies for enhanced understanding of robustness of LAB starter cultures.

BACKGROUND

Food fermentation, which involves the conversion of sugars into organic acids or alcohol by microorganisms, is performed since ancient times. For example, evidence has been found for the presence of fermented beverages as early as 6,000 B.C. in a Chinese village (McGovern, et al., 2004). Initially, food fermentation processes were used for preservation of raw food materials and were spontaneous fermentations with microorganisms present in the raw food material or on the tools applied. The endresult of such spontaneous fermentations is rather unpredictable due to variation in the microbiota of the raw material and handling of the food material. By applying back-slopping methods, where part of the

DOI: 10.4018/978-1-5225-5363-2.ch007

fermentation endproduct is used to initiate a new fermentation, fermentation processes became more controlled (Leroy, 2004; Smid & Hugenholtz, 2010). Industrialization of food fermentation processes in the late nineteenth century led to the use of starter cultures, which were added to the raw food material, resulting in a safer and more stable product. Starter cultures are generally classified as defined or undefined. Undefined starter cultures have a complex composition of strains and originate from traditional fermentations, whereas defined starter cultures are composed of a limited amount of known strains, which are typically isolated from undefined starters (Smid, et al., 2014). Initially, these starter cultures were propagated daily, which is time-consuming and can result in shifts in the strain composition and loss of plasmid-encoded properties (Leroy, 2004; Smid, et al., 2014). The development of concentrated starter cultures, containing high amounts of one or multiple strains in frozen or dried form, enabled the direct inoculation of fermentation processes and has largely solved these issues (Leroy, 2004; Silva, Freixo, Gibbs, & Teixeira, 2011).

The majority of the commercially available bacterial starter cultures contain lactic acid bacteria (LAB) (Santivarangkna, Kulozik, & Foerst, 2007). LAB have a long history in the fermentation of various foods, e.g., vegetables, meat, cereals and milk (Leroy, 2004). Their capability to rapidly acidify food materials by the formation of lactic acid and other organic acids prevents spoilage of the food. Moreover, the formation of aroma-compounds and exopolysaccharides contributes substantially to the taste and texture of the fermented endproduct (Leroy, 2004). The group of LAB includes the genera *Lactobacillus*, *Leuconostoc*, *Pediococcus*, *Streptococcus* and *Lactococcus*. *Lactococcus lactis* is one of the most extensively used LAB in food fermentation, mainly in the production of cheese, butter and buttermilk (Leroy, 2004). Due to its importance for the dairy industry, it became one of the best studied gram-positive bacteria (Kok, Buist, Zomer, van Hijum, & Kuipers, 2005).

Historically, LAB starter culture strains have been primarily selected for their flavor-forming, texture-enhancing and fast-acidifying characteristics. Nowadays robustness is also of high industrial relevance, as this characteristic directly impacts on the profit margins of starter culture producers (Bron & Kleerebezem, 2011; Santivarangkna, Kulozik, & Foerst, 2008). Starter culture strains encounter various stresses during processing and storage as well as during the food fermentation process. The production of starter cultures generally involves freezing (cold stress) or drying (heat or cold stress combined with dehydration stress), whereas during cheese making, starter cultures are exposed to osmotic stress due to high salt concentrations and acid stress during fermentation (Smith, 2010). Viability of starter culture strains is essential for an adequate contribution to the fermentation process and, therefore, robustness, which is defined as the ability to remain viable and active during harsh conditions, is an important characteristic (Bron & Kleerebezem, 2011; Santivarangkna, et al., 2008). Recently, the stress physiology in LAB has been extensively reviewed by Papadimitriou *et al.* (Papadimitriou, et al., 2016); this chapter will focus on the use of molecular understanding to improve the robustness of LAB strains.

Hundreds of LAB genome sequences are currently available in the public domain (Douillard & de Vos, 2014; Laroute, et al., 2017). Moreover, high throughput screening methods have enabled the scientific community to generate large functional datasets as well, both from individual strains in large sets of fermentation conditions and from large panels of strains. This chapter describes how the correlation of genomic and phenotypic data can lead to a better understanding of the strain-specific diversity in robustness characteristics at the genomics level, largely focusing on *Lactococcus lactis*, the best studied and most applied LAB species for cheese fermentations. Moreover, the authors detail how this understanding will facilitate appropriate selection of strains with desired phenotypes and further improvement of industrially applied strains.

THE ROLE OF GENOMICS TECHNOLOGIES TO UNDERSTAND AND IMPROVE ROBUSTNESS OF STARTER CULTURES

Improvement of starter cultures has initially focused on the quest for strains with favorable characteristics in available strain collections. High diversity in functional properties such as bacteriocin production (de Vuyst, 1994) and proteolytic activity (Boutrou, Sepulchre, Gripon, & Monnet, 1998) has been demonstrated within the group of industrial *L. lactis* strains, which are primarily isolated from dairy products. Studies including *L. lactis* strains isolated from other habitats have demonstrated that even more distinct phenotypes in e.g. flavor formation exist within the species (Ayad, Verheul, de Jong, Wouters, & Smit, 1999; Bachmann, et al., 2009). Also robustness has been demonstrated to be a highly diverse phenotype among LAB strains (Dijkstra, Setyawati, et al., 2014; van Bokhorst-van de Veen, van Swam, Wels, Bron, & Kleerebezem, 2012). For example, within a collection of 39 *L. lactis* strains from various sources a large variation (up to 5 log units) in heat and oxidative stress survival was observed (Dijkstra, Setyawati, et al., 2014). Moreover, a large diversity of survival during spray drying of *L. lactis* cultures has been observed, indicating the importance of selection of starter culture strains for robustness characteristics along with acidifying and flavor-forming properties (Dijkstra, Setyawati, et al., 2014).

A better understanding at the genomics level of the observed diversity in functional characteristics among strains is bound to facilitate appropriate selection of strains with desired phenotypes. Between 2001 and 2007 complete genome sequences of three dairy-derived *L. lactis* strains (IL1403 (Bolotin, et al., 2001), SK11 (Makarova, et al., 2006) and MG1363 (Wegmann, et al., 2007)) became publicly available, followed by the genome sequences of the two non-dairy isolates KF147 and KF282 in 2010 (Siezen, et al., 2010). These whole genome sequences enabled a comparative genome hybridization (CGH) approach, in which the genetic content of 39 *L. lactis* strains were compared based on available *L. lactis* genomes using multi-strain DNA micro-array technology (Bayjanov, et al., 2009; Siezen, et al., 2011). This provided a gene presence/absence matrix of a substantial amount of strains, supporting the prediction of the involvement of genes in specific phenotypes by comparing the presence/absence of genes with the observed phenotype, an approach coined genotype-phenotype matching. As an example of the power of this approach, growth of *L. lactis* on the plant dissacharide melibiose could be associated with a large gene cluster encoding sugar transporter and regulatory functions (Bayjanov, Starrenburg, van der Sijde, Siezen, & van Hijum, 2013). Also other LAB genotype-phenotype comparisons were applied, which resulted in the identification of genes in *Lactobacillus plantarum* WCFS1 that modulate the immune response of dendritic cells (Meijerink, et al., 2010) and the identification of a gene involved in mannose adhesion (Pretzer, et al., 2005). Gene deletion mutants were constructed to confirm the role of the associated genes in the studied phenotype. In these studies fifty percent of the genes that were associated with the phenotype caused an alteration in the phenotype upon deletion. This demonstrates that not all associated genes are directly involved in the observed phenotype and gene deletion is required to definitely establish the role of the associated genes. In a study of gastro-intestinal robustness of *L. plantarum* strains, no genes were significantly associated with the robustness phenotype (van Bokhorst-van de Veen, van Swam, et al., 2012), whereas in a study of heat and oxidative stress survival of *L. lactis* strains, several (~50) genes could be associated with robustness (Dijkstra, Setyawati, et al., 2014). In this study gene deletion mutants could not confirm the role of these genes in robustness, possibly because complimentary genes that take over the function originally performed by the gene targeted by deletion were present (Dijkstra, Setyawati, et al., 2014). This demonstrates that despite the success in specific cases genotype-phenotype matching cannot reveal the genes involved in all phenotypes. The role of genes associated with complex phenotypes such as stress survival appear to be more difficult to be confirmed by gene deletion than those associated with relatively simple phenotypes such as mannose-specific adhesion, for which gene deletion rendered a complete shift in phenotype (Pretzer, et al., 2005). Moreover, certain phenotypic differences cannot be explained by the presence or absence of specific genes but concern differences in the regulation of genes that are present in all strains, e.g. flavor forming enzyme activity (Bachmann, et al., 2009). These differentially regulated omnipresent genes can intrinsically never be identified as important for phenotypes by applying genotype-phenotype matching, but require a different approach termed transcriptome-phenotype matching.

Comparison of gene expression levels and corresponding phenotypes induced by various fermentation conditions in transcriptome-phenotype matching studies can pinpoint genes associated with these phenotypes. This method has been successfully applied in LAB in studies of the gastro-intestinal survival of *L. plantarum* WCFS1 (Bron, et al., 2012; van Bokhorst-van de Veen, Lee, et al., 2012) and the heat and oxidative stress survival of four *L. lactis* strains (Dijkstra, Alkema, et al., 2014; Dijkstra, et al., 2016). Association of genes with robustness towards heat and oxidative stress of *L. lactis* strains appeared highly strain-specific (Dijkstra, Alkema, et al., 2014; Dijkstra, et al., 2016). In *L. lactis* strain MG1363 oxidative stress survival was linked to the expression levels of the *metC-cysK* operon (Dijkstra, Alkema, et al., 2014). Involvement of this operon in robustness was confirmed by culturing MG1363 in absence of cysteine, which resulted in a 1.000-fold increased survival during oxidative stress (Dijkstra, Alkema, et al., 2014), making this a concrete example of how improvements in fermentation conditions can enhance industrially relevant phenotypes.

The large impact of fermentation conditions on robustness phenotypes as observed in these transcriptome-phenotype matching studies demonstrates that fermentation conditions during production of starter cultures should be carefully chosen (Bron, et al., 2012; Dijkstra, Alkema, et al., 2014; Dijkstra, et al., 2016). Earlier studies on (pre-)adaptation as a strategy for improvement of specific phenotypes of starter culture strains have already demonstrated that exposure to specific conditions during growth may result in improvement of a desired phenotype. During adaptation changes in the transcriptome that are advantageous for the desired phenotype are provoked. To enhance robustness phenotypes, adaptation generally involves exposure to a mild stress before exposure to a lethal dose of the same (pre-adaptation) or a different stress (cross-adaptation). For example, a brief exposure to 42°C during growth induces the heat shock proteins GroEL and DnaK in *L. lactis* and results in increased survival after subsequent exposure to 50°C as compared with cells that were continuously cultured at 30°C (Whitaker & Batt, 1991). Furthermore, adaptation has been successfully applied to *L. lactis* strains to increase acid, salt, oxidative or ethanol stress survival (Duwat, Ehrlich, & Gruss, 1999; O'Sullivan & Condon, 1997). However, adaptation only results in short term improvement of the desired phenotype, and, therefore, adaptation conditions should be applied consistently.

To achieve permanent improvement of existing strains, in the past, primarily random mutagenesis tools (rendering random DNA mutations by exposure to mutagens such as UV radiation or specific chemicals) and metabolic engineering (optimization of metabolic routes by genetic modification for production of a valuable component) were applied. Acquiring a strain with a desired mutation after random mutagenesis requires screening of a large number of mutants to enable selection of the mutant with the desired phenotype which can be a laborious process. Furthermore, random mutagenesis often results in undesired additional mutations that might affect other important characteristics of the strain. The use of targeted gene mutation prevents the occurrence of these additional mutations. However, consumer acceptance concerning these genetically modified strains is low and regulatory issues disallow the use of these strains in food fermentation (Derkx, et al., 2014).

Other methods described to obtain improved strains are experimental evolution and selection techniques, in which strains are subjected to conditions that are selective towards spontaneous mutants with a specific phenotype, for instance a higher growth rate or the ability to survive in these conditions (Bachmann, Pronk, Kleerebezem, & Teusink, 2014; Derkx, et al., 2014). These strains are considered natural and, therefore, can be applied in food fermentation without any regulatory issues, in contrast to mutants obtained via random or targeted mutagenesis. The development of next-generation sequencing tools enabled the characterization of these optimized mutants by genome re-sequencing and determination of the transcriptome by sequencing and aid the understanding of the underlying mechanisms responsible for the desired phenotypes. Experimental evolution and selection experiments in *L. lactis* strains have demonstrated that adaptation of the plant-derived isolate KF147 to a milk environment is possible and involves changes in nitrogen metabolism and down-regulation of genes involved in the degradation of plant material (Bachmann, Starrenburg, Molenaar, Kleerebezem, & van Hylckama Vlieg, 2012). Another example of experimental evolution is an emulsion-based serial propagation strategy that enables selection of *L. lactis* MG1363 derivatives with increased growth yield. Subsequent genome resequencing demonstrated that the altered phenotype in this derivative was caused by changes in the expression levels of a glucose transporter (Bachmann, et al., 2013). Another experimental evolution study demonstrated that the ability of *L. lactis* strain MG1363 to grow on lactose could be obtained by a single mutation in the promotor region of a normally silent cellobiose transporter gene (Solopova, et al., 2012). Selection through repeated exposure to heat stress allowed isolation of heat-robust derivatives of *L. lactis* strain SK11 (Dijkstra, et al.). Moreover, the heat-robust derivatives displayed a significantly increased survival during spray drying (Dijkstra, et al.). The transcriptome of independently obtained robust derivatives all revealed the up-regulation of a zinc transport operon, suggesting a yet unknown link between metal regulation and heat stress survival (Dijkstra, et al.). This demonstrates that genomic and/or transcriptomic analysis of experimentally improved strains can aid in increasing the knowledge on robustness phenotypes of LAB starter culture strains.

FUTURE RESEARCH DIRECTIONS

The ongoing development and subsequent lower costs of genome sequencing techniques has propelled the availability of an ever increasing number of LAB genomes. The extension of data on gene absence and presence in strain collections will further enhance the strength of the genome-phenotype approach by increasing the probability of pinpointing genes involved in robustness or other desired phenotypes of LAB starter culture strains. Natural competence, i.e. the ability to take up extracellular DNA fragments from the environment, has recently been established in *L. lactis* (David, et al., 2017; Mulder, Wels, Kuipers, Kleerebezem, & Bron, 2017) and this will significantly decrease the effort required to construct mutants, making the confirmation of gene-phenotype associations an even more attractive route. Moreover, readily available genome and transcriptome sequencing techniques should support continuation of setting up transcriptome-phenotype matching studies as well as analysis of experimentally improved strains, supporting further understanding of robustness phenotypes of starter culture strains. Assessment of genome sequences to confirm the absence of virulence and/or antibiotic resistance genes is a straightforward method to limit safety risks in food products. Taken together, the availability of genome sequence data appears to be the industry standard prior to application of novel strains in food fermentations. Targeted mutagenesis is currently not accepted in the food market but CRISPR-Cas technology, using Cas9 protein and a guide RNA enables targeted, highly efficient alterations of genome sequences (Sander & Joung, 2014) and might offer a way to remove undesired genes (either from a safety perspective or alternatively from a metabolic rerouting point of view), bringing us one step closer to the ultimate plug bug in which phenotypes can be combined at will.

CONCLUSION

Genomics technologies have been demonstrated to support our understanding of robustness phenotypes. Genotype- and transcriptome-phenotype matching approaches might not directly provide improved strains that can be applied in food fermentation according to food legislation. Nevertheless, they do support a better understanding of observed differences in phenotypes and can enhance the selection of strains with a desired phenotype or fermentation conditions resulting in a desired phenotype (Figure 1). Furthermore, genomic analysis of improved strains obtained via experimental evolution and selection approaches aid our understanding of robustness phenotypes, which supports the (improved) application of a broader range of LAB strains.

Figure 1. Schematic overview of the workflow to obtain optimally robust starter cultures (indicated by the arrows) and the (genomics) technologies that can support this workflow (indicated by the grey boxes)

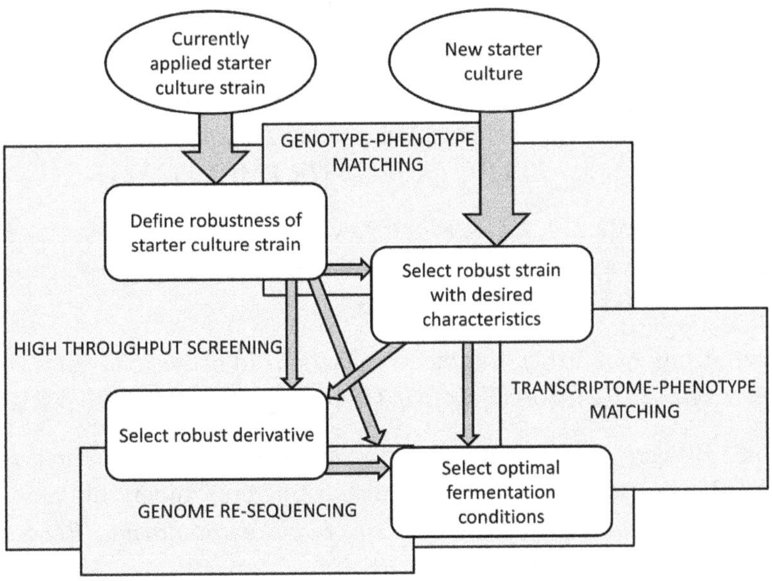

REFERENCES

Ayad, E. H. E., Verheul, A., de Jong, C., Wouters, J. T. M., & Smit, G. (1999). Flavour forming abilities and amino acid requirements of *Lactococcus lactis* strains isolated from artisanal and non-dairy origin. *International Dairy Journal*, *9*(10), 725–735. doi:10.1016/S0958-6946(99)00140-5

Bachmann, H., Fischlechner, M., Rabbers, I., Barfa, N., Branco dos Santos, F., Molenaar, D., & Teusink, B. (2013). Availability of public goods shapes the evolution of competing metabolic strategies. *Proceedings of the National Academy of Sciences of the United States of America*, *110*(35), 14302–14307. doi:10.1073/pnas.1308523110 PMID:23940318

Bachmann, H., Pronk, J. T., Kleerebezem, M., & Teusink, B. (2014). Evolutionary engineering to enhance starter culture performance in food fermentations. *Current Opinion in Biotechnology*, *32C*, 1–7. PMID:25269887

Bachmann, H., Starrenburg, M. J., Dijkstra, A., Molenaar, D., Kleerebezem, M., Rademaker, J. L., & van Hylckama Vlieg, J. E. (2009). Regulatory phenotyping reveals important diversity within the species *Lactococcus lactis*. *Applied and Environmental Microbiology*, *75*(17), 5687–5694. doi:10.1128/AEM.00919-09 PMID:19592532

Bachmann, H., Starrenburg, M. J., Molenaar, D., Kleerebezem, M., & van Hylckama Vlieg, J. E. (2012). Microbial domestication signatures of *Lactococcus lactis* can be reproduced by experimental evolution. *Genome Research*, *22*(1), 115–124. doi:10.1101/gr.121285.111 PMID:22080491

Bayjanov, J. R., Starrenburg, M. J., van der Sijde, M. R., Siezen, R. J., & van Hijum, S. A. (2013). Genotype-phenotype matching analysis of 38 *Lactococcus lactis* strains using random forest methods. *BMC Microbiology*, *13*(1), 68. doi:10.1186/1471-2180-13-68 PMID:23530958

Bayjanov, J. R., Wels, M., Starrenburg, M., van Hylckama Vlieg, J. E., Siezen, R. J., & Molenaar, D. (2009). PanCGH: A genotype-calling algorithm for pangenome CGH data. *Bioinformatics (Oxford, England)*, *25*(3), 309–314. doi:10.1093/bioinformatics/btn632 PMID:19129208

Bolotin, A., Wincker, P., Mauger, S., Jaillon, O., Malarme, K., Weissenbach, J., ... Sorokin, A. (2001). The complete genome sequence of the lactic acid bacterium *Lactococcus lactis* ssp. *lactis* IL1403. *Genome Research*, *11*(5), 731–753. doi:10.1101/gr.GR-1697R PMID:11337471

Boutrou, R., Sepulchre, A., Gripon, J. C., & Monnet, V. (1998). Simple Tests for Predicting the Lytic Behavior and Proteolytic Activity of Lactococcal Strains in Cheese. *Journal of Dairy Science*, *81*(9), 2321–2328. doi:10.3168/jds.S0022-0302(98)70121-3

Bron, P. A., & Kleerebezem, M. (2011). Engineering lactic acid bacteria for increased industrial functionality. *Bioengineered Bugs*, *2*(2), 80–87. doi:10.4161/bbug.2.2.13910 PMID:21636994

Bron, P. A., Wels, M., Bongers, R. S., van Bokhorst-van de Veen, H., Wiersma, A., Overmars, L., ... Kleerebezem, M. (2012). Transcriptomes reveal genetic signatures underlying physiological variations imposed by different fermentation conditions in *Lactobacillus plantarum*. *PLoS One*, *7*(7), e38720. doi:10.1371/journal.pone.0038720 PMID:22802930

David, B., Radziejwoski, A., Toussaint, F., Fontaine, L., Henry de Frahan, M., Patout, C., van Dillen, S., Boyaval, P., Horvath, P., Fremaux, C., & Hols, P. (2017). Natural DNA transformation is functional in *Lactococcus lactis* ssp. *cremoris* KW2. *Appl. Environ. Microbiol.*

de Vuyst, L. (1994). Nisin production variability between natural *Lactococcus lactis* subsp. *lactis* strains. *Biotechnology Letters*, *16*(3), 287–292. doi:10.1007/BF00134627

Derkx, P. M., Janzen, T., Sorensen, K. I., Christensen, J. E., Stuer-Lauridsen, B., & Johansen, E. (2014). The art of strain improvement of industrial lactic acid bacteria without the use of recombinant DNA technology. *Microbial Cell Factories*, *13*(Suppl 1), S5. doi:10.1186/1475-2859-13-S1-S5 PMID:25186244

Dijkstra, A. R., Alkema, W., Starrenburg, M., Hugenholtz, J., van Hijum, S., & Bron, P. A. (2014). Fermentation-induced variation in heat and oxidative stress phenotypes of *Lactococcus lactis* MG1363 reveals transcriptome signatures for robustness. *Microbial Cell Factories*, *13*(1), 148. doi:10.118612934-014-0148-6 PMID:25366036

Dijkstra, A. R., Alkema, W., Starrenburg, M. J., Hugenholtz, J., van Hijum, S. A., & Bron, P. A. (2016). Strain-Dependent Transcriptome Signatures for Robustness in *Lactococcus lactis*. *PLoS One*, *11*(12), e0167944. doi:10.1371/journal.pone.0167944 PMID:27973578

Dijkstra, A. R., Setyawati, M. C., Bayjanov, J. R., Alkema, W., van Hijum, S. A., Bron, P. A., & Hugenholtz, J. (2014). Diversity in robustness of Lactococcus lactis strains during heat stress, oxidative stress, and spray drying stress. *Applied and Environmental Microbiology*, *80*(2), 603–611. doi:10.1128/AEM.03434-13 PMID:24212574

Dijkstra, A. R., Starrenburg, M. J., Todt, T. J., van Hijum, S. A., Hugenholtz, J., & Bron, P. A. *Isolation of a heat-resistant subpopulation of* Lactococcus lactis *SK11 with enhanced spray drying robustness.* Submitted for publication.

Douillard, F. P., & de Vos, W. M. (2014). Functional genomics of lactic acid bacteria: From food to health. *Microbial Cell Factories*, *13*(Suppl 1), S8. doi:10.1186/1475-2859-13-S1-S8 PMID:25186768

Duwat, P., Ehrlich, S. D., & Gruss, A. (1999). Effects of metabolic flux on stress response pathways in *Lactococcus lactis*. *Molecular Microbiology*, *31*(3), 845–858. doi:10.1046/j.1365-2958.1999.01222.x PMID:10048028

Kok, J., Buist, G., Zomer, A. L., van Hijum, S. A., & Kuipers, O. P. (2005). Comparative and functional genomics of lactococci. *FEMS Microbiology Reviews*, *29*(3), 411–433. doi:10.1016/j.fmrre.2005.04.004 PMID:15936843

Laroute, V., Tormo, H., Couderc, C., Mercier-Bonin, M., Le Bourgeois, P., Cocaign-Bousquet, M., & Daveran-Mingot, M. L. (2017). From Genome to Phenotype: An Integrative Approach to Evaluate the Biodiversity of Lactococcus lactis. *Microorganisms*, 5. PMID:28534821

Leroy, F., & De Vuyst, L. (2004). Lactic acid bacteria as functional starter cultures for the food fermentation industry. *Trends in Food Science & Technology*, *15*(2), 67–78. doi:10.1016/j.tifs.2003.09.004

Makarova, K., Slesarev, A., Wolf, Y., Sorokin, A., Mirkin, B., Koonin, E., ... Mills, D. (2006). Comparative genomics of the lactic acid bacteria. *Proceedings of the National Academy of Sciences of the United States of America, 103*(42), 15611–15616. doi:10.1073/pnas.0607117103 PMID:17030793

McGovern, P. E., Zhang, J., Tang, J., Zhang, Z., Hall, G. R., Moreau, R. A., ... Wang, C. (2004). Fermented beverages of pre- and proto-historic China. *Proceedings of the National Academy of Sciences of the United States of America, 101*(51), 17593–17598. doi:10.1073/pnas.0407921102 PMID:15590771

Meijerink, M., van Hemert, S., Taverne, N., Wels, M., de Vos, P., Bron, P. A., ... Wells, J. M. (2010). Identification of genetic loci in *Lactobacillus plantarum* that modulate the immune response of dendritic cells using comparative genome hybridization. *PLoS One*, *5*(5), e10632. doi:10.1371/journal.pone.0010632 PMID:20498715

Mulder, J., Wels, M., Kuipers, O. P., Kleerebezem, M., & Bron, P. A. (2017). Unleashing natural competence in *Lactococcus lactis* by induction of the competence regulator ComX. *Applied and Environmental Microbiology*, *83*(20), e01320-17. doi:10.1128/AEM.01320-17 PMID:28778888

O'Sullivan, E., & Condon, S. (1997). Intracellular pH is a major factor in the induction of tolerance to acid and other stresses in *Lactococcus lactis. Applied and Environmental Microbiology*, *63*, 4210–4215. PMID:9361406

Papadimitriou, K., Alegria, A., Bron, P. A., de Angelis, M., Gobbetti, M., Kleerebezem, M., ... Kok, J. (2016). Stress Physiology of Lactic Acid Bacteria. *Microbiology and Molecular Biology Reviews*, *80*(3), 837–890. doi:10.1128/MMBR.00076-15 PMID:27466284

Pretzer, G., Snel, J., Molenaar, D., Wiersma, A., Bron, P. A., Lambert, J., ... Kleerebezem, M. (2005). Biodiversity-based identification and functional characterization of the mannose-specific adhesin of *Lactobacillus plantarum. Journal of Bacteriology*, *187*(17), 6128–6136. doi:10.1128/JB.187.17.6128-6136.2005 PMID:16109954

Sander, J. D., & Joung, J. K. (2014). CRISPR-Cas systems for editing, regulating and targeting genomes. *Nature Biotechnology*, *32*(4), 347–355. doi:10.1038/nbt.2842 PMID:24584096

Santivarangkna, C., Kulozik, U., & Foerst, P. (2007). Alternative drying processes for the industrial preservation of lactic acid starter cultures. *Biotechnology Progress*, *23*(2), 302–315. doi:10.1021/bp060268f PMID:17305363

Santivarangkna, C., Kulozik, U., & Foerst, P. (2008). Inactivation mechanisms of lactic acid starter cultures preserved by drying processes. *Journal of Applied Microbiology*, *105*(1), 1–13. doi:10.1111/j.1365-2672.2008.03744.x PMID:18266696

Siezen, R. J., Bayjanov, J., Renckens, B., Wels, M., van Hijum, S. A., Molenaar, D., & van Hylckama Vlieg, J. E. (2010). Complete genome sequence of *Lactococcus lactis* subsp. *lactis* KF147, a plant-associated lactic acid bacterium. *Journal of Bacteriology*, *192*(10), 2649–2650. doi:10.1128/JB.00276-10 PMID:20348266

Siezen, R. J., Bayjanov, J. R., Felis, G. E., van der Sijde, M. R., Starrenburg, M., Molenaar, D., ... van Hylckama Vlieg, J. E. (2011). Genome-scale diversity and niche adaptation analysis of *Lactococcus lactis* by comparative genome hybridization using multi-strain arrays. *Microbial Biotechnology*, *4*(3), 383–402. doi:10.1111/j.1751-7915.2011.00247.x PMID:21338475

Silva, J., Freixo, R., Gibbs, P., & Teixeira, P. (2011). Spray-drying for the production of dried cultures. *International Journal of Dairy Technology*, *64*(3), 321–335. doi:10.1111/j.1471-0307.2011.00677.x

Smid, E. J., Erkus, O., Spus, M., Wolkers-Rooijackers, J. C., Alexeeva, S., & Kleerebezem, M. (2014). Functional implications of the microbial community structure of undefined mesophilic starter cultures. *Microbial Cell Factories*, *13*(Suppl 1), S2. doi:10.1186/1475-2859-13-S1-S2 PMID:25185941

Smid, E. J., & Hugenholtz, J. (2010). Functional genomics for food fermentation processes. *Annual Review of Food Science and Technology*, *1*(1), 497–519. doi:10.1146/annurev.food.102308.124143 PMID:22129346

Smith, W. M. (2010). Molecular Mechanisms of Stress Resistance in Lactococcus lactis. Current Research, Technology and Education Topics in Applied Microbiology and Microbial Biotechnology, 2, 1106-1118.

Solopova, A., Bachmann, H., Teusink, B., Kok, J., Neves, A. R., & Kuipers, O. P. (2012). A specific mutation in the promoter region of the silent cel cluster accounts for the appearance of lactose-utilizing *Lactococcus lactis* MG1363. *Applied and Environmental Microbiology*, *78*(16), 5612–5621. doi:10.1128/AEM.00455-12 PMID:22660716

van Bokhorst-van de Veen, H., Lee, I. C., Marco, M. L., Wels, M., Bron, P. A., & Kleerebezem, M. (2012). Modulation of *Lactobacillus plantarum* gastrointestinal robustness by fermentation conditions enables identification of bacterial robustness markers. *PLoS One*, *7*(7), e39053. doi:10.1371/journal.pone.0039053 PMID:22802934

van Bokhorst-van de Veen, H., van Swam, I., Wels, M., Bron, P. A., & Kleerebezem, M. (2012). Congruent Strain Specific Intestinal Persistence of *Lactobacillus plantarum* in an Intestine-Mimicking In Vitro System and in Human Volunteers. *PLoS One*, *7*(9), e44588. doi:10.1371/journal.pone.0044588 PMID:22970257

Wegmann, U., O'Connell-Motherway, M., Zomer, A., Buist, G., Shearman, C., Canchaya, C., ... Kok, J. (2007). Complete genome sequence of the prototype lactic acid bacterium *Lactococcus lactis* subsp. *cremoris* MG1363. *Journal of Bacteriology*, *189*(8), 3256–3270. doi:10.1128/JB.01768-06 PMID:17307855

Whitaker, R. D., & Batt, C. A. (1991). Characterization of the Heat Shock Response in *Lactococcus lactis* subsp. *lactis*. *Applied and Environmental Microbiology*, *57*, 1408–1412. PMID:16348482

Chapter 8
Probiotic Microorganisms and Encapsulation Method Approaches

Seydi Yıkmış
Namık Kemal University, Turkey

Harun Aksu
Istanbul University, Turkey

Mehmet Alpaslan
Namık Kemal University, Turkey

Osman Şimşek
Namık Kemal University, Turkey

ABSTRACT

Nowadays, interest in probiotics, which are useful and necessary for healthy life, is rapidly increasing, and studies on the beneficial effects of probiotics on human health continue intensely. Every year, increasing efforts to prevent cancer, which has been anticipated, has increased the interest in probiotics and therefore synbiotics. Encapsulation methods are one of the most important protection methods currently used to ensure the viability of probiotics and their effectiveness. Especially milk and dairy products are used for many purposes such as increasing the shelf life, increasing the nutritional value, providing digestibility, shortening the ripening period, improving taste and aroma substances. The use of the microencapsulation technique alone can improve probiotic vitality. Combining microencapsulation with various food processing technologies is thought to help improve the vitality of probiotics in production and storage. In this chapter, probiotic microorganisms and encapsulation applications are explored.

DOI: 10.4018/978-1-5225-5363-2.ch008

INTRODUCTION

Probiotics and Prebiotics

Probiotics, which are meant for life in Greek, can be described as living microorganisms that balance the intestinal flora when taken in certain amounts with nutrients and affect the host's health positively. Another name given to probiotics is 'biogenics' (Holzapfel & Schillinger, 2002).

Definitions used in the literature for probiotics:

- The probiotic term, consisting of two parts, "Pro" and "biota", means "for life" and is the antithesis of antibiotics (Patterson & Burkholder, 2003).
- Microorganisms added to foods that affect the host physiology positively by organizing the microorganism and nutritional balance in the intestines, and regulating systemic and mucosal immunity (Coşkun, 2006).
- Probiotics, nonpathogenic-living organisms, are able to reach to the colon since they are resistant to digestion against to the acidic environments of stomach. Thus it can positively affect the human health by influencing microbiota.
- Probiotics are live-food additives that affect the balance of the intestinal microorganisms of the host. (Watson, Preedy, Monedero, & Rodríguez-Díaz, 2016).
- Apart from simple nutritive properties, they are living microorganisms that are beneficial to the host's health when taken in sufficient quantities (Shah & Lankaputhra, 1997).
- Probiotics are organisms and substances contributing to the intestinal microorganism balance (Watson, Preedy, Likotrafiti, & Rhoades, 2016).

Prebiotics are indigestible components that increase the number and activities of beneficial bacteria living in the gut and the effect of probiotics. The classical definition of the prebiotic is that non-digestible nutrient components which have impacts on the host's health positively by increasing the number and type of bacteria in the colon. This term was first used by Gibson and Roberfroid in 1995. However, the emergence of prebiotics dates back to the 1950s. György and colleagues described a bifidogenic factor that selectively stimulates the proliferation of bifidobacteria. (Gyorgy, Mello, Torres, & Barness, 1953). Prebiotic food is a food product that contains a prebiotic component. The prebiotic components are oligosaccharides or polysaccharides which are mainly found in the carbohydrate group and generally function as soluble fibers. Prebiotics usually are found in the form of fructose and galactose polymers (Coşkun, 2006). There are four main groups of common prebiotics commonly used in Europe: inulin, fructooligosaccharides, lactulose, galactooligosaccharides. The most common oligosaccharides added to foods are fructooligosaccharides, galactooligosaccharides and polydextrose. Prebiotics affect the composition and activity of intestinal microbiota positively, by regulating intestinal motility, increasing the absorption and bioavailability of minerals, such as calcium and magnesium, and preventing the proliferation of pathogenic microorganisms. In order to define a food component as aprebiotic, it must have the following properties:

- Must be resistant to digestion,
- Must be hydrolyzed by bacteria present in the colon,

- Stimulate the growth of beneficial bacteria in the colon
- There must be positive effects on the host's health. (Agostoni et al., 2004; Manning & Gibson, 2004).

The positive effects of probiotic microorganisms on human health were first introduced by Nobel Prize-winning Russian researcher, Elie Metchnikoff in 1908. Years ago, Metchnikoff realized that Bulgarian villagers had lived longer. When he investigated the reaoson for that, he found that they consumed plenty of yoğurt. When he examined the yoğurt, he found live bacteria on it and he called it as *Lactobacillus bulgaricus* (Taşdemir, 2017). Probiotics plays an important role in the gastrointestinal system physiology. These non-pathogenic live microorganisms are using to support the immune system, to regulate the intestinal system, to improve digestion and absorbsion of food components including vitamins and minerals, to inhibit pathogenic bacteria and viruses, to prevent tumor formation, to prevent diarrhea, to reduce lactose tolerance (Tamime & Marshall, 1997; Vinderola, Prosello, Ghiberto, & Reinheimer, 2000).

In order to consider a microorganism as a probiotic, the following properties need to be evaluated.

- To have non-invasive and non-carcinogenic properties for the host,
- To be human origin,
- To be resistant to stomach acid and bile,
- To attach to the intestinal epithelium cell,
- To have clinical effects on health,
- To colonize temporary in the gastrointestinal tract,
- To be able to adapt to natural flora,
- To be suitable for experimental studies (a stable strain, able to survive, culturally obtainable, resistant to oxygen, lyophilized),
- To be able to produce antimicrobial substances (bacteriocins, hydrogen peroxide etc) (Coşkun, 2005; Fioramonti, Theodorou, & Bueno, 2003).

The most studied microorganizms considered as probiotics in the worldwide are *Lactobacillus* (*Lactobacillus acidophilus, L. casei, L. buchneri, L. fermentum, L. delbrueckii* subsp. *bulgaricus, L. plantarum, L. reuteri, L. johnsonii, L. rhamnosus, L. salivarius, L. crispatus, L. curvatus*), *Bifidobacterium* (*B. bifidum, B. adolescentis, B. animalis, B. infantis, B. longum* and *B. breve*), *Enterococcus* (*E. faecium and E. faecalis) and Streptococcus, S. intermedius* and *S. diacetylactis*) (Doğan, 2012; Duwat, Cesselin, Sourice, & Gruss, 2000).

The products formed by the use of probiotics and prebiotics are called synbiotics. With synbiotic application, probiotic bacteria can survive longer and can proliferate better in the colon. The aim of the using synbiotic products is to increase the beneficial effects of bacteria on host's health. *In vitro* study findings support this idea that using the synbiotic applications have more advantages than either prebiotic or probiotics alone (Bruzzese, Volpicelli, Squaglia, Tartaglione, & Guarino, 2006; Ewaschuk & Dieleman, 2006).

MECHANISMS OF PROBIOTICS

There are many mechanisms explains how probiotics have protective effects on intestinal system including as follows:

- İnhibition of proliferation of pathogenic bacteria by producing antimicrobial substances,
- Blocking attachment reagons of pathogenic bacteri in the intestinal system,
- Competition for the nutrients,
- Destruction of toxin receptors,
- Functional stimulation of the immune system (Rastall et al., 2005).

ANTIBACTERIAL EFFECT

Probiotic bacteria strains secrete substances including hydrogen peroxide, organic acid, bacteriocin, inhibit the proliferation of pathogenic microorganisms. Hydrolytic enzymes, such as β-galactosidases increases the production of free and short chain fatty acids, lactic acid, propionic acid and butyric acid. These changes cause decreases in pH in the intestinal lümen and eventually reduces in the number of pathogenic bacteria. Thus, they create a favorable environment for the proliferation of beneficial microorganisms (Ouwehand, Kirjavainen, Shortt, & Salminen, 1999; Servin & Coconnier, 2003). Probiotics reduce the absorption of toxic substances such as indole, amine, and ammonia from the intestines, or limit the toxicity of these toxic substances in products. Probiotic bacteria also have the ability to synthesize metabolites that can neutralize some bacterial toxins (Guerin-Danan et al., 1998). Probiotics compete with pathogenic bacteria for mucus layer and limited number of epithelial cells. At the same time, probiotics inhibit the growth of pathogenic bacteria by consuming the nutrients they need to breed bacteria. They produce antimicrobial substances such as hydrogen peroxide, organic acid and bacteriocin. They can alter intestinal microflora by regulating fatty acid profile in the colon. Probiotic strains have been found to convert linoleic acid into antiinflammatory and anticarcinogenic conjugated linoleic acid (Morita et al., 2002).

ADHESIVE MECHANISM

It is thought that probiotics form a barrier against the pathogenic microorganisms in the intestinal system, which reduces the degree of epithelial cell attachment with these microorganisms. Lactic acid bacteria have various surface determinants that allow adhesion to intestinal epithelial cells. Microbial depletion of lactic acid bacteria is associated with passive forces, electrostatic relationships, hydrophobic and steric forces, specific structures covered by lipoteichoic acid and lectins (Servin & Coconnier, 2003). In the adhesion study with *Lactobacillus gasseri*, it has been found that the binding of proteins and carbohydrates is necessary for the binding of divalent cations (Ca + 2). *Lactobacilli* are composed of different combinations of proteins and carbohydrates on the surface of the bacteria that bind to the human intestine. mechanism. The binding of *bifidobacteria* to proteins is mediated by a compound in the protein structure. It is shown that adhesive-like proteins of *bifidobacteria* are species-specific (Bernet, Brassart, Neeser, & Servin, 1993; Önal, Beyatlı, & Aslım, 2005).

EFFECTS ON THE IMMUNE SYSTEM

Probiotic bacteria such as *Bifidobacter* and *Lactobacillus* have been proven to reduce redox potential in the intestinal tract by studies. Probiotic bacteria metabolize carbohydrates and proteins by transferring substrates. They produce electron-accepting metabolites in the absence of oxygen at the phosphorylation site. The fermentation is effected by forming the resulting acetate and lactate (Chichlowski, 2006; Nayir, 2008). The effects of probiotics on the immune system include enhancing antibody production and activity of natural killer (NK) cells, modulating the nuclear factor kappa-B (NFkB) pathway and inducing T cell apoptosis. Probiotics generally induce intestinal anti-inflammatory cytokine production such as interleukin-10 (IL-10) and transforming growth factor-beta (TGF-beta), while tumor necrosis factor-alpha (TNF-α), interferon- (IL-8), which inhibit proinflammatory cytokine production (Maassen et al., 2000). Bacterial flora plays a role in the reduction of Th2 activity and in the formation of oral tolerance. It has been shown that CpG motifs in bacterial DNA increase Th1 differentiation, polyclonal B cell activation, antigen specification IgA production, interleukin (IL) -12 and interferon (IFN) -gamma secretion. The immunomodulator stimulates the production of cytokines IL-10 and TGF-beta stimuli. It also activates monocytes and macrophages (20-24). The formation and continuation of intestinal and systemic tolerance depends on the characteristic structure of the intestinal tract and immunomodulatory cytokines (IL-10, TGF-beta). Microorganisms found in intestinal microflora contribute positively to the effect of these cells (von der Weid, Bulliard, & Schiffrin, 2001).

ENCAPSULATION

The encapsulation; can be defined as packing solid, liquid or gaseous food components, enzymes, cells and other substances, microorganisms, protein or carbohydrate based coating material. Apart from food applications, encapsulation is also used in chemistry, agriculture, feed, medicine, pharmacy, veterinary medicine, biotechnology. The use of the encapsulation process in food additives is common (Cho, Shim, & Park, 2003; Gouin, 2004; Madene, Jacquot, Scher, & Desobry, 2006; Poncelet, n.d.) . Microencapsulation is a technique in which capsules (microcapsules) having a size in the range of micrometers to millimeters are wrapped with one or more coating materials (wall material) of the periphery of an active material (core material). The microcapsules are simply spherical, with a homogeneous wall around them. The substance or mixture contained in the microcapsule is expressed as a core, an internal phase or a filler, whereas the outer wall is called a shell, a coating, a wall material, or a membrane (Gharsallaoui, Roudaut, Chambin, Voilley, & Saurel, 2007).

The basic principles of encapsulation technologies are divided into two:

1. The active ingredient is mixed in a polymer solution and dispersed as fine droplets, the droplets are solidified by techniques such as gelation, drying, cooling, coevaporation.
2. The solid powder particles are mixed in the fluidized bed, the coating solution is applied by spraying, solidified by drying or cooling.

Various techniques can be used to coat the active ingredient in the encapsulation. These techniques can be examined under three headings:

1. Physical processes: spray drying, extrusion, fluidized bed coating and spray cooling
2. Physicochemical processes: simple or complex coacervation and trapping in liposomes
3. Chemical processes: Interface polymerisation and molecular inclusion (Quintanilla-Carvajal et al., 2010).

Depending on the particle size of the encapsulated products nanoparticles, microcapsules and micro-reactors can be defined. The use of microencapsules, food products, makes food more healthful, tasty and more usable. In the food industry, it can be listed the benefits of microencapsulated materials as follows;

- Easy transport of active material
- İmprove stability in the final process and during the process (less evaporation of volatile substances, less degradation or reaction with components such as oxygen or water)
- İnactivation of active substance in food processing systems
- Developing safety (reducing burning of volatile components such as flavor)
- Adjustable properties of active ingredients (particle size, structure, oil or water solubility, color)
- Taste masking
- Controlled release (access to target)
- Creation of visible and structural effects (Desai & Jin Park, 2005).

Encapsulation may be divided into sub-varieties based on size of the droplets formed. These are;

1. Nanoconjugation (200 nm = less than 0.2 μm); different liquids that are absolutely insoluble within each other, that is, heterogeneous phases, are droplets of multiple phase nano size, created by physical difficulty (Rao & McClements, 2011).
2. Macroencapsulation (greater than 5 μm); Macroencapsulation is the encapsulation of a large mass of islets in a diffusion ring (M. Qi et al., 2004).
3. Microencapsulation (0.2-5 μm); Microencapsulation; Is a technology that is formed by wrapping single or multiple coating materials on the active material. This technique, which is used to improve the specific properties of the products and to extend their shelf life, is preferred in the food industry in order to coat the liquid droplets, solid particles or gas components with the coating materials in food purity (Koç, Sakin, & Kaymak-Ertekin, 2010).

In encapsulation technology, many active substances can be protected by encapsulation for their intended use. The most commonly used active ingredients in encapsulation applications are:

- Vitamins and minerals
- Enzymes and proteins
- Organic acids
- Probiotics and prebiotics
- Essential oils
- Sweeteners, preservatives, colorants, aromas
- Fatty acids (ω-3, conjugated linoleic acid)
- Carotenoids (β-carotene, lycopene)

- Antioxidants (Tocopherols, flavonoids, polyphenols) (Ardebili, Pecht, Ardebili, & Pecht, 2009; Botelho, Canas, & Lameiras, 2017; Oliver & Augustin, 2009)(Ardebili, Pecht, Ardebili, & Pecht, 2009; Botelho, Canas, & Lameiras, 2017; Oliver & Augustin, 2009).

ENCAPSULATION METHODS

Spray Drying

Spray drying is one of the oldest methods of encapsulating active materials (Barbosa-Cánovas, 2005). The active material to be coated is kept in a protective polymer material and the dispersion of this material in the coating solution is prepared. The active material and the wall material are sprayed towards the drying chamber and the heated air is removed by centrifugation to allow the sample to be dried (Jiménez-Martín, Gharsallaoui, Pérez-Palacios, Ruiz Carrascal, & Antequera Rojas, 2015). In the spray drying process it is intended to provide microbiological stability of the products, to prevent water and water activity, to prevent chemical and / or microbiological deterioration, to reduce storage and transportation costs and to protect the specific properties of the products. Spray drying technology is described as the most commonly used encapsulation method because of its easy equipment foundation and the fact that the production costs are lower than many of the meta-traits (Gökmen, Palamutoğlu, & Sariçoban, 2012).

Polysaccharides such as alginate, carboxymethyl cellulose, guar gum and chitosan can also be used as coating materials in spray drying methods such as sodium caseinate, whey protein, soy protein, but their low water solubility and cost are disadvantageous for this method (Gouin, 2004). The size of the obtained capsules is 10-100 micrometers. The disadvantages of this method are that the variety of water-soluble coating material used is limited and also cause the deterioration of the thermosensitive bioactive substances at high temperature (Fang & Bhandari, 2010).

Carriers are natural gums (gum arabic, alginates, carrageenans), proteins (milk proteins, soy proteins, gelatins), carbohydrates (maltodextrin and cellulose derivatives) and oils (waxes and 14 emulsifiers). The most commonly used coating material in microencapsulation by spray drying is the arabic gamma or the acacia gamma. Capsules obtained by conventional spray drying release the active ingredients as soon as they come into contact with the water. Spray drying is the most common method of microencapsulation of oils and aromas (Carneiro, Tonon, Grosso, & Hubinger, 2013).

Coacervation

The coacervation method is phase separation from one or more hydrocolloid starting solutions and the newly formed coazervate phase is suspended or emulsified in the same reaction medium around the active ingredient. In this method, not only a colloid (such as gelatin) substance can be used, but also gelatin and gamma acacia (arabic) combination can be used in more complex processes (Fang & Bhandari, 2010). Coacervation is often used in the encapsulation of aroma oils, but it is also a useful tool for fish oils, vitamins, preservatives and enzymes. In addition to having advantages over other encapsulation methods, this method has some problems, such as the high cost of commercialization, the complicated process, and the necessity of using glutaraldehyde, which is restricted in legal regulation of many countries as secondary crustacean material (Gouin, 2004).

Creatıng Emulsion

The emulsion is obtained by kinetically dispersing two phases (oil and water) which are stable and not intermixing, in the form of very small droplets of one of the liquids in the other. High-speed mixers and homogenizers are used in emulsification. Water-soluble ingredients may be encapsulated with water / oil emulsions or double emulsions such as water / oil / water. Oil / water emulsions affect the taste by changing the volume of the aqueous phase and thus the concentration of the taste molecules in the water. Lycopene, beta-carotene, sterols and oil-soluble vitamins can be added in powder form by drying in oil / water emulsions or by spray drying, freeze drying methods (Zuidam & Shimoni, 2010).

The diameter of the obtained droplets is generally within the range of 0.1 - 100µm. Systems in which oil droplets are dispersed in the aqueous phase are called oil (oil / water) emulsions and systems in which water droplets are dispersed in the oil phase are called water (water / oil) emulsions. In addition to these, it is possible to obtain them in multilayer emulsions. In addition, some emulsifiers may be used to enhance the stability of the emulsions (Fang & Bhandari, 2010).

Freeze Drying

Freeze drying is also a method used in the dehydration of thermosensitive components and aromas, also known as lyophilization. The sample is frozen at -90 to -40 °C and dried by sublimation under low pressure with low pressure (Zuidam & Shimoni, 2010).

Freeze drying takes place in three stages; Freezing, basic drying phase and second drying phase. During the ice cream phase; Transforming the fresh water into ice crystals in freezing or freezing, in the basic drying phase; The removal of ice crystals from the product by sublimation, and the removal of the bound water in the food during the second drying phase. Advantages of freeze drying method; The loss of aroma is very low, the reproducibility of the product obtained is very good, and the losses are minimal due to the movement of the dissolving substances in the food. However, there are disadvantages such as high cost and long operation time. As mentioned above, microencapsulation process in food industry is usually carried out by using spray drying method because of low production cost. However, the freeze drying method, which uses low temperatures in the microencapsulation of food products sensitive to temperature (such as fish oil, flavoring substances), may be an alternative (Heinzelmann, Franke, Jensen, & Haahr, 2000).

Dripping Gelation (Ionic Gelation) Method

Dripping or ionic gelation is a practical and economical method. However, it is preferred in laboratory applications because it is not suitable for product production in very large quantities. In this method, the active ingredient is homogenized by means of the polymer solution and the dispersing agent (such as CaCl 2) is instilled by syringe to obtain spherical gel particles. The method of dripping is a method of obtaining calcium alginate particles wherein the crosslinking is based on the principle of sodium in the sodium alginate molecule with the calcium in the CaCl2 molecule.

Gelation formation of the biopolymer solution is observed by dripping to another solution. In addition to alginate, microparticles can also be obtained by dropping the pectin solution into the calcium solution (ionic gelation), by dropping the chitosan solution into the tripolyphosphate solution (ionic gelation), by dropping the whey protein into hot liquid (hot gelation) or by dropping the gelatin solution into cold liquid (cold gelation) (Matalanis, Jones, & McClements, 2011; Shi, He, Teh, Morsi, & Goh, 2011).

It has been reported that some technologies such as the syringe-extrusion system developed in recent years (providing air jet, electrical, potential, vibrating units and alginate flow) make it possible to commercialize alginate particles containing encapsulated food ingredients by producing high amounts of alginate particles (Gouin, 2004).

Fluid Bed Coating

Fluid bed coating, It was discovered by D.E.Wurster in the 1950s and is also known as the Wurster process. The pharmaceutical industry has been using the fluid bed coating method for the purpose of film formation, flavor masking, enhancing the stability of the obtained products and acting in the desired region for a long time since the coating of medicines (Jyothi et al., 2010).

The solid particles to be encapsulated are delivered from the lower part to the coating chamber with heated or cooled air and transported to the upper part of the chamber by the air in motion. Particles carried to the top of the oven are gathered toward the walls of the room for low airflow and return. The coating material dissolved in a dissolved or evaporated solvent is atomized to form a haze cloud in the air chamber, and the coating of the suspended particles in the air is ensured. With the help of the airflow supplied from the bottom of the oven, a fairly uniform coating of particles moving upwards is ensured. The controlled constant airflow of temperature and moisture content allows the evaporation of the solvent from the molten lipid coating material or the dissolved coating material that corresponds to the warm and hard particles. With the aid of the small porous filters located on the top of the oven, the small particles are returned to the air suspension and a further coating is obtained. The final product from the air chamber is warmed and passed through a final drying process before packaging (Barbosa-Cánovas, 2005; Guignon, Duquenoy, & Dumoulin, 2002).

The fluidized bed coating method is used to encapsulate nutrients such as vitamin C, B vitamins, iron sulfate, iron fumarate, sodium ascorbate, potassium chloride and various vitamins / minerals, which are used as nutritional supplements in the food industry. In the meat industry, various food acids are encapsulated with a fluid bed technique to improve color and aroma (DeZarn, 1995; Nedovic, Kalusevic, Manojlovic, Levic, & Bugarski, 2011).

Liposomal Dispensing Method

Liposomes are used in the pharmaceutical, personal care, chemical and food industries to encapsulate both hydrophobic and hydrophilic compounds; are bilayer spherical lipid vesicles composed of polar lipids. Polar lipids are dispersed in a polar environment such as water (Desai & Jin Park, 2005; Teschke & Souza, 2002). Due to its lipid and aqueous phases, liposomes are used for encapsulation and release of water-soluble, oil-soluble and amphilic materials. The main advantages of their use are that the encapsulated material can be stored up to the appropriate time and the rate of release can be controlled. Bioactive agents encapsulated in liposomes can be absorbed from the gastrointestinal tract in such a way that they can be protected from the digestive tract and can provide bioactivity and bioavailability in the intestines (Jafari, Assadpoor, He, & Bhandari, 2008; Sultana et al., 2000). It has been observed that compounds such as antimicrobial, antioxidant and vitamins which are encapsulated in the studies are more resistant to processes such as pasteurization, drying, cooking, frying (Barenholz, 2001; Mozafari et al., 2006).

Coating Materials Used in Encapsulation

The choice of coating materials varies according to the size and shape of the capsule, its stability and permeability properties. The composition of the coating is the main determinant of the method used to enhance the performance of the particulate ingredients and the functional properties of the microcapsules (Barbosa-Cánovas, 2005). There are microcapsules in different shapes, single, multi-wall, irregular, multicore and matrix. The first step in the microencapsulation process is to select the appropriate coating material (capsule). As the coating materials, substances such as gelatin, pectin, starch, kappa-carrageenan, agar, whey, etc. which can form a film are used, such as sugars, gums, proteins, natural and modified polysaccharides, oils or synthetic polymers (Dubey & Dubey, 2009; W. Qi et al., 2006; Sultana et al., 2000).

The carrier system should have the following characteristics;

- Effectively encapsulate and trap the functional agent of the desired amount.
- It should be compatible with the surrounding food or beverage matrix, should not cause a negative effect on product appearance, rheology, taste and shelf life.
- Be able to control the release of functional agent
- Be resistant to environmental influences that may occur during storage, handling, handling and processing.
- Functional agent protected from chemical impairments in order to remain active
- It should be prepared using low cost process applications with the ingredients in the generally accepted as safe (GRAS) status.
- Should not adversely affect the bioavailability of encapsulated material (Desai & Jin Park, 2005).

CURRENT ENCAPSULATION APPLICATIONS IN DAIRY TECHNOLOGY

In recent years, a system has been developed and implemented in order to encapsulate the substrate with the desired properties for the development of taste and aroma at the desired levels in the cheese. The microcapsules obtained in this system are added before the supercoagulation to ensure that flavor and aroma forming products are formed in the capsules during ripening of the cheese.

By combining two types (endopeptidase and exopeptidase) of enzymes, the microencapsulation technique and the reduction of the maturation period, the maturation time was reduced and the storage costs could be reduced. These applications, which are realized with the help of microencapsulation technique and which give positive results with the use of encapsulated enzymes, have been further developed and started to be used in almost all types of cheese today (Kınık, Kavas, & Yılmaz, 2003). In cheese ripening, yeast, natural enzymes of the milk, starter bacteria and enzymes, and non-starter bacteria play a role. The trapped enzymes in the microcapsules are physically separated from the substrate in the bulk and during the clot construction phase, the enzyme in the serum is released only by cleavage of the capsule and performs the maturation process (Anjani, Kailasapathy, & Phillips, 2007). In a study conducted, microencapsulation was applied to the enzyme aminopeptidase added to accelerate the ripening of cheddar cheese. Alginate-chitosan-calcium chloride (1.6-0.1-0.1% w / w) complex was chosen as the coating material for this. The flavor, aroma and textural values of cheese containing encapsulated enzyme were higher than the other samples. In addition, the total amount of free amino acid during storage was determined in samples containing most encapsulated enzymes, followed by free enzymes (Peker & Arslan, 2011).

A number of methods have been developed to remove these adverse effects on lactic acid bacteria during the production of ice cream and frozen products. The most important of these methods is to arrest and encapsulate lactobacilli in a simple and inexpensive method of calcium alginate gel (Kınık et al., 2003).

Probiotic microorganisms are trying to preserve their vitality for a long time in frozen products. However, the cell membranes of probiotics are damaged by the mechanical stress of the ice crystals formed outside or inside the cell during freezing, and therefore they can lose their vitality (Akın, Akın, & Kırmacı, 2007). In one study, the effects of microencapsulation on two probiotic bacteria (*Lactobacillus casei* (Lc-01) and *Bifidobacterium lactis* (Bb-12)) added to freezing were investigated. The product containing microencapsulated probiotic bacteria was kept at -20 °C for 180 days and the survival rates of probiotics were monitored. This ratio was 5.1×10^9 and 4.1×10^9 cfu / ml on day 1 and 4.2×10^6 and 1.1×10^7 cfu / ml after 180 days, respectively. After probiotic bacteria were microencapsulated with calcium alginate, bacterial viability was monitored based on the same storage temperature and time, and over 30% of the previous values were counted as viable bacteria (Homayouni, Azizi, Ehsani, Yarmand, & Razavi, 2008).

L. acidophilus and *B. lactis* and β-D-galactosidase activity were monitored at -20 °C during a 24-week storage. The number of *L. acidophilus* decreased by 2.52 log for free cells after 24 weeks, while 2.06 log and 2.27 log decreased for the same encapsulated state, unfermented and fermented ice cream, respectively (Kailasapathy & Sultana, 2003).

In one study, the number of *L. acidophilus LA-5* free cells in fermented milk beverages with a pH value lower than 4.2 fell below 10^6 CFU / ml after 1 week of storage. A similar case was seen after 2 weeks in *Bifidobacteria* after storage. In the encapsulated cells, the viability of *L. acidophilus* and *Bifidobacterium* was found to be above 10^6 CFU / ml even after 42 days of storage (Mortazavian, Khosrokhvar, Rastegar, & Mortazaei, 2010).

Commercial *Lactobacillus rhamnosus* GG is stabilized in an emulsion-based formulation with whey protein and starch and has not found any difference in bacterial viability loss after spray drying or freeze-drying (Ying et al., 2010).

The viability of *L. acidophilus* and *Bifidobacteria* cells was enhanced by the addition of L-cystine, whey protein concentrate, acid casein hydrolysis and triptone. Casein and whey protein hydrolysates reduced the rate of development of probiotic bacteria *L. acidophilus La-5* and *L. rhamnosus Lr-35* in fermented milk during the production stages, but the viability of these bacteria increased during storage (Lucas, Sodini, Monnet, Jolivet, & Corrieu, 2004).

In one study, it was determined that Cheddar cheese was more effective than probiotic microorganisms *Lactobacilli* and *Enterococci* as food carriers in the digestive system (Lee & Salminen, 2009).

The microencapsulated 8 probiotic bacteria strains were compared in terms of acidity, bile salts and heat tolerance. Microencapsulated probiotic bacteria were found to survive longer when compared to free cells (control group). A decrease of 6.51 log CFU / ml was observed in the viability of the control group whereas a decrease of 3.36 log CFU / ml was found in the microencapsulated strains (Ding & Shah, 2007).

It was determined that the type and concentration of the agent (10 and 30 g/g) utilized and the fat percentage (25 and 75 g/g) utilized by the spray drying method were more than 90% for all systems in investigating the effects on microencapsulation efficiency and efficiency (Young, Sarda, & Rosenberg, 1993).

In one study, viable cell counts of *Lactobacillus bulgaricus* bacteria encapsulated with sodium alginate were investigated in frozen milky desserts. According to the findings, cells larger than 30 µ in diameter were found to be more viable than those in average 15 µ (Sheu, Marshall, & Heymann, 1993).

Polyglycerol monostearate (PGMS) was used as coating material in microencapsulation of the iron components. The sizes of the microcapsules are between 2-5 µm. All samples were lost iron during storage. At higher temperatures (20 °C), more iron is separated from microcapsules than at lower temperatures (4 °C) (Kwak, K. M. Yang, & Ahn, 2003).

A wide variety of microencapsulation techniques are used in the food industry. These techniques are effectively employed in production and development of functional foods. In addition, these technologies also promote the successful transport of bioactive components into the digestive tract (Champagne & Fustier, 2007). Bacteria protected by encapsulation have a protection rate of 80-95% in yoghurt. During the encapsulation process, the bacterial cells are not damaged much and the enzymatic activities are not inhibited. It has been determined in recent years that immobilization of probiotic bacteria and prebiotic substances together in capsule form further increases the resistance of probiotic cells to environmental factors (Peker & Arslan, 2011).

Microencapsulated *Lactobacillus casei* examined the effect of different alginate concentrations (2%, 3% and 4%) on the viability of the NCDC-298 strain. As the alginate concentration increased, the resistance of the bacteria to gastrointestinal conditions was increased and it was found to have a positive effect on viability (Mandal, Puniya, & Singh, 2006).

Emulsion encapsulation has increased the antibacterial effect of the environment in the enhancement of the antibacterial effect of isoeugenol by emulsion encapsulation. Encapsulation increased antibacterial efficacy in carrot juice, but not in milk. It was determined that the chitosan coating of the emulsions caused aggregation (Krogsgård Nielsen et al., 2016).

The level of viability of the artificial gastrointestinal system and the colonies of *L. acidophilus* and *B. bifidum* cultures coated with calcium alginate-corn starch mixture was investigated. As a result of the research, they found that the use of corn starch increased the viability of bacteria in the coating comparedto those coated without starch; Found that the coating process had no significant effect in increasing the viability of the bacteria in high acid and bile salt environments (Sultana et al., 2000).

In one study, *Lactobacillus delbrueckii subsp. bulgaricus* and *Streptococcus thermophilus*, yogurt starter cultures were used encapsulated in alginate-taro starch matrix at different levels. No significant difference was found between the applications in terms of sensory qualities. However, yogurt produced from 2-weekly stored 4% alginate-1.5% taro starch encapsulated lactic acid bacteria was found to be acceptable due to similarity with correct sweetness grade and acidity grade and control procedure. It is believed that taro starch can in fact be used in conjunction with alginate to encapsulate yogurt starter cultures and can protect or enhance the lactic acid bacterial cell viability without affecting the activity of the cultures (Tan, Dizon, Barraquio, & Mercado, 2016).

In the casein encapsulation study of fermentic acid ethyl ester to prevent non-flavor formation in UHT(Ultra High Temperature) milk. The encapsulated ferulic acid ethyl ester has affected the formation of 2-acetyl-2-thiazoline by 32.8-63.2% of the UHT stream after 30 days of storage at 30 °C at the level of 0.18-1.42 mg / mL. Therefore, caseinata encapsulated ferulic acid ethyl ester could potentially be used to improve the quality of life (Guan & Zhong, 2017).

In a study of microencapsulation, the study of the effect of *L. acidophilus* and *B. bifidum* on the viability of cassava cheese during maturation, *L. acidophilus* and *B. bifidum* kashar cheese, which were covered with the extrusion and emulsion techniques, and microbiological, biochemical and sensory properties of kashar cheese were investigated for 90 days. As a result of storage, the viability of probiotic cells was observed to be preserved to a large extent by microencapsulation (Özer, Kirmaci, Şenel, Atamer, & Hayaloğlu, 2009).

In the encapsulation study to promote the reduction of Maillard reaction products in UHT milk with ascorbic acid encapsulation, ascorbic acid was encapsulated with a lipid coating and the effects were tested after the application of a laboratory UHT. Encapsulation promoted delayed release of ascorbic acid and a reduction in the formation of maillard reaction products. It was found that the increase in total lysine uptake up to 45% in the pellet containing encapsulated ascorbic acid, while the decrease in Nε- (carboxymethyl) -L-lysine, Nε- (carboxyethyl) -L-lysine and furoisin decrease between 10 and 53% (Troise et al., 2016).

In one study, *Bifidobacterium bifidum* alginate, pectin and pestle water were coated with gel. Cells that did not undergo the plating process lose their vitality, whereas those in the plating process were able to maintain their viability up to 2 hours at pH 2.5 (Guérin, Vuillemard, & Subirade, 2003).

Lactobacillus sakei spp. found that encapsulation of bacteriocin in phosphatidylcholine and 1,2-dioleoyloxy-3-trimethylammonium-propane / phosphatidylcholine nanovesicles did not affect antimicrobial activity *in vitro* and *in situ* by *Listeria monocytogenes* inhibition by liposomal nanovesicles containing bacteriocins produced by *Lactobacilus sakei 2a* (Malheiros, Cuccovia, & Franco, 2016).

In studying the antilisterial activity of phosphatidylcholine nanospheres containing milk extract of garlic extract (encapsulated in phosphatidylcholine liposomes), nanoliposomes containing garlic extract inhibited various *Listeria spp* species. Milano nanotechnology suggests that it may be a useful strategy for the transport of natural antimicrobials in foodstuffs (Pinilla, Noreña, & Brandelli, 2017).

They found that the optimal combination of microencapsulation materials to be used in the microencapsulation of probiotics for gastrointestinal conditions is 3% sodium alginate mixed with the most suitable 1% peptide and 3% fructooligosaccharide (Chen, Chen, & Lin, 2006).

CONCLUSION

Probiotics have been incorporated into dietary systems other than traditional fermented dairy products, and many probiotic products are now available on the market shelves. Since the viability of probiotics during production and storage is influenced by various factors, it can be difficult to provide the amount of probiotic that can provide beneficial effects to consumers' health. Encapsulation techniques that have found a wide application areas for many years but have not found much use in the food industry have started to be used widely in the food industry in recent years. In this context, the use of microencapsulation techniques, cell-protective agents, the addition of food-promoting food components, oxygen-barrier packaging materials, modification of the antioxidants and storage medium significantly improve and protect the viability of microorganisms. The choice of appropriate encapsulation material is very important and determines the effectiveness of the process. In addition, there is a growing interest in the use of synbiotic products today. When the probiotics reach the intestinal tract, the prebiotics they contain use probiotic bacteria to retain their viability and retain their colons. The use of the microencapsulation technique alone can improve probiotic vitality. Combining microencapsulation with various food processing technologies will help improve the vitality of probiotics in production and storage. Health benefits of probiotics isolated from natural fermented products should be supported by clinical trials. As a result of all these achievements, consumers have come to the conclusion that it is possible to offer products that are more nutritious, reliable, and have a long shelf life.

REFERENCES

Agostoni, C., Axelsson, I., Goulet, O., Koletzko, B., Michaelsen, K. F., & Puntis, J. W. L. ESPGHAN Committee on Nutrition. (2004). Prebiotic oligosaccharides in dietetic products for infants: a commentary by the ESPGHAN Committee on Nutrition. *Journal of Pediatric Gastroenterology and Nutrition, 39*(5), 465–73. Retrieved from http://www.ncbi.nlm.nih.gov/pubmed/15572882

Anjani, K., Kailasapathy, K., & Phillips, M. (2007). Microencapsulation of enzymes for potential application in acceleration of cheese ripening. *International Dairy Journal, 17*(1), 79–86. doi:10.1016/j.idairyj.2006.01.005

Ardebili, H., Pecht, M. G., Ardebili, H., & Pecht, M. G. (2009). Encapsulation Process Technology. In Encapsulation Technologies for Electronic Applications (pp. 129–179). Academic Press. doi:10.1016/B978-0-8155-1576-0.50007-3

Barbosa-Cánovas, G. V. (2005). *Food powders : physical properties, processing, and functionality.* Kluwer Academic/Plenum Publishers.

Barenholz, Y. (2001). Liposome application: Problems and prospects. *Current Opinion in Colloid & Interface Science, 6*(1), 66–77. doi:10.1016/S1359-0294(00)00090-X

Bernet, M. F., Brassart, D., Neeser, J. R., & Servin, A. L. (1993). Adhesion of human bifidobacterial strains to cultured human intestinal epithelial cells and inhibition of enteropathogen-cell interactions. *Applied and Environmental Microbiology, 59*(12), 4121–4128. Retrieved from http://www.ncbi.nlm.nih.gov/pubmed/8285709 PMID:8285709

Botelho, G., Canas, S., & Lameiras, J. (2017). Development of phenolic compounds encapsulation techniques as a major challenge for food industry and for health and nutrition fields. In Nutrient Delivery (pp. 535–586). Academic Press. doi:10.1016/B978-0-12-804304-2.00014-7

Bruzzese, E., Volpicelli, M., Squaglia, M., Tartaglione, A., & Guarino, A. (2006). Impact of prebiotics on human health. *Digestive and Liver Disease, 38*, 283–287. doi:10.1016/S1590-8658(07)60011-5 PMID:17259092

Carneiro, H. C. F., Tonon, R. V., Grosso, C. R. F., & Hubinger, M. D. (2013). Encapsulation efficiency and oxidative stability of flaxseed oil microencapsulated by spray drying using different combinations of wall materials. *Journal of Food Engineering, 115*(4), 443–451. doi:10.1016/j.jfoodeng.2012.03.033

Champagne, C. P., & Fustier, P. (2007). Microencapsulation for the improved delivery of bioactive compounds into foods. *Current Opinion in Biotechnology, 18*(2), 184–190. doi:10.1016/j.copbio.2007.03.001 PMID:17368017

Chen, K.-N., Chen, M.-J., & Lin, C.-W. (2006). Optimal combination of the encapsulating materials for probiotic microcapsules and its experimental verification (R1). *Journal of Food Engineering, 76*(3), 313–320. doi:10.1016/j.jfoodeng.2005.05.036

Chichlowski, M. (2006). *Effect of probiotic consortium on level and mechanism of intestine function.* Academic Press.

Cho, Y. H., Shim, H. K., & Park, J. (2003). Encapsulation of Fish Oil by an Enzymatic Gelation Process Using Transglutaminase Cross-linked Proteins. *Journal of Food Science, 68*(9), 2717–2723. doi:10.1111/j.1365-2621.2003.tb05794.x

Coşkun, T. (2005). Fonksiyonel besinlerin sağlığımız üzerine etkileri. *Çocuk Sagligi ve Hastaliklari Dergisi, 48*, 69–84. Retrieved from http://www.cshd.org.tr/uploads/pdf_CSH_133.pdf

Coşkun, T. (2006). Pro-, Pre- ve Sinbiyotikler. *Çocuk Sagligi ve Hastaliklari Dergisi, 49*, 128–148. Retrieved from http://www.cshd.org.tr/uploads/pdf_CSH_204.pdf

Desai, K. G. H., & Jin Park, H. (2005). Recent Developments in Microencapsulation of Food Ingredients. *Drying Technology, 23*(7), 1361–1394. doi:10.1081/DRT-200063478

DeZarn, T. J. (1995). *Food ingredient encapsulation: an overview* (S. J. Risch & G. A. Reineccius, Eds.). Washington, DC: American Chemical Society. doi:10.1021/bk-1995-0590.ch007

Doğan, M. (2012). Probiyotik Bakterilerin Gastrointestinal Sistemdeki Etki Mekanizması. *Gıda Teknolojileri Elektronik Dergisi, 7*(1), 20–27.

Dubey, R., & Dubey, R. (2009). Microencapsulation Technology and Applications. *Defence Science Journal, 59*(1), 82–95. doi:10.14429/dsj.59.1489

Duwat, P., Cesselin, B., Sourice, S., & Gruss, A. (2000). Lactococcus lactis, a bacterial model for stress responses and survival. *International Journal of Food Microbiology, 55*(1–3), 83–86. doi:10.1016/S0168-1605(00)00179-3 PMID:10791722

Ewaschuk, J. B., & Dieleman, L. A. (2006). Probiotics and prebiotics in chronic inflammatory bowel diseases. *World Journal of Gastroenterology, 12*(37), 5941–5950. doi:10.3748/wjg.v12.i37.5941 PMID:17009391

Fang, Z., & Bhandari, B. (2010). Encapsulation of polyphenols – a review. *Trends in Food Science & Technology, 21*(10), 510–523. doi:10.1016/j.tifs.2010.08.003

Fioramonti, J., Theodorou, V., & Bueno, L. (2003). Probiotics: What are they? What are their effects on gut physiology? *Best Practice & Research. Clinical Gastroenterology, 17*(5), 711–724. doi:10.1016/S1521-6918(03)00075-1 PMID:14507583

Gharsallaoui, A., Roudaut, G., Chambin, O., Voilley, A., & Saurel, R. (2007). Applications of spray-drying in microencapsulation of food ingredients: An overview. *Food Research International, 40*(9), 1107–1121. doi:10.1016/j.foodres.2007.07.004

Gökmen, S., Palamutoğlu, R., & Sariçoban, C. (2012). Gıda Endüstrisinde Enkapsülasyon Uygulamaları. *Electronic Journal of Food Technologies Gıda Teknolojileri Elektronik Dergisi Electronic Journal of Food Technologies, 7*(71), 36–5036. Retrieved from http://www.teknolojikarastirmalar.com/pdf/tr/02_12_7_1_121_723.pdf

Gouin, S. (2004). Microencapsulation. *Trends in Food Science & Technology, 15*(7–8), 330–347. doi:10.1016/j.tifs.2003.10.005

Guan, Y., & Zhong, Q. (2017). Encapsulation of ferulic acid ethyl ester in caseinate to suppress off-flavor formation in UHT milk. *Food Chemistry*, *237*, 532–537. doi:10.1016/j.foodchem.2017.05.140 PMID:28764031

Guérin, D., Vuillemard, J.-C., & Subirade, M. (2003). Protection of bifidobacteria encapsulated in polysaccharide-protein gel beads against gastric juice and bile. *Journal of Food Protection*, *66*(11), 2076–2084. doi:10.4315/0362-028X-66.11.2076 PMID:14627286

Guerin-Danan, C., Chabanet, C., Pedone, C., Popot, F., Vaissade, P., Bouley, C., … Andrieux, C. (1998). Milk fermented with yogurt cultures and Lactobacillus casei compared with yogurt and gelled milk: influence on intestinal microflora in healthy infants. *The American Journal of Clinical Nutrition, 67*(1), 111–7. Retrieved from http://www.ncbi.nlm.nih.gov/pubmed/9440384

Guignon, B., Duquenoy, A., & Dumoulin, E. D. (2002). Fluid Bed Encapsulation of Particles: Principles and Practice. *Drying Technology*, *20*(2), 419–447. doi:10.1081/DRT-120002550

Gyorgy, P., Mello, M. I., Torres, F. E., & Barness, L. A. (1953). Growth promotion in rats by crude concentrates of the bifidus factor. *Proceedings of the Society for Experimental Biology and Medicine. Society for Experimental Biology and Medicine, 84*(2), 464–7. Retrieved from http://www.ncbi.nlm.nih.gov/pubmed/13121074

Heinzelmann, K., Franke, K., Jensen, B., & Haahr, A.-M. (2000). Protection of fish oil from oxidation by microencapsulation using freeze-drying techniques. *European Journal of Lipid Science and Technology*, *102*(2), 114–121. doi:

Holzapfel, W. H., & Schillinger, U. (2002). Introduction to pre- and probiotics. *Food Research International*, *35*(2–3), 109–116. doi:10.1016/S0963-9969(01)00171-5

Homayouni, A., Azizi, A., Ehsani, M. R., Yarmand, M. S., & Razavi, S. H. (2008). Effect of microencapsulation and resistant starch on the probiotic survival and sensory properties of synbiotic ice cream. *Food Chemistry*, *111*(1), 50–55. doi:10.1016/j.foodchem.2008.03.036

Jafari, S. M., Assadpoor, E., He, Y., & Bhandari, B. (2008). Encapsulation Efficiency of Food Flavours and Oils during Spray Drying. *Drying Technology*, *26*(7), 816–835. doi:10.1080/07373930802135972

Jiménez-Martín, E., Gharsallaoui, A., Pérez-Palacios, T., Ruiz Carrascal, J., & Antequera Rojas, T. (2015). Volatile compounds and physicochemical characteristics during storage of microcapsules from different fish oil emulsions. *Food and Bioproducts Processing*, *96*, 52–64. doi:10.1016/j.fbp.2015.07.005

Jyothi, N. V. N., Prasanna, P. M., Sakarkar, S. N., Prabha, K. S., Ramaiah, P. S., & Srawan, G. Y. (2010). Microencapsulation techniques, factors influencing encapsulation efficiency. *Journal of Microencapsulation*, *27*(3), 187–197. doi:10.3109/02652040903131301 PMID:20406093

Kailasapathy, K., & Sultana, K. (2003). Survival and [beta]-D-galactosidase activity of encapsulated and free. *Journal of Dairy Technology*, *58*(223). Retrieved from https://search.proquest.com/docview/199499876/fulltextPDF/7DEDFF93D6784502PQ/1?accountid=25088

Koç, M., Sakin, M., & Kaymak-Ertekin, F. (2010). Mikroenkapsülasyon ve Gıda Teknolojisinde Kullanımı Microencapsulation and its Applications in Food Technology. *Pamukkale Üniversitesi Mühendislik Bilimleri Dergisi, 16*(1), 77–86. Retrieved from https://www.journalagent.com/pajes/pdfs/PAJES_16_1_77_86.pdf

Krogsgård Nielsen, C., Kjems, J., Mygind, T., Snabe, T., Schwarz, K., Serfert, Y., & Meyer, R. L. (2016). Enhancing the antibacterial efficacy of isoeugenol by emulsion encapsulation. *International Journal of Food Microbiology, 229*, 7–14. doi:10.1016/j.ijfoodmicro.2016.04.002 PMID:27089032

Kwak, H. S., K. M. Yang, & Ahn, J. (2003). *Microencapsulated Iron for Milk Fortification*. Academic Press. 10.1021/JF030199+

Lee, Y. K., Yuan, K., & Salminen, S. (2009). Handbook of probiotics and prebiotics (2nd ed.). John Wiley & Sons.

Lucas, A., Sodini, I., Monnet, C., Jolivet, P., & Corrieu, G. (2004). Probiotic cell counts and acidification in fermented milks supplemented with milk protein hydrolysates. *International Dairy Journal, 14*(1), 47–53. doi:10.1016/S0958-6946(03)00147-X

Maassen, C. B., van Holten-Neelen, C., Balk, F., den Bak-Glashouwer, M. J., Leer, R. J., Laman, J. D., … Claassen, E. (2000). Strain-dependent induction of cytokine profiles in the gut by orally administered Lactobacillus strains. *Vaccine, 18*(23), 2613–23. Retrieved from http://www.ncbi.nlm.nih.gov/pubmed/10775795

Madene, A., Jacquot, M., Scher, J., & Desobry, S. (2006). Flavour encapsulation and controlled release - a review. *International Journal of Food Science & Technology, 41*(1), 1–21. doi:10.1111/j.1365-2621.2005.00980.x

Malheiros, P. S., Cuccovia, I. M., & Franco, B. D. G. M. (2016). Inhibition of Listeria monocytogenes in vitro and in goat milk by liposomal nanovesicles containing bacteriocins produced by Lactobacillus sakei subsp. sakei 2a. *Food Control, 63*, 158–164. doi:10.1016/j.foodcont.2015.11.037

Mandal, S., Puniya, A. K., & Singh, K. (2006). Effect of alginate concentrations on survival of microencapsulated Lactobacillus casei NCDC-298. *International Dairy Journal, 16*(10), 1190–1195. doi:10.1016/j.idairyj.2005.10.005

Manning, T. S., & Gibson, G. R. (2004). Prebiotics. *Best Practice & Research. Clinical Gastroenterology, 18*(2), 287–298. doi:10.1016/j.bpg.2003.10.008 PMID:15123070

Matalanis, A., Jones, O. G., & McClements, D. J. (2011). Structured biopolymer-based delivery systems for encapsulation, protection, and release of lipophilic compounds. *Food Hydrocolloids, 25*(8), 1865–1880. doi:10.1016/j.foodhyd.2011.04.014

Morita, H., He, F., Fuse, T., Ouwehand, A. C., Hashimoto, H., & Hosoda, M. … Kurisaki, J. (2002). Adhesion of lactic acid bacteria to caco-2 cells and their effect on cytokine secretion. *Microbiology and Immunology, 46*(4), 293–7. Retrieved from http://www.ncbi.nlm.nih.gov/pubmed/12061632

Mortazavian, A. M., Khosrokhvar, R., Rastegar, H., & Mortazaei, G. R. (2010). Effects of dry matter standardization order on biochemical and microbiological characteristics of freshly made probiotic Doogh (Iranian fermented milk drink). *Italian Journal of Food Science, 22*, 98–102.

Mozafari, M. R., Flanagan, J., Matia-Merino, L., Awati, A., Omri, A., Suntres, Z. E., & Singh, H. (2006). Recent trends in the lipid-based nanoencapsulation of antioxidants and their role in foods. *Journal of the Science of Food and Agriculture, 86*(13), 2038–2045. doi:10.1002/jsfa.2576

Nayir, S. M. (2008). *Sütün yoğurda dönüşümü sırasında içerdiği feneolik antioksidan maddelere probiyotik bakteri etkisinin incelenmesi. Yüksek Lisans Tezi.* Trakya Üniversitesi. Retrieved from http://www.teknolojikarastirmalar.com/pdf/tr/02_2012_7_1_121_721.pdf

Nedovic, V., Kalusevic, A., Manojlovic, V., Levic, S., & Bugarski, B. (2011). An overview of encapsulation technologies for food applications. *Procedia Food Science, 1*, 1806–1815. doi:10.1016/j.profoo.2011.09.265

Oliver, C. M., & Augustin, M. A. (2009). Using dairy ingredients for encapsulation. In Dairy-Derived Ingredients (pp. 565–588). Academic Press. doi:10.1533/9781845697198.3.565

Önal, D., Beyatlı, Y., & Aslım, B. (2005). Probiyotik Bakterilerin Epitel Yüzeylere Yapışması. *Orlab On-Line Mikrobiyoloji Dergisi, 3*(9), 1–10. Retrieved from www.mikrobiyoloji.org/pdf/702050901.pdf

Ouwehand, A. C., Kirjavainen, P. V., Shortt, C., & Salminen, S. (1999). Probiotics: Mechanisms and established effects. *International Dairy Journal, 9*(1), 43–52. doi:10.1016/S0958-6946(99)00043-6

Özer, B., Kirmaci, H. A., Şenel, E., Atamer, M., & Hayaloğlu, A. (2009). Improving the viability of Bifidobacterium bifidum BB-12 and Lactobacillus acidophilus LA-5 in white-brined cheese by microencapsulation. *International Dairy Journal, 19*(1), 22–29. doi:10.1016/j.idairyj.2008.07.001

Patterson, J., & Burkholder, K. (2003). Application of prebiotics and probiotics in poultry production. *Poultry Science, 82*(4), 627–631. doi:10.1093/ps/82.4.627 PMID:12710484

Peker, H., & Arslan, S. (2011). Mikroenkapsülasyon ve süt teknolojisinde kullanım alanları. *Akademik Gıda, 9*(6), 70–80.

Pinilla, C. M. B., Noreña, C. P. Z., & Brandelli, A. (2017). Development and characterization of phosphatidylcholine nanovesicles, containing garlic extract, with antilisterial activity in milk. *Food Chemistry, 220*, 470–476. doi:10.1016/j.foodchem.2016.10.027 PMID:27855927

Poncelet, D. (n.d.). Microencapsulation: fundamentals, methods and applications. In Surface Chemistry in Biomedical and Environmental Science (pp. 23–34). Springer Netherlands. doi:10.1007/1-4020-4741-X_3

Qi, M., Gu, Y., Sakata, N., Kim, D., Shirouzu, Y., Yamamoto, C., ... Inoue, K. (2004). PVA hydrogel sheet macroencapsulation for the bioartificial pancreas. *Biomaterials, 25*(27), 5885–5892. doi:10.1016/j.biomaterials.2004.01.050 PMID:15172501

Qi, W., Ma, J., Yu, W., Xie, Y., Wang, W., & Ma, X. (2006). Behavior of microbial growth and metabolism in alginate–chitosan–alginate (ACA) microcapsules. *Enzyme and Microbial Technology, 38*(5), 697–704. doi:10.1016/j.enzmictec.2005.10.003

Quintanilla-Carvajal, M. X., Camacho-Díaz, B. H., Meraz-Torres, L. S., Chanona-Pérez, J. J., Alamilla-Beltrán, L., Jimenéz-Aparicio, A., & Gutiérrez-López, G. F. (2010). Nanoencapsulation: A New Trend in Food Engineering Processing. *Food Engineering Reviews, 2*(1), 39–50. doi:10.100712393-009-9012-6

Rao, J., & McClements, D. J. (2011). Food-grade microemulsions, nanoemulsions and emulsions: Fabrication from sucrose monopalmitate & lemon oil. *Food Hydrocolloids*, *25*(6), 1413–1423. doi:10.1016/j.foodhyd.2011.02.004

Rastall, R. A., Gibson, G. R., Gill, H. S., Guarner, F., Klaenhammer, T. R., Pot, B., ... Sanders, M. E. (2005). Modulation of the microbial ecology of the human colon by probiotics, prebiotics and synbiotics to enhance human health: An overview of enabling science and potential applications. *FEMS Microbiology Ecology*, *52*(2), 145–152. doi:10.1016/j.femsec.2005.01.003 PMID:16329901

Servin, A. L., & Coconnier, M.-H. (2003). Adhesion of probiotic strains to the intestinal mucosa and interaction with pathogens. *Best Practice & Research. Clinical Gastroenterology*, *17*(5), 741–754. doi:10.1016/S1521-6918(03)00052-0 PMID:14507585

Shah, N. P., & Lankaputhra, W. E. V. (1997). Improving viability of Lactobacillus acidophilus and Bifidobacterium spp. in yogurt. *International Dairy Journal*, *7*(5), 349–356. doi:10.1016/S0958-6946(97)00023-X

Sheu, T.-Y., Marshall, R. T., & Heymann, H. (1993). Improving Survival of Culture Bacteria in Frozen Desserts by Microentrapment. *Journal of Dairy Science*, *76*(7), 1902–1907. doi:10.3168/jds.S0022-0302(93)77523-2 PMID:8345127

Shi, P., He, P., Teh, T. K. H., Morsi, Y. S., & Goh, J. C. H. (2011). Parametric analysis of shape changes of alginate beads. *Powder Technology*, *210*(1), 60–66. doi:10.1016/j.powtec.2011.02.023

Sultana, K., Godward, G., Reynolds, N., Arumugaswamy, R., Peiris, P., & Kailasapathy, K. (2000). Encapsulation of probiotic bacteria with alginate–starch and evaluation of survival in simulated gastrointestinal conditions and in yoghurt. *International Journal of Food Microbiology*, *62*(1–2), 47–55. doi:10.1016/S0168-1605(00)00380-9 PMID:11139021

Tamime, A. Y., & Marshall, V. M. E. (1997). Microbiology and technology of fermented milks. In B. A. Law (Ed.), *Microbiology and Biochemistry of Cheese and Fermented Milk* (pp. 153–192). Boston, MA: Springer US; doi:10.1007/978-1-4613-1121-8_3

Tan, H. A. D., Dizon, E. I., Barraquio, V. L., & Mercado, S. M. (2016). Efficacy of Alginate-Taro (Colocasia esculenta L. Schott) Starch for Yoghurt Starter Culture Encapsulation. *Journal of Science, Engineering and Technology*, *4*(0), 1–10. Retrieved from http://ijterm.org/index.php/jset/article/view/159

Taşdemir, A. (2017). Probiyotikler, Prebiyotikler ve Sinbiyotikler. *Kastamonu Sağlık Akademisi, 2*(1), 71–88. Retrieved from http://dergipark.gov.tr/download/article-file/305079

Teschke, O., & de Souza, E. F. (2002). Liposome Structure Imaging by Atomic Force Microscopy: Verification of Improved Liposome Stability during Adsorption of Multiple Aggregated Vesicles. *Langmuir*, *18*(17), 6513–6520. doi:10.1021/la025689v

Troise, A. D., Vitiello, D., Tsang, C., Fiore, A., Glomb, M. A., Fogliano, V., ... Eisenbrand, G. (2016). Encapsulation of ascorbic acid promotes the reduction of Maillard reaction products in UHT milk. *Food & Function*, *7*(6), 2591–2602. doi:10.1039/C6FO00151C PMID:27240727

Vinderola, C. G., Prosello, W., Ghiberto, T. D., & Reinheimer, J. A. (2000). Viability of probiotic (Bifidobacterium, Lactobacillus acidophilus and Lactobacillus casei) and nonprobiotic microflora in Argentinian Fresco cheese. *Journal of Dairy Science*, *83*(9), 1905–1911. doi:10.3168/jds.S0022-0302(00)75065-X PMID:11003217

von der Weid, T., Bulliard, C., & Schiffrin, E. J. (2001). Induction by a Lactic Acid Bacterium of a Population of CD4+ T Cells with Low Proliferative Capacity That Produce Transforming Growth Factor and Interleukin-10. *Clinical and Vaccine Immunology; CVI*, *8*(4), 695–701. doi:10.1128/CDLI.8.4.695-701.2001 PMID:11427413

Watson, R. S., Preedy, V. R., Likotrafiti, E., & Rhoades, J. (2016). *Probiotics, Prebiotics, Synbiotics, and Foodborne Illness*. Probiotics, Prebiotics, and Synbiotics. doi:10.1016/B978-0-12-802189-7.00032-0

Watson, R. S., Preedy, V. R., Monedero, V., & Rodríguez-Díaz, J. (2016). Intestinal Microbiota and Susceptibility to Viral Infections: Role of Probiotics. In Probiotics, Prebiotics, and Synbiotics (pp. 813–826). Academic Press. doi:10.1016/B978-0-12-802189-7.00062-9

Ying, D. Y., Phoon, M. C., Sanguansri, L., Weerakkody, R., Burgar, I., & Augustin, M. A. (2010). Microencapsulated Lactobacillus rhamnosus GG Powders: Relationship of Powder Physical Properties to Probiotic Survival during Storage. *Journal of Food Science*, *75*(9), E588–E595. doi:10.1111/j.1750-3841.2010.01838.x PMID:21535593

Young, S. L., Sarda, X., & Rosenberg, M. (1993). Microencapsulating Properties of Whey Proteins. 1. Microencapsulation of Anhydrous Milk Fat. *Journal of Dairy Science*, *76*(10), 2868–2877. doi:10.3168/jds.S0022-0302(93)77625-0

Zuidam, N. J., & Shimoni, E. (2010). Overview of Microencapsulates for Use in Food Products or Processes and Methods to Make Them. In *Encapsulation Technologies for Active Food Ingredients and Food Processing* (pp. 3–29). New York, NY: Springer New York; doi:10.1007/978-1-4419-1008-0_2

Chapter 9
Transglutaminase Applications in Dairy Technology

H. Ceren Akal
Ankara University, Turkey

Celalettin Koçak
Ankara University, Turkey

H. Barbaros Özer
Ankara University, Turkey

ABSTRACT

Consumers' expectations from a dairy product have changed dramatically during the last two decades. People are now more eager to purchase more nutritious dairy foods with improved sensory characteristics. Dairy industry has made many efforts to meet such expectations and numerious production strategies and alternatives have been developed over the years including non-thermal processing, membrane applications, enzymatic modifications of milk components, and so on. Among these novel approaches, transglutaminase (TG)-mediated modifications of milk proteins have become fairly popular and such modifications in dairy proteins offer many advantages to the dairy industry. Since late 1980s, a great number of researches have been done on TG applications in milk and dairy products. Especially, milk proteins-based edible films and gels from milk treated with TG have found many application fields at industrial level. This chapter reviews the characteristics of microbial-origin TG as well as its mode of action and recent developments in TG applications in dairy technology.

INTRODUCTION

Enzymatic modifications of food proteins to create a tailor-made end product have been of interest to food industry and food scientists for many decades (Gerard, 2002; Jaros, et al., 2006a; Buchert et al., 2010). During the last two decades, efforts to optimize the yield, shelf-life and quality characteristics of foods by enzymatic modifications of food ingredients have been intensified. These efforts have led to introduction of a large number of commercial enzymes into food industry. Among those enzymes,

DOI: 10.4018/978-1-5225-5363-2.ch009

transglutaminase (TGase; protein-glutamine γ-glutamyltransferase, EC 2.3.2.13) has been widely investigated by researchers and its commercial forms have reached a great market success (Kuraishi et al., 2001). TGase was identified about 50 years ago in guinee pig liver as cytoplasmic TGase 2 (Sarkar et al., 1957). TGase is an enzyme naturally present in most animal tissues and body fluids (Sharma et al., 2001). It has also been discovered in microorganisms including *Streptoverticillium mobaraense, S. ladakanum, S. cinanmoneum, Physaru polysepharum* and *Bacillus subtilis* (Ando et al., 1989; Klein et al., 1992; Tsai et al., 1996; Duran et al., 1998; Kobayashi et al., 1998). TGase extracted from tissues or body fluids of animals (*i.e*., cattle, swine and fish) have been investigated for their suitability for food productions at industrial scale. Factor XIII, a type of TGase extracted from blood of cattle and swine at slaughter, was found to be unsuitable for food applications since this enzyme needs thrombin for its activation. In the presence of thrombin red pigmentation often occurs and limits animal origin TGase for food applications (Motoki & Seguro, 1998). With the improvements in transgenesis procedures, mass production of microbial TGase with low cost and controlled enzyme activity is now possible (Lerner & Matthias, 2015). For this purpose, host microorganisms used were *Eschericia coli* (guinea pig liver TGase, *Streptoverticillium* TGase, fish TGase) (Ikura et al., 1980; Takehana et al., 1994; Yasueda et al., 1995), yeasts (human factor XIIIa) (Bishop et al., 1990) and *Streptomyces* spp. (*Streptoverticillium* TGase) (Washizu et al., 1994). In 1980s, guniee pig liver enzyme was widely investigated for the feasibility of modifying food proteins for industrial applications (Motoki & Nio, 1983; Motoki et al., 1984; 1986, 1987a, 1987b, Nio et al., 1985, 1986). Although these studies yielded promising results, limited supply of guinee pig liver and the social and technical barriers against its use in food applications have encouraged researches to find alternative sources. A great deal of research has been dedicated to screen enzyme-producing strains of microorganisms to find out natural and suitable source(s) of TGase. The TGase-like enzymes secreted by the microorganisms were screened for their ability to form glutamine-lysine bonding between food proteins, which is characteristic for TGase (Nonaka et al., 1989). Then the enzyme extracted from a variant of *Streptoverticillium mobaraense* was found to be suitable food applications (Washizu et al., 1994). Later, other microorganisms were also screened for production of microbial TGase (mTGase). Readers may refer to Kieliszek & Misiewicz (2014) for current list of microorganisms screened for mTGase production.

TGase catalyses an acyl transfer reaction between γ-carboxyamide groups of peptide-bound glutamine residues (acyl donor) and the primary amino groups glutamine and lysine residues (Liu & Damodaran, 1999). Ammonia is released as a by-product of this reaction and could be used as a marker to monitor the reaction (Kellerby et al., 2006). TGase is able to catalyse three different types of reactions depending on the reaction conditions. These are i) acyl transfer reaction, ii) cross-linking (polymerisation) reaction between glutamine and lysine and iii) deamidation (Zhu & Tramper, 2008; Han et al., 2009). These reactions lead to the formation of new intra- and intermolecular bonds between proteins, which further modify the structure and functionality of proteins, *i.e.* solubility, water holding capacity, emulsification capacity, rennetability, thermal stability, ethanol stability etc. (Dickinson, 1997; Lorenzen & Schlimme, 1998; Motoki & Seguro, 1998; Lorenzen, 2000; O'Sullivan et al., 2002; Huppertz & de Kruif, 2007a). mTGase is an extracellular enzyme of the class of transferases (Yokoyama et al., 2004) and show no requirement to cofactors (Macedo & Sato, 2005). Although mTGase may show some activity at pH 4 and 9, its pH optimum is between 5 and 8. The optimum activity temperature of mTGase is around 55 °C and the enzyme losses its activity at 70 °C for a few minutes (Yokoyama et al., 2004). TGase is accepted as GRAS by FDA since 1998 and food aid by EFSA and is widely used in dairy, meat and bakery industries (Romeih & Walker, 2017). This chapter will only focus on the dairy applications of mTGase.

MODE OF ACTION OF TGASE ON MILK PROTEINS

The polymerization of milk proteins by mTGase has been well studied (Sharma et al., 2001; O'Sullivan et al., 2002; Rodriguez-Nogales, 2006). The caseins are good substrate for TGase due to their open structure, being more pronounced for κ-casein due to its location near the outside of the micelle (Rodriguez-Nogales, 2005; de Kruif et al., 2002). On the other hand, depending on the source of TGase, the reactivity of κ-casein towards the enzyme may vary. For example, susceptibility of κ-casein for a guniea pig liver TGase was found to be lower than that of α_{s1}-casein and β-casein (Christensen et al. 1996). However, κ-casein was the most susceptible casein fraction to crosslinking by mTGase in unheated milk (Sharma et al., 2001). If the milk was preheated (at 85 °C for 15 min), the susceptibility of κ-casein to mTGase remained comparable. This was further attributed to high reactivity of caseinomacropeptide (CMP) part (hydrophilic part) of κ-casein to mTGase (Tolkach & Kulozik, 2005). Sharma et al. (2001) showed that, in unheated milk, TGase had a small effect on individual milk proteins in skim milk. On contrast, in preheated skim milk, TGase-mediated cross-linking, deamidation and/or amine incorporation were observed and rapid crosslinking was evident during the first 30 min of reaction. Same authors also reported that monomeric forms of both κ-casein and β-casein decreased in the presence of TGase indicating more susceptibility of these two casein fractions to TGase action. Overall, the protein structure is a determinative factor for TGase-triggered cross-linking of caseins and Na-caseinate is more prone to crosslinking by TGase than native caseins (Bönisch et al., 2004). Ikura et al. (1992) demonstrated that cross-linking of α_s- and β-casein by TGase did not affect the functional properties of caseins. On the other hand, solubility of crosslinked α_s-casein showed a pH-dependency and was higher than native casein at pHs ranging from 4 to 6. The solubility of κ-casein was not affected by crosslinking. The heat-stability of casein micelle cross-linked by TGase improved due probably to the cross-linking of polyelectrolyte brush on micelle surface, which is critical for micelle dissociation (Huppertz & de Kruif, 2007a). In an earlier study, Mizuno et al. (1999) found that the glass transition temperature of casein was elevated by mTGase. Micellar calcium phosphate plays a critical role in maintaining the integrity of TGase-treated casein micelle. TGase-treated casein micelle free from micellar calcium phosphate had higher level of sedimentation, as reported by Moon et al. (2009). This was linked to excessive intra-micellar TGase-triggered cross-linking of casein molecules which led to the formation of sedimentable covalently bond caseins. On the other hand, Huppertz & de Kruif (2008) found that the integrity of enzymatically modified structure of casein micelle was not influenced by the removal of micellar calcium phosphate. TGase-triggered cross-linking of casein micelles results in casein nanogel particles containing covalently linked casein network. Nanogels (1-1000 nm) may have potential in drug delivery as well as coatings and composite manufacture. Crosslinking by TGase also limits the mobility of casein micelle (Mounsey et al., 2005).

The globular whey proteins, unless denatured, are poor substrates for TGase (Traorè & Meunier, 1992; Schorsch et al., 2000; Sakamoto et al., 1994; Christensen et al., 1996; Han & Damodaran, 1996; Sharma et al., 2002; Gauche et al., 2008). Protein polymerization triggered by TGase increases linearly with increasing degree of whey protein denaturation by heat (Lorenzen, 2000). Faergemand et al. (1997) showed that whey proteins require a reductant, such as dithioerythritol or pH-induced structural change, to facilitate the cross-link reaction by TGase. However, some researchers have demonstrated that β-lactoglobulin in its native state may well be affected by the TGase (Sharma et al., 2001; Nieuwenhuizen et al., 2004; Rodriguez-Nogales, 2006). Overall, if the whey proteins are subjected to mTGase in single-protein systems, dimers and polymers may occur without being structurally modified. However, if the whey proteins are in multi-protein systems like milk, especially β-lactoglobulin remains intact unless denatured. Han & Damodaran (1996) showed that due to thermodynamic incompatibility of β-lactoglobulin and β-casein, these two proteins cannot simultaneously be bound at the active site of TGase in milk system. Modifications of lysine and glutamine residues of α-lactalbumin by TGase show a strong structure- and temperature-dependency. For example, at 30 °C, no modification in lysine and glutamine residues was observed in holo α-lactalbumin, but modifications in lysine 13, 16, 108 and 114, and glutamine 39 and 43 were evident in apo α-lactalbumin (Niewenhuizen et al., 2003). On contrary, at 50 °C, four lysine residues were modified by TGase in holo α-lactalbumin.

Although the susceptibility of casein fractions to TGase is high in model systems, in milk system this susceptibility may be fairly low. While, in heat treated milk, the casein polimerization by TGase was observed, in unheated milk this reaction was almost absent (Sharma et al., 2001; Mautner et al., 1999). This situation was attributed to the presence of indigenous TGase inhibitors in milk (De Jong et al., 2002, 2003). The inhibitor described by De Jong et al. (2002) had a molecular mass of 200-250 Da and affected TGase independently from the source of enzyme (*i.e.*, microbial, guinea pigs or human plasma) and is easily eliminated by heat treatment. Pre-heating conditions for optimal TGase-mediated protein crosslinking were reported as >72 °C (Mautner et al., 1999) or 84,5 °C for 60 min (Rodriguez-Nogales, 2006). However, intense heat treatment to milk is sometimes undesirable for dairy technology. It is known that high heat treatment adversely affects the rennetability of milk. Therefore, strategies alternative to high heat treatment to milk were developed to increase TGase activity in milk. In an earlier approach, Miwa et al. (2002) added food grade glutathione (GSH) to raw milk and incubated milk at 40 °C for 3 h in the presence of mTGase at a level of 2 U/g protein. The authors showed that the susceptibility of especially κ-casein to TGase increased remarkably and concentration of high molecular weight compounds increased accordingly. The mechanism of GSH on micellar cross-linking by TGase was further investigated by Bönisch et al. (2007a). Authors demonstrated that GSH affected the interactions between TGase and TGase inhibitors and in the presence of GSH heat treatment beyond pasteurization norm (72 °C for 15 s) was not necessary. Other alternatives to increase the TGase activity in milk are removing inhibitors through membrane separation techniques (Bönisch et al., 2006) and UHT treatment of milk (Bönisch et al., 2004).

It is a well-established fact that Ca^{2+}-independent mTGase produced by *Streptoverticillum mobaraense* has a better ability to polymerase proteins than guinea pig liver TGase, possibly due to the higher thermostability (37 °C) of the former enzyme (Nonaka et al., 1996; Imm et al., 2000). While mTGase is completely independent from Ca^{+2}, guinea pig liver origin TGase generally has an absolute requirement for Ca^{+2} (Aeschlimann & Paulsson, 1994). Among the strong inhibitors of mTGase are Cu^{+2}, Zn^{+2}, Pb^{+2} and Li^{+} (Yokoyama et al., 2004). It is hypothesised that heavy metals bind the thiol groups of cysteine residue, which is a part of active site of TGase.

TGase-mediated protein modifications improve the gelling, renneting, foaming and emulsifying properties of milk proteins and decrease gel permeability. These modifications in functional properties of milk proteins allow development of dairy products with improved texture, yield and heat stability (Benjakul et al., 2008; Agyare et al., 2009; Giosafatto et al., 2012; Guo et al., 2013). TGase has an ability to alter the properties of casein micelles and protect or improve the integrity of the micelle against conditions at which micelle is normally dissociated, *i.e.* treatment with urea, sodium dodecyl sulfate, ethanol, sodium citrate or acidification (Lam et al., 2017). TGase also effectively improves the thermal stability of whey proteins at neutral pHs (Tanimoto & Kinsella, 1988; Wang, 2013; Zhong, et al., 2013). Agyare & Damodaran (2010) showed that the thermal stability (T_d) of β-lactoglobulin in TGase-treated whey protein isolate (WPI) was improved without any major structural changes but precipitation was observed at pH 4.0-4.5 due to a reduction in hydrophilic-hydrophobic balance of the protein surface as a result of elimination of the positive charge of lysyl residues. More compact structure of β-lactoglobulin results in higher thermal stability in this milk protein fraction (Tang & Ma, 2007). The modulation in thermal stability of proteins may be attributed to the partial unfolding of protein molecules and the subsequent re-arrangement of conformation. Modulation of functional properties of milk proteins including surface hydrophobicity, emulsifying activity and foam stability with fatty acids has been known for some years (Toledano & Magdassi, 1997). The fatty-acylation of proteins can be achieved either by chemically or enzymatically. TGase may be used to acylate the proteins and alter their functional properties. Whey proteins are among the most studied biopolymers because of their properties to form emulsions and nanoemulsions (Lee et al., 2011), nanofibrils (Loveday et al., 2011), hydrogels (Betz et al., 2012), and nanostructures (Chen & Subirade, 2006) for delivering nutraceuticals and pharmaceuticals. Turbidity in beverages containing whey protein isolates is the major challenge of beverage industry. Pre-treatment of whey protein isolate with mTGase may be an option to produce a transparent beverage with added salts such as sports drinks (Wang et al., 2012).

The effects of TGase on the gelation mechanisms and characteristics of acid gels have been studied in detail. Faergemand and Qvist (1997) demonstrated that skim milk gels treated with TGase had a finer protein network with thin strands between the particles than that of the untreated gels. Later, this finding was supported by Anema et al. (2005) who investigated the gel characteristics of heated, unheated, TGase-treated or high pressure applied skim milk formed through acidification by glucono-δ-lactone. Application of TGase alone or in combination with high pressure yielded higher storage modulus (G') values than the gels heated or unheated. This effect was attributed to the introduction in the numbers of intermolecular crosslinks. High hydrostatic pressure to casein micelle solution at 250 and 300 MPa resulted in an initial rapid increase in light transmission followed by partial reversal in light transmission due probably to re-association of micellar fragments (Huppetz & Smiddy, 2008). The disruption and possible re-association of casein micelle was partly prevented by employing TGase prior to high pressure application, as reported by Huppetz & Smiddy (2008). On the other hand, combination of high hydrostatic pressure (i.e. 400-600 MPa) with TGase-mediated cross-linking produced higher levels of protein oligomerisation in β-lactoglobulin gels, as reported by Lauber et al. (2001). Same authors showed that weak oligomerisation by TGase was evident when TGase was added after high pressure application. In general, acid-type gels are weaker when the rate of acidification is accelerated. Pre-treatment of milk with TGase prior to acidification causes a stiffer structure at lower rates of acidification but the degree of stiffness decreases when the acidification is faster (Jacob et al., 2011). Therefore, it is fair to assume that enzymatically cross-linked protein structure act as steric hindrances depending on the gelation time span. On the other hand, prolonged incubation time of acid gels treated with TGase or increased TGase concentration results in reduced gel firmness (Jaros et al., 2014). Jaros et al. (2006b, 2010) proposed that proper rearrangements of oligomerised proteins during acidification are restricted to some extent due to decreases in total bonds. Similar observations were made by Eissa & Khan (2006) who showed that the hydrophobic interactions in denatured whey protein were restricted by TGase. In addition, with the increase in aggregate size above a critical value, the proteins forming gel matrix becomes less flexible (Jaros et al., 2014). In TGase-treated acid casein gels, an adverse relationship between dimer size and gel stiffness at given isopeptide content were reported by Raak et al. (2017a). Recently, mechanical behaviours of casein gels cross-linked by TGase were comprehensively studied by Nieuwland et al. (2014). They showed that with the increase in cross-linking, gradual decrease in length scale of gel was evident. Same authors demonstrated that the elasticity of casein gels was determined at a length scale of casein micelle and fracture properties were determined at a length scale within casein micelle. At maximum casein gel stiffness, the amount of water entrapped in modulated gel matrix was minimum (Nieuwland et al., 2014). Stiffness of TGase-cross-linked acid casein gels is largely determined by degree of acidification. Below pI values, almost all κ- and α_s-casein were retained in the gel matrix cross-linked by TGase; however, these casein fractions were released in the untreated gels under the same conditions. This indicates that cross-linking of caseins retains charged molecules within the matrix, leading to more electrostatic interactions and eventually more protein interactions (Raak et al., 2017b). Raak et al. (2017b) also showed that uncross-linked β-casein plays a crucial role in preventing gel structure weakening in acid casein gels.

Depending on the gelation pH, temperature and TGase concentration, the gelation profile of caseinomacropeptide varies. At higher gelation temperature and pH (*i.e.* 90 ℃, pH 4.5) the concentration of TGase required for a proper gelation was around 1 U/g protein, whereas at lower gelation temperature and pH (*i.e.* 70 ℃, pH 3.0) this was around 20 U/g protein (Kocaman, 2016).

Chemically acidified sodium caseinate gels treated with TGase were reported to have more homogeneous structure and higher gel stability compared with their untreated counterparts (Myllärinen et al., 2007). Song & Zhao (2013) studied the impacts of oligochitosan glycosylation and cross-linking by TGase on rheological, gelling and emulsifying properties of sodium caseinate. In comparison with the caseinate, glycosylated and cross-linked caseinate had lower level of reactable amino groups, higher apparent viscosity and decreased emulsifying activity index. The gelling time of the resulting gel was also shorter than the untreated gels.

In an earlier study, Faergemand et al. (1997) demonstrated that TGase-catalysed cross-linking in milk gels chemically acidifed by glucono-δ-lactone resulted in stronger structures with lower permeability than untreated gels. Viscoelasticity of TGase-treated acid gels were reported to increase with the increase in enzyme dosages (Nonaka et al. 1992). On the other hand, at excessively high TGase dosages (i.e. 15 U/g protein), the gels became softer and fragile (Sakamoto et al., 1994). The breaking strength and elastic moduli of TGase-treated gels were higher than that of heat-set gels (Dickinson, 1997). In another study, the surface rheological characteristics of enzymatically cross-linked milk proteins were investigated and a dramatic increase in surface shear viscosity of caseinate from 4 $mN.m^{-1}$ to 1000 mNm^{-1} in a few hours of enzymatic reaction was noted (Faergemand et al., 1997). Similarly, the dilatational elasticities of caseinate and β-lactoglobulin gels were reported to increase (Faergemand & Murray, 1998). Recently, Ardelean et al. (2013) compared the characteristics of acid gels from cow's and goat's milk treated with TGase. They found that although the TGase-treated goat's milk gel was much weaker than that made from cow's milk, the improvement in gel stiffness by TGase was evident. Same authors postulated that enrichment of dry matter of goat's milk with cow's skim milk powder yielded gel characteristics similar to cow's milk gels treated under same conditions (Ardelean et al. 2012). Re-arrangement of protein network in acid gels at low pH values often leads to contraction of protein strands and eventually wheying off. In case of use of TGase in modification of acid milk gels, protein network is almost fixed at the early stages of gelation, limiting the network rearrangements at the later stages of gelation and cold storage (Ercili-Cura et al., 2013). This eventually results in less wheying off from the gel matrix.

As stated earlier, the rate of TGase reactions on milk proteins is largely determined by protein structure. Although caseins are good substrates for TGase, the rates of crosslinking of individual casein fractions vary due to differences in their conformation. Lysine is critical amino acid for TGase-triggered cross-linking reactions and β-,α_{s1}-, α_{s2}- and κ-caseins have different numbers of lysine residues (11, 14, 24, and 9 residues, respectively). In addition, β-casein has 21 glutamine residues, making this fraction more available for TGase (Swaisgood, 1982). It is well known that β- and α_{s1}-casein are good emulsifiers, being more pronounced in β-casein owing to its high steric stability provided by its charged N-terminal region (Dickinson, 1999; Macierzanka et al., 2011). Macierzanka et al. (2011) investigated the TGase-mediated crosslinking kinetics of sodium caseinate after emulsification in detail. They found that when sodium caseinate was adsorbed to an oil-in-water interface in an emulsion, the accessibilities of the casein fractions were changed, leading to reduced rates of cross-linking by TGase. TGase-triggered inter-droplet crosslinking leads to conversion of liquid emulsions to particulate gels (Chen et al., 2006). The spatial distribution of proteins in a gelled matrix plays a key role on the accessibility of enzyme (Grossmann et al., 2017). Zeeb et al. (2013) demonstrated that surface distances between single droplets were reduced due to close packing. There are two conditions for a stable emulsion gel based on TGase-mediated globular milk proteins: quick adsorbtion of unfolded proteins onto oil droplets and high susceptibility of unfolded proteins to TGase (Kang et al., 2003). The flexibility of proteins, particularly in relation to disulfide bond reduction, may be one of the important factors determining the properties of emulsion gel.

Faergemand et al. (1998) compared the emulsifying properties of milk proteins cross-linked with mTGase before and after emulsification. The authors found that creaming stability of the emulsion system in which proteins were enzymatically stabilized was increased. This was attributed to the modifications in adsorbed layer of emulsion or increased viscosity of the system. Increase in emulsion stability of TGase-treated β-casein was reported by Liu and Damodaran (1999). This was attributed to improved steric stability of the emulsion system resulting from branched nature of TGase-catalysed polymers, or increased cohesive interactions between polymerised β-casein molecules. Similarly, β-lactoglobulin-stabilized emulsion gels by TGase were found to have weaker frequency dependency and stronger strain dependency than that of heat-set β-lactoglobulin gels (Dickinson & Yamamoto, 1996). Whey protein isolate is best known with its low oxygen permeability and water vapour transmission rate essential for packaging of sensitive foods (Schmid et al., 2014). Packaging-related properties of whey films cross-linked with TGase were first studied by Schmid et al. (2014). The authors showed that the oxygen permeability of the enzymatically-modified whey protein isolate gels decreased by up to a factor of five, and water vapour transmission rate decreased by up to a factor two. This information is critical regarding replacement of petrochemical based barriers with whey protein based films.

Rheological, agregation and emulsification properties of whey protein isolate treated with TGase were reported to be improved and TGase-treated whey protein isolate gelified faster than untreated ones (Yildirim & Hettiarachchy, 1998; Qi et al., 2015). Here the ratio of enzyme to substrate is of critical importance in physical characteristics of TGase-treated WPI gels. Truong et al. (2004) showed that excessive intra- and interchain crosslinking formed by TGase could cause formation of polymers that were too large for effective network development. TGase may also be used in immobilized state in modification of properties of whey protein isolate-based gels. This method offers a number advantageous including elimination of a downstream inactivation process, recovery of enzyme and preventing excess cross-linking (Wilcox & Swaisgood, 2002). Wilcox et al. (2002) developed a system in which purification and immobilization of mTGase were achieved simultaneously which eliminates the heat inactivation step of enzyme activity after the reaction has ended. The system contains a combination of recombinant DNA and immobilized enzyme technologies. Immobilization of TGase was achieved by selective adsorption of the enzyme on biotinylated controlled pore-glass. Nanotechnology has been widely investigated for their potential use in production of nanoscale food materials with unique functional characteristics. Zhang

& Zhong (2009) used microemulsions as nanoreactor to produce nanoscale particles of whey proteins. When these nanoparticles were subjected to heat treatment, turbidity in the solution was evident, whereas the nanoparticles crosslinked with TGase gave a clear solution which makes it potentially available for whey protein-based beverage development.

In general, the effect of milk fat on the rate of cross-linking of caseins is negligible but TGase caused an increase in the stability of milk fat globule against coalescence (Hinz et al., 2007). On the other hand if the TGase-treated whole milk was homogenized, the distribution of milk proteins over the surface of homogenized fat globules was less than that of untreated milk (Hinz et al., 2007). This finding is in harmony with O'Sullivan et al. (2002) and Smiddy et al. (2006) who demonstrated that TGase-treated casein micelles were not affected by mechanical or chemical disruptions.

TGase may even further be used as a tool to add amino acids to the end product with improved nutritional value (Nonaka et al., 1996; Ali et al., 2010; Damodaran & Agyare, 2013). TGase may also be involved in reduction of antigenic properties of milk proteins, especially whey proteins. TGase was demonstrated to be able to incorporate glucosamine and oligochitosan into caseinate solution (Jiang & Zhao, 2011, 2012). In the presence of glucosamine and oligochitosan, the whey protein concentrate could be glycated and cross-linked under the catalytic action of TGase. The glycated and cross-linked products led to more open secondary structure of proteins leading to reduction in *in vitro* antigenicity (Zhang et al., 2016).

A number of studies have explored the potential applications of mTGase in dairy foods. Today mT-Gase is widely used in the manufacture of dairy-based edible films, yogurt, cheese, ice cream and other dairy products.

APPLICATION OF TGASE IN DAIRY BASED EDIBLE FILMS

Development of film formulations for food applications has long been of interest of food industry. Owing to its low cost and suitability to large scale productions, gelatin is widely used as a natural polymer for film formulations often in combination with a plasticiser such as glycerol (Vanin et al., 2005). However, it is not suitable for foods with high mositure levels due to high hygroscopic nature of gelatin (Gómez-Guillén et al., 2011). TGase may well be employed to improve the water resistance, mechanical, thermal, and morphological properties of gelatin films through cross-linking gelatin films (Kołodziewska & Piotrowska, 2007; Wang et al., 2015; Yi et al., 2006). TGase also yields a smoother and more compact gelatin film through modifying α-helix structure of gelatin especially at higher glycerol concentrations (Liu et al., 2017). Mixtures of casein and gelatin were also investigated for their suitability for production of edible film. Chambi & Grosso (2006) showed that casein:gelatin (75:25) mixture treated with TGase had the lowest water vapour permeability than the gels made either from casein or gelatin alone. Although the overall gel structures of the TGase-treated mixed gel were modified, the tensile strength of the resulting gel was not changed. Unlikely, the mechanical properties of the α_{s1}-casein gels treated with TGase increased remarkably (Motoki et al., 1987b). Treatment of mixture of whey proteins and chitosan (Di Pierro et al., 2006) and carboxymethylated chitosan (CMC) (in ratio of 75:25 or 50:50) (Jiang et al., 2016) with TGase resulted in improvement of mechanical and water vapour barrier properties of composite gels which may be used as an alternative edible packaging material. Whey protein-pectin films treated with TGase (8 U/ g of whey protein isolate) were reported to succeed to prevent conversion of foods such as biscuits from a glassy state to a rubbery state (Marquez et al., 2014). Oh et al. (2004)

investigated the possibility of incorporating zein hydrolysate into dairy-based edible films cross-linked with TGase. Combination of addition of zein hydrolysate and TGase treatment yielded a dairy-based gel with improved flexibility and reduced level of plasticiser without impairing the water vapour permeability.

APPLICATION OF TGASE IN YOGURT MANUFACTURE

Consumers demand towards reduced or non-fat yogurts has increased dramatically during the last two decades. Reducing or completely removing milk fat eventually leads to weakness in yogurt gel matrix, which is often criticised by the consumers. To overcome this handicap, various production strategies including addition of natural or synthesised gums, dairy-based powders or enzymatic modifications of milk proteins have been developed. Among these strategies, incorporation of mTGase seems to be the most promising application in regard to improving texture in the end product. Up to date, many studies have been conducted on mTGase application in yogurt and other fermented milk products (Faergemand & Qvist, 1997; Lauber et al., 2000; Lorenzen et al., 2002; Jaros et al., 2007; Özer et al., 2007; Domagała et al., 2013). The common conclusion drawn from these studies is that mTGase improves texture and water holding capacity of non- or reduced-fat yogurts by forming additional covalent bondings. The major outcomes of these studies are summarized in Table 1. Ercili-Cura et al. (2013) extensively studied the structural mechanism leading to improved water retention in acid milk gels by mTGase. Decreases in both tan δ values and backscattered light intensities of the mTGase-treated gels during acidification indicated that gel matrix was formed by smaller compartments, which were homogeneously distributed over the matrix. Authors also demonstrated that spontaneous whey syneresis at higher gelation temperatures (over 30 °C) was reduced by mTGase. Majority of protein cross-linking is completed at early stages of acidification leading to limiting further network rearrangements and contractions. This eventually limits the whey separation during the later stages of gel formation and probably cold storage (Ercili-Cura et al., 2013). This view was later supported by Darnay et al. (2015) who investigated the nanostructural changes in mTGase-treated low fat yogurt gels during fermentation stage by small angle neutron scattering (SANS). Authors showed that mTGase was capable of largely cross-linking milk proteins within the first 1 hour of fermentation. Pre-treatment of mTGase at 40 °C for 60 min was demonstrated to accelerate the gelation of both cow's and goat's milks (Ardelean et al., 2013), being more pronounced in the former. It has long been known that protein concentration, structure and polymorphism are the determinative factors for gelation kinetics of milk (Park et al., 2007; Uniacke-Lowe et al., 2010). Ardelean et al. (2013) found that under the same mTGase treatment conditions, while nearly 31% of cow's milk proteins cross-linked, this was only 9.6% in goat's milk proteins.

In order to reduce the processing time, mTGase is employed simultaneously with yogurt starter culture and it keeps its reactivity along with storage period. It is expected that, the gelation kinetics and matrix properties of the acid casein gels made from milks pre-treated with mTGase prior to heat inactivation and treated with mTGase simultaneously with starter culture after heat treatment would be different. Bönisch et al. (2007b) investigated the possibility of using mTGase and starter culture simultaneously. Authors showed that after 120 min incubation at 40 °C, no residual enzyme activity was found below pH 4.0 and mTGase did not affect with the fermentation time. Most of the enzymatic cross-linking of proteins was reported to be completed between pH 6.6 and 5.7.

mTGase may interfere with the metabolic activities of starter cultures. It was reported that acetaldehyde production by yogurt starter bacteria was partly hindered by mTGase (Özer et al., 2007; Öner et

Table 1. Summary of the studies on the effects of mTGase on yogurt

Conditions	Outcomes	Sources
mTGase from *Streptomyces lydicus* employed at 41 °C for 60 min prior to acidification of milk by glucono-δ-lactone (enzyme:protein ratio 0.4%)	Finer protein network with 4- to 6-fold stiffer and 2- to 3-fold lower permeability compared with untreated gels.	Færgemand & Qvist (1997)
Pre-incubation of skim milk with mTGase at 40 °C for 150 min followed by heat inactivation (enzyme concentration 3 U/g protein-Activa®MP)	Breaking strength of skimmilk yogurt increased from 550 cN to 920 cN with increasing pre-incubation time.	Lauber et al. (2000)
Pre-incubation of whole milk with mTGase at 40 °C for 120 min followed by heat inactivation at 80 °C for 1 min or concomitant use of mTGase with starter culture (enzyme:substrate ratio 1:2000-Activa®MP)	Pre-incubation resulted in prolonged fermentation while simultaneously use of mTGase with starter culture had no effect on fermentation time. Independent from mTGase addition sequence, post-acidification was lower. Enzyme treated yogurts were firmer with reduced syneresis and creamy structure. Regarding odour and consistency the enzyme treated yogurts were less "yogurt specific".	Lorenzen et al. (2002)
Pre-incubation of whole milk with mTGase at 40 °C for 120 min followed by heat inactivation at 80 °C for 2 min (enzyme concentration 3 U/g protein-Activa®MP)	Final G' (storage modulus) of the enzyme-treated gels increased. When high hydrostatic pressure and mTGase were used together, the gel strength was higher than when only pressure or mTGase was employed.	Anema et al. (2005)
Semi-skimmed UHT milk (1.5% milk fat) was treated with mTGase (enzyme concentration 3 U/g protein-Activa®MP) at 40 °C for 1 h. Enzyme was inactivated by heating at 85 °C for 10 min. UHT milk was acidified by GDL	Enzyme-treated gels had higher elasticity and rupture force. Increased enzyme concentration or extended incubation time led to lower number of bonds due probably to restriction of matrix rearrangement resulting in lower gel firmness.	Jaros et al. (2006b)
Pre-treatment of skimmed raw milk at 40 °C for 120 min (enzyme concentrations varying from 0 to 0.5 g/L of milk-Activa®MP), followed by heat inactivation at 90 °C for 5 min.	Growth slowing effect of mTGase on yogurt starter culture was evident, being more pronounced at higher mTGase concentrations. A dose-dependent improvement in yogurt texture was reported. Production of acetaldehyde was hindered. The physical and sensory properties of non-fat yogurt was acceptable up to a mTGase level of 0.3 g/L.	Özer et al. (2007)
mTGase and starter culture were used simultaneously (enzyme concentration 0.6 U/g protein-Activa®MP).	No residual enzyme activity below pH 4.0 at 40 °C for 120 min exposure time. No effect of enzyme on fermentation time. 80% of protein cross-linking completed between pH 6.6-5.7. Throughout cold storage, apparent viscosity increased, yielding a coarse structure.	Bönisch et al. (2007b)
Pre-treatment of skimmilk at 42 °C for 180 min (enzyme concentrations varying between 0-3 U/g milk protein-Activa®MP or Activa®YG), followed by heat inactivation at 95 °C for 3 min.	mTGase containing glutathione (GSH) yielded higher apparent viscosity and protein polymerisation than mTGase containing no GSH. After 6-wk storage at 4 °C, no changes in the rheological properties of the stirred yogurts were evident.	Bönisch et al. (2007c)
Pre-treatment of reconstituted skimmed milk at 30 °C for 105 min (enzyme concentration 1.2 U/g milk protein-Activa®MP), followed by heat inactivation at 80 °C for 10 min.	Viscosity of the stirred yogurts improved at low shear rates or shear stresses, resulting from enhanced interactions between cross-linked protein aggregates. Shear stress required for breakdown of superaggregates was higher when the proteins were cross-linked prior to acidification.	Jaros et al. (2007)
Pre-treatment of raw milk at 40 °C for 2 h before fermentation (enzyme concentration 1.0 U/g milk protein), followed by heat inactivation at 80 °C for 1 min.	Enzyme-treated set yogurts had higher gel strength and lower whey syneresis. Acetaldehyde concentrations in mTGase-treated yogurts were lower than the untreated yogurts.	Öner et al. (2008)
Three different yogurts were made: • Yogurt from milk • Yogurt from milk of which 20% was substituted by liquid milk whey • Yogurt from milk of which 30% was substituted by liquid milk whey mTGase was added after heat treatment of yogurt milk at 95 °C for 5 min (enzyme concentration 0.5 U/g milk protein)	Consistency indices of the mTGase treated samples increased. Yogurt made from milk of which 20% was replaced with whey had similar physical characteristics when mTGase was used.	Gauche et al. (2009)
Three different mTGase applications were examined: • Pre-treatment of heat treated milk at 42 °C for 90 min before fermentation (no enzyme inactivation) • Pre-treatment of heat treated milk at 42 °C for 90 min before fermentation (enzyme inactivated at 80 °C for 5 min) • Concomitant addition of mTGase and starter culture (no enzyme inactivation) (enzyme:protein ratio 1:20)	mTGase active set yogurts had higher water holding capacity than mTGase inactivated or control samples. Enrichment of protein or dry matter content of milk may well be excluded to obtain a satisfactory physical quality set yogurt when mTGase is employed.	Yüksel & Erdem (2010)
Casein solution (27 g/kg) or raw milk were pre-incubated with mTGase (enzyme concentration 3 U/g protein), followed by inactivation of enzyme chemically (N-ethylmaleimide, 8 mmol/L) or thermally at 85 °C for 3 min.	Up to a certain level of protein oligomerisation (70% for casein solution and 25% for milk), gel stiffness and firmness increased in the enzyme-treated samples. Above these levels of polymerisation, the rheological behaviours showed an opposite characteristics, along with increased whey separation. Dimeric and trimeric caseins were found to contribute to the enhanced physical properties of milk protein gels.	Jaros et al. (2010)
mTGase was used simultaneously with starter culture. Incubation was carried out at 37 or 42 °C (enzyme concentration: 0.04%-Activa®MP)	The levels of immunoreactivity of individual proteins were affected by mTGase and incubation temperatures. At 37 °C, β-lactoglobulin immunoreactivity of yogurts were absent while at 42 °C this protein fraction was present at low level. At 42 °C, κ-casein was not detected by ELISA. mTGase also significantly affected the sensory properties of the yogurts.	Wróblewska et al (2011)

continued on following page

Table 1. Continued

Conditions	Outcomes	Sources
Five different mTGase applications were examined: • Concomitant use of mTGase and starter culture • Pre-treatment of raw milk with mTGase at 50 °C for 10 min or 1 h • Pre-treatment of raw milk with mTGase at 60 °C for 10 min or 1 h • Treatment of pasteurized milk with mTGase at 50 °C for 10 min or 1 h • Treatment of pasteurized milk with mTGase at 60 °C for 10 min or 1 h (enzyme concentration: 1 U/g protein- Activa®MP)	Enzyme addition after pasteurization improved the gel strengths and reduced the whey syneresis in the yogurts. Enzymatically cross-linked proteins were more evenly distributed over the gel network.	Şanlı et al. (2011)
UHT goat's milk and UHT cow's milk were enriched with low-heat bovine skimmilk powder. The samples were inoculated with mTGase and incubated for 0.5 to 7 h at 40 °C (enzyme concentration: 1.8 U/g protein- Activa®MP). Enzyme reaction was stopped by adding N-ethylmaleimide (8 mmol/L) or by heat treatment at 85 °C for 10 min.	Enzyme-treated goat's milk gels acidified by glucono-δ-lactone at 30 °C significantly higher gel stiffness and lower whey syneresis than the untreated goat's milk gels. Combination of dry matter enrichment and mTGase application yielded goat's milk gels similar to cow's milk gels. This combination may well be used in goat's milk yogurt with satisfactory physical characteristics.	Ardelean et al. (2012)
Pre-treatment of pasteurized milk with mTGase at 40 °C for 120 min (enzyme concentrations: 0.74, 1.29 or 1.85 U/g protein- Activa®MP), followed by heat inactivation at 75 °C for 1 min.	The physical properties of mTGase-treated yogurts used for strained yogurt (*ca.* 21% dry matter level) production were improved. The textural properties of the strained yogurts made from mTGase treated yogurts were not affected by mTGase treatment. With the increase in mTGase level, the degree of proteolysis decreased. Microbiological and sensory properties of the strained yogurts were not affected by mTGase application.	Dinkçi (2012)
mTGase obtained from *Streptoverticillium mobaraense* (DSM-40847) were added into heat treated milk either concomitantly with starter culture or prior to starter addition (at 40 °C for 60 min) in free or immobilized state, followed by heat inactivation at 90 °C for 5 min (enzyme concentration: 45 or 75 U/l)	mTGase contributed to the physical properties of the yogurt samples significantly. Yogurts treated with mTGase at a level of 45 U/l yielded physical properties similar to mTGase free yogurts enriched with skimmilk powder at a level of 3%. Immobilized mTGase had no remarkable effect on yogurts.	Mahmood & Sebo (2012)
Thermally (at 85 °C for 30 min) and high pressure (600 MPa at 55 °C for 10 min) treated milks were inoculated with mTGase at 42 °C for 180 min, followed by thermal inactivation of enzyme at 80 °C for 1 min (enzyme concentration: 2.2 U/g protein- Activa®YG)	Yogurts from HP treated milks (independent from mTGase application) were creamier than those from thermally treated milks. Sensorial and physical qualities of yogurts were improved by HP plus mTGase combination. Dry matter fortification or stabilizer addition may well be excluded when HP+mTGase combination is applied to yogurt milk.	Tsevdou et al. (2013)
Pre-treatment of heated goat's milk (at 85 °C for 15 min) with mTGase at 5 °C for 16 h, followed by heat inactivation of enzyme at 80 °C for 1 min (enzyme concentrations: 1, 2 or 3 U/ g protein- Activa®MP)	Overall, mTGase positively affected the physical (firmness and whey separation) and sensory characteristics of yogurts made from goat's milk throughout storage period of 14 days, being more pronounced at mTGase treatment at a level of 2 U/g protein.	Domagała et al. (2013)
Pre-treatment of heat treated goat's milk (at 85 °C for 15 min) with mTGase at 40 °C for 120 min, followed by heat inactivation of enzyme at 80 °C for 1 min (enzyme concentrations: 0.5, 1.0, 2.0 or 4.0 U/g protein- Activa®MP)	Labneh (concentrated yogurt with dry matter level of *ca.* 36%) made from mTGase-treated goat's milk had more compact structure than the untreated samples.	Şanlıdere-Aloğlu & Öner (2013)
Pre-treatment of raw cow's milk with mTGase at 50 °C for 60 min (no enzyme inactivation, enzyme concentration: 1.0 U/g protein- Activa®MP)	mTGase did not affect the acidity development in yogurts but proteolytic activities of the yogurt starter culture were adversely affected by mTGase. The firmness and whey separation values of the mTGase-treated yogurts (dry matter level of 11%) were similar to non-treated yogurts (dry matter content of 14.8%).	Şanlı et al. (2014)
Treatment 1: Pre-treatment of milk with mTGase at 40 °C for 2 h, followed by heat inactivation of enzyme at 80 °C for 1 min *Treatment 2:* Addition of starter culture, probiotics and mTGase simultaneously (enzyme concentrations: 0.02, 0.06, 0.12%, w/w)	The gel structure was stronger in the probiotic yogurt with no mTGase inactivation (treatment 2) than the other yogurts. Highest hysteresis loop area was obtained in the sample with the highest mTGase level in inactivated form.	Iličić et al (2014)
mTGase and starter culture were used simultaneously to make low fat set-yogurt (enzyme concentration: 0.5 or 1.0 U/g protein)	Treatment of mTGase did not result in any difference in sensory properties of yogurts but syneresis was remarkably prevented by mTGase.	Pakseresht et al. (2017)

al., 2008). On the other hand, no concrete correlation between the colony counts of starter bacteria and mTGase treatment was evident. Wróblewska et al. (2013), for example, reported no remarkable difference in the colony counts of mTGase-treated fermented milks stored 4 weeks. Farnsworth et al. (2006) investigated the survivability of probiotic cultures (*Lb. acidophilus*, *Bifidobacterium* spp. and *Lb. casei* subsp. *casei*) in goat's milk yogurt treated with mTGase.

With the increase in mTGase levels from 0 to 1 U/g protein, the physical qualities of yogurts improved accordingly and the numbers of probiotic bacteria remained almost stable throughout storage period of 8 weeks. Similar results were reported by Pavunc et al. (2011) who found no marginal effect of mTGase on viability of *Lactobacillus helveticus* M92 in yogurt during 28 days of storage. As a novel approach,

Guyot & Kulozik (2011) investigated the effect of mTGase-treated milk powders on characteristics of skimmilk yogurt. The authors manufactured skimmilk powder from milk treated with mTGase at concentrations of 0.5, 1.0, 3.0 or 10.0 U/g protein. Enzyme activity was allowed for 120 min at 40 °C and then was stopped by heating at 85 °C for 2 min. It was concluded that with the addition of mTGase-modified skimmilk powder, half of the protein addition was required to obtain an equivalent viscosity compared to the control yogurt.

mTGase may also be successfully employed in the manufacture of Ayran (Turkish drinking yogurt). Şanlı et al. (2013) reported that Ayran made from mTGase-treated milk had higher viscosity and lower phase separation than the control sample. Similarly, Wróblewska et al. (2013) studied the effects of mTGase on the immunoreactive and sensory properties of fermented milk beverages. Authors found no reaction between yogurt proteins and antibodies towards α-lactalbumin and β-lactoglobulin. Sensory profiles of the fermented milks incubated at 37 °C and 42 °C showed remarkable differences, as revealed by quantitative descriptive analysis.

Although majority of the mTGase researches have been carried out on yogurt made from cow's or goat's milk, effectiveness of mTGase on yogurts from other milk types has also been investigated. Abou-Soliman et al. (2017), for example, investigated the physicochemical, microstructural and rheological properties of yogurt made from mTGase-treated camel-milk. For a satisfactory quality of camel-milk yogurt, dry matter fortification or stabilizer supplementation is the must. However, in case of use of mT-Gase simultaneously with starter culture at a level of 0.4% (w/w) yielded camel milk yogurt with fairly improved physical properties. mTGase may also be used as an alternative to dry matter fortification of fat-free buffalo milk with acceptable physical quality. Supplementation of skimmed buffalo milk with mTGase (at a level of 1 U/g protein) and buttermilk powder (at a level of 1 or 2%, w/w) was recommended by Romeih et al., (2014). Zhang et al. (2012) used mTGase purified from high salt medium in the production of yak-milk yogurt. As expected, treatment of mTGase at ratios varying from 0.25 to 1.0 U/g protein enhanced the rheological properties of yak-milk yogurt, being more pronounced at higher mTGase levels. More importantly, mTGase treatment resulted in lowered wet yak hair and sweat odor, which both are defined as quality defects mostly associated with yak milk products.

Vast majority of TGase used in researches or industrial production of yogurt is originated from *Streptoverticillium mobaraense*. During the last few years, TGase from *Zea mays* has been investigated for dairy applications (Li et al., 2017). Li et al. (2015) demonstrated that *Zea mays* origin TGase had a capability of producing stronger acid gels with higher firmness, consistency and index of viscosity. More recently, a group of researchers from China characterised recombinant *Zea mays* TGase expressed in *Pichia pastoris* and *Eschericia coli* (Li et al., 2014, 2015). Both TGase yielded yogurts with improved rheological properties and recombinant TGase from *Zea mays* was recommended to substitue for mT-Gase in the production of yogurt.

APPLICATION OF TGASE IN CHEESE-MAKING

The application of mTGase in cheese manufacturing seems promising, since similar results in terms of increased gel strength and reduced syneresis like those found in applications in yogurt can be expected (Özer et al., 2012). mTGase has also been shown to potentially increase the yield in cheese (Bönisch et al., 2008; Cozzolino et al., 2003). Unlike yogurt, in cheese manufacturing, the conditions of optimum for rennet contradict to mTGase, resulting in partly reduced rennet activity. As stated earlier, mTGase is more active on denatured whey proteins than their undenatured form due to their globular structure. To increase mTGase activity on whey proteins, milk should be pre-heat treated. Pre-heat treatment of milk also helps inhibiting mTGase inhibitors present in milk. On the other hand, the heat-triggered whey protein-casein interactions through disulphide bridges weaken rennet activity on milk proteins. Lorenzen et al. (2000) explains this situation as "surface sealing effect" of denatured whey proteins. However, this theory was later excluded by O'Sullivan et al. (2002) who demonstrated that mTGase did not interfere with the release of caseinomacropeptide (CMP) from whey protein free solution. The authors postulated that mTGase plays active role in the primary phase of rennet coagulation but not in the secondary phase. One should bear in mind that O'Sullivan et al. (2002) employed size exclusion-high performance liquid chromatography to evaluate the hydrolysis products of non-cross linked proteins. This technique is insufficient to analyse degree of hydrolysis of cross-linked proteins, as stated by Huppertz and de Kruif (2007b). In order to beter understand the influence of mTGase on the primary and secondary phases of renneting of milk, Huppertz and de Kruif (2007b) monitored the rennet floculation time by Berridge method. The authors detected no monomeric κ-casein in renneted milk after treatment of mTGase at 30 °C for 4 to 24 hours. They further demonstrated by diffusing wave spectroscopy that mTGase interfered with the secondary phase of renneting, leading to partly inhibiton of rennet-induced coagulation of milk. Yüksel et al. (2011) evaluated the possible effects of mTGase on renneting process on Berridge substrate and showed that both primary and secondary phases of renneting were affected by mTGase treatment. They proposed that mTGase should be used 5 min after rennet addition to obtain a satisfactory level of coagulation. As a well known fact that raw milk contains mTGase inhibitors and for an optimum mTGase activity these inhibitors should be eliminated in a way. Although pre-heat treatment of milk is an effective way to inhibit the mTGase inhibitors, because of the reasons discussed above briefly, a solution alternative to pre-heat treatment was investigated. Bönisch et al. (2008) used mTGase preparation containing glutathione (GSH) in renneting of milk without employing heat treatment beyond pasteurization. These authors also confirmed that enzymatic cross-linking of milk proteins interferes with both the primary and secondary phases of renneting. When mTGase was used simultaenously with rennet, formation of additional covalent bonds stabilizes the gel network, leading to more whey retention, as shown by Bönisch et al. (2008). Retention of whey proteins in the cheese curd leads to an increase in the cheese yield; however, the optimization of the release of CMP is essential for a cheese with desired textural properties. Özer et al. (2012) investigated the optimization conditions of mTGase and rennet in milk when they were employed concomitantly. The authors studied the interactions of initial milk pH (6.1, 6.3 or 6.5), renneting temperature (30 °C or 34 °C) and mTGase doses (0.6 U/g protein or 1.8 (U/g protein). Overall, at higher renneting temperature and lower mTGase concentration, the yields of rennet gels were lower. The coagulation time (t_c) of the gels decreased with decreased pH (*i.e.* 6.1 or 6.3) and increased renneting temperature. Authors concluded that the optimum initial milk pH, coagulation temperature and mTGase concentration were 6.3, 30 °C and 1.8 U/g protein, respectively. Interestingly, the authors failed to detect gelation at initial milk pH of 6.5 when mTGase concentration was 1.8 U/g protein. Both level of CMP released into serum phase and degree of polymerization during 90 min of coagulation stage showed that mTGase plays role in both phases of renneting, as previously demonstrated by Bönisch et al. (2008). Overall, while the effect of mTGase on primary phase of renneting is related to the restricted release of CMP from micelle which leads to extended coagulation time, the effect of the enzyme on the secdondary phase of renneting is related to higher extend of protein cross-linking which leads to retention of more water in the matrix resulting in weaker gel structure. More recently, Domagała et al. (2015) studied the influence of milk protein cross-linking by mTGase on rennet coagulation time and the gel properties. The authors examined effects of different mTGase concentrations (1, 2 or 3 U/g protein) and different incubation conditions (0 h, 40 °C for 2 h, 25 °C for 4 h or 5 °C for 16 h) on rennet coagulation time and gel characteristics. In line with Özer et al. (2012), they found that at higher mTGase concentration (3 U/g protein) and higher incubation temperature (at 40 °C for 2 h) rennet coagulation time increased and the gel firmness was the lowest. Microstructural evaluations revealed that the mTGase-treated gels had smaller casein micelles with thinner chains and smaller clusters than the untreated rennet gel. Domagała et al. (2015) recommended using mTGase and rennet simultaneously in cheese-making.

The effects of mTGase on properties of processed cheese were studied by De Sá & Bordignon-Luiz (2010). The authors compared the sequence of addition of mTGase to milk based on physical properties of the end product. Regarding syneresis index, wheying off and flow behaviour of the cheeses, they recommended to add mTGase to milk 7 min after rennet addition. The total solids content of mTGase-treated Turkish white brined cheese (Beyaz peynir) was lower than the control cheese but the protein-in dry matter level in the former cheese was higher (Özer et al., 2013). On the other hand, mTGase (added at a level of 1.8 U/g protein at 30 °C or 34 °C) remarkably hindered the development of proteolysis in the same cheese. After 90 days of storage at 4 °C, the levels of total free amino acids (TFAA) were remarkably lower in the enzymatically modified cheeses. This finding was also supported by lower concentrations of hydrophilic, intermediate and hydrophobic peptides in the mTGase-treated cheeses as well. On contrary, large deformation textural properties of the enzyme-treated cheeses were greater than the control cheeses (higher hardness and cohesiveness values and stable springiness values in the former samples). The moisture-adjusted yield values of the enzymatically-modified cheeses were 18.9% (renneted at 30 °C) and 17.1% (renneted at 34 °C) compared with the control cheeses (17.4% and 14.9%, in the same order). Similarly, Hu et al. (2013) reported increases in protein and fat recoveries in low fat Cheddar cheese made from mTGase-treated milk. At the early stage of ripening, the degree of proteolysis was higher than the control cheese but with extended ripening, the level of proteolysis decreased along with harder structure and lower meltability in the end product. In a novel approach to Edam cheese- making, Aaltonen et al. (2014) standardized the protein level of milk to 12% by UF (with 10 kDa molecular cut-off) and treated the retentate with mTGase (at a level of 132 U/kg). mTGase-treated retentate was incubated at 17 °C for 2 h and retentates with and without mTGase were diluted by raw milk to protein and fat levels of 4.2% and 3.3%, respectively. Authors reported a 4% increase in yield of Edam cheese made from mTGase-treated UF-milk, with no sign of differences in the protein profiles and organoleptic properties of the cheeses compared to the control cheeses. Yield and textural characteristics of Panela cheese supplemented with dairy or vegetable proteins and treated with TGase were studied by Salinas-Valdés et al. (2015). Soybean or peanut proteins were added to milk to increase protein level by 50 or 100%. The mixtures were pasteurized and then treated with TGase (at a level of 0.12%). The highest cheese yield was obtained from cheese made from milk supplemented with plant proteins (both soybean and peanut) at a level of 100% and treated with TGase. Combination of soybean protein supplementation and TGase-treatment yielded the highest hardness, cohesiveness, chewiness and springincss values. Soy cream cheese treated with mTGase was reported to exhibit more solid-like behaviour at 25 °C and less elastic at 4 °C compared with dairy cream cheese (Lim et al., 2011). Owing to its nutritonal value, soy cream cheese may well be an alternative to dairy-based cream cheese as long as the rheological properties of the former are developed by enzymatic means.

The possibility of using mTGase in the manufacture of goat's whey cheese with slightly increased yield and no impairement in sensory profiles of the end product was demonstrated by Karzan et al. (2016). A novel cheese made from mTGase-treated milk named "cross-linked cheese" was developed by Di Pierro et al. (2010). Authors compared two different stages of mTGase addition: concomitantly with rennet and following cutting the curd. Sequence of mTGase treatment affected the protein content of the cheese significantly, being much higher in the latter application (39.5% *vs.* 33.8%). As expected, mTGase treatment limited the development of proteolysis independent from sequence of mTGase addition. Rennet coagulation time of milk was affected remarkably by the stage of adding mTGase (Cozzolino et al., 2003). Fortification of milk with WPI followed by mTGase treatment led to increase in moisture content, cheese yield and protein-in-dry matter levels of low fat Iranian cheese as reported by Sayadi et al. (2013).

Ultrafiltration of cheese whey is a common way of recovering whey proteins and reducing the biological pollution potential of cheese wastes. Major challenge of this application is the membrane fouling caused by smaller molecular weight β-lactoglobulin and α-lactalbumin (Tolkach & Kulozik, 2006; Norazman et al. 2013). Therefore, increasing molecular weight of whey proteins by enzymatic cross-linking may be a solution to increase rate of recovery of whey proteins from UF modules. This hypotesis was studied extensively by Wen-qiong et al. (2017) who set the optimal conditions for whey protein recovery as follows: TGase concentration, 40 U/g whey protein; time and temperature of catalytic reaction, at 40 ºC for 60 min; pH of solution, 5.0. Under these conditions, the recovery rate of was increased by 15-20% with 10% reduction in lactose rejection rate and 30-40% increase in relative permeate flux rate compared with untreated control solution. Enzymatic cross-linking of whey proteins also resulted in decreases in total resistance and cake resistance values due probably to increased protein particle size along with reduced zeta potential of the modified proteins.

OTHER DAIRY APPLICATIONS OF TGASE

It is known that caseins can be gelled by mTGase without heat treatment or acidification as long as the substrate concentration is high enough (*i.e.* above 8%) (Dickinson & Yamamoto, 1996). This makes mTGase-triggered hydrogels based on casein suitable for microencapsulation of heat-labile biological materials such as enzymes or living probiotic microorganisms (Heidabach et al., 2009). A novel microencapsulation method based on enzymatic cross-linking of caseins was developed by Heidabach et al. (2009). Briefly, mixture of caseinate suspension and probiotic culture was treated with mTGase (at a concentration of 10 U/g protein). The mixture was then added into vegetable oil to form droplets. The droplets were gelled at 30 or 40 °C for 2 to 3 hours during emulsifying process. Such a matrix successfully protected probiotic bacteria under simulated gastric digestion. Same matrix was reported to protect *Lactobacillus paracasei* subsp. *paracasei* F19 and *Bifidobacterium bifidum* Bb12 during freeze drying (Heidabach et al., 2010). Li et al. (2016) microencapsulated *Lactobacillus rhamnosus* GG (LGG) in soy proteins matrix cross-linked by mTGase. The microencapsulation yield was 67.4% and water activity of the microencapsulated LGG and free LGG were the same after freze drying process. The cross-linked soy protein-based matrix efficiently protected LGG under simulated gastric juice (pH 2.5 and 3.6), intestinal juice (0.3% and 2% bile salt) and storage at 4 °C compared with unprotected LGG cells.

The optimum conditions for TGase activity in ice cream production were determined by using response surface methodology by Rossa et al. (2011). TGase was added into milk prior to adition of ice cream mix and was deactivated thermally at 80 °C for 2 min. TGase improved the consistency index of ice cream, which favored a pseudoplastic behavior. The maximum response was noted at 4 U/g protein mTGase concentration and at 56.8 °C for 90 min reaction conditions.

REFERENCES

Aaltonen, T., Huumonen, I., & Myllärinen, P. (2014). Controlled transglutaminase treatment in Edam cheese-making. *International Dairy Journal, 38*(2), 179–182. doi:10.1016/j.idairyj.2013.12.004

Abou-Soliman, N. H. I., Sakr, S. S., & Awad, S. (2017). Physico-chemical, microstructural and rheological properties of camel-milk yogurt as enhanced by microbial transglutamianse. *Journal of Food Science and Technology, 54*(6), 1616–1627. doi:10.100713197-017-2593-9 PMID:28559621

Aeschlimann, D., & Paulsson, M. (1994). Transglutaminases: Protein cross-linking enzymes in tissues and body fluids. *Thrombosis and Haemostasis, 73*, 256–263. PMID:7914385

Agyare, K. K., Addo, K., & Xiong, Y. L. L. (2009). Emulsifying and foaming properites of transglutaminase-treated wheat gluten hydrolysateas influenced by pH, temperature and salt. *Food Hydrocolloids, 23*(1), 72–81. doi:10.1016/j.foodhyd.2007.11.012

Agyare, K. K., & Damodaran, S. (2010). pH-stability and thermal properties of microbial transglutaminase-treated whey protein isolate. *Journal of Agricultural and Food Chemistry, 58*(3), 1946–1953. doi:10.1021/jf903530d PMID:20088508

Ali, N. A., Ahmed, S. H., Mohamed, E. A., Ahmed, I. A. M., & Babiker, E. E. (2010). Effect of transglutaminase cross linking on the functional properties as a function of NaCl concentration of legumes protein isolate. *International Journal of Biological and Life Sciences, 6*(1), 8–13.

Ando, H., Adachi, M., Umeda, K., Matsuura, A., Nonaka, M., Uchio, R., ... Motoki, M. (1989). Purification and characteristics of a novel transglutaminase derived from microorganisms. *Agricultural and Biological Chemistry, 53*, 2613–2617.

Anema, S. G., Lauber, S., Lee, S. K., Henle, T., & Klostermeyer, H. (2005). Rheological propertis of acid gels prepared from pressure- and transglutaminase-treated skim milk. *Food Hydrocolloids, 19*(5), 879–887. doi:10.1016/j.foodhyd.2004.12.001

Ardelean, A. I., Jaros, D., & Rohm, H. (2013). Influence of microbial transglutaminase cross-linking on gelation kinetics and texture of acid gels made from whole goats and cows milk. *Dairy Science & Technology, 93*(1), 63–71. doi:10.100713594-012-0092-3

Ardelean, A. I., Otto, C., Jaros, D., & Rohm, H. (2012). Transglutaminase treatment to improve physical properties of gels made from enriched goat milk. *Small Ruminant Research, 106*(1), 47–53. doi:10.1016/j.smallrumres.2012.05.002

Benjakul, S., Phatcharat, S., Tammatinna, A., Visessanguan, W., & Kishimura, H. (2008). Improvement of gelling properties of lizardfish mince as influenced by microbial transglutaminase and fish freshness. *Journal of Food Science, 73*(6), S239–S246. doi:10.1111/j.1750-3841.2008.00813.x PMID:19241566

Betz, M., Hörmansperger, J., Fuchs, T., & Kulozik, U. (2012). Swelling behaviour, charge and mesh size of thermal protein hydrogels as influenced by pH during gelation. *Soft Matter, 8*(8), 2477–2485. doi:10.1039/c2sm06976h

Bishop, P. D., Teller, D. C., Smith, R. A., Lasser, G. W., Gilbert, T., & Seale, R. L. (1990). Expression, purification and characterization of human factor XIII in *Saccharomyces cerevisiae*. *Biochemistry, 29*(7), 1861–1869. doi:10.1021/bi00459a028 PMID:2184890

Bönisch, M. P., Heidebach, T. C., & Kulozik, U. (2008). Influence of transglutaminase protein cross-linking on the rennet coagulation of casein. *Food Hydrocolloids, 22*(2), 288–297. doi:10.1016/j.foodhyd.2006.11.015

Bönisch, M. P., Huss, M., Lauber, S., & Kulozik, U. (2007b). Yoghurt gel formation by means of enzymatic protein cross-linkling during microbial fermentation. *Food Hydrocolloids, 21*(4), 585–595. doi:10.1016/j.foodhyd.2006.07.002

Bönisch, M. P., Huss, M., Weitl, K., & Kulozik, U. (2007c). Transglutaminase cross-linking of milk proteins and impact on yoghurt gel properties. *International Dairy Journal, 17*(11), 1360–1371. doi:10.1016/j.idairyj.2007.01.019

Bönisch, M. P., Lauber, S., & Kulozik, U. (2004). Effect of ultra-high temperature treatment on the enzymatic cross-linking of micellar casein and sodium caseinate by transglutaminase. *Journal of Food Science, 69*(8), E398–E404. doi:10.1111/j.1365-2621.2004.tb09902.x

Bönisch, M. P., Lauber, S., & Kulozik, U. (2007a). Improvement of enzymatic cross-linking of casein micelles with transglutaminase by glutathione addition. *International Dairy Journal, 17*(1), 3–11. doi:10.1016/j.idairyj.2006.01.007

Bönisch, M. P., Tolkach, A., & Kulozik, U. (2006). Inactivation of an indigenous transglutaminase inhibitor in milk serum by means of UHT-treatment and membrane separation techniques. *International Dairy Journal, 16*(6), 669–678. doi:10.1016/j.idairyj.2005.08.014

Buchert, J., Ercili-Cura, D., Ma, H., Gasparetti, C., Monogioudi, E., Faccio, G., ... Kruus, K. (2010). Crosslinking food proteins for improved functionality. *Annual Review of Food Science and Technology, 1*(1), 113–138. doi:10.1146/annurev.food.080708.100841 PMID:22129332

Chambi, H., & Grosso, C. (2006). Edible films produced with gelatin and casein cross-linked with transglutaminase. *Food Research International, 39*(4), 458–466. doi:10.1016/j.foodres.2005.09.009

Chen, L., Romandetto, G. E., & Subirade, M. (2006). Food protein-based materials as nutraceutical delivery system. *Trends in Food Science & Technology, 17*(5), 272–283. doi:10.1016/j.tifs.2005.12.011

Chen, L., & Subirade, M. (2006). Alginate-whey protein granular microspheres as oral delivery vehicles for bioactive compounds. *Biomaterials, 27*(26), 4646–4654. doi:10.1016/j.biomaterials.2006.04.037 PMID:16714058

Christensen, B. M., Sørensen, E. S., Højrup, P., Petersen, T. E., & Rasmussen, L. K. (1996). Localization of potential transglutaminase cross-linking sites in bovine caseins. *Journal of Agricultural and Food Chemistry, 44*(7), 1943–1947. doi:10.1021/jf9602131

Cozzolino, A., Di Pierro, P., Mariniello, L., Sorrentino, A., Masi, P., & Porta, R. (2003). Incorporation of whey proteins into cheese curd by using transglutamianse. *Biotechnology and Applied Biochemistry, 38*(3), 289–295. doi:10.1042/BA20030102 PMID:12906711

Damodaran, S., & Agyare, K. K. (2013). Effect of microbial transglutaminase treatment on thermal stability and pH-solubility of heat-shocked whey protein isolate. *Food Hydrocolloids*, *30*(1), 12–18. doi:10.1016/j.foodhyd.2012.04.012

Darnay, L., Len, A., Koncz, A., Friedrich, L., & Rosta, L. (2015). Small angle neutron scattering study of nanostructural changes in microbial transglutaminase-treated low-fat yogurt during fermentation. *Food Science and Biotechnology*, *24*(6), 2125–2128. doi:10.100710068-015-0282-x

De Jong, G. A. H., Boumans, J. W. L., & Winjgaards, G. (2002). *Food grade transglutaminase inhibitor and uses thereof.* European Patent 1201136.

De Jong, G. A. H., Winjgaards, G., & Koppelman, S. J. (2003). Transglutaminase inhibitor from milk. *Journal of Food Science*, *68*(3), 820–825. doi:10.1111/j.1365-2621.2003.tb08249.x

De Kruif, C. G., Tuinier, R., Holt, C., Timmins, P. A., & Rollema, H. S. (2002). Physicochemical study of κ-casein and β-casein dispersions and the effect of cross- linking by transglutaminase. *Langmuir*, *18*(12), 4885–4891. doi:10.1021/la025543w

De Sá, E. M. F., & Bordignon-Luiz, M. (2010). The effect of transglutamianse on the properties of milk gels and processed cheese. *International Journal of Dairy Technology*, *63*(2), 243–251. doi:10.1111/j.1471-0307.2010.00568.x

Di Pierro, P., Chico, B., Villalonga, R., Mariniello, L., Damiano, A. E., Masi, P., & Porta, R. (2006). Chitosan-whey protein based edible films produced in the absence or presence of transglutaminase: Analysis of their mechanical and barrier properties. *Biomacromolecules*, *7*(3), 74–749. doi:10.1021/bm050661u PMID:16529409

Di Pierro, P., Mariniello, L., Sorrentino, A., Valeria, C., Giosafatto, L., Chianese, L., & Porta, R. (2010). Transglutaminase-induced chemical and rheological properties of cheese. *Food Biotechnology*, *24*(2), 107–120. doi:10.1080/08905431003784465

Dickinson, E. (1997). Enzymatic crosslinking as a tool for food colloid rheology control and interfacial stabilization. *Trends in Food Science & Technology*, *8*(10), 334–339. doi:10.1016/S0924-2244(97)01067-4

Dickinson, E. (1999). Caseins in emulsions: Interfacial properties and interactions. *International Dairy Journal*, *9*(3-6), 305–312. doi:10.1016/S0958-6946(99)00079-5

Dickinson, E., & Yamamoto, Y. (1996). Rheology of milk protein gels and protein-stabilized emulsion gels cross-linked with transglutaminase. *Journal of Agricultural and Food Chemistry*, *44*(6), 1371–1377. doi:10.1021/jf950705y

Dinkçi, N. (2012). The influence of transglutaminase treatment on functional properties of strained yoghurt. *Journal of Animal and Veterinary Advances*, *11*(13), 2238–2246. doi:10.3923/javaa.2012.2238.2246

Domagała, J., Najgebauer-Lejko, D., Wieteska-Śliwa, I., Sady, M., Wszołek, M., Bonczar, G., & Filipczak-Fiutak, M. (2015). Influence of milk protein cross-linking by transglutaminase on the rennet coagulation time and the gel properties. *Journal of the Science of Food and Agriculture*, *96*(10), 3500–3507. doi:10.1002/jsfa.7534 PMID:26585165

Domagała, J., Wszołek, M., Tamime, A. Y., & Kupiec-Teahan, B. (2013). The effect of transglutamianse concentration on the texture, syneresis and microstructure of set-type goat's milk yoghurt during the storage period. *Small Ruminant Research, 112*(1-3), 154–161. doi:10.1016/j.smallrumres.2012.12.003

Duran, R., Junqua, M., Schmitter, J. M., Gancet, C., & Goulas, P. (1998). Purification, characterisation, and gene cloning of transglutaminase from *Streptoverticillium cinnamoneum* CBS 683.68. *Biochimie, 80*(4), 313–319. doi:10.1016/S0300-9084(98)80073-4 PMID:9672751

Eissa, A. S., & Khan, S. A. (2006). Modulation of hydrophobic interactions in denatured whey proteins by transglutaminase enzyme. *Food Hydrocolloids, 20*(4), 543–547. doi:10.1016/j.foodhyd.2005.07.005

Ercili-Cura, D., Lille, M., Legland, D., Gaucel, S., Poutanen, K., Partaen, R., & Şantto, R. (2013). Structural mechanisms leading to improved water retention in acid milk gels by transglutaminase. *Food Hydrocolloids, 30*(1), 419–427. doi:10.1016/j.foodhyd.2012.07.008

Faergemand, M., & Murray, B. S. (1998). Interfacial dilatational properties of milk proteins cross-linked by transglutaminase. *Journal of Agricultural and Food Chemistry, 46*(3), 884–890. doi:10.1021/jf970733b

Faergemand, M., Otte, J., & Qvist, K. B. (1997). Enzymatic crosslinking of whey proteins by a Ca^{+2}-independent microbial transglutaminase from *Streptomyces lydicus*. *Food Hydrocolloids, 11*(1), 19–25. doi:10.1016/S0268-005X(97)80006-9

Faergemand, M., Otte, J., & Qvist, K. B. (1998). Emulsifying properties of milk proteins cross-linked with microbial transglutaminase. *International Dairy Journal, 8*(8), 715–723. doi:10.1016/S0958-6946(98)00111-3

Faergemand, M., & Qvist, K. B. (1997). Transglutaminase: Effect on rheological properties, microstructure and permeability of set style acid skim milk gel. *Food Hydrocolloids, 11*(3), 287–292. doi:10.1016/S0268-005X(97)80058-6

Farnsworth, J. P., Li, J., Hendricks, G. M., & Guo, M. R. (2006). Effects of transglutaminase treatment on functional properties and probiotic culture survivability of goat milk yogurt. *Small Ruminant Research, 65*(1-2), 113–121. doi:10.1016/j.smallrumres.2005.05.036

Gauche, C., Tomazi, T., Barreto, P. L. M., Ogliari, P. J., & Bordignon-Ruiz, M. T. (2009). Physical properties of yoghurt manufactured with milk whey and transglutaminase. *Lebensmittel-Wissenschaft + Technologie, 42*(1), 239–243. doi:10.1016/j.lwt.2008.05.023

Gauche, C., Vieira, J. T. C., Ogliari, P. J., & Bordignon-Luiz, M. T. (2008). Crosslinking of milk with whey proteins by transglutaminase. *Process Biochemistry, 43*(7), 788–794. doi:10.1016/j.procbio.2008.04.004

Gerard, J. A. (2002). Protein–protein crosslinking in food: Methods, consequences, applications. *Trends in Food Science & Technology, 13*(12), 391–399. doi:10.1016/S0924-2244(02)00257-1

Giosafatto, C. V. L., Rigby, N. M., Wellner, N., Ridout, M., Husband, F., & Mackie, A. R. (2012). Microbial transglutaminase-mediated modification of ovalbumin. *Food Hydrocolloids, 26*(1), 261–267. doi:10.1016/j.foodhyd.2011.06.003

Gómez-Guillén, M. C., Giménez, B., López-Caballero, M. E., & Montero, M. P. (2011). Functional and bioactive properties of collagen and gelatin from alternative sources: A review. *Food Hydrocolloids*, *25*(8), 1813–1827. doi:10.1016/j.foodhyd.2011.02.007

Grossmann, L., Wefers, D., Bunzel, M., Weiss, J., & Zeeb, B. (2017). Accessibility of transglutaminase to induce protein crosslinking in gelled food matrices-influence of network structure. *Lebensmittel-Wissenschaft + Technologie*, *75*, 271–278. doi:10.1016/j.lwt.2016.09.005

Guo, J., Jin, Y. C., Yang, X. Q., Yu, S. J., Yin, S. W., & Qi, J. R. (2013). Computed microtomography and mechanical property analysis of soy protein porous hydrogel prepared by homogenizing and microbial transglutaminase crosslinking. *Food Hydrocolloids*, *31*(2), 220–226. doi:10.1016/j.foodhyd.2012.10.023

Guyot, C., & Kulozik, U. (2011). Effect of transglutaminase-treated milk powders on the properties of skim milk yoghurt. *International Dairy Journal*, *21*(9), 628–635. doi:10.1016/j.idairyj.2010.10.010

Han, M., Zhang, Y., Fei, Y., Xu, X., & Zhou, G. (2009). Effect of microbial transglutaminase on NMR relaxometry and microstructure of pork myofibrillar protein gel. *European Food Research and Technology*, *228*(4), 665–670. doi:10.100700217-008-0976-x

Han, X. Q., & Damodaran, S. (1996). Thermodynamic compatibility of substrate protiens affects their cross-linking by transglutaminase. *Journal of Agricultural and Food Chemistry*, *44*(5), 1211–1217. doi:10.1021/jf950569x

Heidabach, T., Först, P., & Kulozik, U. (2009). Transglutamşnase-induced caseinate gelation for themicroencapsulation of probiotic cells. *International Dairy Journal*, *19*(2), 77–84. doi:10.1016/j.idairyj.2008.08.003

Heidabach, T., Först, P., & Kulozik, U. (2010). Influence of casein-based microencapsulation on freeze-drying and storage of probiotic cells. *Journal of Food Engineering*, *98*(3), 309–316. doi:10.1016/j.jfoodeng.2010.01.003

Hinz, K., Huppertz, T., Kulozik, U., & Kelly, A. L. (2007). Influence of enzymatic cross-linking on milk fat globules and emulsifying properties of milk proteins. *International Dairy Journal*, *17*(4), 289–293. doi:10.1016/j.idairyj.2006.05.001

Hu, Y., Ge, K., Guo, H., Luo, J., Wang, F., & Ren, F. (2013). Effect of transglutaminase on yield, compositional and functional properties of low-fat Cheddar cheese. *Food Science and Technology Research*, *19*(3), 359–367. doi:10.3136/fstr.19.359

Huppertz, T., & de Kruif, C. D. (2007a). Ethanol stability of casein micelles cross-linked with transglutaminase. *International Dairy Journal*, *17*(5), 436–441. doi:10.1016/j.idairyj.2006.05.005

Huppertz, T., & de Kruif, C. D. (2008). Structure and stability of nanogel particles prepared by internal crosslinking of casein micelles. *International Dairy Journal*, *18*(5), 556–565. doi:10.1016/j.idairyj.2007.10.009

Huppertz, T., & de Kruif, C. G. (2007b). Rennet-induced coagulation of enzymatically cross-linked casein micelles. *International Dairy Journal*, *17*(5), 442–447. doi:10.1016/j.idairyj.2006.06.026

Huppertz, T., & Smiddy, M. A. (2008). Behaviour of partially cross-linked casein micelles under high pressure. *International Journal of Dairy Technology*, *61*(1), 51–55. doi:10.1111/j.1471-0307.2008.00370.x

Ikura, K., Kometani, T., Sasaki, R., & Chiba, H. (1980). Cross-linking of soy bean 7S and 11S proteins by transglutaminase. *Agricultural and Biological Chemistry*, *44*, 2979–2984.

Ikura, K., Sasaki, R., & Motoki, M. (1992). Use of transglutaminase in quality-improvement and processing of food proteins. *Comments in Agricultural and Food Chemistry*, *2*(6), 389–407.

Iličić, M. D., Milanović, S. D., Carić, M. D., Dokić, L. P., & Kanurić, K. G. (2014). Effect of transglutaminase on texture and flow properties of stirred probiotic yoghurt during storage. *Journal of Texture Studies*, *45*(1), 13–19. doi:10.1111/jtxs.12038

Imm, J. Y., Lian, P., & Lee, C. M. N. (2000). Gelation and water binding properties of transglutaminase-treated skim milk powder. *Journal of Food Science*, *65*(2), 200–205. doi:10.1111/j.1365-2621.2000.tb15979.x

Jacob, M., Nöbel, S., Jaros, D., & Rohm, H. (2011). Physical properties of acid milk gels: Acidification rate significantly interacts with cross-linking and heat treatment of milk. *Food Hydrocolloids*, *25*(5), 928–934. doi:10.1016/j.foodhyd.2010.09.003

Jaros, D., Heidig, C., & Rohm, H. (2007). Enzymatic modification through microbial transglutaminase enhances the viscosity of stirred yogurt. *Journal of Texture Studies*, *38*(2), 179–198. doi:10.1111/j.1745-4603.2007.00093.x

Jaros, D., Jacob, M., Otto, C., & Rohm, H. (2010). Excessive cross-linking of caseins by microbial transglutaminase and its impact on physical properties of acidified milk gels. *International Dairy Journal*, *20*(5), 321–327. doi:10.1016/j.idairyj.2009.11.021

Jaros, D., Partschefeld, C., Henle, T., & Rohm, H. (2006a). Transglutaminase in dairy products: Chemistry, physics and applications. *Journal of Texture Studies*, *37*(2), 13–155. doi:10.1111/j.1745-4603.2006.00042.x

Jaros, D., Pätzold, J., Schwarzenbolz, U., & Rohm, H. (2006b). Small and large deformation rheology of acid gels from transglutaminase treated milks. *FOBI*, *1*, 124–132.

Jaros, D., Schwarzenbolz, U., Raak, N., Löbner, J., Henle, T., & Rohm, H. (2014). Cross-linking with microbial transglutaminase: Relationship between polymerisation degree and stiffness of acid casein gels. *International Dairy Journal*, *38*(2), 174–178. doi:10.1016/j.idairyj.2013.10.011

Jiang, S. J., Zhang, X., Ma, Y., Tuo, Y., Qian, F., Fu, W., & Mu, G. (2016). Characterization of whey protein-carboxymethylated chitosan composite films with and without transglutaminase treatment. *Carbohydrate Polymers*, *153*, 153–159. doi:10.1016/j.carbpol.2016.07.094 PMID:27561482

Jiang, S. J., & Zhao, X. H. (2011). Transglutaminase-induced cross-linking and glucosamine conjugation of casein and some functional properties of the modified product. *International Dairy Journal*, *21*(4), 198–205. doi:10.1016/j.idairyj.2010.12.004

Jiang, S. J., & Zhao, X. H. (2012). Cross-linking and glucosamine conjugation of casein by transglutaminase and the emulsifying property and digestibility in vitro of the modified product. *International Journal of Food Properties*, *15*(6), 1286–1299. doi:10.1080/10942912.2010.521274

Kang, Y. N., Kim, H., Shin, W. S., Woo, G., & Moon, T. W. (2003). Effect of disulfide bond reduction on bovine serum albumin-stabilized emulsion gel formed by microbial transglutaminase. *Journal of Food Science*, *68*(7), 2215–2220. doi:10.1111/j.1365-2621.2003.tb05749.x

Karzan, T. M., Nawal, H. S., & Ashna, T. A. (2016). The effect of microbial transglutaminase enzyme on some physicochemical and sensory properties of goat's whey cheese. *International Food Research Journal*, *23*, 688–693.

Kellerby, S. S., Gu, Y. S., McClements, D. J., & Decker, E. A. (2006). Lipid oxidation in a menhaden oil-in-water emulsion stabilized by sodium caseinate cross-linked with transglutaminase. *Journal of Agricultural and Food Chemistry*, *54*(26), 10222–10227. doi:10.1021/jf062143w PMID:17177563

Kieliszek, M., & Misiewicz, A. (2014). Microbial transglutaminase and its application in the food industry. A review. *Folia Microbiologica*, *59*(3), 241–250. doi:10.100712223-013-0287-x PMID:24198201

Klein, J. D., Guzman, E., & Kuehen, G. D. (1992). Purification and partial characterization of transglutaminase from *Physarum polycephalum*. *Journal of Bacteriology*, *174*(8), 2599–2605. doi:10.1128/jb.174.8.2599-2605.1992 PMID:1348244

Kobayashi, E., Hashiguchi, K., Yikozeki, K., & Yamanaka, S. (1998). Molecular cloning of the transglutaminase gene from *Bacillus subtilis* and its expression in *Escherichia coli*. *Bioscience, Biotechnology, and Biochemistry*, *62*(6), 1109–1114. doi:10.1271/bbb.62.1109 PMID:9692191

Kocaman, E. (2016). *Effects of ph and transglutaminase enzyme on gelation of caseinomacropeptide* (M.Sc. Thesis). Istanbul Technical University, Istanbul, Turkey.

Kołodziewska, I., & Piotrowska, B. (2007). The water vapour permeability, mechanical properties and solubility of fish gelatin-chitosan films modified with transglutamşnase or 1-ethyl-3-(3 –dimethylaminopropyl) carbodiimide (EDC) and plasticized with glycerol. *Food Chemistry*, *103*(2), 295–300. doi:10.1016/j.foodchem.2006.07.049

Kuraishi, C., Yamazaki, K., & Susa, Y. (2001). Transglutaminase: Its utilization in the food industry. *Food Reviews International*, *17*(2), 221–246. doi:10.1081/FRI-100001258

Lam, E., Holt, C., Edwards, P., McKinnon, I., Otter, D., Li, N., & Hemar, Y. (2017). Effect of transglutaminase treatment on the physico-chemical properties of skim milk with added ethylendiaminetetraacetic acid. *Food Hydrocolloids*, *69*, 329–340. doi:10.1016/j.foodhyd.2017.02.002

Lauber, S., Henle, T., & Klostermeyer, H. (2000). Relationship between the crosslinking of caseins by transglutaminase and the gel strength of yoghurt. *European Food Research and Technology*, *210*(5), 305–309. doi:10.1007002170050554

Lauber, S., Noack, I., Klostermeyer, H., & Henle, T. (2001). Oligomerization of β-lactoglobulin by microbial transglutaminase during high pressure treatment. *European Food Research and Technology*, *213*(3), 246–247. doi:10.1007002170100368

Lee, S. J., Choi, S. J., Li, Y., Decker, E. A., & McClements, D. J. (2011). Protein-stabilized nanoemulsions and emulsions: Comparison of physicochemical stability, lipid oxidation, and lipase digestibility. *Journal of Agricultural and Food Chemistry*, *59*(1), 415–427. doi:10.1021/jf103511v PMID:21133433

Lerner, A., & Matthias, T. (2015). Possible association between celiac diseaase and bacterial transglutaminase in food processing: A hypothesis. *Nutrition Reviews*, *73*(8), 544–552. doi:10.1093/nutrit/nuv011 PMID:26084478

Li, C., Wang, C. L., Sun, Y., Li, A. L., Liu, F., & Meng, X. C. (2016). Microencapsulation of *Lactobacillus rhamnosus* GG by transglutaminase cross-linked soyprotein isolate to improve survival in simulated gastrointestinal conditions and yoghurt. *Journal of Food Science*, *81*(7), M1726–M1734. doi:10.1111/1750-3841.13337 PMID:27228279

Li, H., Cui, Y., Zhang, L., Luo, X. L., Fan, R., Xue, C., ... Han, X. (2015). Production of a transglutaminase from *Zea mays* in *Eschericia coli* and its impact on yoghurt properties. *International Journal of Dairy Technology*, *68*(1), 54–61. doi:10.1111/1471-0307.12165

Li, H., Cui, Y., Zhang, L., Zhang, L., & Yu, J. (2017). Optimization of recombinant *Zea mays* transglutaminase production and its influence on the functional properties of yoghurt. *Food Science and Biotechnology*, *26*(3), 723–730. doi:10.100710068-017-0083-5

Li, H., Zhang, L., Cui, Y., Luo, X. L., Xue, C., Wang, S., ... Han, X. (2014). Characterization of recombinant *Zea mays* transglutaminase expressed in Pichia pastoris and its impact on full and non-fat yoghurts. *Journal of the Science of Food and Agriculture*, *94*(6), 1225–1230. doi:10.1002/jsfa.6402 PMID:24105803

Lim, T. J., Easa, A. M., Karim, A. A., Bhat, R., & Liong, M. T. (2011). Development of soy-based cream cheese via the addition of microbial transglutamianse, soy protein isolate and maltodextrin. *British Food Journal*, *113*(9), 1147–1172. doi:10.1108/00070701111174587

Liu, F., Chiou, B. S., Avena-Bustillos, R. J., Zhang, Y., Li, Y., McHugh, T., & Zhong, F. (2017). Study of combined effects of glycerol and transglutaminase on properties of gelatin films. *Food Hydrocolloids*, *65*, 1–9. doi:10.1016/j.foodhyd.2016.10.004

Liu, M., & Damodaran, S. (1999). Effect of Transglutaminase-catalysed polymerization of beta-casein on its emulsifying properties. *Journal of Agricultural and Food Chemistry*, *47*(4), 1514–1519. doi:10.1021/jf981030c PMID:10564008

Lorenzen, P. C. (2000). Renneting properties of transglutaminase-treated milk. *Milchwissenschaft. Milk Science International*, *54*, 563–566.

Lorenzen, P. C., Neve, H., Mautner, A., & Schlimme, E. (2002). Effect of enzymatic cross-linking of milk proteins on functional properties of set-style yoghurt. *International Journal of Dairy Technology*, *55*(3), 152–157. doi:10.1046/j.1471-0307.2002.00065.x

Lorenzen, P. C., & Schlimme, E. (1998). Properties and potential fields of application of transglutamianse preparations in dairying. In *IDF Bulletin 332* (pp. 47–53). Brussels: International Dairy Federation.

Loveday, S. M., Su, J., Rao, M. A., Anema, S. G., & Singh, H. (2011). Effect of calcium on the morphology and functionality of whey protein nanofibrils. *Biomacromolecules*, *12*(10), 3780–3788. doi:10.1021/bm201013b PMID:21894942

Macedo, J., & Sato, H. (2005). Propriedades e aplicações da transglutaminase microbiana em alimentos. *Alimentos e Nutrição*, *16*(4), 413–419.

Macierzanka, A., Bordron, F., Rigby, N., Mills, E. N. C., Lille, M., Poutanen, K., & Mackie, A. R. (2011). Transglutaminase cross-linking kinetics of sodium caseinate is changed after emulsification. *Food Hydrocolloids*, *23*(5), 843–850. doi:10.1016/j.foodhyd.2010.07.010

Mahmood, W. A., & Sebo, N. H. (2012). Improvement of yogurt properties by microbial transglutaminase. *Jordan Journal of Agricultural Sciences*, *8*, 333–341.

Marquez, R. G., Di Pierro, P., Esposito, M., Mariniello, L., & Porta, R. (2014). Application of transglutaminase-crosslinked whey protein /pectin films as water barrier coatings in fried and baked foods. *Food and Bioprocess Technology*, *7*, 47–455.

Mautner, A., Meisel, H., Lorenzen, P. C., & Schlimme, E. (1999). Determination of the dipeptide ε-(γ-glutamyl)lysine in transglutaminase crosslinked proteins by amino acid analysis. *Kieler Milchwirtschaftliche Fosrhungsberichte*, *51*, 155–163.

Miwa, N., Kumazawa, Y., Nakagoshi, H., & Sakaguchi, S. (2002). *Methods for modifying raw material milk and dairy product prepared by using the modified raw material milk*. European patent 1 197 152 A2.

Mizuno, A., Mitsuiki, M., & Motoki, M. (1999). Glass transition of casein as affected by transglutaminase. *Journal of Food Science*, *64*(5), 796–799. doi:10.1111/j.1365-2621.1999.tb15914.x

Moon, J. H., Hong, Y. H., Huppertz, T., Fox, P. F., & Kelly, A. L. (2009). Properties of casein micelles cross-linked by transglutaminase. *International Journal of Dairy Technology*, *62*(1), 27–32. doi:10.1111/j.1471-0307.2008.00442.x

Motoki, M., Aso, H., Seguro, K., & Nio, N. (1987a). α_{s1}-casein film prepared using transglutaminase. *Agricultural and Biological Chemistry*, *51*, 993–996.

Motoki, M., & Nio, N. (1983). Crosslinking between different proteins by transglutaminase. *Journal of Food Science*, *48*(2), 561–566. doi:10.1111/j.1365-2621.1983.tb10790.x

Motoki, M., Nio, N., & Takinami, K. (1984). Functional properties of food proteins polymerized by transglutaminase. *Agricultural and Biological Chemistry*, *48*, 1257–1261.

Motoki, M., Nio, N., & Takinami, K. (1987b). Functional properties of heterologous polymer prepared by Transglutaminase between milk casein and soybean globulin. *Agricultural and Biological Chemistry*, *51*, 237–239.

Motoki, M., & Seguro, K. (1998). Transglutaminase and its use for food processing. *Trends in Food Science & Technology*, *9*(5), 204–210. doi:10.1016/S0924-2244(98)00038-7

Motoki, M., Seguro, K., Nio, N., & Takinami, K. (1986). Glutamine-specific deamidation of α_{s1}-casein by transglutaminase. *Agricultural and Biological Chemistry*, *50*, 3025–3030.

Mounsey, J. S., O'Kennedy, B. T., & Kelly, P. M. (2005). Influence of transglutaminase treatment on properties of micellar casein and products made therefrom. *Le Lait*, *85*(4-5), 405–418. doi:10.1051/lait:2005028

Myllärinen, P., Buchert, J., & Autio, K. (2007). Effect of transglutaminase on rheological properties and microstructure of chemically acidified sodium caseinate gels. *International Dairy Journal, 17*(7), 800–807. doi:10.1016/j.idairyj.2005.10.031

Nieuwenhuizen, W. F., Dekker, H. L., de Koning, L. J., Gröneveld, T., de Koster, C. G., & de Jong, G. A. (2003). Modification of glutamine and lysine residues in holo and apo alpha-lactalbumin with microbial transglutaminase. *Journal of Agricultural and Food Chemistry, 51*(24), 7132–7139. doi:10.1021/jf0300644 PMID:14611184

Nieuwenhuizen, W. F., Dekker, H. L., Gröneveld, T., Koster, C. G., & Jong, G. A. H. (2004). Transglutaminase-mediated modification of glutamine and lysine residues in native bovine β-lactoglobulin. *Biotechnology and Bioengineering, 85*(3), 248–258. doi:10.1002/bit.10898 PMID:14748079

Nieuwland, M., Bouwman, W. G., Bennik, M. L., Silletti, E., & de Jongh, H. H. J. (2014). Characterizing length scales that determine the mechanical behavior of gels from crosslinked casein micelles. *Food Biophysics, 10*(4), 416–427. doi:10.100711483-015-9399-y

Nio, N., Motoki, M., & Takinami, K. (1985). Gelation of casein and soybean globulins by transglutaminase. *Agricultural and Biological Chemistry, 49*, 2283–2286.

Nio, N., Motoki, M., & Takinami, K. (1986). Gelation mechanisms of protein solutions by transglutamianse. *Agricultural and Biological Chemistry, 48*, 851–855.

Nonaka, M., Matsuura, Y., & Motoki, M. (1996). Incorporation of lysine dipeptides into α_{s1}-casein by Ca^{+2}-independent microbial transglutaminase. *Bioscience, Biotechnology, and Biochemistry, 60*(1), 131–133. doi:10.1271/bbb.60.131

Nonaka, M., Sakamoto, H., Toiguchi, S., Kawajiri, H., Soeda, T., & Motoki, M. (1992). Sodium caseinate and skim milk gels formed by incubation with microbial transglutaminase. *Journal of Food Science, 57*(5), 1214–1218. doi:10.1111/j.1365-2621.1992.tb11302.x

Nonaka, M., Tanaka, H., Okiyama, A., Motoki, M., Ando, H., Umeda, K., & Matsuura, A. (1989). Polymerization of several proteins by Ca^{+2}-independent transglutaminase derived from microorganisms. *Agricultural and Biological Chemistry, 53*(10), 2619–2623.

Norazman, N., Wu, W., Li, H., Wasinger, V., Zhang, H., & Chen, V. (2013). Evaluation of chemical cleaning of UF membranes fouled with whey protein isolates via analysis of residual protein components on membrane surface. *Separation and Purification Technology, 103*, 241–250. doi:10.1016/j.seppur.2012.10.039

O'Sullivan, M. M., Kelly, A. L., & Fox, P. F. (2002). Effect of transglutamianse on the heat stability of milk: A possible mechanism. *Journal of Dairy Science, 85*(1), 1–7. doi:10.3168/jds.S0022-0302(02)74045-9 PMID:11860101

Oh, J. H., Wang, B., Field, D. P., & Aglan, H. A. (2004). Characteristics of edible films made from dairy proteins and zein hydrolysate cross-linked with transglutaminase. *International Journal of Food Science & Technology, 39*(3), 287–294. doi:10.1111/j.1365-2621.2004.00783.x

Öner, Z., Karahan, A. G., Aydemir, S., & Sanlidere-Aloglu, H. (2008). Effect of transglutaminase on physicochemical properties of set-style yogurt. *International Journal of Food Properties*, *11*(1), 196–205. doi:10.1080/10942910701286346

Özer, B., Guyot, C., & Kulozik, U. (2012). Simultaneous use of transglutaminase and rennet in milk coagulation: Effect of initial milk pH and renneting temperature. *International Dairy Journal*, *24*(1), 1–7. doi:10.1016/j.idairyj.2011.10.002

Özer, B., Hayaloğlu, A. A., Yaman, H., Gürsoy, A., & Şener, L. (2013). Simultaneous use transglutaminase and rennet in white-brined cheese production. *International Dairy Journal*, *33*(2), 129–134. doi:10.1016/j.idairyj.2013.02.001

Özer, B., Kirmaci, H. A., Oztekin, S., Hayaloglu, A., & Atamer, M. (2007). Incorporation of microbial transglutaminase into non-fat yogurt production. *International Dairy Journal*, *17*(3), 199–207. doi:10.1016/j.idairyj.2006.02.007

Pakseresht, S., Tehrani, M. M., & Razavi, S. A. M. (2017). Optimization of low-fat set-type yoghurt: Effect of altered whey protein to casein ratio, fat content and microbial transglutaminase on rheological and sensorial properties. *Journal of Food Science and Technology*, *54*(8), 2351–2360. doi:10.100713197-017-2675-8 PMID:28740292

Park, Y. M., Juarez, M., Ramos, M., & Haenlein, G. F. W. (2007). Physico-chemical characteristics of goat and sheep milk. *Small Ruminant Research*, *68*(1-2), 88–113. doi:10.1016/j.smallrumres.2006.09.013

Pavunc, A. L., Beganović, J., Kos, B., Buneta, A., Beluhan, S., & Šušković, J. (2011). Influence of microencapsulation and transglutaminase on viability of probiotic strain *Lactobacillus helveticus* M92 and consistency of set yoghurt. *International Journal of Dairy Technology*, *64*(2), 254–261. doi:10.1111/j.1471-0307.2010.00647.x

Qi, W., Chen, C., Liu, M., Yu, G., Cai, X., Guo, P., ... Mei, S. (2015). Whey protein isolate modified by transglutaminase aggregation and emulsion gel properties. *Global Conference on Polymer and Composite Materials*, 1-11. 10.1088/1757-899X/87/1/012034

Raak, N., Jaros, D., & Rohm, H. (2017a). Cross-linking with microbial transglutaminase: Isopeptide bonds and polymer size as drivers for acid casein gel stiffness. *International Dairy Journal*, *66*, 49–55. doi:10.1016/j.idairyj.2016.10.015

Raak, N., Rohm, H., & Jaros, D. (2017b). Enzymatic cross-linking of casein facilitates gel structure weakening ınduced by overacidification. *Food Biophysics*, *12*(2), 261–268. doi:10.100711483-017-9483-6

Rodriguez-Nogales, J. M. (2005). Enzymatic cross-linking of ewe's milk proteins by transglutaminase. *European Food Research and Technology*, *221*(5), 692–699. doi:10.100700217-005-0041-y

Rodriguez-Nogales, J. M. (2006). Enhancement of transglutaminase-induced protein cross-linking by preheat treatment of cows' milk: A statistical approach. *International Dairy Journal*, *16*(1), 26–32. doi:10.1016/j.idairyj.2005.01.003

Romeih, E., Abdel-Hamid, M., & Awad, A. A. (2014). The addition of buttermilk powder and transglutaminase improves textural and organoleptic properties of fat-free buffalo yogurt. *Dairy Science & Technology*, *94*(3), 297–309. doi:10.100713594-014-0163-8

Romeih, E., & Walker, G. (2017). Recent advances on microbial transglutaminase and dairy application. *Trends in Food Science & Technology*, *62*, 133–140. doi:10.1016/j.tifs.2017.02.015

Rossa, P. N., De Sá, E. M. F., Burin, V. M., & Bordignon-Luiz, T. M. (2011). Optimization of microbial transglutaminase activity in ice cream using response surface methodology. *LWT-Food Sceince and Technology, 44*, 29-34.

Sakamoto, H., Kumizawa, Y., & Motoki, M. (1994). Stregth of protein gels prepared with microbial transglutaminase as related to reaction conditions. *Journal of Food Science*, *59*(4), 866–871. doi:10.1111/j.1365-2621.1994.tb08146.x

Salinas-Valdés, A., Millán, J. D. R., Serna-Saldívar, S. O., & Chuck-Hernández, C. (2015). Yield and textural characteristics of Panela cheeses produced with dairy-vegetable protein (soybean or peanut) blends supplemented with transglutaminase. *Journal of Food Science*, *80*(12), S2950–S2956. doi:10.1111/1750-3841.13126 PMID:26550775

Şanlı, T., Sezgin, E., Deveci, O., Şenel, E., & Benli, M. (2011). Effect of using transglutaminase on physical, chemical and sensory properties of set-type yoghurt. *Food Hydrocolloids*, *25*(6), 1477–1481. doi:10.1016/j.foodhyd.2010.09.028

Şanlı, T., Sezgin, E., Şenel, E., & Benli, M. (2013). The effect of transglutaminase on some physicochemical and sensory properties of the Turkish drinking yoghurt Ayran. *International Journal of Dairy Technology*, *66*(3), 410–415. doi:10.1111/1471-0307.12045

Şanlı, T., Sezgin, E., Şenel, E., & Benli, M. (2014). The effects of transglutaminase, exopolysaccharide-producing starter culture, sensory and texture properties of low-fat set yoghurt. *International Journal of Dairy Technology*, *67*, 237–244. doi:10.1111/1471-0307.12127

Şanlıdere-Aloğlu, H., & Öner, Z. (2013). The effect of treating goat's milk with transglutaminase on chemical, structural and sensory properties of labneh. *Small Ruminant Research*, *109*(1), 31–37. doi:10.1016/j.smallrumres.2012.10.005

Sarkar, N. K., Clarke, D. D., & Waelsch, H. (1957). An enzymatically catalyzed incorporation of amines into proteins. *Biochimica et Biophysica Acta*, *25*, 451–452. doi:10.1016/0006-3002(57)90512-7 PMID:13471608

Sayadi, A., Madadlou, A., & Khosrowshahi, A. (2013). Enzymatic cross-linking of whey proteins in low fat Iranian white cheese. *International Dairy Journal*, *29*(2), 88–92. doi:10.1016/j.idairyj.2012.10.006

Schmid, M., Sängerlaub, S., Wege, L., & Stäbler, A. (2014). Properties of transglutaminase crosslinked whey protein isolate coatings and cast films. *Packaging Technology & Science*, *27*(10), 799–817. doi:10.1002/pts.2071

Schorsch, C., Carrie, H., & Norton, I. T. (2000). Cross-linking casein micelles by a microbial transglutaminase: Influence of cross-links in acid-induced gelation. *International Dairy Journal*, *11*, 785–793.

Sharma, R., Lorenzen, P. C., & Qvist, K. B. (2001). Influence of transglutaminase treatment of skim milk on the formation of ε-(γ-glutamyl)lysine and the susceptibility of individual proteins towards crosslinking. *International Dairy Journal*, *11*(10), 785–793. doi:10.1016/S0958-6946(01)00096-6

Sharma, R., Zakora, M., & Qvist, K. B. (2002). Susceptibility of an industrial α-lactalbumin concentrate to cross-linking by microbial transglutaminase. *International Dairy Journal, 12*(12), 1005–1012. doi:10.1016/S0958-6946(02)00122-X

Smiddy, M. A., Martin, J. E. G. H., Kelly, A. L., de Kruif, C. G., & Huppertz, T. (2006). Stability of casein micelles cross-linked by transglutaminase. *Journal of Dairy Science, 89*(6), 1906–1914. doi:10.3168/jds.S0022-0302(06)72258-5 PMID:16702254

Song, C. L., & Zhao, X. H. (2013). Rheological, gelling and emulsifying properties of a glycosylated and cross-linked caseinate generated by transglutaminase. *International Journal of Food Science & Technology, 48*(12), 2595–2602. doi:10.1111/ijfs.12255

Swaisgood, H. E. (1982). Chemistry of milk proteins. In P. F. Fox (Ed.), *Developments in dairy chemistry* (Vol. 1, pp. 1–59). London: Applied Science Publishers.

Takehana, S., Washizu, K., Ando, K., Koikeda, S., Takeuchi, K., Matsui, H., ... Takagi, H. (1994). Chemical synthesis of the gene for microbial transglutaminase from *Streptoverticillium* and its expression in *Escherica coli. Bioscience, Biotechnology, and Biochemistry, 58*(1), 88–92. doi:10.1271/bbb.58.88 PMID:7765335

Tang, C.-H., & Ma, C.-H. (2007). Modulation thermal stability of β-lactoglobulin by transglutaminase treatment. *European Food Research and Technology, 225*(5-6), 649–652. doi:10.100700217-006-0460-4

Tanimoto, S.-Y., & Kinsella, J. E. (1988). Enzymatic modifications of proteins: Effects of transglutaminase cross-linking on some properties of β-lactoglobulin. *Journal of Agricultural and Food Chemistry, 36*(2), 281–285. doi:10.1021/jf00080a010

Toledano, O., & Magdassi, S. (1997). Formation of surface active gelatin by covalent attachment of hydrophobic chains. *Journal of Colloid and Interface Science, 193*(2), 172–177. doi:10.1006/jcis.1997.5030 PMID:9344517

Tolkach, A., & Kulozik, U. (2005). Fractionation of whey proteins and caseinomacro-peptide by means of enzymatic crosslinking and membrane separation techniques. *Journal of Food Engineering, 67*(1-2), 13–20. doi:10.1016/j.jfoodeng.2004.05.058

Tolkach, A., & Kulozik, U. (2006). Transport of whey proteins through 0.1 mm ceramic membrane: Phenomena, modeling and consequences for concentration or diafiltration processes. *Desalination, 199*(1-3), 340–341. doi:10.1016/j.desal.2006.03.183

Traorè, F., & Meunier, J. C. (1992). Crosslinking activity of placental F XIIIa on whey proteins and caseins. *Journal of Agricultural and Food Chemistry, 40*(3), 399–402. doi:10.1021/jf00015a007

Truong, V., Clare, D. A., Catignani, G. L., & Swaisgood, H. (2004). Cross-linking and rheological changes of whey proteins treated with microbial transglutaminase. *Journal of Agricultural and Food Chemistry, 52*(5), 1170–1176. doi:10.1021/jf034397c PMID:14995116

Tsai, G.-J., Lin, S.-M., & Jiang, S.-T. (1996). Transglutaminase from *Streptoverticillium ladakanum* and application to minced fish product. *Journal of Food Science, 61*(6), 1234–1238. doi:10.1111/j.1365-2621.1996.tb10968.x

Tsevdou, M. S., Eleftheriou, E. G., & Taoukis, P. S. (2013). Transglutamianse treatment of thermally and high pressure processed milk: Effects on the properties and storage stability of set yoghurt. *Innovative Food Science & Emerging Technologies, 17*, 144–152. doi:10.1016/j.ifset.2012.11.004

Uniacke-Lowe, T., Huppertz, T., & Fox, P. (2010). Equine milk proteins: Chemistry, structure and nutritional significance. *International Dairy Journal, 20*(9), 609–629. doi:10.1016/j.idairyj.2010.02.007

Vanin, F. M., Sobral, P. J. A., Menegalli, F. C., Carvalho, R. A., & Habitante, A. M. Q. B. (2005). Effect of plasticizers and their concentrations on thermal and functional properties of gelatin-based films. *Food Hydrocolloids, 19*(5), 899–907. doi:10.1016/j.foodhyd.2004.12.003

Wang, W. (2013). *Whey proteins cross-linked by transglutaminase or glycated with maltodextrin: physicochemical bases of the improved heat stability* (PhD Thesis). University of Tenesse.

Wang, W., Zhong, Q., & Hu, Z. (2012). Nanoscale understanding of thermal aggregation of whey protein pretreated by transglutaminase. *Journal of Agricultural and Food Chemistry, 61*(2), 435–446. doi:10.1021/jf304506n PMID:23252670

Wang, Y., Liu, A., Ye, R., Wang, W., & Li, X. (2015). Transglutamianse-induced crosslinking of gelatin-calcium carbonate composite films. *Food Chemistry, 166*, 414–422. doi:10.1016/j.foodchem.2014.06.062 PMID:25053075

Washizu, K., Ando, K., Koikeda, S., Hirose, S., Matsuura, A., Akagi, H., ... Takeuchi, K. (1994). Molecular cloning of the gene for microbial transglutamianse *Streptoverticillium* and its expression in *Streptomyces lividans*. *Bioscience, Biotechnology, and Biochemistry, 58*(1), 82–87. doi:10.1271/bbb.58.82 PMID:7765334

Wen-qiong, W., Lan-wei, Z., Xue, H., & Yi, L. (2017). Cheese whey protein recovery by ultrafiltration through transglutamianse (TG) catalysis whey protein cross-linking. *Food Chemistry, 215*, 31–40. doi:10.1016/j.foodchem.2016.07.057 PMID:27542447

Wilcox, C. P., Clare, D. A., Valentine, V. W., & Swaisgood, H. (2002). Immobilization and utilization of the recombinant fusion proteins trypsin-streptavidin and streptavidin-transglutaminase for modification of whey protein isolate functionality. *Journal of Agricultural and Food Chemistry, 50*(13), 3723–3730. doi:10.1021/jf011603c PMID:12059150

Wilcox, C. P., & Swaisgood, H. E. (2002). Modification of the rheological properties of whey protein isolate through the use of an immobilized microbial transglutaminase. *Journal of Agricultural and Food Chemistry, 50*(20), 5546–5551. doi:10.1021/jf0117154 PMID:12236677

Wróblewska, B., Kaliszewsk-Suchodoła, A., Kołakowski, P., Pawlikowska, K., & Troszyńska, A. (2013). The effect of microbial transglutaminase on the immunoreactive and sensory properties of fermented milk beverages. *International Journal of Food Science & Technology, 48*(5), 1007–1017. doi:10.1111/ijfs.12054

Wróblewska, B., Kaliszewska, A., Kołakowski, P., Pawlikowska, K., & Troszyńska, A. (2011). Impact of transglutaminase reaction on the immunoreactive and sensory quality of yoghurt starter. *World Journal of Microbiology & Biotechnology, 27*(2), 215–227. doi:10.100711274-010-0446-z

Yasueda, H., Nakanishi, K., Kumazawa, Y., Nagase, K., Motoki, M., & Matsui, H. (1995). Tissue-type transglutaminase from red sea bream (*Pagrus major*) sequence analysis of the cDNA and functional expression in *Eschericia coli. European Journal of Biology*, *232*(2), 411–419. doi:10.1111/j.1432-1033.1995.tb20826.x PMID:7556189

Yi, J. B., Kim, Y. T., Bae, H., Whiteside, W. S., & Park, H. J. (2006). Influence of transglutamianse-induced cross-linking on properties of fish gelatin films. *Journal of Food Science*, *71*(9), E376–E383. doi:10.1111/j.1750-3841.2006.00191.x

Yildirim, M., & Hettiarachchy, N. S. (1998). Properties of films produced by cross-linking whey proteins and 11S globulin using transglutaminase. *Journal of Food Science*, *63*(2), 248–252. doi:10.1111/j.1365-2621.1998.tb15719.x

Yokoyama, K., Nio, N., & Kikuchi, Y. (2004). Properties and applications of microbial transglutaminase. *Applied Microbiology and Biotechnology*, *64*(4), 447–454. doi:10.100700253-003-1539-5 PMID:14740191

Yüksel, Z., Avcı, E., & Erdem, Y. K. (2011). Modification of the rennet process in Berridge substrate by transglutamianse. *International Journal of Dairy Technology*, *64*(3), 365–371. doi:10.1111/j.1471-0307.2011.00667.x

Yüksel, Z., & Erdem, Y. K. (2010). The influence of transglutaminase on functional properties of set yoghurt. *International Journal of Dairy Technology*, *63*(1), 86–97. doi:10.1111/j.1471-0307.2009.00539.x

Zeeb, B., Grossmann, L., & Weiss, J. (2013). Accessibility of transglutaminase to induce protein cross-linking in gelled food matrices-Impact of membrane structure. *Food Biophysics*, 1–8.

Zhang, L., Zhang, L., Yi, H., Du, M., Ma, C., Han, X., ... Zhang, Y. (2012). Enzymatic characterization of transglutaminase from *Streptomyces mobaraensis* DSM 40587 in high salt and effect of enzymatic cross-linking of yak milk proteins on functional properties of stirred yogurt. *Journal of Dairy Science*, *95*(7), 3559–3568. doi:10.3168/jds.2011-5125 PMID:22720914

Zhang, W., & Zhong, Q. (2009). Microemulsions as nanoreactors to produce whey protein nanoparticles with enhanced heat stability by sequential enzymaticc-linking and thermal pretreatments. *Journal of Agricultural and Food Chemistry*, *57*(19), 9181–9187. doi:10.1021/jf902045u PMID:19807163

Zhang, Y.-H., Liu, J.-Q., Xu, D., & Zhao, X.-H. (2016). Impacts of glucosamine/ oligochitosan glycation and cross-linking by transglutaminase on the structure and *in vitro* antigenicity of whey proteins. *International Journal of Dairy Technology*, *69*(2), 169–176. doi:10.1111/1471-0307.12246

Zhong, Q., Wang, W., Hu, Z., & Ikeda, S. (2013). Sequential preheating and transglutaminase pretreatments improve stability of whey protein isolate at pH 7.0 during thermal sterilization. *Food Hydrocolloids*, *31*(2), 306–316. doi:10.1016/j.foodhyd.2012.11.018

Zhu, Y., & Tramper, J. (2008). Novel applications for microbial transglutaminase beyond food processing. *Trends in Biotechnology*, *26*(10), 559–565. doi:10.1016/j.tibtech.2008.06.006 PMID:18706723

Chapter 10
Role of Microbial Cultures and Enzymes During Cheese Production and Ripening

Şebnem Öztürkoğlu Budak
Ankara University, Turkey

Celalettin Koçak
Ankara University, Turkey

Peter A. Bron
NIZO Food Research BV, The Netherlands

Ronald P. de Vries
Westerdijk Fungal Biodiversity Institute, The Netherlands & Utrecht University, The Netherlands

ABSTRACT

Many different kinds of cultures, enzymes, and methods are used during the production and ripening of a variety of cheese types. In this chapter, the importance, types, and applications of microbial cultures during cheese production are discussed. Moreover, an overview of the important role of enzymatic systems, either derived from these cultures or directly added to the milk fermentation, is presented. The main biochemical events including glycolysis, lipolysis, and proteolysis during cheese ripening are explained, focusing on their end products, which contribute to the development of the overall aroma of cheese.

INTRODUCTION

Cheese ripening is a complex process that influences the chemical, physical, microbiological, rheological and textural properties of cheese. The most eminent outcome of cheese ripening is the differential development of the flavor and textural characteristics of the huge variety of cheeses that are produced throughout the world. How the cheese ripens and how flavor/texture development progresses, depends mainly on the inoculated starter bacteria and bacteria naturally occurring in milk (Beresford & Williams, 2004; Settanni & Moschetti, 2010). Several aspects are important in cheese production and ripening, and contribute to the specific properties of each cheese type (El Soda, Madkor, & Tong, 2000). These aspects are discussed below.

DOI: 10.4018/978-1-5225-5363-2.ch010

There are two different types of starter cultures used in cheese productions, defined and undefined starter cultures. Defined starter cultures are usually composed of one or more strains with known characteristics. The individual strains in defined starter cultures are usually isolated from undefined complex starter cultures (Smid et al., 2014). Several genera of lactic acid bacteria (LAB), including *Lactobacillus delbrueckii* subsp. *lactis*, *Lactococcus lactis* subsp. *lactis*, *Lactococcus lactis* subsp. *cremoris*, *Leuconostoc* spp., *Streptococcus thermophilus* and *Lactobacillus helveticus* (Cogan, 2014), can be considered as defined starter LAB, whereas the undefined starter cultures usually originated from cheese production plants, environments or farms (Smid et al., 2014). During artisanal production runs, undefined starter LAB are typically added by the transfer of an inoculum from an old batch of cheese to the new batch (a process termed back-slopping) to initiate the fermentation with the activity of the indigenous microbiota present in the inoculum. However, during industrial production, consistent and controlled processes are required in order to ensure a safe and standardized final product (Erkuş et al., 2013). A significant difference between defined starter cultures and undefined complex starters is their sensitivity to bacteriophages. During cheese production, defined starter cultures are generally more vulnerable to bacteriophage attack than undefined complex starters (Stadhouders, 1986). This property as well as their function as a source for the isolation of new dairy strains, explains the interest in undefined complex starter cultures (Erkuş et al., 2013).

This review describes the diversity and evolution of the microbiota during cheese ripening, and how these ever-changing microbial players contribute to the flavor of the fermentation end-product via their proteolytic and lipolytic capacities.

TYPES OF CULTURES USED DURING CHEESE PRODUCTION AND RIPENING

The microorganisms important for cheese production and ripening can originate from milk or they can be intentionally or unintentionally added to cheese production.

Primary Cultures

These microorganisms are carefully selected and deliberately added to milk or cheese during production. Primary cultures are also called 'starter bacteria', 'lactic cultures' or starter lactic acid bacteria (LAB), because of their properties to initiate the production of lactic acid from lactose (Parente & Cogan, 2004).

Primary cultures are frequently different species of several genera of lactic acid bacteria (LAB), including *Lactococcus lactis* subsp. *cremoris, Lactococcus lactis* subsp. *lactis, Leuconostoc* sp., *Streptococcus thermophilus, Lactobacillus delbrueckii* subsp. *lactis*, and *Lactobacillus helveticus*. These cultures are added to the milk for lactic acid production and for the formation of biochemical changes, which create the specific aroma and flavor characteristics during ripening (Collins, McSweeney, & Wilkinson, 2003; Ong & Shah, 2008; Singh, Drake, & Cadwallader, 2003). Intracellular enzymes of LAB, such as proteases and lipases that are released into the cheese matrix due to cell lysis by mechanical, physical and chemical factors, have been shown to participate in the breakdown of peptides, amino acids and fatty acids (Parente & Cogan, 2004). The pH decrease that developed after the addition of primary cultures is also required for rennet activity during coagulation. pH also influences cell lysis, and this release of a vast range of enzymatic activities is the principal factor enabling starter cultures to accelerate the cheese ripening. Primary cultures also contribute to the microbial safety of cheese by producing antimicrobial compounds and by competing with spoilage and pathogenic microorganisms. However, in some traditional produced cheeses no primary culture is added. Instead, these cheeses are depend on adventitious growth of lactic acid bacteria already present in milk or contamination from milk or cheese environment (Mullan, 2014).

Defined Strain Starters (DSS)

Defined strains are usually isolates of mixed cultures. In dairy technology raw milk or cheeses made from raw milk are potentially good sources for defined strains starters. Defined strain starters have some disadvantages such as loss of activity and rapid onset of destructive phage infections. For these reasons, defined strains are usually replaced by pairs of phage-unrelated strains that are rotated daily (Parente & Cogan, 2004).

In recent years, there have been many efforts to isolate new strains from natural sources such as raw milk, plants and different types of traditional dairy products. These strains are expected to have the ability to produce acid rapidly and withstand attack from bacteriophage attacks (Parente & Cogan, 2004).

Mixed Strain Starters (MSS)

Mixed strain starters are obtained by a selection of natural starters and are propagated under controlled conditions. Their composition and stability are improved without losing the ability to withstand the phase attacks (Stadhouders & Leenders, 1984). They are widely used for the production of different types of cheeses such as Cottage, Camembert, Mozzarella, Roquefort, Stilton, Gouda, Edam and Emmantal (Parente & Cogan, 2004). These starter cultures may contain undefined mixtures of strains, like natural or traditional starters, which differ in their physiological and technological properties. For this reason they may contain many phage-resistant strains (Stadhouders & Leenders, 1984).

Natural Starters

Many LAB species that have a potential to be used as starter culture have been isolated from cheese made from raw milk or without the addition of a starter culture. These strains are usually natural contaminants of the milk or cheese factory environment and have advantages compared to pure cultures in the light of including phage-resistant strains. Therefore, many attempts are still made to isolate such strains from raw milk, factories and other natural sources (Parente & Cogan, 2004).

Secondary (Adjunct) Cultures

Similar to primary cultures, secondary cultures are also intentionally added to cheese as adjuncts to the primary acid-producing starter cultures (Parente & Cogan, 2004). They have no function in lactic acid production and are only involved in ripening. Their major role is to contribute to the biochemical changes through their diverse enzymatic systems. These cultures are responsible for reducing the bitter and sweet flavor in cheese (Parente & Cogan, 2004).

Secondary cultures can be bacteria, yeast or fungi (El-Soda et al., 2000) and can be incorporated into the milk or applied deliberately on to the surface of the cheese (Fox, Guinee, Cogan, & McSweeney, 2000; Cogan, 2014). This group is more diverse both from a taxonomical and functional point of view.

The foremost species are molds, such as *Penicillium roqueforti*, *Penicillium camemberti*, and *Penicillium freudenreichii* (Parente and Cogan, 2004). *P. roqueforti* is responsible for the veins in Blue cheeses, while *P. camemberti* is responsible for the white fluffy surface of Camembert and Brie (Cogan, 2014). In addition, *Geotrichum candidum*, *Debaryomyces hansenii*, *Kluyveromyces* spp. are examples of yeast species that are deliberately inoculated on the surface of cheeses. *Brevibacterium linens*, *Arthrobacter* spp. and coryneform bacteria are examples of bacterial species that are used as secondary cultures in surface ripened cheeses such as Tilsit, Munster and Limburger.

Adventitious (Unintentionally Added) Microorganisms

Non-starter lactic acid bacteria (NSLAB) are naturally occurring microorganisms present in milk or beneficial contaminants that originated from cheese production environments. They display a high tolerance to the hostile environment and strongly influence the biochemistry of curd maturation, contributing to development of the final characteristics of the cheese (Settanni & Moschetti, 2010). The presence of NSLAB in cheese was reported to lead to an increase in the level of small peptides and amino acids in cheese (El Soda et al., 2000) and consequently they were reported to play a role in secondary events, such as flavor formation (Bhowmik & Marth, 1990) in certain cheese varieties, particularly artisanal cheeses (Beresford, Fitzsimons, Brennan, & Cogan, 2001). Among NSLAB, *Lactobacillus* strains are the most common and can be found in relatively high numbers, with *L. casei*, *L. paracasei*, *L. plantarum* and *L. curvatus* being the predominant species. In addition, non-*Lactobacillus* species of NSLAB include Pediococci (*Pediococcus acidilactici*, *Pediococcus pentosaceus*) and Enterococci (*Enterococcus durans*, *Enterococcus faecalis*, *Enterococcus faecium*), but they are typically present in relatively low numbers (Chamba & Irlinger, 2004; Fox, McSweeney, & Lynch, 1998).

In addition to their favorable effects, NSLAB also introduce variability into the ripening process that can result in fluctuations in the final characteristics (Franciosi, Settanni, Carlin, Cavazza, & Poznanski, 2008). Due to these properties, significant differences were reported between cheeses produced at the same factory on different days and between cheeses from different vats of the same day (Fitzsimons, Cogan, Condon, & Beresford, 1999). To minimize microbial variability during the ripening process, desired NSLAB could be selected and used as a secondary culture (Settanni & Moschetti, 2010). However, their effects to human health should also be considered when selecting strains as secondary cultures (Settanni & Moschetti, 2010).

Apart from NSLAB, secondary microbiota might also be adventitious in some cases, depending on the selective conditions such as pH, humidity, temperature and water activity, although they are described as selected microorganisms that are inoculated to milk or curd (Fox et al., 2000).

ROLE OF ENZYMES DURING CHEESE RIPENING

In general, the main flavor forming pathways in cheese are catalyzed by enzymes originating from: (i) milk (ii) coagulant, (iii) starter LAB, (iv) secondary/adjunct culture, (v) NSLAB, and (vi) exogenous enzymes. These enzymes play a principal role in the formation of amino acids and fatty acids that contribute directly to the basic taste of cheese and indirectly to cheese flavor because of their role as precursors for aroma components (McSweeney, 2004).

Indigenous (Native) Milk Enzymes

These enzymes originate mainly from leakage of blood or immune system components into the milk. The key indigenous protease enzymes of milk include plasmin (milk alkaline protease) and cathepsin D (milk acid proteinase). The hydrolysis of milk proteins by proteases affects the texture and flavor of dairy products. This can have either beneficial or detrimental effects, depending on the extent of hydrolysis and type of dairy product (Fox, McSweeney,n & Wallace, 1993). Plasmin and cathepsin D act on β- and α_{S1}-casein, respectively. Plasmin occurs in milk along with its precursor plasminogen (inactive form of plasmin) and is associated with the casein micelles (Bastian & Brown, 1996). Cathepsin D originates from the lysosomes of somatic cells (Wilkinson & Kilcawley, 2005). Similar to indigenous proteases, lipoprotein lipase (LPL) is the indigenous milk lipase (Fox & Stepaniak, 1993) and originates from blood. It contributes to the lipolysis and off-flavor formation in cheese. LPL is rather nonspecific for the type of fatty acid but is specific for sn1 and sn3 positions of mono-, di- and triglycerides. Therefore, lipolysis in milk leads to preferential release of short- and medium-chain fatty acids, which in milk triglycerides are esterified predominantly at the sn3 position.

Coagulant

Addition of milk clotting enzymes such as chymosin and pepsin are able to hydrolyze peptide bonds and are responsible for the coagulation during cheese production (Visser, 1993) by the specific cleavage of the Phe105-Met106 bond of κ-casein, which destabilizes the casein micelle structure (Yegin et al., 2011). The ready-to-use forms of these enzymes are known as rennet (Koçak & Güzel-Seydim, 2011). Enzymatic milk coagulation is a two-phase process. In the first phase, the enzyme hydrolyzes the κ-casein molecule and splits the protein into two fragments: hydrophobic para-κ-casein and a hydrophilic macropeptide. The second phase consists of the coagulation of the casein micelles that have been destabilized by proteolytic hydrolysis (Claverie-Martin & Vega-Hernàndez, 2007).

Other than chymosin from mammalian species (mainly bovine), there are also commercially available enzymes produced by *Rhizomucor* spp., which have gained wide industrial acceptance as substitutes for the bovine chymosin (Claverie-Martin & Vega-Hernàndez, 2007). However, it is known that fungal rennet preparations may present a degree of residual proteolytic activity that subsequently participates in proteolysis of the caseins during ripening and results in the production of bitter peptides during cheese ripening (Yegin et al., 2011).

Indigenous enzymes and coagulants are not stable against heat treatments. Inactivation of indigenous protease (Farkye & Imafidon, 1995) and lipase (Deeth & Fitz-Gerald, 1983) was observed during heat treatments of milk or curd. It was reported that heating above 65°C resulted in reduced activity in plasmin (Metwalli, De Jongh, & Van Boekel, 1998). Another study (Deeth & Fitz-Gerald, 1983) demonstrated that activity of LPL in raw milk cheeses was extensively inactivated at 72°C for 15 seconds. Rennet enzymes can also be inactivated at the curdling stage and are inhibited by added salt. Hayes, Oliveira, McSweeney, & Kelly (2002) reported that chymosin activity showed minimal inactivation during the production of Swiss-type cheese, whereas after cooking residual chymosin activity in whey decreased to approximately 10%. It was also reported that residual chymosin activity decreased during ripening (Hayes et al., 2002). Therefore, especially for ageing cheeses, it might be considered that the most significant factor during ripening is the presence of microbial enzymes released by primary (starter LAB) and secondary cultures added by cheese makers, as well as NSLAB present in milk or contaminants from environment (McSweeney, 2004; Upadhyay, McSweeney, Magboul, & Fox, 2004).

Proteases

Proteases can be divided into four classes: cysteine proteases, serine proteases, aspartic proteases and metalloproteases. Each class has a characteristic set of functional amino acid residues arranged in a particular configuration to form the active site. Proteases can also be subdivided into two major groups based on their ability to cleave N- or C-terminal peptide bonds (exopeptidase) or internal peptide bonds (endopeptidase) (Yegin et al., 2011).

Aspartic (acid) protease, having an endopeptidase property with two aspartic acid residues, is the most important protease group in cheese technology because of its milk clotting effect during cheese production (Claverie-Martin & Vega-Hernàndez, 2007). The commercial milk-clotting enzyme (rennet mainly in the form of chymosin) also belongs to the group of aspartic proteases. Most aspartic proteases show maximal activity at low pH values (~3–5) and are inhibited by pepstatin A (Sumantha, Larroche, & Pandey, 2006). However, the cell wall associated proteases of LAB are serine proteases having a molecular weight of 145kDa and a pH optimum of 5.5-6.2 (Fox et al., 2000).

ENZYMES FROM MICROBIAL SOURCES

Enzymes of Starter LAB and NSLAB

The starter LAB reach maximum numbers of living cells shortly after cheese production, followed by a gradual decrease during ripening as a consequence of lysis. In contrast, NSLAB are typically present at low numbers in the beginning of the cheese production process, but increase during the months of ripening and ultimately dominate the microbiota of long-ripened cheeses (Fox et al., 2000; Beuvier & Buchin, 2004). The proteases in LAB are anchored to the cell membrane and protrude through the cell wall (Fox et al., 2000). The proteolytic system of *Lactococcus* was described as cell envelope-associated proteases (CEPs) and are responsible for the formation of many small peptides in cheese and the release of amino acids (Fox & McSweeney, 1996). In most cheese varieties, LAB are the major contributors to flavor development through the action of their intracellular enzyme systems (Wilkinson & Kilcawley, 2005).

Similar to starter LAB, NSLAB strains were also shown to contain an active proteolytic system that contributes the cheese ripening (Broome, Krause, & Hickey, 1990). They appear to contribute little to primary proteolysis in Cheddar cheese but do contribute to the release of free amino acids (Fox et al., 1998). Therefore, studies performed on the proteolytic system of NSLAB showed the possibility of their application as adjunct cultures to enhance proteolysis or to generate specific flavor characteristics in cheese (Klein & Lortal, 1999).

Enzymes of Secondary Cultures

Most cheese varieties, particularly traditional cheeses made from raw milk, have a secondary part in their microbiota which is important for functionalities other than acid production (Fox et al., 2000; Gupta, Gupta, & Rathi, 2004). Among fungi, Penicillia are good producers of extracellular enzymes, such as lipases and proteases (Hamlyn et al., 1981). *P. roqueforti* and *P. camemberti* are reported to secrete aspartyl and metalloproteases and specifically hydrolyze α_{s1}- and β- caseins. Similarly, extracellular and intracellular aminopeptidases purified from *Brevibacterium linens* are particularly active on

α_{S1}- and β- caseins (Rattray & Fox, 1997, 1999). *Pseudomonas* spp. is another bacterial strain that was reported to have proteolytic activity (Schokker & van Boekel, 1997). Propionibacterium spp. are weakly proteolytic but strongly peptidolytic.

Exogenous (Adjunct) Enzymes

The main aim of employing exogenous enzymes in cheese industry is to accelerate cheese ripening. Furthermore, the use of these enzymes provides high organoleptic properties (El Soda et al., 2000). Commercial protease and lipase preparations (Wilkinson & Kilcawley, 2005) or selected strains can be added to milk or cheese directly to produce the relevant enzymes (El Soda et al., 2000). In contrast to NSLAB, adjuncts are specifically selected and intentionally added to cheese or milk to balance the textural and flavor profiles that are otherwise hard to attain (Wilkinson & Kilcawley, 2005). Despite their positive effects, addition of these enzymes can also stimulate the bitter taste in cheese (Wilkinson & Kilcawley, 2005).

BIOCHEMICAL EVENTS DURING CHEESE RIPENING

The biochemical alterations in cheese during ripening may be grouped into primary and secondary events. Primary events include the metabolism of residual lactose (glycolysis) and of lactate and citrate, and breakdown of fat (lipolysis) and protein (proteolysis). Lactose is broken down to glucose and galactose. Glucose is efficiently used for the production of ATP and lactate, whereas galactose accumulates in dairy products due to partial lactose fermentation by lactic acid bacteria (Neves et al., 2010). Following the primary events, secondary events occur including the catabolism of fatty acids and amino acids. Products of secondary biochemical events directly influence the development of many volatile flavor compounds (McSweeney & Sousa, 2000). Primary and secondary biochemical events are presented in Figure 1.

Glycolysis

Glycolysis includes the fermentation of lactose to lactic acid by lactic acid bacteria and it is an essential primary reaction in the manufacture of all cheese varieties. Glycolysis is the fastest reaction that occurs during ripening of cheese. In many cheese varieties glycolysis is terminated with the transition of lactose to lactic acid, by the inhibitory effect of salt. However, in some cheese varieties this process partially continues during cheese ripening (Fox et al., 1993; Fox & McSweeney, 1996).

The three main metabolic processes that occur during glycolysis are lactose, lactic acid and citrate metabolisms. During lactose metabolism the largest proportion of lactose (98%) in milk is removed with separated whey during cheese production. However, the residual lactose in cheese curd is rapidly metabolized to lactate and/or lactic acid during the early stages of ripening by the activity of LAB or NSLAB, which decreases the pH from pH 6.5-6.6 to 6.2-6.4. The extent of lactate metabolism depends mainly on the salt-in-moisture (S/M) content in the curd. At low S/M content on low populations of NSLAB, residual lactose is converted mainly to L-lactate by LAB. Whereas, at high populations of NSLAB, considerable amounts of D-lactate are formed by conversion of L-lactate (Fox et al., 2000). Moreover, secondary microbiota contributes to lactate metabolism in some cheese varieties, such as surface mold-ripened, surface smear-ripened and Blue cheese. The metabolism of lactate is reported to

Figure 1. Principal biochemical mechanisms occurring during cheese ripening (Reproduced from Fox et al., 2004)

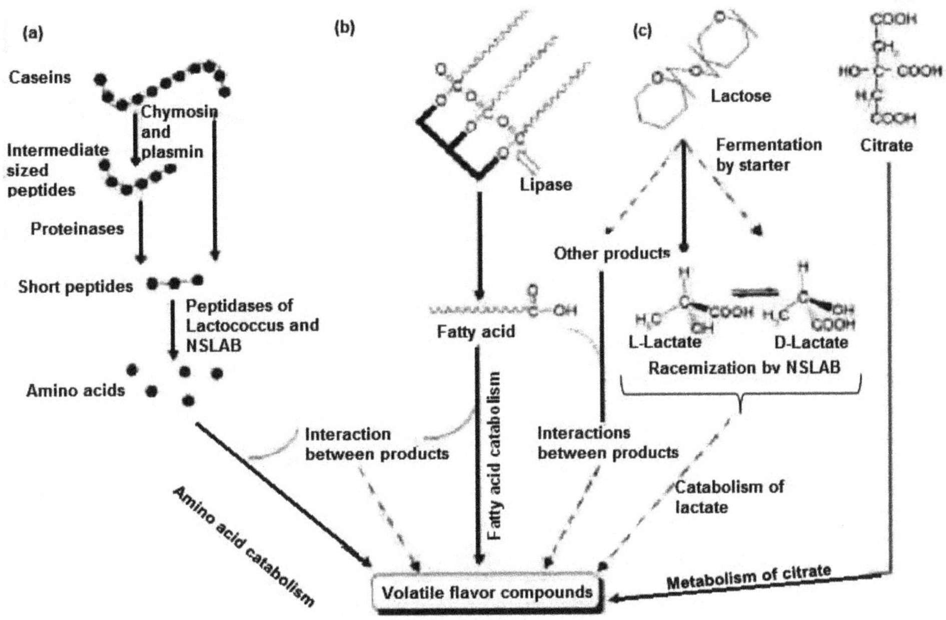

be very extensive in these varieties. The secondary microbiota that colonize and dominate the cheeses rapidly metabolize lactate to CO_2 and H_2O and cause an increase in pH. When lactate is exhausted, secondary cultures metabolize the proteins, producing NH_3, which diffuses inside the cheese and increase the pH. The elevated pH stimulates the activity of native or microbial proteases which contributes to proteolysis in the further stages (McSweeney & Fox, 2004).

During glycolysis, moreover, citrate metabolism is performed by (Cit^+) microorganisms and mainly important for Dutch cheese with holes (eyes). The citrate found in milk is approximately 1.8 g/L and 94% of it is in soluble form, which is also lost with whey during production. The remaining 6% is in colloidal form, and is metabolized to metabolic products such as acetic acid, diacetyl and CO_2 by *Leuconostoc* spp. and (Cit^+) *Lactococcus* spp. (Fox & McSweeney, 1996; Parente & Cogan, 2004).

Lipolysis (Breakdown of Fat)

Lipolysis is defined as the hydrolysis of ester bonds in triglycerides, resulting in the generation of fatty acids and glycerol (Schweiger et al., 2014) (Figure 2). The release of free fatty acids (FFA) is catalyzed by lipases, which could be residual native milk lipoprotein lipase (LPL), rennet preparations containing pregastric esterase (PGE), lipase and esterase of starter LAB and NSLAB, adjunct cultures or secondary fungal cultures and exogenous lipolytic enzymes (Wilkinson & Kilcawley, 2005). Lipolysis and modification of resultant FFAs are significant contributors to overall cheese aroma (McSweeney & Sousa, 2000). However, extensive lipolysis is also considered undesirable in most cheese varieties (Sousa & Malcata, 1997) as it can cause a rancid taste due to an excessive accumulation of volatile FFAs (Collins et al., 2003; Wilkinson & Kilcawley, 2005).

Figure 2. Formation of FFAs from triglyceride during lipolysis (Based on Fox et al., 1993)

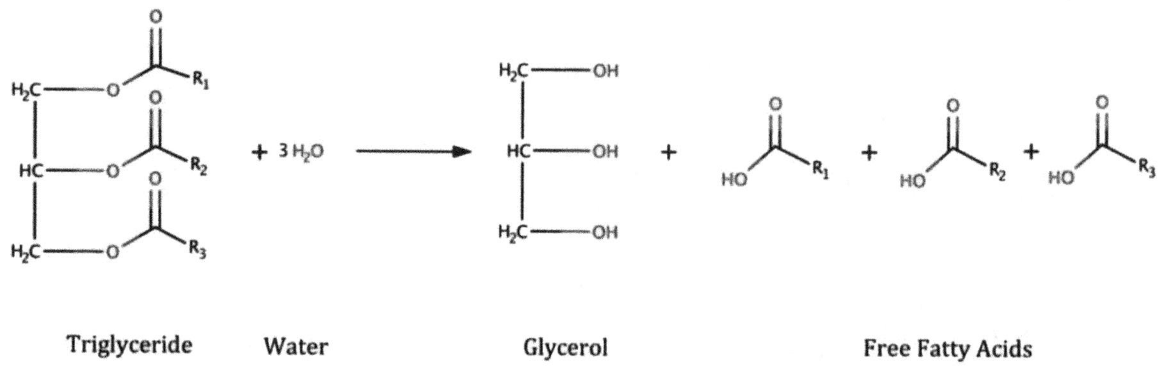

Lipolysis is an important biochemical event that takes place during the ripening of many cheese varieties. Particularly, mold-ripened cheeses and cheeses from goat and sheep milk are characterized by extensive lipolysis (Molimard & Spinnler, 1996). The reason of this is that the triglycerides of ruminant (sheep, goat) milk fat are rich in short-chain fatty acids which have low thresholds and contribute significantly to the flavor of cheese, when liberated. The intensity of lipolysis in mold-ripened cheeses is often related to the lipolytic ability of the secondary microbiota or adjunct culture. The most common lipolytic organisms associated with cheese are *Penicillium* spp. which secrete highly active extracellular lipases. The two lipases secreted by *P. roqueforti* were characterized previously, and it was shown to secrete two lipases, with a pH optimum at 6.0-6.5 and 7.5-8.0, which belong to the acid and alkaline lipases, respectively. Similarly, smear ripened cheese varieties are often characterized by high levels of lipolysis (McSweeney, 2004).

Catabolism of Fatty Acids

FFAs released by lipolysis are used as the precursors for consecutive secondary reactions and are converted to other sapid and aromatic compounds like methyl ketones, lactones and alcohols. The aroma of Blue cheese is dominated by saturated n-methyl ketones including heptan-2-one, nonan-2-one and undecan-2-one. The concentrations of methyl ketones are proportional to the level of lipolysis (Fox et al., 2000).

The catabolism of fatty acids in cheese by *Penicillium* spp. involves 3 main stages:

1. Oxidation of β-ketoacids,
2. Decarboxylation to methyl ketones with one less carbon atom,
3. Reduction of methyl ketones to the corresponding secondary alcohol.

A number of factors affect the rate of methyl ketone production, including temperature, pH, physiological state of the microorganism and the ratio of the fatty acids' concentration to the dry weight of spores. Both resting spores and fungal mycelium are capable of producing methyl ketones, but the rate of production of methyl ketones does not directly depend on the concentrations of FFA precursors. In fact, high concentrations of FFA are toxic to *P. roqueforti* (Fox et al., 2000).

Lactones are cylic esters resulting from the intramolecular esterification of a hydroxy acid through the loss of water to form a ring structure. Lactones possess a strong aroma, which are important for the overall aroma of cheese, despite not being specifically cheese-like. δ- and γ-lactones are found in cheese. They may have originated spontaneously from their corresponding hydroxy acids, which were released from triglycerides in a previous stage or produced from keto acids released by lipolysis, followed by reduction to hydroxy acids. δ-lactones have very low flavor thresholds and their concentrations correlate with age and flavor intensity. Moreover, they have been reported to contribute to the aroma of Cheddar cheese (Fox et al., 2000).

Proteolysis (Breakdown of Protein)

Proteolysis is the hydrolysis of cheese proteins leading to the formation of medium and small soluble peptides and free amino acids from large water-insoluble peptides (comparable in size to intact caseins). It is the most complex of the three primary biochemical events during cheese ripening. Proteolysis is widely considered to be the most important process for the development of sensory and textural properties of cheese (Fox, 1989; McSweeney, 2004), except for Blue cheese which is dominated by the mechanism of lipolysis (Fox et al., 2000). Unfortunately, it also contributes to off-flavors in cheeses (Upadhyay et al., 2004). Proteolysis contributes to cheese ripening through four pathways:

1. Direct contribution to flavor via the formation of peptides and amino acids or indirect contribution via catabolism of amino acids to amines, acids, thiols, thioesters etc.,

Figure 3. Catabolism of fatty acids and the transition of products to volatile flavor compounds in cheese (Reproduced from Molimard & Spinnler, 1996)

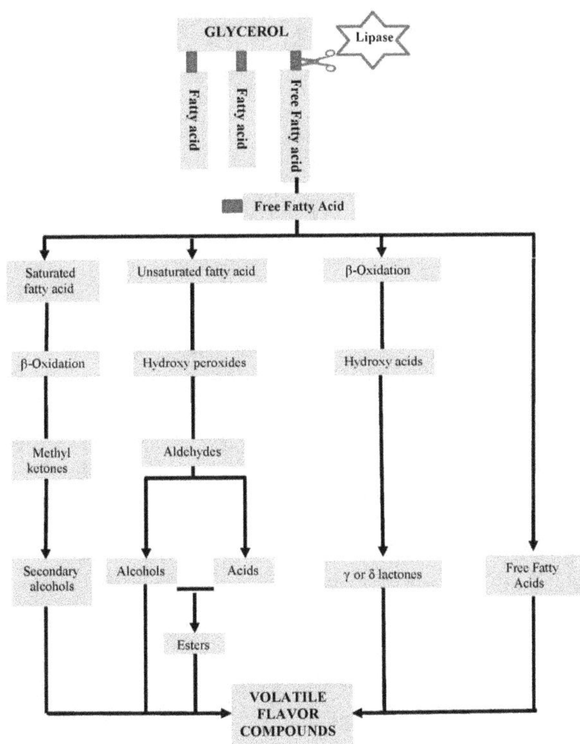

2. Higher release of sapid compounds during mastication,
3. Alterations in pH via the formation of $NH_{3,}$
4. Modifications of the texture of cheese due to the breakdown of the protein network, increase in pH and greater water binding by the newly-formed amino and carboxyl groups (Fox, 1989; McSweeney, 2004).

Proteolysis is catalyzed by proteases and peptidases from residual coagulant, native milk, (especially plasmin), starter LAB, NSLAB, and in certain varieties other secondary microbiota. Particularly, extensive proteolysis occurs in surface mold- and surface smear- ripened cheeses because of the very potent proteolytic system of their secondary cultures (Fox et al., 2000). *B. linens* are reported to have both intracellular and extracellular proteases, but mostly extracellular proteases were purified and characterized, all of which are described as serine protease (Hayashi, Revell, & Law, 1990).

Catabolism of Amino Acids

Amino acid catabolism, which is particularly significant in mold- and smear ripened cheese varieties, is an indicator of extensive proteolysis (Fox et al., 2000). As shown in Figure 4, catabolism of amino acids involves transamination, decarboxylation, desulfuration, deamination and hydrolysis of amino acid side chains, leading to the production of a wide array of compounds, including carboxylic acids, amines, NH_3, CO_2, aldehydes, alcohols, thiols and other sulphur compounds, phenols and hydrocarbons. All of these are particularly important in many cheese varieties. Decarboxylation involves the conversion of amino acid to the corresponding amine with loss of CO_2. Deamination results in the formation of NH_3

Figure 4. Amino acid catabolism and formation of cheese flavor (Reproduced from Molimard & Spinnler, 1996)

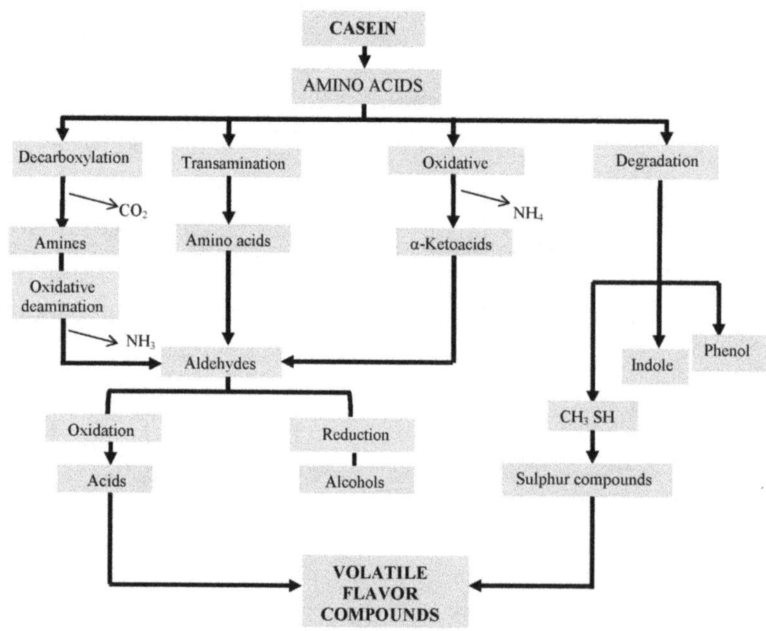

and α-ketoacids. Further, transamination results in the formation of other aminoacids by the action of transaminases. Aldehydes formed by the above processes can then be oxidized to acids or reduced to the corresponding alcohols. In addition, sulphur containing compounds are another volatile group produced from aminoacids (Yvon & Rijnen, 2001; Fox et al., 2000).

Aminotransferases secreted from microorganisms are specific to different groups of amino acids, such as branched-chain (Leu, Ile, Val), aromatic (Phe, Tyr, Trp), sulphuric (Cys, Met) or acidic (Asp). Further, a specific α-keto acid is produced from each amino acid by the help of aminotransferases (Yvon and Rijnen, 2001). Starter LAB have a limited capacity for synthesis of amino acids. However, as mentioned above, the number of essential amino acids is strain-dependent (Ayad, Verheul, de Jong, Wouters, & Smit, 1999). Therefore, it can be considered that both NSLAB and secondary cultures need fewer amino acids to grow and produce α- keto acid decarboxylase enzymes than the industrial starter LAB strains (van Kranenburg et al., 2002). Wild isolates of *L. lactis subsp. cremoris* and *L. lactis* subsp. *lactis* strains are reported to require only 1-3 amino acids. The absence of some amino acid biosynthethic pathways in NSLAB and secondary cultures might be a consequence of their adaptation to dairy products. In milk, amino acids are readily available by the proteolysis of caseins. Wild strains are not naturally associated with a rich environment such as milk, which makes them more dependent on synthesis of their own amino acids compared to industrial strains (van Kranenburg et al., 2002).

Effect of Microorganisms and Enzymes on the Characteristics of Divle Cave Cheese: A Case Study

The wealth of information on the metabolic and enzymatic processes leading to components important for the flavor of cheeses is heavily biased towards industrial cheeses, whereas many locally produced, artisanal cheeses have not been researched at all. As an example, Divle Cave cheese is a semi-hard cheese made from unpasteurized and semi-skimmed ewe's milk which is traditionally produced in spring when milk is abundant, in Karaman, a rural region of Turkey. The south of this region consists of a big cavity among the mountains, named Divle cave, at a depth of 70 meters. Cheeses are produced without using any starter culture or fungal spores and are ripened in goatskin bags in Divle cave for 4 months. Divle Cave cheese has its own characteristics mainly due to the ripening period in the cave where the temperature is 8-10ºC and the humidity is 85-90%. This semi-hard cheese has a distinctive taste, flavor and crumbly texture.

The sensory characteristics of cheeses cannot be dissociated from the diversity of their microbiota and microbial activities. Divle Cave cheese was researched for the determination of the effects of the microbial community and enzymes secreted by the individual strains of this microbiota on the chemical and sensory properties of cheese. The bacteria, filamentous fungi and yeasts present in the middle and at the end of ripening were identified by molecular techniques. Moreover, protease and lipase enzyme production by these strains was also described. The production flow chart of Divle Cave cheese is given in Figure 5.

Complex Microbiota of Divle Cave Cheese

A complex microbiota with species that belong to the classes of Bacilli (9), Gammaproteobacteria (3) and Actinobacteria (11), yeasts (5), and filamentous fungi (18) was detected in Divle Cave cheese during the ripening period. Besides the wide diversity determined in cheese microbiota, particularly the

Figure 5. Production flow chart of Divle Cave cheese

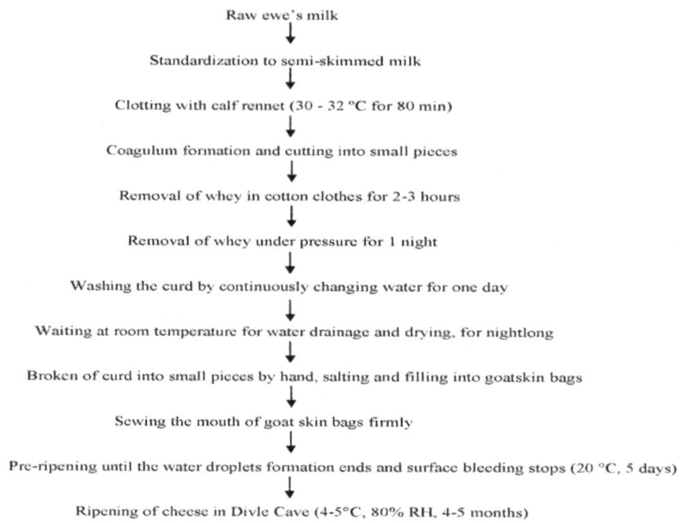

diversity among filamentous fungi was remarkable, considering that 15 different species of *Penicillium* were identified. The increase of pH in the second half of the ripening process favored the growth of fungi and Actinobacteria, and interestingly, no LAB were isolated except for *L. paraplantarum*. During the progress of ripening, the broad microbial diversity decreased in the cheese core, which is likely due to the anaerobic environment and the evolution of the chemical composition inside the cheese. In contrast to lower counts determined inside the cheese, particularly yeast and filamentous fungi on the outside of the cheese showed a gradual increase throughout the whole process and are detected at high levels at the end of ripening (Ozturkoglu-Budak, Figge, Houbraken, & de Vries, 2016a). This confirmed the high alcohols, ketones, and esters determined as the major volatile groups during volatile profiling analyses, which are probably produced by the abundant yeast and filamentous fungi in Divle cheese (Ozturkoglu-Budak et al., 2016b). Another strong effect on the cheese flavor caused by the presence of a highly abundant and diverse fungal community was thought to be the formation of methyl ketones by their enzymatic activities (Molimard & Spinnler, 1996). These compounds are produced by the oxidation of fatty acids and related secondary alcohols produced from FFA, and play an important role in determining the characteristic flavor components in Blue cheese (Gallois & Langlois, 1990). Since methyl ketones have typical odors detectable at a low perception threshold, these compounds were also reported as significant contributors to the aroma of Camembert cheese (Di Cagno et al., 2003; Izco & Torre, 2000) and Blue cheeses (Hayaloglu, Fox, Guven, & Cakmakci, 2007).

Another interesting finding of this study is that no pathogenic microorganisms were determined in the cheese at the end (120th day) of ripening. Although we detected coliforms, staphylococci and *Enterobacteriaceae* until the 90th day of ripening, but disappeared afterwards (Ozturkoglu-Budak, 2016d). Therefore, it could be considered that the microbiological safety of Divle Cave cheese was ensured through the 90-days ripening process, even though it is produced from raw ewes' milk. This result demonstrated that the microbiological risk of cheese can be reduced by a suitable ripening process. The ability of traditional cheeses to combat pathogens is related to native antipathogenic strains or microbial consortia rather than natural non-microbial inhibitors from milk (Montel et al., 2014). The inhibition seems to be associated with qualitative and quantitative composition of the above mentioned strains and the effects of their metabolites on other microorganisms. Particularly LAB, the dominant indigenous microorganisms of raw milk cheese, produce pathogen-inhibiting substances such as bacteriocins, organic acids and hydrogen peroxide (Yoon, Lee, & Choi, 2016).

Contributors to the Ripening and Flavor Development of Divle Cave Cheese

A relationship was also described between the identified microbiota and the temporal development of the volatile compounds during production and ripening of Divle Cave cheese. In addition to their role in inhibition of food-borne pathogens, the naturally existing microbial community also develops the flavor and sensory characteristics of a specific cheese variety, by its enzyme profiles along with the native enzymes of milk and rennet. Analyses of SPME-GC-MS revealed a total of 110 volatile compounds derived from Divle Cave cheese during production and ripening comprising 13 acids, 18 alcohols, 12 ketones, 15 esters, 12 terpenes, 7 aldehydes, 18 hydrocarbons and 15 miscellaneous compounds. Volatile compound analyses also identified the most abundant or key flavor compounds, likely contributing to the characteristic aroma and unique taste of this artisanal raw ewe's milk cheese. Key flavor compounds, such as 2-butanone, 2-butanol and ethyl butanoate, that are specific on sensory properties of Divle Cave cheese were found to belong to the volatile compounds which are known as characteristically produced by some of the determined fungal strains (Ozturkoglu-Budak et al., 2016b).

According to volatile profile analyses, carboxylic acids, secondary alcohols, methyl ketones, and ethyl esters were detected both frequently and at high concentration at different stages of the ripening process (Ozturkoglu-Budak et al., 2016b). A high alcohol amount in the beginning of production until the 60th day of ripening could originate from the microbial populations of LAB, enterococci, and yeast derived from raw milk (Bontinis, Mallatou, Pappa, Massouras, & Alichanidis, 2012). However, when ripening progressed a decrease was observed in alcohol amounts, which could be the result of the esterification reaction between alcohols and fatty acids (Delgado, González-Crespo, Cava, García-Parra, & Ramírez, 2010). This observation correlates with the increasing ester amounts towards the end of ripening in Divle Cave cheese. In addition, the high fat content of ewe's milk (4.58%), corresponding to high fat content of cheese (16.5-19.5%) during ripening, might have promoted the presence of high amounts of lipolysis products. Carboxylic acids are not only aroma components, but can also serve as precursors for the formation of esters, methyl ketones, alcohols and lactones (Collins et al., 2003). Due to their low detection thresholds, particularly short- and medium-chain carboxylic acids are thought to contribute significantly to the aroma profile of cheeses (Tavaria, Silva Ferreira, & Malcata, 2004). Therefore in Divle Cave cheese, lipolysis was hypothesized as the main biochemical reaction for production of carboxylic fatty acids, which can be accepted as the initiator compound for the flavor development (Ozturkoglu-Budak, 2016d).

To answer this question, lipolysis and proteolysis levels were determined at specific points beginning from raw milk until the end of ripening by using GC and HPLC, respectively. Within this scope, accumulated FFA and small peptides are considered as representative of the degree of lipolysis and proteolysis. As expected, the lipolysis and proteolysis degree generally increased during the whole ripening period and a substantial accumulation of FFA and peptides was observed at the end of ripening. Relatively high levels of proteolysis and lipolysis determined at the end of ripening could be explained by the indigenous and secondary microbiota as well as native proteases, lipases and esterases derived from raw milk.

The FFA and peptide accumulation was due to contributions of the secondary microbiota producing the catabolic enzymes that we isolated from cheese, as well as native proteases, lipases and esterases present in raw milk. The action of proteases causes the degradation of hydrophobic peptides and release of low molecular weight peptides and amino acids (Gonzales De Llano, Ramos, Polo, Sanz, & Martinez-Castro, 1995), as was also demonstrated in Divle Cave cheese (Ozturkoglu-Budak, 2016d). The strains potentially contributing to ripening were selected according to the protease and lipase production of individual strains on sterilized Divle Cave cheese, prepared from irradiated cheese. Yeasts, fungal and bacterial strains were inoculated on these culture media individually and incubated for growth and enzyme production at conditions adapted from authentic conditions. The intense lipolysis observed (Ozturkoglu-Budak, 2016d) is in agreement with the data shown by (Desmasures, 2014), in which mold-ripened cheeses are characterized by an intense lipolysis as a result of the extracellular lipases particularly secreted from *Penicillium camemberti, P. roqueforti,* and *G. candidum*.

Lipases were mainly secreted by yeast and fungal strains in Divle Cave cheese. *Y. lipolytica* was identified as the best contributor to cheese ripening with regard to lipase production, particularly early in the ripening process. Fungal strains of *M. racemosus, P. commune, P. biforme, P. brevicompactum* and *P. cavernicola* were the best lipase producing strains after *Y. lipolytica*. In addition, most of the *Penicillium* spp. were also responsible for proteolysis during ripening. *P. brevicompactum* had the highest protease production followed by *P. corylophilum* and *P. chrysogenum*. Among the bacterial strains, indigenous and secondary microbiota including *M. luteus, B. antiquum, B. stratosphericus, P. proteolytica* and *P. glacincola* were considered to be important for the ripening process (Ozturkoglu-Budak, Wiebenga, Bron, & de Vries, 2016c). Unexpectedly, we detected lower proteolytic and lipolytic activities of LAB. This may be caused by the fact that they are intracellular enzymes in these bacteria that are not released into the cheese matrix without cell lysis (El Soda et al., 2000). The majority of the proteases secreted by most yeasts and filamentous fungi were found to belong to aspartic (acid) proteases. Milk clotting enzymes also belong to this group of enzymes (Claverie-Martin & Vega-Hernàndez, 2007). This finding indicates the importance of proteases produced by yeast and fungal strains for cheese ripening. In contrast, inhibition studies revealed that the bacterial proteases mainly belonged to the serine protease class (Ozturkoglu-Budak et al., 2016c).

FFA content determined by lipolysis analyses revealed similarities with the carboxylic acid content of volatile profile results, in which SCFA such as butyric (C4:0) and pentanoic (C5:0) acids were detected at the end of ripening. Among FA, SCFA (C4:0-C8:0) and MCFA (C10:0-C14:0) are known to contribute significantly to the aroma profile of cheeses due to their low detection thresholds (Kraggerud, Skeie, Høy, Røkke, & Abrahamsen, 2008; Pinho, Peres, & Ferreira 2003; Tavaria et al., 2004). In particular the more volatile SCFA were reported to directly affect the aroma of cheese in many aged cheeses (Curioni & Bosset, 2002). Therefore, our similar results reiterate the significant contribution of these compounds to the aroma of Divle Cave cheese. Interestingly, the MCFA amount was higher in Divle Cave cheese than SCFA, despite the broad volatile diversity. The reason for this could be esterification of SCFA with alcohols (Delgado et al., 2010). A positive outcome of this is the prevention of rancid taste which can stem from excessive butyric acid accumulation (Ozturkoglu-Budak, 2016d).

Towards the end of ripening, the breakdown of accumulated branched-chain amino-acids leads to the production of flavor compounds including aldehydes, alcohols and acids (McSweeney, 2004). The enzymatic conversion of amino acids such as leucine, isoleucine and valine is catalyzed by aminotransferases (AT) and they formed the carboxylic acids of 3-methyl butanoic (isovaleric acid), 2-methyl butanoic and 2-methyl propanoic acid (isobutyric acid), respectively (Ardö, 2006; Smit, Smit, & Engels, 2005).

3-methylbutanoic acid was the most abundant branched-chain fatty acid in Divle Cave cheese (Ozturk-oglu-Budak, 2016d). 3-methylbutanoic acid gives rancid, cheese-like and sweet flavor to sheep cheeses (Bergamini, Wolf, Perotti, & Zalazar, 2010; Marilley & Casey, 2004; Yvon & Rijnen, 2001). 2-methyl propanoic acid (isobutyric acid) gives a very similar aromatic note to Camembert and Cheddar cheeses with the aroma of butyric acid, which is a short-chain fatty acid arising mainly from lipolysis (Yvon & Rijnen, 2001). The high amount of branched chain acids synthesized from amino acids are associated with severe proteolysis (Carbonell, Nunez, & Fernandez-Garcia, 2002), which is similar to our finding in Divle Cave cheese. Moreover, this excessive degradation of proteins in long ripened cheeses causes the formation of bitter peptides. This bitterness in cheese was described as a necessary consequence of proteolysis in previous reports (Lemieux & Simard, 1991). Therefore, a better understanding of the origin of these compounds is needed in order to control and accelerate the formation of cheese flavor.

CONCLUSION

The type of work described in this paper encompasses both the microbial profiling and the assessment of the temporal protease and lipase activities during the cheese production and ripening processes. It highlights that the native and secondary microbiota and their enzymes are associated with more complex volatile profiles and higher scores for specific sensory attributes. Ultimately, dedicated research efforts on specific individual members of the cheese microbiota will include full genome sequencing and proteolysis and lipolysis profiling. This would allow the scientific community to better understand the specific enzymes and/or metabolic routes involved in the metabolites important for the flavor of cheeses. Moreover, from a more applied point of view, it should be noted that such molecular understanding could also lead to the more frequent employment of individual microbial strains as adjuncts in cheese production as a more economical alternative for the employment of a purified enzyme.

REFERENCES

Ardö, Y. (2006). Flavor formation by amino acid catabolism. *Biotechnology Advances*, *24*(2), 238–242. doi:10.1016/j.biotechadv.2005.11.005 PMID:16406465

Ayad, E. H. E., Verheul, A., de Jong, C., Wouters, J. T. M., & Smit, G. (1999). Flavor forming abilities and amino acid requirements of *Lactococcus lactis* strains isolated from artisanal and non-dairy origin. *International Dairy Journal*, *9*(10), 725–735. doi:10.1016/S0958-6946(99)00140-5

Bastian, E. D., & Brown, R. J. (1996). Plasmin in Milk and Dairy Products: An Update. *International Dairy Journal*, *6*(5), 435–457. doi:10.1016/0958-6946(95)00021-6

Beresford, T. P., Fitzsimons, N. A., Brennan, N. L., & Cogan, T. M. (2001). Recent advances in cheese microbiology. *International Dairy Journal*, *11*(4-7), 259–274. doi:10.1016/S0958-6946(01)00056-5

Beresford, T. P., & Williams, A. (2004). The microbiology of cheese ripening. In P. F. Fox, McSweeney, P. L. H., Cogan, T. M. & Guinee, T. P. (Ed.), Cheese: Chemistry, Physics and Microbiology: Vol 1. General Aspects (pp. 287-318). London: Elsevier. doi:10.1016/S1874-558X(04)80071-X

Bergamini, C. V., Wolf, I. V., Perotti, M. C., & Zalazar, C. A. (2010). Characterisation of biochemical changes during ripening in Argentinean sheep cheeses. *Small Ruminant Research*, *94*(1-3), 79–89. doi:10.1016/j.smallrumres.2010.07.004

Beuvier, E., & Buchin, S. (2004). Raw milk cheeses. In P. F. Fox, McSweeney, P. L. H., Cogan, T. M. & Guinee, T. P. (Ed.), Cheese: Chemistry, Physics and Microbiology: Vol 1. General Aspects (pp. 319-345). London: Elsevier. doi:10.1016/S1874-558X(04)80072-1

Bhowmik, T., & Marth, E. H. (1990). Role of Micrococcus and Pediococcus species in cheese ripening. *Journal of Dairy Science*, *73*(4), 859–866. doi:10.3168/jds.S0022-0302(90)78740-1

Bontinis, T. G., Mallatou, H., Pappa, E. C., Massouras, T., & Alichanidis, E. (2012). Study of proteolysis, lipolysis and volatile profile of a traditional Greek goat cheese (Xinotyri) during ripening. *Small Ruminant Research*, *105*(1-3), 193–201. doi:10.1016/j.smallrumres.2012.01.003

Broome, M. C., Krause, D. A., & Hickey, M. W. (1990). The use of non-starter lactobacilli in Cheddar cheese manufacture. *Australian Journal of Dairy Technology*, *45*, 67–73.

Carbonell, M., Nunez, M., & Fernandez-Garcia, E. (2002). Evolution of the volatile components of ewe raw milk La Serena cheese during ripening. Correlation with flavour characteristics. *Le Lait*, *82*(6), 683–698. doi:10.1051/lait:2002042

Chamba, J. F., & Irlinger, F. (2004). Secondary and adjunct cultures. In M. P. L. H. Fox P. F., Cogan T. M., & Guinee T. P. (Ed.), Cheese: Chemistry, Physics and Microbiology (Vol. 1, pp. 191-206). London: Elsevier Academic Press. doi:10.1016/S1874-558X(04)80068-X

Claverie-Martin, F., & Vega-Hernàndez, M. C. (2007). Aspartic Proteases Used in Cheese Making. In J. Polaina & A. P. MacCabe (Eds.), *Industrial Enzymes: Structure, Function and Applications* (pp. 207–219). Dordrecht: Springer Netherlands. doi:10.1007/1-4020-5377-0_13

Cogan, T. M. (2014). Srater cultures: Employed in Cheesemaking. In C. A. B. L. Tortorello (Ed.), *Encyclopedia of Food Microbiology* (2nd ed.; pp. 508–514). Oxford, UK: Academic Press. doi:10.1016/B978-0-12-384730-0.00322-0

Collins, Y. F., McSweeney, P. L. H., & Wilkinson, M. G. (2003). Lipolysis and free fatty acid catabolism in cheese: A review of current knowledge. *International Dairy Journal*, *13*(11), 841–866. doi:10.1016/S0958-6946(03)00109-2

Curioni, P. M. G., & Bosset, J. O. (2002). Key odorants in various cheese types as determined by gas chromatography-olfactometry. *International Dairy Journal*, *12*(12), 959–984. doi:10.1016/S0958-6946(02)00124-3

Deeth, H. C., & Fitz-Gerald, C. H. (1983). Lipolytic enzymes and hydrolytic rancidity in milk and milk products. In P. E. Fox (Ed.), *Developments in Dairy Chemistry-2: Lipids* (pp. 195–240). London: Applied Science Publishers. doi:10.1007/978-94-010-9231-9_6

Delgado, F. J., González-Crespo, J., Cava, R., García-Parra, J., & Ramírez, R. (2010). Characterisation by SPME–GC–MS of the volatile profile of a Spanish soft cheese P.D.O. Torta del Casar during ripening. *Food Chemistry*, *118*(1), 182–189. doi:10.1016/j.foodchem.2009.04.081

Desmasures, N. (2014). Cheese: Mold-Ripened Varieties. In M. L. Tortorello (Ed.), *Encyclopedia of Food Microbiology* (2nd ed.; pp. 409–415). Oxford, UK: Academic Press. doi:10.1016/B978-0-12-384730-0.00060-4

Di Cagno, R., Banks, J., Sheehan, L., Fox, P. F., Brechany, E. Y., Corsetti, A., & Gobbetti, M. (2003). Comparison of the microbiological, compositional, biochemical, volatile profile and sensory characteristics of three Italian PDO ewes' milk cheeses. *International Dairy Journal, 13*(12), 961–972. doi:10.1016/S0958-6946(03)00145-6

El Soda, M., Madkor, S. A., & Tong, P. S. (2000). Adjunct Cultures: Recent Developments and Potential Significance to the Cheese Industry. *Journal of Dairy Science, 83*(4), 609–619. doi:10.3168/jds.S0022-0302(00)74920-4 PMID:10791774

Erkuş, O., de Jager, V. C. L., Spus, M., van Alen-Boerrigter, I. J., van Rijswijck, I. M., Hazelwood, L., ... Smid, E. J. (2013). Multifactorial diversity sustains microbial community stability. *The ISME Journal, 7*(11), 2126–2136. doi:10.1038/ismej.2013.108 PMID:23823494

Farkye, N. Y., & Imafidon, G. I. (1995). *Thermal denaturation of indigenous milk enzymes. In Heat-induced changes in milk* (pp. 331–348). Brussels, Belgium: Int. Dairy Fed.

Fitzsimons, N. A., Cogan, T. M., Condon, S., & Beresford, T. (1999). Phenotypic and genotypic characterization of non-starter lactic acid bacteria in mature cheddar cheese. *Applied and Environmental Microbiology, 65*, 3418–3426. PMID:10427029

Fox, P. F. (1989). Proteolysis during cheese manufacture and ripening. *Journal of Dairy Science, 72*(6), 1379–1400. doi:10.3168/jds.S0022-0302(89)79246-8

Fox, P. F., Guinee, T. P., Cogan, T. M., & McSweeney, P. L. H. (2000). *Fundamentals of Cheese Science.* Gaithersburg, MD: Aspen.

Fox, P. F., Law, J., McSweeney, P. L. H., & Wallace, J. (1993). *Biochemistry of Cheese Ripening. Cheese: Chemistry, Physics and Microbiology* (Vol. 1). London: Chapman & Hall.

Fox, P. F., & McSweeney, P. L. H. (1996). Proteolysis in cheese during ripening. *Food Reviews International, 12*(4), 457–509. doi:10.1080/87559129609541091

Fox, P. F., McSweeney, P. L. H., & Lynch, C. M. (1998). Significance of non-starter lactic acid bacteria in Cheddar cheese. *Australian Journal of Dairy Technology, 53*, 83–89.

Fox, P. F., & Stepaniak, L. (1993). Enzymes in cheese technology. *International Dairy Journal, 3*(4-6), 509–530. doi:10.1016/0958-6946(93)90029-Y

Franciosi, E., Settanni, L., Carlin, S., Cavazza, A., & Poznanski, E. (2008). A Factory-Scale Application of Secondary Adjunct Cultures Selected from Lactic Acid Bacteria During Puzzone di Moena Cheese Ripening. *Journal of Dairy Science, 91*(8), 2981–2991. doi:10.3168/jds.2007-0764 PMID:18650274

Gallois, A., & Langlois, D. (1990). New results in the volatile odorous compounds of French cheeses. *Le Lait, 70*(2), 89–106. doi:10.1051/lait:199028

Gonzales De Llano, D., Ramos, M., Polo, C., Sanz, J., & Martinez-Castro, I. (1995). Evolution of the Volatile Components of an Artisanal Blue Cheese During Ripening. *Journal of Dairy Science*, *73*(7), 1676–1683. doi:10.3168/jds.S0022-0302(90)78842-X

Gupta, R., Gupta, N., & Rathi, P. (2004). Bacterial lipases: An overview of production, purification and biochemical properties. *Applied Microbiology and Biotechnology*, *64*(6), 763–781. doi:10.100700253-004-1568-8 PMID:14966663

Hamlyn, P. F., Bradshaw, R. E., Mellon, F. M., Santiago, C. M., Wilson, J. M., & Peberdy, J. F. (1981). Efficient protoplast isolation from fungi using commercial enzymes. *Enzyme and Microbial Technology*, *3*(4), 321–325. doi:10.1016/0141-0229(81)90007-7

Hayaloglu, A. A., Fox, P. F., Guven, M., & Cakmakci, S. (2007). Cheeses of Turkey: 1. Varieties ripened in goat-skin bags. *Le Lait*, *87*(2), 79–95. doi:10.1051/lait:2007006

Hayashi, K., Revell, D. F., & Law, B. A. (1990). Effect of Partially Purified Extracellular Serine Proteinases Produced by *Brevibacterium linens* on the Accelerated Ripening of Cheddar Cheese. *Journal of Dairy Science*, *73*(3), 579–583. doi:10.3168/jds.S0022-0302(90)78705-X

Hayes, M. G., Oliveira, J. C., McSweeney, P. L. H., & Kelly, A. L. (2002). Thermal inactivation of chymosin during cheese manufacture. *The Journal of Dairy Research*, *69*(02), 269–279. doi:10.1017/S0022029902005472 PMID:12222804

Izco, J. M., & Torre, P. (2000). Characterisation of volatile flavour compounds in Roncal cheese extracted by the 'purge and trap' method and analysed by GC–MS. *Food Chemistry*, *70*(3), 409–417. doi:10.1016/S0308-8146(00)00100-X

Klein, N., & Lortal, S. (1999). Attenuated starters: An efficient means to influence cheese ripening—a review. *International Dairy Journal*, *9*(11), 751–762. doi:10.1016/S0958-6946(99)00155-7

Koçak, C., & Güzel-Seydim, Z. B. (2011). Kazein Kimyası ve Sütün Pıhtılaşma Mekanizması. In Peynir Biliminin Temelleri (pp. 53-78). İzmir-Türkiye: Sidas.

Kraggerud, H., Skeie, S., Høy, M., Røkke, L., & Abrahamsen, R. K. (2008). Season and ripening temperature influence fatty acid composition and sensory properties of semi-hard cheese during maturation. *International Dairy Journal*, *18*(8), 801–810. doi:10.1016/j.idairyj.2007.11.003

Lemieux, L., & Simard, R. E. (1991). Bitter flavour in dairy products. 1. A review of the factors likely to influence its development, mainly in cheese manufacture. *Le Lait*, *71*(6), 599–636. doi:10.1051/lait:1991647

Marilley, L., & Casey, M. G. (2004). Flavours of cheese products: Metabolic pathways, analytical tools and identification of producing strains. *International Journal of Food Microbiology*, *90*(2), 139–159. doi:10.1016/S0168-1605(03)00304-0 PMID:14698096

McSweeney, P. L. H. (2004). Biochemistry of cheese ripening: Introduction and overview. In Cheese: Chemistry, Physics and Microbiology: Vol 1. General Aspects, (pp. 347-360). London: Elsevier.

McSweeney, P. L. H., & Fox, P. F. (2004). Metabolism of residual lactose and of lactate and citrate. In Cheese: Chemistry, Physics and Microbiology: Vol 1. General Aspects (pp. 361-372). London: Elsevier. doi:10.1016/S1874-558X(04)80074-5

McSweeney, P. L. H., & Sousa, M. J. (2000). Biochemical pathways for the production of flavour compounds in cheeses during ripening: A review. *Le Lait*, *80*(3), 293–324. doi:10.1051/lait:2000127

Metwalli, A. M., De Jongh, H. H. J., & Van Boekel, M. J. S. (1998). Heat inactivation of bovine plasmin. *International Dairy Journal*, *8*(1), 47–56. doi:10.1016/S0958-6946(98)00017-X

Molimard, P., & Spinnler, H. E. (1996). Review: Compounds involved in the flavor of surface mold-ripened cheeses: Origins and properties. *Journal of Dairy Science*, *79*(2), 169–184. doi:10.3168/jds.S0022-0302(96)76348-8

Montel, M.-C., Buchin, S., Mallet, A., Delbes-Paus, C., Vuitton, D. A., Desmasures, N., & Berthier, F. (2014). Traditional cheeses: Rich and diverse microbiota with associated benefits. *International Journal of Food Microbiology*, *177*, 136–154. doi:10.1016/j.ijfoodmicro.2014.02.019 PMID:24642348

Mullan, W. M. A. (2014). Importance of Selected Genera. Encyclopedia of Food Microbiology, 3, 515-521.

Neves, A. R., Pool, W. A., Solopova, A., Kok, J., Santos, H., & Kuipers, O. P. (2010). Towards enhanced galactose utilization by Lactococcus lactis. *Applied and Environmental Microbiology*, *76*(21), 7048–7060. doi:10.1128/AEM.01195-10 PMID:20817811

Ong, L., & Shah, N. P. (2008). Release and identification of angiotensin-converting enzyme-inhibitory peptides as influenced by ripening temperatures and probiotic adjuncts in Cheddar cheeses. *Lebensmittel-Wissenschaft + Technologie*, *41*(9), 1555–1566. doi:10.1016/j.lwt.2007.11.026

Ozturkoglu-Budak, S. (2016d). *Dynamics of complex microbiota and enzymes in Divle Cave cheese and their biochemical consequences* (PhD thesis). Utrecht University, The Netherlands.

Ozturkoglu-Budak, S., Figge, M. J., Houbraken, J., & de Vries, R. P. (2016a). The diversity and evolution of microbiota in traditional Turkish Divle Cave cheese during ripening. *International Dairy Journal*, *58*, 50–53. doi:10.1016/j.idairyj.2015.09.011

Ozturkoglu-Budak, S., Gursoy, A., Aykas, D. P., Koçak, C., Dönmez, S., de Vries, R. P., & Bron, P. A. (2016b). Volatile compound profiling of Turkish Divle Cave cheese during production and ripening. *Journal of Dairy Science*, *99*(7), 1–12. doi:10.3168/jds.2015-10828 PMID:27108178

Ozturkoglu-Budak, S., Wiebenga, A., Bron, P. A., & de Vries, R. P. (2016c). Protease and lipase activities of fungal and bacterial strains derived from an artisanal raw ewe's milk cheese. *International Journal of Food Microbiology*, *237*, 17–27. doi:10.1016/j.ijfoodmicro.2016.08.007 PMID:27541978

Parente, E., & Cogan, T. M. (2004). Starter cultures: General aspects. In Cheese: Chemistry, Physics and Microbiology (Vol. 1, pp. 123-147): Academic Press.

Pinho, O., Peres, C., & Ferreira, I. M. (2003). Solid-phase microextraction of volatile compounds in "Terrincho" ewe cheese. Comparison of different fibers. *Journal of Chromatography. A*, *1011*(1-2), 1–9. doi:10.1016/S0021-9673(03)01066-5 PMID:14518757

Rattray, F. P., & Fox, P. F. (1997). Purification and characterization of an intracellular esterase from Brevibacterium linens ATCC 9174. *International Dairy Journal*, *7*(4), 273–278. doi:10.1016/S0958-6946(97)00013-7

Rattray, F. P., & Fox, P. F. (1999). Aspects of Enzymology and Biochemical Properties of Brevibacterium linens Relevant to Cheese Ripening: A Review1. *Journal of Dairy Science*, *82*(5), 891–909. doi:10.3168/jds.S0022-0302(99)75308-7 PMID:10342227

Schokker, E. P., & van Boekel, M. A. J. S. (1997). Production, purification and partial characterization of extracellular proteinase from Pseudomonas fluorescens 22 F. *International Dairy Journal*, *7*(4), 265–271. doi:10.1016/S0958-6946(97)00008-3

Schweiger, M., Eichmann, T. O., Taschler, U., Zimmermann, R., Zechner, R., & Lass, A. (2014). Measurement of Lipolysis. *Methods in Enzymology*, *538*, 171–193. doi:10.1016/B978-0-12-800280-3.00010-4 PMID:24529439

Settanni, L., & Moschetti, G. (2010). Non-starter lactic acid bacteria used to improve cheese quality and provide health benefits. *Food Microbiology*, *27*(6), 691–697. doi:10.1016/j.fm.2010.05.023 PMID:20630311

Singh, T. K., Drake, M. A., & Cadwallader, K. R. (2003). Flavor of Cheddar Cheese: A Chemical and Sensory Perspective. *Comprehensive Reviews in Food Science and Food Safety*, *2*(4), 166–189. doi:10.1111/j.1541-4337.2003.tb00021.x

Smid, E. J., Erkus, O., Spus, M., Wolkers-Rooijackers, J. C. M., Alexeeva, S., & Kleerebezem, M. (2014). Functional implications of the microbial community structure of undefined mesophilic starter cultures. *Microbial Cell Factories*, *13*(Suppl 1), S2. doi:10.1186/1475-2859-13-S1-S2 PMID:25185941

Smit, G., Smit, B. A., & Engels, W. J. M. (2005). Flavour formation by lactic acid bacteria and biochemical flavour profiling of cheese products. *FEMS Microbiology Reviews*, *29*(3), 591–610. doi:10.1016/j.fmrre.2005.04.002 PMID:15935512

Sousa, M. J., & Malcata, F. X. (1997). Effects of processing conditions on the caseinolytic activity of crude extracts of Cynara cardunculus L. *Food Science & Technology International*, *2*(4), 255–263. doi:10.1177/108201329600200408

Stadhouders, J. (1986). The control of cheese starter activity. *Netherlands Milk and Dairy Journal*, *40*, 155–173.

Stadhouders, J., & Leenders, G. J. M. (1984). Spontaneously developed mixed-strain cheese: Their behaviour toward phages and their use in the Dutch cheese industry. *Netherlands Milk and Dairy Journal*, *38*, 157–181.

Sumantha, A., Larroche, C., & Pandey, A. (2006). Microbiology and industrial biotechnology of food-grade proteases: A perspective. *Food Technology and Biotechnology*, *44*, 211–220.

Tavaria, F. K., Silva Ferreira, A. C., & Malcata, F. X. (2004). Volatile free fatty acids as ripening indicators for Serra da Estrela cheese. *Journal of Dairy Science*, *87*(12), 4064–4072. doi:10.3168/jds.S0022-0302(04)73548-1 PMID:15545367

Upadhyay, V. K., McSweeney, P. L. H., Magboul, A. A. A., & Fox, P. F. (2004). Proteolysis in cheese during ripening. In Cheese: Chemistry, Physics and Microbiology: Vol 1. General Aspects (pp. 391-433). London: Elsevier. doi:10.1016/S1874-558X(04)80076-9

van Kranenburg, R., Kleerebezema, M., van Hylckama Vliega, J., Ursing, B. M., Boekhorst, J., Smit, A. B., ... Siezen, R. J. (2002). Flavour formation from amino acids by lactic acid bacteria: Predictions from genome sequence analysis. *International Dairy Journal*, *12*(2-3), 111–121. doi:10.1016/S0958-6946(01)00132-7

Visser, S. (1993). Proteolytic Enzymes and Their Relation to Cheese Ripening and Flavor: An Overview. *Journal of Dairy Science*, *76*(1), 329–350. doi:10.3168/jds.S0022-0302(93)77354-3

Wilkinson, M. G., & Kilcawley, K. N. (2005). Mechanisms of incorporation and release of enzymes into cheese during ripening. *International Dairy Journal*, *15*(6-9), 817–830. doi:10.1016/j.idairyj.2004.08.021

Yegin, S., Fernandez-Lahore, M., Jose Gama Salgado, A., Guvenc, U., Goksungur, Y., & Tari, C. (2011). Aspartic proteinases from Mucor spp. in cheese manufacturing. *Applied Microbiology and Biotechnology*, *89*(4), 949–960. doi:10.100700253-010-3020-6 PMID:21127856

Yoon, Y., Lee, S., & Choi, K.-H. (2016). Microbial benefits and risks of raw milk cheese. *Food Control*, *63*, 201–215. doi:10.1016/j.foodcont.2015.11.013

Yvon, M., & Rijnen, L. (2001). Cheese flavour formation by amino acid catabolism. *International Dairy Journal*, *11*(4-7), 185–201. doi:10.1016/S0958-6946(01)00049-8

Chapter 11
Microbial Non-Coagulant Enzymes Used in Cheese Making

Ekaterini Moschopoulou
Agricultural University of Athens, Greece

ABSTRACT

In this chapter, the use of microbial non-coagulant proteases, microbial lipases, and microbial transglutaminase in the cheese making procedure is discussed. Microbial proteases and lipases have been used for over 30 years to accelerate cheese ripening and consequently to enhance the cheese flavor development by increasing proteolysis and lipolysis level in a shorter time. They are commercially produced by bacteria and fungi species. Transglutaminase is a relative new enzyme, which catalyzes the cross-linking of peptide bonds and helps to improve the cheese texture and to increase the cheese yield. Today, cheeses from almost all cheese categories are produced using these enzymes.

INTRODUCTION

Enzymes play a significant role in cheese production throughout process, i.e. from milk coagulation to cheese ripening. Depending on the technology applied, the enzymes involved in cheese production may be indigenous, i.e. coming from the cheese milk like plasmin, lipoprotein lipase and several others, or exogenous. Exogenous are those added for clotting the cheese milk like rennet enzymes and milk coagulants or those added for other purposes such as accelerating of ripening, improving the cheese textural characteristics, developing new cheese products etc. Cheese ripening is a process of three biochemical pathways, i.e. proteolysis, lipolysis and glycolysis that occur with the aid of starter and non-starter microorganisms and their enzymes, the residual milk-clotting enzyme, the indigenous milk enzymes and the added enzymes. This review deals with the exogenous enzymes of microbial origin like non-milk coagulant microbial proteases, lipases and transglutaminase, which are used in cheese production.

DOI: 10.4018/978-1-5225-5363-2.ch011

BACKGROUND

It is commonly accepted that the most important exogenous enzymes in cheese industry are the milk-clotting enzymes. The rennet enzymes as well as the milk coagulants, i.e. enzymes from non-ruminant animals, plants, microorganisms or biotechnologically invented, belong to aspartic group of proteases (EC 3.4.23). Although microbial coagulants are aspartic proteases they have special molecular characteristics and specificity (Moschopoulou, 2017). The exogenous non-coagulant enzymes used in cheese making for various purposes are mainly proteases and lipases and in some cases transglutaminase. Production, characterization and use of exogenous microbial enzymes in dairy processing have been extensively reviewed (Neelakantan et al., 1999; Jooyandeh et al., 2009; El-Hofi et al., 2011; Feijoo –Siota et al., 2014; Garcia et al., 2016; Romeih & Walker 2017).

Proteases are generally classified according to their optimum pH range as:

1. Acidic,
2. Neutral or
3. Alkaline proteases

and according to their catalytic mechanism of action, as

1. Serine proteases (EC 3.4.21),
2. Serine carboxyl proteases (EC 3.4.16),
3. Cysteine proteases (EC 3.4.22),
4. Metallo proteases I (EC 3.4.24) or metallo carboxyl proteases (EC 3.4.17).

They are also classified according to their specificities for amino acid sequences to be cleaved (Garcia et al., 2016). Microbial proteases are bacterial or fungal, having different molecular characteristics and mechanism of action. Bacterial neutral proteases may exhibit activity at the pH range 6.5-8.5, and at 10-70°C (optimum temperature at about 55°C) and are characterized by their high affinity for hydrophobic amino acids pairs. Some of the neutral proteases belong to the metallo-protease type and require divalent metal ions for their activity, while those belonging to serine proteases are not affected by chelating agents. Bacterial alkaline proteases present activity at high pH values e.g. pH 9.0, have broad substrate specificity and show optimal temperature at about 60°C. Fungi produce a big variety of proteolytic enzymes. For example, *Aspergillus oryzae* produces acid, neutral and alkaline proteases, which are active at a wide pH range, i.e. pH 4-11 and show broad substrate specificity. Fungal acid proteases are mostly used in milk clotting, while fungal neutral proteases and metallo-proteases combined with bacterial proteases are used to reduce bitterness in cheeses. They hydrolyze hydrophobic amino acid bonds, are active at pH 7.0 and are inhibited by chelating agents (Rao et al., 1998; Tavano, 2017). Production of microbial enzymes is made either by submerged fermentation or by solid-state fermentation. Bacterial commercial proteases are mostly produced by submerged fermentation, whereas fungal proteases are produced by solid-state fermentation (Aguilar et al., 2008).

Lipases (EC 3.1.1.3) are hydrolytic enzymes that act on ester bonds. These enzymes existed widely in bacteria, yeasts and fungi. Fungi are recognized as the best lipase sources and are used preferentially for industrial applications. The most important fungal species that produce lipases are *Candida rugosa*, *Candida antarctica*, *Rhizomucor miehei*, *Aspergillus niger*, *Aspergillus oryzae* and *Geotrichum candidum*

(Birschbach, 1994; Jooyandeh et al., 2009). Molecular characteristics and mechanism of action differ among the various microbial lipases. Most of microbial lipases present pH optima in the pH range 4-9 and preferentially liberate free fatty acids (FFA) from the 1 and 3 positions of triglycerides (Birschbach, 1994; Celligoi et al., 2017; El-Hofi et al., 2011; Jooyandeh et al., 2009). Ha & Lindsay (1993) studied the hydrolysis of ruminant milk fats by eleven lipases from animal and microbial origin and demonstrated that kid goat lipase, *P. roqueforti, A. niger* and *R. arrhizus* lipases produced high amounts of medium-chain branched fatty acids. In contrast, *Candida cylindracea* lipase yielded high amounts of volatile n-chain fatty acids non selectively and only small quantities of volatile branched-chain fatty acids. Fungal lipases can be produced by solid-state fermentation (Kumar & Ray, 2014).

The enzyme transglutaminase (EC 2.3.2.13) (TGase) modifies food proteins by forming both inter- and intra- molecular isopeptide bonds in various food proteins by cross-linking the amino acid residues of protein bound glutamine (acyle donor) and lysine (acyle receptor) (Romeih & Walker, 2017). Microbial TGase (MTGase) was firstly isolated from *Streptoverticillium sp.* in 1989 and has been recognized as GRAS, i.e. generally recognized as safe, substance for human nutrition. The optimum pH activity of MTGase ranges between 5 and 7. At pH 6.0 and at temperature range between 40°C and 50°C shows its full activity, while it is inactivated by heat treatment at 70°C for 5 min or at 75°C for 1 min. Microbial sources and molecular characteristics of MTGase as well as its application in various dairy products have been outlined in detail (Kieliszek & Misiewicz, 2014; Jaros et al., 2006; Ozrenk, 2006; Romeih & Walker, 2017). Currently, MTGase has potential applications in food industry such as dairy, bakery and meat products, and edible films for improving their firmness, viscosity, elasticity and water-holding capacity. In the dairy sector, MTGase is used for several reasons, i.e. to incorporate whey proteins in cheese (Masotti et al., 2017), to modify milk proteins by cross-linking of caseins for increased cheese yield (Domagala et al., 2016), to catalyze the whey protein cross-linking for protein recovery during ultrafiltration of whey (Wen-qionq et al., 2017) and to increase viscosity and gel strength of yoghurt products (Romeih & Walker, 2017).

MICROBIAL ENZYMES USED AS AGENTS FOR ACCELERATING CHEESE RIPENING

For most of the cheese types, ripening is an expensive and time consuming process and for this reason, efforts for accelerating it have been made towards different directions. Various attempts have been carried out to shorten the cheese ripening period such as:

1. Increase of the ripening temperature,
2. Addition of exogenous enzymes like glycosidase, protease and lipase,
3. Modification of the starters,
4. Development of cheese slurries,
5. Application of high pressure and
6. Combination of the previous methods (Wilkinson, 1999; Wilkinson & Kilcawley, 2005).

The mechanisms of incorporation and release of exogenous enzymes into cheese during ripening have been outlined in detail by Wilkinson & Kilcawley (2005). In summary, there are four patterns of including exogenous enzymes in cheese: a) addition to cheese milk prior to cheesemaking procedure,

b) addition to cheese milk with starter or rennet, c) addition at dry salting and d) addition directly into the cheese block. Addition of enzymes in the free form using the two first ways on inclusion are not technically realistic as about of 90% of the enzyme is lost in whey during curd drainage. Alternatively, addition of exogenous enzymes during dry salting is more efficient than salting in brine, but data for enzyme distribution, solubility, and diffusion rates are rare. Addition of enzymes in the form of aqueous solution directly into Mozzarella cheese block has been made in experimental scale using a high-pressure injector. To overcome the problem of retaining the exogenous enzymes and to control their release in the cheese mass the encapsulation technique has been used recently. Numerous techniques of microbial enzyme encapsulation exist (Lević et al., 2017), but stability and release are affected by several parameters of cheesemaking. The materials used for protease and lipase encapsulation seem to affect the efficiency of the entrapped enzymes. For example, the addition of solution 0.1M calcium chloride and 0.1% (w/v) chitosan improves the hydro-gelling efficiency of alginate polymers and increase the encapsulation efficiency to 70% (Kailasapathy et al., 2006). Moreover, Akin & Güler-Akin (2017) have investigated the κ-carragenan, gellan and sodium alginate used as capsule materials. They have concluded that sodium alginate is the best in the case of accelerating the ripening of Kashar cheese. An alternative method to encapsulate enzymes is the use of liposomes, which are microscopic lipid vesicles consisting of an outer shell or phospholipid and an internal aqueous core. Liposomes seem to have potential advantages over other methods of enzyme encapsulation (Wilkinson & Kilcawley, 2005). Finally, it should be noted that acceleration of glycolysis by exogenous β-galactosidase does not seem to be of great value in acceleration of cheese ripening as lactose hydrolysis mainly occurs at the early stage of ripening.

Application of Exogenous Microbial Enzymes in Accelerating Cheese Ripening

Cheese ripening is a complex series of biochemical events. Primary biochemical events include the metabolism of residual lactose and of lactate and citrate, lipolysis and proteolysis. Secondary biochemical events include the metabolism of fatty acids and amino acids developing many volatile flavor compounds (McSweeney, 2004). Exogenous protease and peptidases are involved in secondary proteolysis of cheese by hydrolyzing the peptides that release from primary proteolysis. Addition of exogenous proteases or lipases complements the indigenous enzymes in cheese and thus the process of proteolysis or lipolysis respectively results in acceleration of ripening. However, although the practice is known over the last four decades, it is applied to certain cheese types due to various technical and legal limitations especially in Europe (Wilkinson & Kilcawley, 2005).

Microbial Proteases

Microbial proteases as agents for accelerating cheese ripening have been studied over 30 years ago. Commercially available proteases either alone or in mixtures showing various specific proteolytic activities are presented in Table 1. They are used to enhance or debitter the cheese flavor, to accelerate the ripening and to produce enzyme-modified cheeses (EMC) in order to develop a variety of cheese flavor profiles for incorporation into processed cheese, cheese spreads, cheese dips, cheese analog and cheese sauces (Kilara & Chandan, 2011). However, excessive amounts of exogenous proteases tend to lead to the development of bitterness and weaken cheese body, which are due either to excessive proteolysis or to tendency specifically to liberate bitter peptides (Law & Wigmore 1982a,b). In general, type and amount

Table 1. Commercial non-coagulant proteases used in cheese industry

Producer	Name of Product	Microbial Source	Application
AB Enzymes	Corolase LAP	*Aspergillus sojae*	Debittering
Amano	Thermoase PC10F	*Bacillus thermoproteolyticus*	Protein hydrolysis
	Peptidase R	*Rhizopus oryzae*	Debittering
Biocatalysts	Flavorpro 192	*Aspergillus oryzae*	Debittering
	Flavorpro Umami	*Aspergillus* sp.	EMC production
	Promod™	*Aspergillus* spp.	EMC production having various flavors
Danisco	Accelase	*Lactococcus lactis*	Accelerating cheese ripening
	Savorase	*L. lactis*	EMC production
	Debitrase	*L. lactis*	Debbitering
DSM	Accelerzyme CPG	*A. niger*	Accelerating cheese ripening
	Accelerzyme NP	*B. amyloliquefaciens*	Accelerating cheese ripening
Medipharm	Enzobact	*Lactobacillus helveticus*	Debbitering and flavor development
Novozymes	Flavourzyme MG/A	*A. oryzae*	Debbitering and flavor development
	Neutrase 0.8L	*Bacillus amylolichefaciens*	Production of cheese flavor

of starter and rennet used as well as bacteria proliferation are the most important factors of producing bitter peptides that are strongly correlated with hydrophobicity (Lemieux & Simard 1991).

Cheddar cheese has been extensively studied with regard the acceleration of its ripening using either traditional, i.e. elevated ripening temperatures, addition of enzymes, addition of cheese slurry or modern methods, i.e. attenuated starters, genetically engineered starters and recombinant enzymes and microencapsulation of ripening enzymes (Azarnia et al., 2006). Although fungal acid protease, bacterial alkaline protease and broad specificity protease may produce bitter cheeses, Law & Wigmore (1982a) showed that the 2-month old cheeses treated with microbial proteases exhibited a flavor intensity equivalent to that of untreated 4-month old cheese. However, the microorganism of origin seems to play role in suitability of a protease. It was demonstrated by the same authors that, commercial preparations of *Bacillus subtilis* neutral protease could accelerate flavor development in Cheddar cheese without defect formation, compared to a fungal acid protease from *A. oryzae* and to a broad specificity protease with aminopeptidase activity produced by *Streptomyces grisens*. In contrast, the alkaline protease substilisin A from *Bacillus licheniformis* caused bitter flavor (Law & Wigmore 1982b). *B. subtilis* neutral protease gene has been expressed in *L. lactis* to produce a biologically active protease, which may help to accelerate cheese ripening (Van De Guchte et al., 1990). Nattokinase, a substilisin-like serine protease produced by *Bacillus natto* has been showed to accelerate primary proteolysis of Cheddar cheese, to increase the free amino acids (FAA) and finally to affect positively the cheese flavor (Upadhyay et al., 2006). On the other hand, Trépanier et al. (1992) reported that addition of Neutrase© in Cheddar cheese in combination with live and heat -socked lactobacilli at various ratios increased cheese fracturability and decreased cheese cohesiveness, proposing a good manufacture process. The proposed protocol consisted of a) addition of live *L. casei-casei L2A* to control the undesirable microflora, b) addition of heat-shocked cells of the same species at a ratio 1.0% and c) addition of Neutrase© at a ratio $<1.0\times10^{-5}$ AU per g of cheese in order to increase by 60% the flavor intensity. Two partially purified fractions (Pro-C and Pro-D-E) of serine protease produced by *Brevibacterium linens* were compared to commercial bacterial metalloproteinase (Neutrase) in accelerating ripening of Cheddar cheese and the results showed no significant differences in casein hydrolysis among the cheeses. However, cheese flavor increased significantly in the case of the cheeses made with 26 units of Pro-C and Pro-D-E per kgof curd (Hayashi et al., 1990). The commercial enzymes Accelase AM317, Accelase AHC50 and Accelrzyme CPG were evaluated for Cheddar cheese accelerating ripening (Kilcawley et al., 2012). Results showed some adverse rheological effects in cheeses produced by both Accelase blends, which were attributed to excessive primary proteolysis and higher levels of moisture. According to Doolan et al. (2014), exogenous proteases in Cheddar cheese is better to be added during cheese salting than in cheese milk prior to renneting. Furthermore, Kheadr et al (2000) showed that addition of bacterial or fungal proteinase, encapsulated in liposomes with trapping efficiency about 33%, at a ratio up to $1x10^{-5}$ AU or up to $2x10^{-6}$ AU per g of cheese respectively, could short the ripening time of Cheddar cheese. Low fat Cheddar cheese from condensed milk made with lyophilized protease and lipase from *A. oryzae* exhibited accelerated ripening and more typical flavor than the control cheese (Brandsma et al., 1994). In addition, the use of the commercial peptidase from *Aspergillus oryzae* (Flavorzyme®) encapsulated in calcium-alginate emulsion resulted in Cheddar cheese with improved health benefits as the biologically active peptides Leu-Thr-Glu and Asp-Val-Pro-Ser-Glu were released during ripening (Seneweera & Kailasapathy, 2011). Regarding the microbial flora aspect, Bautista & Kroll (1988) studied the survival of some microorganisms in enzyme-accelerated Cheddar cheese by protease (Neutrase 1-5S) and peptidases and concluded that Gram-positive bacteria were not affected and Gram-negative microorganisms were recovered at lower numbers.

Microbial proteases have also been studied in accelerating ripening of Spanish cheeses. Nunez et al. (1991) manufactured ovine Manchego cheese with neutral protease from *Bacillus subtillis* (Neutrase L and Novozym 257) and the matured cheese in the case of Neutrase L was more crumbly and less elastic than control cheese. In addition, the same cheese, in which Neutrase L protease was added at a ratio 0.004AU per L of milk, developed the same flavor intensity in 35 days at 16°C as control cheese did in 90 days at 12°C. Neutral protease from *Bacillus subtilis,* encapsulated with phospholipase C in stimulated release liposomes, was added in cheese milk and compared to acid proteases chymosin and cardoon cyprosins, yielded higher amount of soluble nitrogen in the produced Manchego cheese than the other proteases (Picon et al., 1997). Manchego type cheese from bovine/ovine milk mixture was also manufactured with addition of commercial enzyme preparation of *Aspergillus oryzae* (FlavorAge) at a ratio 50mg per kg of milk displayed significantly shorter ripening time (Fernandez-Garcia et al., 1993). In the same cheese type made from pasteurized ovine cheese milk, cysteine proteinase from *Micrococcus sp. INIA528* was added and the produced cheese had intense flavor in 36 days of ripening instead of the 60 days of ripening of control cheese (Moherano et al., 1998). Izco et al. (2000) added to the cheese milk a mixture of commercial exogenous enzymes (Lipomod™ 187 and Promod™ 215) at a ratio 10g per 100L of milk to manufacture ovine Ossau-Iraty type cheese. Results showed that after 4 months of ripening, proteolysis of cheese was 20% higher than the control cheese and the produced FAA such as asparagine, serine and proline had covered the bitter taste.

Regarding other cheese types, the addition of commercial proteases (Accelase AM250M® and Promod™ 215P in Pasta filata type cheese, after curd stretching, intensified the proteolysis degree without affecting the gross composition of the cheese, as it was compared to the cheese made without addition of exogenous enzymes (Zaharia, 2012). Feta cheese ripening time could be successfully shortened from 80 days to 40 days, using an optimum level of neutral or acid microbial proteases in order to increase the TCA-soluble nitrogen compounds about 35% in 40 days (Vafopoulou et al., 1989). Furthermore, it has been reported that proteolytic enzymes purified from *Brevibacterium linens* were used in accelerated ripening of surface ripening cheese such as Danbo, Limburger and Gruyere (Hayashi, 1996). In addition, crude cytoplasmic extract from *Pseudomonas fluorescens* combined with nonstarter lactic acid bacteria reduced from 60 days to 28 days the ripening period of Pecorino Umbro cheese without flavor defects (Corsetti at al., 1998).

It is worth noting that a possible synergistic effect of proteases with lipases seems to occur. Addition of proteases combined with lipases has been reported to increase cheese proteolysis more than the protease alone (Kheadr et al., 2003). Two different fungal lipase from *R. miehei* or *A. niger* alone or in combination with fungal protease from *Asp. oryzae* and bacterial neutral protease from *B. subtilis* (Neutrase) in accelerating ripening of ovine or bovine Manhego type cheese was investigated by Fernandez-Garcia et al. (1994a). It was reported that the combination of lipases and proteases increased proteolysis whereas the addition of only lipase decreased it, while *B. subtilis* induced bitterness.

More recently many efforts were made towards the discovery of new microbial proteases. Calasso et al. (2015) used freeze-dried cell-free extracts from *Lactobacillus casei LC01*, *Weissela cibaria IXFS*, *Xafnia alvei Moller ATCC510015* and *Debaryomyces hansenii LCF-558* in ovine cheese and demonstrated that gross composition of cheese and microbial growth were not affected. In contrast both, the secondary proteolysis and the level of volatile compounds increased. Furthermore, neutral protease from *Penicillium candidum (PCA 1/TT031)* experimentally purified under solid state fermentation was suggested for accelerating cheese ripening (Alhelli et al., 2016). Finally, there is high number of patents, which deal with a) the production of recombinant strains of microorganisms for producing aminopeptidase or carboxypeptidase and b) the description of their use in accelerating ripening of special cheese varieties (Feijoo-Siota et al., 2014).

Microbial Lipases

Lipolysis degree during cheese ripening varies among the different cheese types. For example, up to 2% of the fatty acids were liberated in 12 months old Cheddar cheese, whereas it was up to 25% in Blue cheese type (Garcia et al., 2016). Furthermore, lipolysis is more important in some cheese varieties, as it considered being the vital key of their special flavor. For this reason, exogenous pregastric esterases originated from small ruminants with an s_n-3 stereospecificity are often used in some Italian cheeses. Ripening process with exogenous lipases is about 2 to 5 times faster than without. However, excessive lipolysis may cause flavor defect, i.e. rancidity and other off-flavors, as well as reduction in cheese yield. Some commercial lipases of microbial origin are presented in Table 2.

Exogenous microbial lipases have been used in several cheese varieties. Cheddar cheese, made with commercial enzymes (neutral bacterial protease, Flavorzyme and fungal protease and lipase (Palatase M) individually encapsulated in liposomes and added at different combinations of lipase with the other enzymes, presented in all treatments shorter ripening time and higher flavor intensity, compared to control cheese (Kheadr et al., 2003). When Palatase M and Lipase 50 were encapsulated in liposomes and added in Cheddar cheese at levels up to 0.5 lipase unit per g of milk fat, a better flavor intensity was perceived. Higher levels of lipase addition caused a pronounced soapy flavor after 2-3 months of ripening (Kheadr et al., 2002).

The Spanish hard cheese Manchego has also been manufactured with individual addition of a commercial microbial lipase from *A. niger* (Palatase), a microbial protease from *B. subtilis* (Neutrase) and a β-galactosidase from *Klyveromyces fragilis* (Lactozym). It was showed that the concentrations

Table 2. Commercial microbial lipases used in cheese industry

Producer	Name of Product	Microbial Source	Application
Amano	Lipase AK	*Pseudomonas fluorescens*	EMC production
Biocatalysts	Lipomod™ (various types)	*Fungal*	EMC production with various cheese flavors
Chr. Hansen	Spiceit®	*R. miehei*	Short ripened Italian type cheeses
Novozymes	Palatase	*R. miehei*	Hydrolysis of small fatty acids in the carboxyl acid part
	Resinase	*As. oryzae*	Ester hydrolysis of triglycerides

of FFA, nitrogen fractions and FAAs in the ripened cheeses were increased by lipase, protease and β-galactosidase respectively (Fernandez-Garcia et al. 1988). Moreover, Manchego-type cheese ripening was also accelerated by addition of 50mg Flavor Age (a blend of proteases, peptidases and lipase from *A. oryzae*) per kg of cheese milk. The released FFA were C4:0-C10:0 however the enhanced flavor in this work was less importance as reported by the authors, compared to the achieved amino acids liberation (Fernandez-Garcia et al., 1994b). It is worth noting that the use of microbial lipase in some cheese types is not always successful, compared to animal lipases. Addition of commercial lipases of different origin and lamb rennet paste was studied for flavor development in Idiazabal, an ovine raw milk Spanish cheese. The results showed that the use of fungal lipases increased the concentration of C16:0 and C18:1 FFA, whereas animal lipases (pregastric esterases) and lamb rennet paste increased the concentration of short-chain FFAs. With respect to sensory characteristics, cheese made with animal lipase or lamb rennet paste received higher organoleptic values than those of cheese made with fungal lipase, as they had induced a more typical flavor of Idiazabal cheese (Barron et al., 2004; Hernandez et al., 2005). In this context, Barron et al. (2004) reported that fungal lipase added to the milk released mainly medium- and long-chain FFA whereas pregastric lipase released preferentially short-chain FFA and concluded that the latter was more suitable for traditionally-flavor cheese.

Lipases in combination with sodium dodecylsulphate (SDS, 0.075%) enhanced the flavor of Blue cheese and improved the distribution of *Penicillium roqueforti* throughout the cheese mass without increasing proteolysis as SDS alone (Farag et al., 1992). Aminifar & Eman-Djomeh (2014) investigated the commercial microbial lipase (Palatase from *R. Miehei*) in Lighvan cheese. The authors reported that pH, acidity and gross composition of cheese were not affected, and FFA content increased considerably, whereas bitterness was not markedly noted. Addition of microbial lipase (Piccantase A) in cheese milk was recommended for accelerating the flavor development in Tulum cheese for a short ripening period (Yilmaz et al., 2005). Mihalic cheese ripening was also successfully accelerated using the fungal lipase from *R. miehei* in combination with a *Bacillus subtilis* protease (neutral protease) (Ozcan & Kurdal, 2012).

Concerning soft cheese types, flavor of Domiati cheese made from reconstituted milk was more intensified when the commercial microbial lipase (Piccantase B) from *Rhizomucor miehei* was added to the cheese milk, compared to the cheese inoculated with animal lipase at the same ratio. The optimal ratio of addition was limited to 0.025% (Ashour et al. 1986). Yazdanpanah et al. (2015) used commercial lipase from *Aspergillus niger* encapsulated or in free form in production of ultrafiltrated Feta-type cheese. The enzyme was completely released in the cheese mass during the first 15 days of ripening, but the highest lipolysis degree was detected in the cheese fortified with the commercial lipase in free form.

Additionally, Kashar cheese ripened with lipase enzymes encapsulated in κ-carragenan exhibited higher lipolysis compared to the cheese treated with lipase encapsulated in gellan or sodium alginate (Akin et al., 2012). In contrast, in same cheese type, κ-carragenan capsules caused poor textural properties and low sensory scores, suggesting that sodium alginate capsules are the most suitable (Akin & Güler-Akin, 2017). It is worth noting that results from experiments on the use of the same type of lipase in the same cheese are contradictory, suggesting that every commercial enzyme is not suitable for every type of cheese. This fact is attributed to the differences that may come out from the microorganism of origin, production procedure, heat treatment of milk, etc. For example, Kasar cheese was manufactured with addition of commercial fungal lipase to the cheese milk at a ratio 5 or 15ml of enzyme preparation to 100 L of milk and displayed noticeable flavor at the beginning of ripening and later excessive rancidity with the increasing amounts of enzyme (Koçak et al., 1996). In contrast, another fungal lipase used in the same cheese and at the same ratio did not accelerate the release of fatty acids, compared to control cheese throughout ripening (Güler et al. 2004).

Recently, an attempt has been made to exploring the potential of new lipases in cheese industry. The gene lp_1760 was cloned and tributyrin esterase Lp_1760 from *Lactobacillus plantarum* was produced. The genelp_1760 encoded a putative esterase lipase in *E. coli* Y3L21 (DE3), which presented high activity in the presence of NaCl, lactic acid and calcium chloride, suggesting its potential use in cheese manufactures (Esteban-Torres et al., 2014). Mase et al. (2013) produced a new lipase enzyme from the strain 39-A of the yeast *Cryptococcus flavescens*, which hydrolyzed the milk fat triglycerides similarly to pregastric esterases. Addition of this lipase in Mozzarella cheese showeda favorable cheese flavor, suggesting that this lipase has potential to replace pregastric esterases in the cheese industry. Moreover, a lipase purified from *Sporidiobolus pararosens* was also suggested as potential substitute of pregastric esterase (Mase et al. 2011). Furthermore, *Serratia marcescens MBB05*, a strain for a cost effective enzyme production, produced a lipase enzyme that is suitable for Cheddar cheese ripening (Shakila Begam et al., 2012).

MICROBIAL TRANSGLUTAMINASE

Microbial transglutaminase (MTGase) is a relative new enzyme, which, helps to increase the cheese yield and to improve the cheese texture by cross-linking of peptide bonds. The most known commercial TGase products are ACTIVA® by AJINOMOTO Company and Probind® by BDF Ingredients Company.

Several authors investigated the use of MTGase in the manufacture of cheese, where a number of patents concerning the application of MTGase in cheese production are also available (Kuraishi et al., 2001; Kieliszek & Misiewicz, 2014). More recently, Romeih & Walker (2017) reported the proposed three ways for producing cheese with MTGase: 1) addition of active enzyme in cheese milk prior to heat treatment, i.e. pasteurization, inactivating it in this way, 2) addition at the same time with milk clotting enzyme and 3) addition of active enzyme after curd cutting step. In this context, Domagala et al. (2016) demonstrated that textural modifications and improved yield of rennet cheese could be better achieved with the simultaneous addition of rennet and MTGase in the cheese milk. The degree of cross-linked proteins depends on the MTGase amount, the incubation time and temperature. Excessive protein cross-linking leads to deterioration of textural properties (Mazuknaite et al., 2013). According to Gustaw et al. (2008), the conditions of gel formation of casein MTGase/rennet system is a compromise between the conditions required for rennet induced casein aggregation and those controlling the MTGase induced intra- and inter-micellar cross-linking. Furthermore, Cozzolino et al. (2003) demonstrated that addition of MTGase in the cheese milk at the same time with rennet or after cutting the obtained coagulum, caused significant protein lost in whey.

Various attempts have been carried out utilizing MTGase to increase cheese yield and enhance the cheese quality particularly the low-fat types. Low-fat Cheddar cheese treated with MTGase had significantly improved the protein and the fat contents as well as cheese yield, and showed higher hardness and lower meltability than control cheese at the end of ripening (Hu et al., 2013). Trappist cheese, a popular Hungarian cheese was produced from bovine milk with 2.8%, or 3.5% or 5% fat content with MTGase. All cheeses showed modifications in texture, but cheese with low fat content showed significantly higher springiness and cohesiveness values than the 3.5% and 5% fat cheeses. The latter, however, were higher scored with regard to sensory characteristics (Darnay et al., 2017a). Moreover, partially treatment with MTGase of pasteurized milk with 3.5% and 5% fat milk affected the color as well as the yield of the produced cheeses, and improved their hardness and their protein contents (Darnay et al., 2017b).

Edam cheese made with MTGase showed increased yield 4% without reducing its hardness and without affecting the whey protein profile and organoleptic characteristics (Aaltonen et al., 2014).

White-brined cheese, produced by simultaneous addition of MTGase and rennet in the cheese milk at two different temperatures, showed higher yield value, moisture and protein-in –dry matter contents than control cheese. Regarding the cheese texture, cheeses treated with MTGase exhibited higher hardness and springiness than the control cheese (Özer et al., 2013). Sayadi et al. (2013) produced low fat Iranian white cheese, adding TGase at the same time with rennet in the cheese milk and with or without fortification with whey protein isolate (WPI). Cheese treated with MTGase without WPI showed higher yield and moisture content and decreased fractureability than cheese enriched with WPI, while the latter showed higher protein recovery and hardness than the former. Additionally, Gustaw et al. (2008) demonstrated that, depending on the combination way of milk clotting, i.e. only MTGase, or rennet and MTGase, or heat treatment of cheese milk, MTGase altered the syneresis and the mechanical properties of UF-Feta and Dambo cheeses. Fresh probiotic cheese, produced after treating the cheese milk with MTGase under different conditions, presented higher yield, lower syneresis, better consistency, and general better sensory properties (Radošević et al., 2007).

With respect to acid induced coagulation, Mazuknaite et al. (2013) manufactured Lithuanian cottage cheese, an acid curd cheese, using MTGase added to cheese milk and incubating for different times. The results showed that the highest cheese yield was achieved by acid-gel treated with 3.6 U of MTGase per g of protein at 40°C for 60 min. Tvarog, another acid curd cheese, was also made by treating the cheese milk with a) active MTGase subsequently inactivated by heating and b) with active TGase. Cheeses did not differ each other as far as moisture content concerned, while the cheese produced with inactivated enzyme scored higher for its sensory attributes than cheeses treated with the active enzyme or without enzyme (Dmytrów et al., 2010).

Regarding other cheese varieties, goat whey cheese showed increased yield by about 1.0% when using MTGase at a ratio 8U per mg of whey protein. The MTGase treated cheese showed also higher hardness and no change in other properties like color and consistency, compared to control cheese (Karzan et al., 2016). Also, processed cheese made with milk gels treated with TGase after rennet addition, presented reduced syneresis index and increased consistency index (De sa et al., 2010). Panela cheese was manufactured with dairy-vegetable protein (soybean or peanut protein) mixtures treated with MTGase (Salinas-Valdes et al., 2015). The authors concluded that MTGase increased the moisture content, the yield and the ratio between final and initial solids in the case of soybean cheeses. Moreover, cheese with soybean showed the highest chewiness and cohesiveness values, while cheeses with peanut protein showed the highest hardness values. Di Pierro et al. (2010) isolated TGase from *Streptoverticillium mobaraense* and manufactured a 'crosslinked cheese', which exhibited significantly increased moisture content and yield, reduced proteolysis and increased hardness.

Finally, Lim et al. (2011) developed a non-dairy cream cheese product using MTGase, soy protein isolate and maltodextrin and reported that its quality and characteristics were similar to those of dairy cream cheese.

FUTURE RESEARCH DIRECTIONS

Microbial exogenous enzymes added to milk during cheese making process could be further improved with regard to their characteristics. Since microbial proteases and peptidases can hydrolyze casein to produce peptides, more research is needed towards the production of such enzymes that will be able to produce bioactive peptides with anti-hypertensive and anti-inflammatory activity, developing in this way 'bio-functional cheese products'. Therefore, new microbial enzyme technologies should be conducted to meet the challenge of increasing the healthy compounds and nutritional value of cheese. In addition, effort and attempts for producing commercial MTGase with low cost should be made as it is still considered an expensive enzyme.

CONCLUSION

Microbial non-coagulant proteases, microbial lipases have advantages over their counterparts from animal sources, because they are easily produced at low cost, their availability is inexhaustible and their characteristics can be improved using biotechnology tools. These enzymes together with microbial transglutaminase reduce the cost of cheese production by shortening the ripening time and increasing the yield. From this point of view, microbial non-coagulant enzymes are useful tools in cheesemaking technology.

REFERENCES

Aaltonen, T., Huumonen, I., & Myllärinen, P. (2014). Controlled transglutaminase treatment in Edam cheese making. *International Dairy Journal*, *38*(2), 179–182. doi:10.1016/j.idairyj.2013.12.004

Aguilar, C. N., Gutiérrez-Sánchez, G., Rado-Barragán, P. A., Rodríguez-Herrera, R., Martínez-Hernandez, J. L., & Contreras-Esquivel, J. C. (2008). Perspectives of solid-state fermentation for production of food enzymes. *American Journal of Biochemistry and Biotechnology*, *4*(4), 354–366. doi:10.3844/ajbbsp.2008.354.366

Akin, M. S., & Güler-Akin, M. B. (2017). Effects of capsule materials on the textural and sensory characteristics of Kashar cheese ripened with encapsulated lipase and protease. *Bulgarian Journal of Agricultural Science*, *23*(2), 330–337.

Akin, M. S., Güler-Akin, M. B., Kirmaci, H. A., Atasoy, A. F., & Türkoglu, H. (2012). The effects of lipase-encapsulating carriers on the accelerated ripening of Kashar cheese. *International Journal of Dairy Technology*, *65*(2), 243–249. doi:10.1111/j.1471-0307.2012.00821.x

Alhelli, A. M., Manap, M. Y. A., Mohammed, A. S., Mirhosseini, H., Suliman, E., Shad, Z., ... Hussin, A. S. M. (2016). Response surface methodology modelling of an aqueous two-phase system for purification of protease from *Penicillium candidum (PCA 1/TT031)* under solid-state fermentation and its biochemical characterization. *International Journal of Molecular Sciences*, *17*(11), 1872. doi:10.3390/ijms17111872 PMID:27845736

Aminifar, M., & Emam-Djomeh, Z. (2014). Changes of texture, microstructure and free fatty acid contents of Lighvan cheese during accelerated ripening with lipase. *Journal of Agricultural Science and Technology*, *16*(1), 113–123.

Ashour, M. M., Abdel Baky, A. A., El Neshawy, A. A., & Salem, O. M. (1986). Improving the quality of Domiati cheese made from recombined milk. *Food Chemistry*, *20*(2), 85–96. doi:10.1016/0308-8146(86)90145-7

Azarnia, S., Robert, N., & Lee, B. H. (2006). Biotechnological methods to accelerate Cheddar cheese ripening. *Critical Reviews in Biotechnology*, *26*(3), 121–143. doi:10.1080/07388550600840525 PMID:16923531

Barron, L. J. R., Hernández, I., Bilbao, A., Flanagan, C. E., Nájera, A. I., Virto, M., ... De Renobales, M. (2004). Changes in lipid fractions and sensory properties of Idiazabal cheese induced by lipase addition. *The Journal of Dairy Research*, *71*(3), 372–379. doi:10.1017/S0022029904000135 PMID:15354585

Bautista, L., & Kroll, R. (1988). Survival of some non-starter bacteria in naturally ripened and enzyme-accelerated Cheddar cheese. *The Journal of Dairy Research*, *55*(4), 597–602. doi:10.1017/S0022029900033380 PMID:3148643

Birschbach, P. (1994). Origins of lipases and their characteristics. *Bulletin of the International Dairy Federation*, (294): 7–10.

Brandsma, R. L., Mistry, V. V., Anderson, D. L., & Baldwin, K. A. (1994). Reduced Fat Cheddar Cheese from Condensed Milk. 3. Accelerated Ripening. *Journal of Dairy Science*, *77*(4), 897–906. doi:10.3168/jds.S0022-0302(94)77024-7

Calasso, M., Mancini, L., Di Cagno, R., Cardinali, G., & Gobbetti, M. (2015). Microbial cell-free extracts as sources of enzyme activities to be used for enhancement flavor development of ewe milk cheese. *Journal of Dairy Science*, *98*(9), 5874–5889. doi:10.3168/jds.2015-9362 PMID:26142846

Celligoi, M., Baldo, C., Rodrigues de Melo, M., Gasparin, F., Marques, T., & Barros, M. Lipase. Properties, functions and food applications. In R. C. Ray & C. M. Rosell (Eds.), *Microbial enzyme technology in food applications* (pp. 214–240). Boca Raton, FL: CRC Press.

Corsetti, A., Gobbetti, M., Smacchi, E., De Angelis, M., & Rossi, J. (1998). Accelerated ripening of Pecorino Umbro cheese. *The Journal of Dairy Research*, *65*(4), 631–642. doi:10.1017/S0022029998003161

Cozzolino, A., Di Pierro, P., Mariniello, L., Sorrentino, A., Masi, P., & Porta, R. (2003). Incorporation of whey proteins into cheese curd by using transglutaminase. *Biotechnology and Applied Biochemistry, 38*(3), 289–295. doi:10.1042/BA20030102 PMID:12906711

Darnay, L., Králik, F., Oros, G., Koncz, Á., & Firtha, F. (2017a). Monitoring the effect of transglutaminase in semi-hard cheese during ripening by hyperspectral imaging. *Journal of Food Engineering, 196*, 123–129. doi:10.1016/j.jfoodeng.2016.10.020

Darnay, L., Tóth, A., Salamon, B., Papik, K., Oros, G., Jónás, G., ... Friedrich, L. (2017b). Texture-modifying properties of microbial transglutaminase on 2 popular hungarian products: Trappist cheese and frankfurter. *Acta Alimentaria, 46*(1), 116–122. doi:10.1556/066.2017.46.1.15

De sá, E.M.F., & Bordignon-Luiz, M.T. (2010). The effect of transglutaminase on the properties of milk gels and processed cheese. *International Journal of Dairy Technology, 63*(2), 243-251.

Di Pierro, P., Mariniello, L., Sorrentino, A., Giosafatto, C. V. L., Chianese, L., & Porta, R. (2010). Transglutaminase-induced chemical and rheological properties of cheese. *Food Biotechnology, 24*(2), 107–120. doi:10.1080/08905431003784465

Dmytrów, I., Jasinska, M., & Dmytrów, K. (2010). Effect of microbiological transglutaminase on selected physicochemical properties of tvarog. *Italian Journal of Food Science, 22*(4), 449–460.

Domagała, J., Najgebauer-Lejko, D., Wieteska-Śliwa, I., Sady, M., Wszołek, M., Bonczar, G., & Filipczak-Fiutak, M. (2016). Influence of milk protein cross-linking by transglutaminase on the rennet coagulation time and the gel properties. *Journal of the Science of Food and Agriculture, 96*(10), 3500–3507. doi:10.1002/jsfa.7534 PMID:26585165

Doolan, I. A., Nongonierma, A. B., Kilcawley, K. N., & Wilkinson, M. G. (2014). Partitioning of starter bacteria and added exogenous enzyme activities between curd and whey during Cheddar cheese manufacture. *International Dairy Journal, 34*(1), 159–166. doi:10.1016/j.idairyj.2013.07.005

El-Hofi, M., El-Tanboly, E.-S., & Abd-Rabou, N. S. (2011). Industrial application of lipases in cheese making: A review. *Internet Journal of Food Safety, 13*, 293–302.

Esteban-Torres, M., Mancheño, J. M., de las Rivas, B., & Muñoz, R. (2014). Production and characterization of a tributyrin esterase from *Lactobacillus plantarum* suitable for cheese lipolysis. *Journal of Dairy Science, 97*(11), 6737–6744. doi:10.3168/jds.2014-8234 PMID:25173466

Farag, A.A., Aly, M.E. & El-Alfy, M.B. (1992). Enhancement of Blue cheese flavour using sodium dodecylsulphate and lipase. *Food/Nahrung, 36*(1), 1-7.

Feijoo-Siota, L., Blasco, L., Rodríguez-Rama, J. L., Barros-Velázquez, J., Miguel, T. D., Sánchez-Pérez, A., & Villa, T. G. (2014). Recent patents on microbial proteases for the dairy industry. *Recent Advances in DNA & Gene Sequences, 8*(1), 44–55. doi:10.2174/2352092208666141013231720 PMID:25564028

Fernández-García, E., López-Fandiño, R., & Alonso, L. (1994b). Effect of a food-grade enzyme preparation from *Aspergillus oryzae* on free fatty acid release in Manchego-type cheese from ovine and bovine milk. *Zeitschrift fur Lebensmittel-Untersuchung und -Forschung, 199*(4), 262–264. doi:10.1007/BF01193308

Fernandez-Garcia, E., Lopez-Fandiño, R., Alonso, L., & Ramos, M. (1994a). The Use of Lipolytic and Proteolytic Enzymes in the Manufacture of Manchego Type Cheese from Ovine and Bovine Milk. *Journal of Dairy Science*, *77*(8), 2139–2149. doi:10.3168/jds.S0022-0302(94)77156-3

Fernandez-Garcia, E., Olano, A., Cabezudo, D., Martin-Alvarez, P. J., & Ramos, M. (1993). Accelerated ripening of Manchego type cheese by added commercial enzyme preparation from *Aspergillus oryzae*. *Enzyme and Microbial Technology*, *15*(6), 519–524. doi:10.1016/0141-0229(93)90086-H

Fernández-García, E., Ramos, M., Polo, C., Juárez, M., & Olano, A. (1988). Enzyme accelerated ripening of Spanish hard cheese. *Food Chemistry*, *28*(1), 63–80. doi:10.1016/0308-8146(88)90007-6

Garcia, H. S., Lopez-Hernandez, A., & Hill, C. G. (2016). Enzyme technology - Dairy industry applications. Comprehensive Biotechnology, 4, 567-574.

Güler, Z., Gün, I., & Uraz, T. (2004). The use of fungal lipase to accelerate the ripening of Kasar cheese. *Milchwissenschaft. Milk Science International*, *59*(5-6), 277–279.

Gustaw, W., Mleko, S., & Tomczynska-Mleko, M. (2008). Production of UF-milk cheeses: Effect of rennet and transglutami-nase on coagulation time, texture and syneresis. *Milchwissenschaft. Milk Science International*, *63*(2), 141–144.

Ha, J. K., & Lindsay, R. C. (1993). Release of volatile branched-chain and other fatty acids from ruminant milk fats by various lipases. *Journal of Dairy Science*, *76*(3), 677–690. doi:10.3168/jds.S0022-0302(93)77391-9 PMID:8463483

Hayashi, K. (1996). Accelerated maturation of cheese by proteolytic enzymes produced by *Brevibacterium linens*. *Japan Agricultural Research Quarterly*, *30*(2), 129–137.

Hayashi, K., Revell, D. F., & Law, B. A. (1990). Effect of Partially Purified Extracellular Serine Proteinases Produced by *Brevibacterium linens* on the Accelerated Ripening of Cheddar Cheese. *Journal of Dairy Science*, *73*(3), 579–583. doi:10.3168/jds.S0022-0302(90)78705-X

Hernández, I., De Renobales, M., Virto, M., Pérez-Elortondo, F. J., Barron, L. J. R., Flanagan, C., & Albisu, M. (2005). Assessment of industrial lipases for flavour development in commercial Idiazabal (ewe's raw milk) cheese. *Enzyme and Microbial Technology*, *36*(7), 870–879. doi:10.1016/j.enzmictec.2004.12.032

Hu, Y.-N., Ge, K.-S., Jiang, L., Guo, H.-Y., Luo, J., Wang, F., & Ren, F.-Z. (2013). Effect of transglutaminase on yield, compositional and functional properties of low-fat cheddar cheese. *Food Science and Technology Research*, *19*(3), 359–367. doi:10.3136/fstr.19.359

Izco, J. M., Irigoyen, A., Torre, P., & Barcina, Y. (2000). Effect of added enzymes on the free amino acids and sensory characteristics in Ossau-Iraty cheese. *Food Control*, *11*(3), 201–207. doi:10.1016/S0956-7135(99)00100-0

Jaros, D., Partschefeld, C., Henle, T., & Rohm, H. (2006). Transglutaminase in dairy products: Chemistry, physics, applications. *Journal of Texture Studies*, *37*(2), 113–155. doi:10.1111/j.1745-4603.2006.00042.x

Jooyandeh, H., Kaur, A., & Minhas, K. S. (2009). Lipases in dairy industry: A review. *Journal of Food Science and Technology*, *46*(3), 181–189.

Kailasapathy, K., Anjani, K., & Seneweera, S. (2006). Recent trends in accelerated cheese ripening using microencapsulated enzymes. *Australian Journal of Dairy Technology*, *61*(2), 78–80.

Karzan, T. M., Nawal, H. S., & Ashna, T. A. (2016). The effect of microbial transglutaminase enzyme on some physicochemical and sensory properties of goat's whey cheese. *International Food Research Journal*, *23*(2), 688–693.

Kheadr, E. E., Vuillemard, J.-C., & El Deeb, S. A. (2000). Accelerated Cheddar cheese ripening with encapsulated proteinases. *International Journal of Food Science & Technology*, *35*(5), 483–495. doi:10.1046/j.1365-2621.2000.00398.x

Kheadr, E. E., Vuillemard, J.-C., & El-Deeb, S. A. (2002). Acceleration of Cheddar cheese lipolysis by using liposome-entrapped lipases. *Journal of Food Science*, *67*(2), 485–492. doi:10.1111/j.1365-2621.2002.tb10624.x

Kheadr, E. E., Vuillemard, J. C., & El-Deeb, S. A. (2003). Impact of liposome-encapsulated enzyme cocktails on Cheddar cheese ripening. *Food Research International*, *36*(3), 241–252. doi:10.1016/S0963-9969(02)00166-7

Kieliszek, M., & Misiewicz, A. (2014). Microbial transglutaminase and its application in the food industry. A review. *Folia Microbiologica*, *59*(3), 241–250. doi:10.100712223-013-0287-x PMID:24198201

Kilara, A., & Chandan, R. C. (2011). Enzyme-modified Dairy Ingredients. In R. Chandan & A. Kilara (Eds.), *Dairy Ingredients for Food Processing* (pp. 317–333). Wiley-Blackwell. doi:10.1002/9780470959169.ch12

Kilcawley, K. N., Nongonierma, A. B., Hannon, J. A., Doolan, I. A., & Wilkinson, M. G. (2012). Evaluation of commercial enzyme systems to accelerate Cheddar cheese ripening. *International Dairy Journal*, *26*(1), 50–57. doi:10.1016/j.idairyj.2012.03.015

Koçak, C., Bítlís, A., Gürsel, A., & Avsar, Y. K. (1996). Effect of added fungal lipase on the ripening of Kasar cheese. *Milchwissenschaft. Milk Science International*, *51*(1), 13–16.

Kumar, D. S., & Ray, S. (2014). Fungal lipase production by Solid State Fermentation- An overview. *Journal of Analytical & Bioanalytical Techniques*, *6*(1), 230.

Kuraishi, C., Yamazaki, K., & Susa, Y. (2001). Transglutaminase: Its utilization in the food industry. *Food Reviews International*, *17*(2), 221–246. doi:10.1081/FRI-100001258

Law, B. A., & Wigmore, A. S. (1982a). Accelerated cheese ripening with food grade proteinases. *The Journal of Dairy Research*, *49*(1), 137–146. doi:10.1017/S0022029900022214

Law, B. A., & Wigmore, A. S. (1982b). Microbial proteinases as agents for accelerated cheese ripening. *International Journal of Dairy Technology*, *35*(2), 75–76. doi:10.1111/j.1471-0307.1982.tb02718.x

Lemieux, L., & Simard, R. E. (1991). Bitter flavour in dairy products. I. A review of the factors likely to influence its development, mainly in cheese manufacture. *Le Lait*, *71*(6), 599–636. doi:10.1051/lait:1991647

Levic, S., Dordevic, V., Knezevic-Jugovic, Z., Kalusevic, A., Milasinovic, N., Bugarski, A., & Nedovic, V. (2017). Enzyme encapsulation. Technologies and their applications in food processing. In R. C. Ray & C. M. Rosell (Eds.), *Microbial enzyme technology in food applications* (pp. 469–502). Boca Raton, FL: CRC Press.

Lim, T.-J., Easa, A.-M., Karim, A.-A., Bhat, R., & Liong, M.-T. (2011). Development of soy-based cream cheese via the addition of microbial transglutaminase, soy protein isolate and maltodextrin. *British Food Journal, 113*(9), 1147–1172. doi:10.1108/00070701111174587

Mase, T., Asano, K., Ikeda, Y., Kato, Y., Esaki, H., & Isshiki, S. (2013). Characterization and application of Lipase 39-A from *Cryptococcus flavescens* for cheese flavoring. *Food Science and Technology Research, 19*(1), 89–95. doi:10.3136/fstr.19.89

Mase, T., Sonoda, M., Morita, M., & Hirose, E. (2011). Characterization of a lipase from *Sporidiobolus pararoseus* 25-A which produces cheese flavor. *Food Science and Technology Research, 17*(1), 17–20. doi:10.3136/fstr.17.17

Masotti, F., Cattaneo, S., Stuknytė, M., & De Noni, I. (2017). Technological tools to include whey proteins in cheese: Current status and perspectives. *Trends in Food Science & Technology, 64*, 102–114. doi:10.1016/j.tifs.2017.04.007

Mazuknaite, I., Guyot, C., Leskauskaite, D., & Kulozik, U. (2013). Influence of transglutaminase on the physical and chemical properties of acid milk gel and cottage type cheese. *Journal of Food Agriculture and Environment, 11*(3-4), 119–124.

McSweeney, P. L. H. (2004). Biochemistry of cheese ripening. *International Journal of Dairy Technology, 57*(2/3), 127–144. doi:10.1111/j.1471-0307.2004.00147.x

Mohedano, A. F., Fernández, J., Garde, S., Medina, M., Gaya, P., & Nuñez, M. (1998). The effect of the cysteine proteinase from *Micrococcus sp. INIA 528* on the ripening process of Manchego cheese. *Enzyme and Microbial Technology, 22*(5), 391–396. doi:10.1016/S0141-0229(97)00211-1

Moschopoulou, E. (2017). Microbial milk coagulants. In R. C. Ray & C. M. Rosell (Eds.), *Microbial enzyme technology in food applications* (pp. 199–312). Boca Raton, FL: CRC Press.

Neelakantan, S., Mohanty, A. K., & Kaushik, J. K. (1999). Production and use of microbial enzymes for dairy processing. *Current Science, 77*(1), 143–148.

Nuñez, M., Guillen, A. M., Rodriguez-Marin, M. A., Marcilla, A. M., Gaya, P., & Medina, M. (1991). Accelerated Ripening of Ewes' Milk Manchego Cheese: The Effect of Neutral Proteinases. *Journal of Dairy Science, 74*(12), 4108–4118. doi:10.3168/jds.S0022-0302(91)78605-0

Ozcan, T., & Kurdal, E. (2012). The effects of using a starter culture, lipase, and protease enzymes on ripening of Mihalic cheese. *International Journal of Dairy Technology, 65*(4), 585–593. doi:10.1111/j.1471-0307.2012.00868.x

Özer, B., Hayaloglu, A. A., Yaman, H., Gürsoy, A., & Şener, L. (2013). Simultaneous use of transglutaminase and rennet in white-brined cheese production. *International Dairy Journal, 33*(2), 129–134. doi:10.1016/j.idairyj.2013.02.001

Ozrenk, E. (2006). The use of transglutaminase in dairy products. *International Journal of Dairy Technology*, *59*(1), 1–7. doi:10.1111/j.1471-0307.2006.00220.x

Picón, A., Gaya, P., Medina, M., & Nuñez, M. (1997). Proteinases encapsulated in stimulated release liposomes for cheese ripening. *Biotechnology Letters*, *19*(4), 345–348. doi:10.1023/A:1018350901132

Radošević, V., Tonković, K., Gregurek, L., Kos, B., & Šušković, J. (2007). Production of fresh probiotic cheese with addition of transglutaminase [Proizvodnjasvježegprobiotičkogsira s dodatkomtransglutaminaze]. *Mljekarstvo*, *57*(1), 15–29.

Rao, M. B., Tanksale, A. M., Ghatge, S., & Deshpande, V. V. (1998). Molecular and biotechnological aspects of microbial proteases. *Microbiology and Molecular Biology Reviews*, *62*(3), 597–635. PMID:9729602

Romeih, E., & Walker, G. (2017). Recent advances on microbial transglutaminase and dairy application. *Trends in Food Science & Technology*, *62*, 133–140. doi:10.1016/j.tifs.2017.02.015

Salinas-Valdés, A., De la Rosa Millán, J., Serna-Saldívar, S. O., & Chuck-Hernández, C. (2015). Yield and Textural Characteristics of Panela Cheeses Produced with Dairy-Vegetable Protein (Soybean or Peanut) Blends Supplemented with Transglutaminase. *Journal of Food Science*, *80*(12), S2950–S2956. doi:10.1111/1750-3841.13126 PMID:26550775

Sayadi, A., Madadlou, A., & Khosrowshahi, A. (2013). Enzymatic cross-linking of whey proteins in low fat Iranian white cheese. *International Dairy Journal*, *29*(2), 88–92. doi:10.1016/j.idairyj.2012.10.006

Seneweera, S., & Kailasapathy, K. (2011). Impact of microencapsulated peptidase (*Aspergillus oryzae*) on Cheddar cheese proteolysis and its biologically active peptide profile. *Protein and Peptide Letters*, *18*(7), 741–746. doi:10.2174/092986611795446076 PMID:21342096

Shakila Begam, M., Pradeep, S. F., & Pradeep, B. V. (2012). Production, purification, characterization and applications of lipase from *Serratia marcescens MBB05. Asian Journal of Pharmaceutical and Clinical Research*, *5*(suppl.4), 237–245.

Tavano, O. L. (2017). Protesaes. In R. C. Ray & C. M. Rosell (Eds.), *Microbial enzyme technology in food applications* (pp. 162–179). Boca Raton, FL: CRC Press.

Trépanier, G., Abboudi, M. E., Lee, B. H., & Simard, R. E. (1992). Accelerated Maturation of Cheddar Cheese: Influence of Added Lactobacilli and Commercial Protease on Composition and Texture. *Journal of Food Science*, *57*(4), 898–902. doi:10.1111/j.1365-2621.1992.tb14318.x

Upadhyay, V. K., Kelly, A. L., & Mcsweeney, P. L. H. (2006). Use of nattokinase, a subtilisin-like serine proteinase, to accelerate proteolysis in Cheddar cheese during ripening. *Le Lait*, *86*(3), 227–240. doi:10.1051/lait:2006004

Vafopoulou, A., Alichanidis, E., & Zerfiridis, G. (1989). Accelerated ripening of Feta cheese with heat-shocked cultures or microbial proteinases. *The Journal of Dairy Research*, *56*(2), 285–296. doi:10.1017/S0022029900026492

Van De Guchte, M., Kodde, J., Van Der Vossen, J. M. B. M., Kok, J., & Venema, G. (1990). Heterologous gene expression in *Lactococcus lactis subsp. lactis*: Synthesis, secretion, and processing of the *Baccilius subtilis* neutral protease. *Applied and Environmental Microbiology*, *56*(9), 2606–2611. PMID:2125811

Wen-qiong, W., Lan-wei, Z., Xue, H., & Yi, L. (2017). Cheese whey protein recovery by ultrafiltration through transglutaminase (TG) catalysis whey protein cross-linking. *Food Chemistry*, *215*, 31–40. doi:10.1016/j.foodchem.2016.07.057 PMID:27542447

Wilkinson, M. C., & Kilcawley, K. N. (2005). Mechanisms of incorporation and releases of enzymes into cheeses during ripening. *International Dairy Journal*, *15*(6-9), 817–830. doi:10.1016/j.idairyj.2004.08.021

Wilkinson, M. G. (1999). Acceleration of cheese ripening. In P.F. Fox (Ed.), Cheese: Chemistry, Physics and Microbiology: Vol. 1. General Aspects (pp. 523-556). Gaithersburg, MD: Aspen Publication.

Yazdanpanah, S., Ehsani, M. R., & Mizani, M. (2015). Water mobility in accelerated ripening of UF Feta cheese. *Journal of Agricultural Science and Technology*, *17*(3), 663–674.

Yilmaz, G., Ayar, A., & Akin, N. (2005). The effect of microbial lipase on the lipolysis during the ripening of Tulum cheese. *Journal of Food Engineering*, *69*(3), 269–274. doi:10.1016/j.jfoodeng.2004.08.017

Zaharia, N. S. (2012). Research regarding accelerated ripening of pasta filata cheeses. *The Annals of the University Dunarea de Jos of Galati. Fascicle VI, Food Technology*, *36*(1), 26–38.

KEY TERMS AND DEFINITIONS

Cheese Ripening: The biochemical process that gives to cheese its special sensory, textural, and compositional characteristics.

Lipase: An enzyme that hydrolyzes triglycerides by splitting ester bonds.

Microbial Enzyme: Enzyme, which is secreted from a microorganism.

Protease: An enzyme that hydrolyzes proteins by splitting the peptide bond.

Transglutaminase: An enzyme that creates bonds by cross-linking of peptide bonds.

Chapter 12
Microbial Production of Recombinant Rennet:
Recent Developments

Zumrut Begum Ogel
Konya Food and Agriculture University, Turkey

ABSTRACT

Rennet, traditionally obtained from calves, is non-vegeterian and unethical due to the slaughter of unweaned animals. Chymosin is highly specific to the Phe105-Met106 bond of κ-casein and has low proteolytic activity. Microbial aspartic proteases can partly replace chymosin. However, recombinant DNA technology has allowed chymosin itself to be produced by bacteria, yeast, and molds. Not only rennet from calf, but from animals like goat kid, lamb, buffalo, camel, and others can be used in cheesemaking. Chymosins of these animals can be cloned and successfully expressed in microorganisms and can be employed in the production of novel as well as traditional cheese products from the milk of camel, goat, and even horse and donkey. This chapter outlines the recombinant DNA techniques applied over the past few years to improve the microbial production of recombinant rennet, from animals and plants.

INTRODUCTION

The history of rennet as a milk coagulant goes back *c.* 2500 years when people began to use the stomach of a young animal, either lamb, calf or goat to curdle milk for cheese. Later they started using a brine extract of the stomach which was a kind of semi-purified enzyme (McGee, 2004). This enzyme is now known as rennet which is actually a mixture of enzymes extracted from the inner mucosa of the fourth stomach chamber (the abomasum) of young animals, mainly calf, and is used in cheesemaking. Rennet is a by-product of veal production which is obtained generally from male animals of 1 month to at most 35 weeks of age.

Due to an increasing demand for cheese and due to the inefficient supply of calf rennet and the need for the slaughter of young animals, studies were initiated to develop alternative ways of producing rennet. One way to overcome this problem was to use microbial fermentation to produce the enzyme. Among the

DOI: 10.4018/978-1-5225-5363-2.ch012

enzymes of animal rennet, chymosin is the one which is responsible for the curdling reaction. Therefore, the so-called microbial rennet is either calf chymosin produced by bacteria, yeast or fungi through recombinant DNA technology or the native enzyme of a microorganism, generally a thermophilic fungus, such as *Rhizomucor,* which is capable of curdling milk for cheesemaking.

In recent years, there has been a demand towards the use of the traditional methods of yogurt and cheesemaking among the conscious consumers who prefer to consume natural food. Although traditional methods generally tend to be environmentally friendly, the traditional method for rennet preparation is against animal rights and is non-vegeterian due to the use of the stomachs of suckling young animals. The alternative methods that require genetic manipulations are also generally criticized by the conscious consumers who are against genetically modified organisms (GMOs).

Good recent reviews are available on recombinant chymosin (Kumar, et al. 2010, Yegin & Dekker, 2013) and on microbial milk coagulants (Yegin & Dekker, 2013, Moschopoulou, 2017). In this chapter the emphasis will be on the most recent developments in the field of recombinant milk-coagulant production in order to improve yields by means of novel methods and approaches towards more efficient heterologous expression systems. It is also aimed to discuss the safety of these products, and to focus on possible future developments based on the recent research trends. In this chapter, for the sake of simplicity, the term rennet will be used to indicate any milk coagulant, irrespective of its source.

BACKGROUND

The major milk protein is casein, constituting about 80% of the total milk proteins. The remaining 20% correspond to the so-called whey proteins, including lactalbumin, lactoglobulin and immunoglobulins. The caseins consist of α, β, κ, γ-forms and subforms of them with few amino acid variations. α-, and β-caseins are sensitive to calcium-mediated coagulation by the formation of colloidal calcium phosphate bridges (Coultate, 2002). The different casein proteins are clustered in the form of micelles, where κ–caseins aggregate such that the formation of colloidal calcium phosphate bridges between submicelles is limited due to its highly hydrophilic but non-calcium-binding C-terminal trisaccharide residues. Thereby, κ-casein acts as a micelle stabilizer.

In classical cheesemaking, milk is first coagulated into a smooth semi-solid curd, resembling yogurt, and is next cut in order to remove the whey. This type of milk coagulation is, generally, referred to as curdling to distinguish it from acid coagulation. Unlike yogurt formation which requires acid production and isoelectric point precipitation of caseins, curd can be formed at higher pH values by the action of enzymes. In particular one enzyme is important for curdling. This enzyme is called chymosin.

Chymosin cleaves a specific peptide bond of κ–casein, namely the bond between methionine (Met) 106 and phenylalanine (Phe) 105. This cleavage divides the κ–casein molecule into two peptides. The smaller peptide, named the glycomacropeptide fragment carries carbohydrates. Carbohydrates confer protection to caseins. When the glycopeptide is cleaved, the protective action of κ–casein is lost and this results in formation of calcium bridges destabilizing micelles.

The traditional source of chymosin is rennet obtained from the stomach of young unweaned ruminants; Calf, goat, buffalo, lamb etc., where it is found together with pepsin (10-20%). Although chymosin is also present in adult animals, the ratio of chymosin to pepsin decreases significantly once the animal is weaned. The disadvantage of this is not just a decrease in the amount of chymosin, but an increase in pepsin at the same time, which causes over-proteolysis and the formation of bitterness in cheese.

Due to a shortage of rennet from young calf in the 1970's, studies have been initiated to find alternative sources of chymosin. Chymosin (EC 3.4.23.4) is an aspartic protease and enzymes of similar activity are produced by other organisms, including plants and microorganisms. However, being able to cleave the Met106 and Phe105 bond of κ-Casein and form a curd, is not sufficient to replace chymosin in industrial applications. The enzyme should be specific and active under conditions similar to chymosin, it should not cause undesirable organoleptic properties, production should be economical, it should neither be toxic to humans nor to the lactic acid bacteria used as a starter culture in the product, etc. (Khademi, et al. 2013). Thus, despite the extensive studies on finding a substitute for chymosin among the microbial sources, only three species of filamentous fungi (*Rhizomucor miehei* and *Rhizomucor pusillus (*formerly known as *Mucor pusillus*) and *Cryphonectria parasitica (*formerly known as *Endothia parasitica)*) were successfully commercialized (Khademi, et al. 2013).

Microbial aspartic proteases (EC 3.4.23), in particular that of *R. miehei* (Kumar, et al. 2010), are fairly effective in replacing rennet, however the activity of the protease and resulting cheese quality can never be exactly the same, including yield of cheese and organoleptic properties. Furthermore, although excellent ripened cheeses were obtained by fungal rennet preparations, this required the destabilization of the enzyme by chemical modifications (Yegin & Dekker, 2013).

By the advancements in genetic engineering techniques, it became possible to produce animal chymosin in microorganisms, including bacteria, yeast and molds, simply by cloning and expressing the chymosin gene in microbes. Many successful applications, including *Escherichia coli* K-12, as the host, resulted in chymosin indistinguishable from its original counterpart, and was given GRAS status by the FDA (Food and Drug Administration). Recombinant chymosin is by far more advantageous than the original animal-derived form because of animal rights, ethics, religious feelings, vegetarian approval, as well as ease of cheese manufacture and higher enzyme yields, any time of the year. There are a number of studies comparing cheese made with recombinant chymosin and native calf chymosin. All these studies have clearly shown that the two forms of enzymes have similar milk clotting to proteolysis ratio and differences observed in cheese quality and production parameters are statistically insignificant (Green, et al. 1985, Bines, et al. 1989, Ustunol and Hicks, 1990, Barbano and Rasmussen, 1992). Today, recombinant chymosin, also known as the fermentation-produced chymosin constitutes around 70-80% of the global rennet market (Yegin & Dekker, 2013).

STRUCTURAL AND FUNCTIONAL FEATURES OF RENNET

All aspartic proteases that act as rennet are synthesized in the form of preproenzymes. The pre-sequence corresponds to the signal peptide, which is required for the secretion of the enzyme, and is cleaved after passage through the cell membrane in prokaryotes or the ER (endoplasmic reticulum) membrane in eukaryotes. The pro-sequence encodes a leader peptide which keeps the enzyme in an inactive form (called the zymogen) until conditions are suitable for its activity. The zymogens are generally activated under acidic conditions, and are stable mostly around pH 3-5.5. The leader peptide is cleaved by autocatalytic cleavage (Barholt-Perdersen, et al. 1979).

Calf chymosin gene encodes 381 amino acids, including a 16 amino acids long signal peptide and a 42 amino acids long propeptide. The mature enzyme consists of 323 amino acids (35.6 kDa). The fungal aspartic proteases consist of *around* 320- 360 amino acids. The most widely used commercial fungal rennet is the one of *R. miehei*. *R. miehei* rennet is synthesized as a 430 amino acids long preproenzyme, consisting of 22 amino acids long signal peptide, 47 amino acids long propeptide and the mature protein consists of 361 amino acids (40.5 kDa) (Feijoo-Siota, et al. 2014). The tertiary structures of calf chymosin and *R. miehei* aspartic protease are similar. The enzymes consist of a single polypeptide folded into a globular protein containing 13% helical segments and 48% β-sheets (Newman, et al. 1991). Aspartic proteases that act as rennet contain two catalytic aspartates. In calf chymosin these are the Asp34 and Asp216 residues, occuring in the Asp-Thr-Gly motif, whereas in *R. miehei* these are the Asp60 and Asp 260 residues (Mandujano-Gonzales, et al. 2016). In calf chymosin, these motifs appear to be significant for the specific cleavage of the Phe105-Met106 bond.

Most plant aspartic proteases are within the A1 family but contain an additional PSI (plant-specific insert) region of *c.* 100 amino acids. This PSI region is located between the N- and C-terminal domains and following the two catalytic Asp residues of the animal and fungal aspartic proteases (Yegin & Dekker, 2013). This so-called PSI region is similar to saposins and saposin-like proteins (Chen, et al. 2002). Three intra-molecular disulfide bridges are formed by the characteristic 6 conserved cysteines of saposin-like proteins. These bridges stabilize a "saposin-fold," which facilitates transient or permanent interaction with membranes (Liepinsh, et al. 1997).

From a functional point of view, a rennet-substitute should have a suitable ratio of clotting to proteolytic activity (C/P). This ratio indicates the specificity of the enzyme for the Phe105-Met106 bond of κ-casein and the level of proteolysis. In general, most rennet substitutes have a lower C/P due to a higher level of proteolysis. For instance, while milk clotting activity of bovine chymosin is 100 and proteolytic activity is 25, resulting in a C/P ratio of 4, those of *R. miehei* are 19 and 149, resulting in a C/P of 0.13. Remarkably, camel chymosin has milk-clotting activity of 170, indicating 70% higher specificity to the Phe105-Met106 bond of κ-casein. Proteolysis is the same as bovine chymosin, thus C/P is 6.8 (Szecsi, et al. 2013).

Higher levels of proteolysis causes loss of casein and fat, and bitterness during ripening. Furthermore, proteolytic specificity is also important as it determines the structure and flavour of the ripened cheese. Some cheese varieties are excessively heated, thus causing the inactivation of the coagulating enzymes. However, in others, coagulating enzyme continues to function during curing and ripening. For example, chymosin breaks down one of the caseins, namely αs1-casein much more than other caseins. A high pH optimum for the rennet-substitute is also desirable. Rennet activity is stable and able to coagulate milk even at the pH of milk (pH 6.5- 6.7), although its activity increases with decreasing pH, generally reaching an optimum around pH 5.5. Most pepsins and microbial proteases are denatured at the pH of milk which has been a major difficulty in developing rennet substitutes (Hill, 2017). The denaturation temperature is also important because, first, rennet must be eliminated during cooking in some cheese varieties such as parmesan cheese, in order not to interfere with the ripening process. Second, rennet should be eliminated by pasteurization in order to prevent further proteolysis, especially in certain products made from whey, however, some microbial rennets withstand pasteurization. For example, thermostability was one of the major problems with the enzyme from *R. pusillus* (Yegin & Dekker, 2013). Finally, consistent activity is important to achieve standard quality of cheese. Animal rennet is a mixture of enzymes and the ratio of chymosin to pepsin is variable. However, the recombinant enzyme is pure in that sense, resulting in higher specific activities and allowing better standardisation, hence yielding more consistent quality of products.

RECOMBINANT RENNET

Recombinant rennet production in bioreactors by means of microorganisms appears to be one of the most attractive approaches for rennet production. The source organism, for the gene encoding rennet, may be either animal, plant or microbe. In order to produce recombinant enzymes, the genes have to be isolated and inserted into the host organism to be used for production. One possibility to obtain the genes, is the isolation of mRNA either from suckling animals, or plants or microbes expressing the rennet-substitute. Further, cDNA is synthesized by reverse transcription of RNA. The cDNA population is then used in PCR to amplify, generally, the pro-enzyme encoding region, using primers that are designed to target the pro-enzyme. The amplified fragment is then cloned onto a suitable vector, generally downstream of an efficient secretion signal, if desirable. First, the construct is enriched in *E. coli* and next transformed into a suitable host, which may be again an *E. coli* strain, or another suitable host, either yeast or mold. Once rennet-substitutes are successfully expressed in a heterologous system, they can then be subject to suitable sequence changes that will enhance their production or activity during cheesemaking, or pH or temperature stability, *etc.*

As the enzyme is separated from the host and purified, following the bioreaction, genetically modified organisms (GMOs) are used only as a vehicle and do not appear in the final product. This is different than using the GMO directly as a food source, as in the case of GMO-plants, such as corn. The recombinant enzymes appear to have the same properties as the native enzyme. Nonetheless, for the consumers who do not wish to consume cheese made by using recombinant rennet, it is advised to use acid-coagulated cheese. In Turkey, due to the importance of yogurt as a staple part of the diet, yogurt has also been traditionally used in cheesemaking. This type of cheesemaking is not well known in the scientific literature. Thus, studies can be conducted to understand the mechanism of coagulation during cheesemaking from yogurt and, for standardisation in order to commercially produce such non-rennet cheese varieties for the consumers who wish to consume non-GMO and vegeterian products.

The recent developments in the field of microbially-produced recombinant rennet are summarized below.

MICROBIAL PRODUCTION OF RECOMBINANT RENNET

Microorganisms can be easily cultivated in bioreactors and, thereby, provide a practical means of producing enzymes. There is a huge accumulated knowledge on microbial heterologous expression systems which can be exploited to design efficient ways of producing rennet either by bacteria, yeast or molds. Knowledge on protein structure, folding, processing and secretion mechanisms has allowed the design of host-vector systems that resulted in successful expression systems for rennet, not only from calf but also from goat, lamb and camel.

In general, unicellular organisms, namely most bacteria and yeasts are more resistant to the shear stress caused by mechanical agitation of the bioreactor medium. This is essential not only for increasing the residance time of air bubbles during aerobic cultivations but also for obtaining a homogeneous mixture, which may otherwise decrease productivity. Unlike the unicellular organisms, on the one hand, the filamentous fungi are less shear resistant, thus their mycelia are damaged by mechanical agitation and their filaments may cause fauling of bioreactor spargers and impellers. On the other hand, molds are excellent in the extracellular production of proteins and enzymes, whereas bacteria and yeasts are less efficient in terms of secretion.

Most of these problems have been overcome by developments in genetic engineering and fermentation technology, such as the construction of suitable host-vector systems and the use of pneumatic agitation-based airlift bioreactors or solid-state fermentations.

Production of Rennet by Bacterial Expression Systems

E. coli was the first model organism for recombinant DNA studies, thus it is genetically and physiologically well characterized and there are well-established methods and suitable host-vector systems for its transformation and heterologous protein production. Indeed prochymosin has been successfully expressed in *E. coli* (Nishimori, et al. 1982) and the resulting enzyme could be activated by auto-catalytic cleavage under acidic conditions. The recombinant enzyme was indistinguishable from the native enzyme in terms of its properties. The FDA has approved the safety of the recombinant chymosin from *E. coli* and indicated that the impurities in the preparation did not make the product unsafe.

Nevertheless, *E. coli*, is not efficient in terms of protein secretion. Foreign proteins are often aggregated in the form of insoluble inclusion bodies and the protein has to be solubilized and refolded by denaturation and renaturation. Furthermore, in order to protect the foreign protein from the self-defence system of cells, they need to be camouflaged by constructing protein fusions. The various studies for the efficient expression of chymosin in *E. coli* are summarized by Kumar, et al. (2010). In general, strong promoters are used to trigger high levels of expression and the signal sequences fused to prochymosin are also significant in determining the faith of the product. For example, the prochymosin gene expressed in *Bacillus subtilis* formed an insoluble aggregate when expressed with plasmid pSM316 under the control of the T5 phage promoter and the induction of a two-cistron sequence at the 5' region of the gene. However, the protein was produced extracellularly when the *B. subtilis* subtilisin signal sequence was fused to the 5' end of the prochymosin gene (Parente, et al. 1991).

Lactococcus lactis is a model lactic acid bacterium for genetic engineering studies and it is widely used in the dairy industry and is a GRAS organism. *L. lactis*, therefore, appeares to be a suitable host for the heterologous production of chymosin. Various expression and secretion cassettes have been constructed for the production of prochymosin in *L. lactis* (Simons, et al. 1988; Simons & Blobel 1992). These cassettes contained 1. The *prt*P promoter sequence, belonging to the lactococcal extracellular proteinase which is a key enzyme in casein degradation and, 2. Ribosome binding site (RBS) and various parts of the coding sequence of the SK11 proteinase fused to prochymosin. The precursor form of PrtP contains a 150-residue propeptide and the mature protein is attached to the cell envelope by a C-terminal membrane anchor. A fusion with the coding sequence for the first 7 amino acids of PrtP, resulted in intracellular accumulation of prochymosin (Simons & Blobel, 1992). This intracellular fraction could not be activated at low pH. Fusion of the coding sequences of prochymosin and the first 33 or the first 62 amino terminal residues of PrtP directed the synthesis and secretion of prochymosin. Although the putative signal peptide of PrtP consists of the first 33 amino acids, the secretion efficiency could be increased when the first 62 amino acids were used. This may be due to positively charged residues present at the amino-terminal end of prochymosin. The presence of this charged region immediately after the signal peptide cleavage site is most likely to obstruct efficient secretion of the protein. Indeed, Ogel (1993) has shown that, even a two amino acids change at the N-terminus of galactose oxidase from Arg-Ala into Phe-Phe, can change the final destination of the enzyme from an extracellular form to a cell wall-bound form.

Recently, Luerce, et al. (2014), have cloned the prochymosin B, chymosin A and B genes downstream of the RBS and signal peptide of the major lactococcal secreted protein Usp45, on vector pXYSEC. However, none of the forms were secreted into the extracellular medium. Cell fractionation and detection by anti-chymosin antibodies suggested that prochymosin B was attached to the cell envelope (Luerce, et al. 2014).

Production of Rennet by Yeast Expression Systems

Due to problems related to secretion, post-translational processing and proper folding in bacterial expression systems, the yeast expression systems appear to provide an advantage for the expression of genes of higher eukaryotes. Unlike molds, yeasts are unicellular and easy to cultivate, thereby provide the advantages of the bacterial systems. Yet they are eukaryotic and, therefore, their protein expression, secretion and processing mechanisms resemble those of higher eukaryotes more than the prokaryotic systems.

Among the yeast expression systems, *Saccharomyces cerevisiae* is the eukaryotic model organism in genetic studies, however, in terms of correct processing and efficient secretion of heterologous proteins, better alternatives have been developed. For the recombinant production of rennet, *Pichia pastoris* and *Kluyveromyces lactis* have proven to be better alternatives. Compared to *S. cerevisiae, P. pastoris* has higher heterologous protein expression levels, up to 10-100 fold (Espinoza-Molina, et al. 2016).

P. pastoris has attracted most of the attention in the recent literature mainly due to its advantages including its long safe use, its GRAS status by the FDA, the availability of simple and efficient molecular manipulation methods, and its capability of proper post-translational processing and secretion of heterologous proteins (Jiang, et al. 2012). As for the overexpression of proteins, the methanol induced AOX1 (Alcohol oxidase 1) promoter is generally preferred, as in the case of the recent studies conducted for the expression of the buffalo, yak and camel chymosins in *P. pastoris* (Tyagi, et al. 2017, Luo, et al. 2016, Wang, et al. 2015). However, from an industrial point of view, methanol induction is not practical on a large scale. It is volatile and toxic, causing health problems and environmental pollution. In order to overcome this problem, recently, Jiang, et al. (2012) have expressed the bovine prochymosin B gene under the control of the constitutive glyceraldehyde 3-phosphate dehydrogenase (GAPDH) promoter on vector pGAPZα. These vectors allow secretion of the proteins due to the presence of the *S. cerevisiae* α-factor prepropeptide sequence fused to the 5' end of the cloned genes. The prochymosin produced and secreted by this construct exhibited satisfactory levels of activity during kappa-casein hydrolysis and its optimum conditions for activity were similar to the native calf chymosin (Jiang, et al. 2012).

Goat chymosin is also one of the enzymes expressed successfully in *P. pastoris* (Kumar, et al. 2007). However, the low level of production has limited its use for the characterization studies. The use of bioreactors, for larger scale growth and expression of recombinant goat chymosin, has shown that the scale-up of production can significantly improve the yield of the enzyme (Tyagi, et al. 2016).

Despite of the use of strong promoters, efficient secretion signals, proper post-translational processing, and scale-up, protein synthesis levels in *P. pastoris* can still be enhanced by optimizing codon usage. Codon usage (codon bias) indicates that in different organisms some of the codons of a certain amino acid (if encoded by more than one codon) may be preferred over others, hence creating a bias towards these codons. As a result, the remaining codons are rarely used in the sequence of genes. Studies have shown that highly expressed genes generally contain these preferred codons in their sequence, suggesting that the rate of translation is increased by these preferred codons, possibly due to the abundance of the corresponding tRNAs. One of the recent reviews on the significance of codon bias on heterologous gene expression is by Gustaffson, et al. (2004).

Different organims may have different codon usages, thus affecting the rate of translation in heterologous expression systems if the codon usage of the source organism is different than that of the host organism. It has been shown that codon optimization may enhance productivity up to 50 fold (Espinoza-Molina, et al. 2016).

Due to relatively low levels of chymosin production in *P. pastoris*, Espinoza-Molina et al. (2016) have recently generated a synthetic copy of the *Bos taurus* (Calf) chymosin gene (CYM). The synthetic gene was optimized in terms of codon usage such that the low frequency codons of *P. pastoris* were replaced by the preferred codons. In total 332 nucleotides were modified to optimize 289 codons, however, the amino acid sequence of the protein did not change. With this approach, a 12-fold increase was obtained and the resulting chymosin displayed similar proteolytic properties compared to the native enzyme (Espinoza-Molina et al. 2016).

Kluyveromyces lactis is another well-developed yeast expression system of GRAS status by the FDA, and has been successfully used as an expression host for calf chymosin, which has been commercialized under the trademark MaxiRen® (DSM, The Netherlands). Recently *K. lactis* was used by Almeida, et al. (2015) for the expression of the plant rennet 'cardosin B' from *Cynara cardunculus*. The extracts of *C. cardunculus* have been traditionally used in making some excellent cheeses from goat and sheep milk. However, there is only artisanal production of these cheeses using plant extracts. Industrial production requires high level of production of the enzyme. Thus the authors have developed a recombinant expression system for the possible industrial production and use of the enzyme. Cardosins of *Cynara* have a broader proteolytic activity as compared to chymosin. As a result, cardosins not only cleave the Phe105-Met106 bond, but also cleave other bonds that impart a characteristic flavor to the final product. The procardosin B gene was first cloned downstream of the α-mating factor leader sequence on vector pKLAC1, to promote secretion, and the LAC promoter, to allow induction by lactose. Next, a protein engineering approach to remove the plant-specific domain, followed by codon optimiziation was applied to improve the yield upto 35 fold in *K. lactis*. The resulting enzyme displayed properties similar to the native cardosin B (Almeida, et al. 2015).

Production of Rennet by Mold Expression Systems

Molds are the enzyme factories of nature and are responsible for the degradation and recycling of most of the dead plant material in the soil. Plant materials to be degraded contain large molecular weight polymeric compounds. Thus, molds have to secrete many of their enzymes to the extracellular environment in order to digest these polymers into smaller molecular weight compounds that can enter the cells for further catabolism.

The fact that molds have the natural ability of secreting plenty of extracellular enzymes makes molds excellent host organisms for the extracellular production of heterologous proteins. Bacteria as well as yeasts are less efficient in that sense. However, industrial production prefers extracellular production of heterologous proteins because it is easier, thus cheaper to purify proteins form culture supernatants, instead of going through the elaborative steps of cell lysis and intracellular protein purification.

Among the molds, *Aspergillus* species appear to be unrivaled as a heterologous expression host. There are well-developed host-vector systems for Aspergillus species. Although some Aspergilli may be pathogenic, such as *Aspergillus fumigatus*, others, such as *Aspergillus oryzae* have been used for thousands of years in food fermentations.

As for the recombinant production of calf chymosin, *Aspergillus niger* has been used successfully as an expression host, and the product has been commercialized under the trademark Chy-Max® (Chr. Hansen Denmark). More recently recombinant camel chymosin was also successfully produced by *A. niger* (Kappeler, et al. 2006). Interestingly, camel chymosin has shown seven fold higher clotting to general proteolytic activity (C/P) as compared to bovine chymosin (Kappeler, et al. 2006). Although the open-reading frames of bovine and camel have 87% homology, bovine chymosin is unable to coagulate camel milk, possibly due to differences between the κ-casein structures of the two species (Kappeler, et al. 2006). Yet, camel chymosin is able to coagulate bovine milk as efficiently as bovine chymosin, while the strength of coagulum obtained by coagulating camel milk with camel chymosin is much lower than the coagulum of bovine milk and camel chymosin.

Recently Lanella (2015), has discovered that camel chymosin can effectively coagulate equid milk, including donkey and horse. These studies have led to the commercial production of camel chymosin, which can be used not only in making cheese from bovine milk, but also from camel milk and equid milk. Camel milk is mainly consumed in Eastern Africa, and equid milk is consumed in Mongolia and, in general, in Central Asia. However, donkey milk has become popular recently also in the western countries, especially for those individuals who have allergy against cow's milk (Lanella, 2015).

In the *A. niger* var. *awamori* expression system, prochymosin gene is integrated downstream of the starch-inducible *gla*A promoter and the *gla*A gene from *A. niger*, and is expressed in the form of a fusion protein. The vector utilizes the pyr4 gene of *Neurospora crassa* and a pyrimidine auxotroph of *A. niger var. awamori* as a host. One other preferred promoter for expression in *Aspergillus*, is the constitutive *gpd*A promoter of *A. nidulans*, on vector pAN7-1 (Punt, et al. 1988). Cardoza et al. (2003) have compared the expression of a codon-optimized synthetic copy of bovine chymosin on constitutive and inducible promoters, where *gpd*A was used as the constitutive promoter and the aspergillopepsin promoter *pep*B of *A. awamori*, was used as the inducible promoter. Although *gpd*A allowed constituttive expression and production throughout growth, *pepB* promoter was induced by pH and yielded a very high expression during the growth phase.

CONCLUSION AND FUTURE PROSPECTS

Due to ethical reasons and animal rights, the use of animal-derived rennet has replaced itself with microbial rennet. Microorganisms produce aspartic proteases that resemble animal chymosin, however, animal chymosin itself can be produced by microorganisms using recombinant DNA technology. Almost 60-80% of the rennet used by the cheesemaking industry corresponds to the recombinant animal rennet because of its unique properties that can hardly be met by the microbial coagulants. Recent developments in the area suggest that the calf rennet may lose its former popularity due to the development of efficient production systems for other recombinant animal rennets, including goat, camel, buffalo, yak, and lamb. Furthermore, market shelves may be introduced with novel cheese products not only from cow's milk and goat's milk but from camel and equine milk too. While camel and equine milk have been traditionally consumed for thousands of years in Asia and Africa, the western market is not familiar with such products. For example, Kımız, is a fermented horse milk product consumed by the

Asian Turkish cultures for thousands of years. The consumption of milk from animals other than cow, opens a new area of research for the development of cheese and other fermented products from these sources, which can be studied in terms of their process parameters, functionality, such as the probiotic population, taste and aroma, etc.

For those consumers, who are both vegeterian and are against GMO, cheese production without the use of rennet appears to increase in popularity. Thus, research may also be conducted on cheese produced by acid coagulation or yogurt or other novel methods.

REFERENCES

Almeida, C. M., Gomes, D., Faro, C., & Simoes, I. (2015). Engineering a cardosin B-derived rennet for sheep and goat cheese manufacture. *Applied Microbiology and Biotechnology*, *99*(1), 269–281. doi:10.100700253-014-5902-5 PMID:24986551

Barbano, B. M., & Rasmussen, R. R. (1992). Cheese yield performance of fermentation-produced chymosin and other milk coagulants. *Journal of Dairy Science*, *75*(1), 1–12. doi:10.3168/jds.S0022-0302(92)77731-5 PMID:1592942

Barholt-Perdersen, V., Asbaek-Christensen, K., & Foltmann, B. (1979). Investigations on the activation of bovine prochymosin. *European Journal of Biochemistry*, *94*(2), 573–580. doi:10.1111/j.1432-1033.1979.tb12927.x PMID:34519

Bines, V. E., Young, P., & Law, B. A. (1989). Comparison of cheddar cheese made with recombinant calf chymosin and standard calf rennet. *The Journal of Dairy Research*, *56*(4), 657–664. doi:10.1017/S0022029900029186

Cardoza, R. E., Gutierrez, S., Ortega, N., Casqueiro, J., & Martin, J. F. (2003). Expression of a synthetic copy of the bovine chymosin gene in *Aspergillus awamori* from constitutive and pH-regulated promoters and secretion using two different pre-pro sequences. *Biotechnology and Bioengineering*, *83*(3), 249–259. doi:10.1002/bit.10666 PMID:12783481

Chen, X., Pfeil, J. E., & Gal, S. (2002). The three typical aspartic proteinase genes of *Arabidopsis thaliana* are differentially expressed. *European Journal of Biochemistry*, *269*(18), 4675–4684. doi:10.1046/j.1432-1033.2002.03168.x PMID:12230581

Coultate, T. P. (2002). *Food: The chemistry of its components*. Cambridge, UK: The Royal Society of Chemistry.

Espinoza-Molina, J. A., Acosta-Muniz, C. H., Sepulveda, D. R., Zamudio-Flores, P. B., & Rios-Velasco, C. (2016). Codon optimization of the '*Bos taurus* chymosin' gene for the production of recombinant chymosin in *Pichia pastoris*. *Molecular Biotechnology*, *58*(10), 657–664. doi:10.100712033-016-9965-7 PMID:27394727

Feijoo-Siota, L., Blasco, L., Rodríguez-Rama, J. L., Barros-Velázquez, J., Miguel, T. D., Sánchez-Pérez, A., & Villa, T. G. (2014). Recent patents on microbial proteases for the dairy industry. *Recent Advances in DNA & Gene Sequences*, *8*(1), 44–55. doi:10.2174/2352092208666141013231720 PMID:25564028

Green, M. L. (1985). Cheddar cheesemaking with recombinant calf chymosin synthesized in *Escherichia coli. The Journal of Dairy Research*, *52*(2), 281–286. doi:10.1017/S0022029900024158

Gustaffson, C., Govindarajan, S., & Minshull, J. (2004). Codon bias and heterologous protein expression. *Trends in Biotechnology*, *22*(7), 346–353. doi:10.1016/j.tibtech.2004.04.006 PMID:15245907

Hill, A. (2017). *Cheese technology. University of Guelph*. Dairy Science and Technology Education Series.

Jiang, X. P., Yin, M. L., Chen, P., & Yang, Q. (2012). Constitutive expression, purification, and characterization of bovine prochymosin in *Pichia pastoris* GS115. *World Journal of Microbiology & Biotechnology*, *28*(5), 2087–2093. doi:10.100711274-012-1012-7 PMID:22806030

Kappeler, S. R., van den Brink, H. J. M., Rahbek-Nielsen, H., Farah, Z., Puhan, Z., Hansen, E. B., & Johansen, E. (2006). Characterization of recombinant camel chymosin reveals superior properties for the coagulation of bovine and camel milk. *Biocehmical and Biophysical Reserach Communications*, *342*(2), 647–654. doi:10.1016/j.bbrc.2006.02.014 PMID:16488399

Khademi, F., Abachi, S., Mortazavi, A., Ehsani, M. A., Tabatabaei, M. R., & Malekzadeh, F. A. (2013). Optimization of fungal rennet production by local isolate of *Rhizomucor nainitalensis* under solid substrate fermentation system. *IOSR Journal of Pharmacological and Biological Sciences*, *5*(2), 115–121. doi:10.9790/3008-052115121

Kumar, A., Grover, S., Sharma, J., & Batish, V. K. (2010). Chymosin and other milk coagulants: Sources and biotechnological interventions. *Critical Reviews in Biotechnology*, *30*(4), 243–258. doi:10.3109/07388551.2010.483459 PMID:20524840

Kumar, A., Sharma, J., Grover, S., Mohanty, A. K., & Batish, V. K. (2007). Molecular cloning and expression of goat (*Capra hircus*) prochymosin in *E. coli. Food Biotechnology*, *21*(1), 57–69. doi:10.1080/08905430701191163

Lanella, G. (2015). Equid milk renneting through pure camel chymosin and cheese manufacturing. [DRJAFS]. *Direct Research Journal of Agriculture and Food Science*, *3*(9), 167–172.

Liepinsh, E., Andersson, M., Ruysschaert, J. M., & Otting, G. (1997). Saposin fold revealed by the NMR structure of NK-lysin. *Nature Structural Biology*, *4*(10), 793–795. doi:10.1038/nsb1097-793 PMID:9334742

Luerce, T. D., Azevedo, M. S. P., LeBlanc, J. G., Azevedo, V., Miyoshi, A., & Pontes, D. S. (2014). Recombinant *Lactococcus lactis* fails to secrete bovine chymosine. *Bioengineered*, *5*(6), 363–370. doi:10.4161/bioe.36327 PMID:25482140

Luo, F., Jiang, W. H., Yang, Y. X., Li, J., & Jiang, M. F. (2016). Cloning and expression of yak active chymosin in *Pichia pastoris. Asian-Australasian Journal of Animal Sciences*, *29*(9), 1363–1370. doi:10.5713/ajas.16.0038 PMID:27004812

Mandujano-Gonzales, V., Villa-Tanaca, L., Anducho-Reyes, M. A., & Mercado-Flores, Y. (2016). Secreted fungal aspartic proteases: A review. *Revista Iberoamericana de Micologia*, *33*(2), 76–82. doi:10.1016/j.riam.2015.10.003 PMID:27137097

McGee, H. (2004). *On food and cooking: The science and lore of the kitchen*. New York: Scribner.

Moschopoulou, E. (2017). Microbial milk coagulants. In Microbial enzyme technology in food applications (pp. 199-210). CRC Press, Taylor and Francis Group.

Newman, M., Safro, M., Frazao, C., Khan, G., Zdanov, A., Tickle, I. J., ... Andreeva, N. (1991). X-ray analysis of aspartic proteinases IV. Structure and refinement at 2.2 Å resolution of bovine chymosin. *Journal of Molecular Biology*, *221*, 1295–1309. PMID:1942052

Nishimori, K., Kawagushi, Y., Hidaka, M., & Beppu, T. (1982). Expression of cloned calf prochymosin gene sequence in *Escherichia coli. Gene*, *19*(3), 337–344. doi:10.1016/0378-1119(82)90024-5 PMID:6295885

Ogel, Z. B. (1993). *Molecular characterization of a galactose oxidase gene* (Ph.D. thesis). The University of Leeds, UK.

Parente, D., de Ferra, F., Galli, G., & Grandi, G. (1991). Prochymosin expression in *Bacillus subtilis. FEMS Microbiology Letters*, *61*(2-3), 243–249. doi:10.1111/j.1574-6968.1991.tb04356.x PMID:1903749

Punt, P. J., Dingemanse, M. A., Jacobs-Meijsing, B. J. M., Pouwels, P. H., & van den Hondel, C. A. M. J. J. (1988). Isolation and characterization of the glyceraldehyde-3-phosphate dehydrogenase gene of *Aspergillus nidulans. Gene*, *69*(1), 49–57. doi:10.1016/0378-1119(88)90377-0 PMID:3066699

Simon, S. M., & Blobel, G. (1992). Signal peptides open protein-conducting channels in *Escherichia coli. Cell*, *69*(4), 677–684. doi:10.1016/0092-8674(92)90231-Z PMID:1375130

Simons, G., Rutten, G., Homes, M., & de Vos, W. M. (1988). Production of bovine prochymosin by lactic acid bacteria. In *Proceedings 2nd Netherlands Biotechnology Congress*, 183-187.

Szecsi, P. B., & Harboe, M. (2013). Chymosin. In N. D. Rawlings & G. S. Salvesen (Eds.), *Handbook of proteolytic enzymes* (3rd ed.; Vol. 1). Academic Press. doi:10.1016/B978-0-12-382219-2.00005-3

Tyagi, A., Kumar, A., Mohanty, A. K., Kaushik, J. K., Groveer, S., & Batish, V. K. (2017). Expression of buffalo chymosin in *Pichia pastoris* for application in mozarella cheese. *Lebensmittel-Wissenschaft + Technologie*, *84*, 733–739. doi:10.1016/j.lwt.2017.06.033

Tyagi, A., Kumar, A., Yadav, A. K., Saklani, A. C., Grover, S., & Batish, V. K. (2016). Functional expression of recombinant goat chymosin in *Pichia pastoris* bioreactor cultures: A commercially viable alternate. *Lebensmittel-Wissenschaft + Technologie*, *69*, 217–224. doi:10.1016/j.lwt.2016.01.015

Ustunol, Z., & Hicks, C. L. (1990). Effect of milk-clotting enzymes on cheese yield. *Journal of Dairy Science*, *73*(1), 8–16. doi:10.3168/jds.S0022-0302(90)78639-0

Wang, N., Wang, K. Y., Li, G., Guo, W., & Liu, D. (2015). Expression and characterization of camel chymosin in *Pichia pastoris. Protein Expression and Purification*, *111*, 75–81. doi:10.1016/j.pep.2015.03.012 PMID:25837439

Yegin, S., & Dekker, P. (2013). Progress in the field of aspartic proteinases in cheese manufacturing: Structures, functions, catalytic machanism, inhibition, and engineering. *Dairy Science & Technology*, *93*(6), 565–594. doi:10.100713594-013-0137-2

Chapter 13
Adjunct Cultures in Cheese Technology

Ayşe Gürsoy
Ankara University, Turkey

Nazlı Türkmen
Ankara University, Turkey

ABSTRACT

Cheese ripening involves highly complex biochemical events. Coagulant enzymes as well as the utilized starters play an important role in these events. Two types of starters are used: primary and secondary. The main role of the primary culture, which consists of lactic acid bacteria, is to carry out lactic production during fermentation. They contribute to proteolysis and limited flavor formation with the enzymes they possess. Secondary or adjunct cultures are used to develop the texture and to accelerate the ripening. During the selection of this type of culture, enzyme profiles (i.e., proteolytic and lipolytic activities and their autolyse levels) in cheese are the primary factors to be taken into consideration. Apart from these, the other factors are their positive effects on health, availability, and economy. Adjunct cultures include yeast, molds, and bacteria. Some of the heterofermentative lactobacilli species, in particular weakened strains, are used as adjunct cultures in order to accelerate the ripening and shorten the ripening time in fat-reduced and low-fat cheeses. This chapter explores adjunct cultures in cheese technology.

INTRODUCTION

Conversion of milk into cheese involves a series of complex biochemical reactions. These biochemical reactions are carried out either by indigeneous microflora of milk or by selected defined microorganisms (Beresford et al., 2001). Defined lactic acid bacteria (LAB) which are also called starter culture, are usually added into milk prior to rennetting and they give cheese characteristic aroma-flavour and texture. Fox et al., 2000, Parente & Cogan 2004). The primary function of starter culture is to generate lactic acid but they also have the following functions in cheese (Broome et al., 2003)

1. Production of aroma components such as acetic acid, acetaldehyde and diacetyl,

DOI: 10.4018/978-1-5225-5363-2.ch013

2. Suppresion of the growth of pathogens by producing antimicrobial agents,
3. Alledgely promote human health if the culture carries probiotic properties,
4. Contribution to ripening due to rich enzyme systems.

The cheese starter cultures can be used as primary cultures which are primarily responsible for lactic acid production and formation of characteristic aroma/flavour and texture of the given cheese variety, or secondary (or adjunct) cultures which target to give cheese one or more specific function.

Primary cultures or lactic acid bacteria, are specifically selected microorganisms based on the characteristic aroma/flavor and texture of the cheese to be produced and incorporated into the cheese milk before cheese manufacturing. They are primarily responsible for lowering the milk pH during fermentation and generate required level of lactic acid. They are also rich in specific enzymes which are actively involved in ripening process of cheese, i.e. proteolytic, lipolytic or glycolytic enzmes Beresford et al., 2001, Parente & Cogan, 2004). In the cheese varieties made from raw milk no starter culture is added to milk. Some Spanish and Italian cheese are good examples of these varieties. In these varieties, indigenous or random (contaminant) microorganisms in the milk contribute to acid and aroma formation. Among the lactic acid bacteria species/subspecies used as primary starter cultures are; *Lactococcus lactis (*subsp. *lactis,* subsp. *cremoris,* subsp. *lactis* biovar. *diacetytlactis) Leuconostoc* spp., *Streptococcus thermophilus, Lactobacillus acidophilus, Lb. delbrueckii* subsp *lactis, Lb. delbrueckii* subsp *bulgaricus, Lb. helveticus* and *Lb. fermentum. Lactococcus lactis* spp. are found in almost all cheeses. The thermophilic lactic acid bacteri species such as *S. thermophilus, Lb. delbrueckii* subsp. *bulgaricus* are used in varieties in which high heat treatment is applied during production, i.e. pasta-filata type cheeses. Species of *Lb. casei, Lb. plantarum, Lb. salivarius, Enterococcus faecium, E. faecalis* and even *Staphylococcus* can also be found mostly in artisanal Mediterranean cheeses (Powell et al. 2003, Parente & Cogan, 2004).

Adjunct cultures are not generally able to produce lactic acid and are employed together with primary cultures; however, they play an important role in ripening. Yeasts, molds and bacteria such as *G. candidum, P. roqueforti, P. camemberti, D. hansenii, Brevibacterium linens, Propionibacterium freudenreichii, Corynebacterium, Staphylococcus, Micrococcus*, and heterofermentative lactobacilli are frequently used as adjuncts (Chamba & Irlinger, 2004, Parente & Cogan 2004). Depending on the characteristics of the cheese, ripening may primarily develop on the surface of the cheese or inside the cheese block. For example, the formation of CO_2 in Emmental cheese, the development of blue veins in Blue cheese or the growth of velvet-like molds on the surface of Camembert cheese are the changes caused by the previously mentioned cultures. Though the primary cheese starter culture is used in Camembert cheese, the color, aroma and textural properties of the ripened cheese are formed by the metabolic activity of *P. camemberti*, which is completely used as an adjunct culture.

SECONDARY CULTURES

Most cheese varieties are produced from heat treated milk. This eventually causes changes in the indigeneous flora of raw milk. In order to compensate the heat-triggered changes in cheese milk flora, cheese industry has put an effort in searching of new methods to provide consumers with products both reliable and with high organoleptic quality in a convenient ripening period. Among these methods, the use of adjunct culture seems to be the most promising one (El Soda et al., 2000, Fox et al., 2000).

The use of molds as adjuncts dates back to older times than commercial lactic starter bacteria. The original microflora of raw milk is established by environment of dairy plant, used equipments and even brine. In cheese varieties made from raw milk these microflora give cheese specific characteristics. For example, in some surface-ripening pasta-filata cheeses, passage of surface microflora from ripened cheese to fresh cheese is provided as a result of washing surface ripened cheese with NaCl. Moldy bread crumbs may be added to the curd before blocking in Roquefort cheese production (Chamba & Irlinger, 2004).

Adjunct cultures also contribute to acceleration of ripening. (El Soda et al. 2000, Madkor et al., 2000). The adjunct cultures have no or very weak ability to convert lactose to lactic acid during fermentation. On the other hand, they provide and control the development of aroma, color and texture in product, owing to their unique properties such as halotolerance, ability to grow at low pH levels, lactate utilization, carbon dioxide formation and proteolytic / peptidolytic, lipolytic / esterolytic and amino acid-converting abilities (Rattray 2003).

Microorganisms found in adjunct cultures can be classified as follows:

1. Propionic acid bacteria
2. Heterofermentative Lactobacilli
3. Moulds
4. Yeast
5. Coryneform and Staphylococcus

Propionic Acid Bacteria

Propionic acid bacteria (PAB) has been known since 19th century where fermentation properties of Emmental cheese was studied intensively. PAB are GR +, non-motile, spore-free, anaerobic and microaerophilic pleomorphic rods. Their optimum growth temperature is 25-32 ° C, optimum pH value is 6-7.5. They require pantothenic acid (Vitamin B5) and biotin (vitamin H) for their growth. It is classified under coryneform bacteria group from Actinomycetaceae group but not to be included in lactic acid bacteria group (Frank & Hassan, 1998, Thierry et al., 2011)

They are able to metabolize lactate to propionate, acetat, CO_2 and 1 mole ATP as well as succinic acid and isovaleric acid, pyruvic acid, lactic acid, formic acid, acetoin, diacetyl, dimethylsulphide, acetaldehyde, propionaldehyde, ethanol, propanol and trehalose (Deborde, 2003, Fröhlich-Wayder & Bachmann 2004)

3 mol lactate → 2 mol propionate + 1 mol acetate + 1 mol CO2 + 1 mol ATP

PAB is of soil origin and transported to cattle's rumen by grass and straw which are contaminated by cattle faeces. Usual contamination route of milk by PAB are milking equipment water and uncleaned tools and equipment. However the contamination rate of PAB in raw milk is not usually at the level which is enough to initiate any fermentation (Frank & Hassan, 1998, Deborde, 2003).

Propionibacteria are either milk and dairy products origin or skin origin. However, *Propionibacterium acidifaciens* has been isolated as a new species from the human tooth and has been proposed as a member of the cutaneous group in recent years (Deborde, 2003, Rattray, 2003, Chamba & Irlinger, 2004, Downes & Wade 2009). Those with milk origin are used as adjunct culture, especially in Swiss-type cheeses such as Emmental, Gruyère, Appenzeller, Sbrinz, Jarlsberg, Maasdammer and Samso and ensures cheese-specific nutty sweet flavor and pore formation (Steffen et al.,1993, Deborde 2003, Fröhlich-Wayder & Bachmann 2004). PAB species with food origin are given in Table 1.

Among the dairy-origin *Propionibacteria* which are used as starter, *P. freudenreichii* subsp. *shermanii* is the most important one found in the microflora of Swiss-type (Emmental and Gruyere) cheeses. *P. freudenreichii* subsp. *freudenreichii* is common in the microflora of Comte cheese and *P. jensenii* is found in the microflora of Maasdamar cheese. In Swiss cheese, propionic acid fermentation is initiated by keeping the cheese blocks at 18-22°C, for a period of *ca.*1 month following the production and accordingly bacteria count reaches to 10^8-10^9 cfu/g. During this period, apart from formation of propionic acid, acetic acid and volatile aroma compounds from lactic acid CO_2 formed by heterofermentative fermentation of lactate provides large and regular pores which are characteristic for this variety. If cheese is made from raw milk, smaller and irregular pore formation can be obtained as a result of metabolic activity of indigeneous PAB of raw material. The metabolic activity of PAB is deccelerated by reducing the ambient temperature following the formation of pores (Beresford et al. 2001, Fröhlich-Wayder & Bachmann 2004, Fröhlich-Wayder et al. 2002, Deborde 2003).

PAB have intracellular esterases and lipases and their lipolytic activity is 10-100 times higher than lactic acid bacteriaThis is of critical importance in development of specific taste and aroma in some Swiss-type cheeses. The lipolytic activity of PAB shows a strain-dependent charateristics and *P. freudenreichii* has been reported to be the species having the highest lipolytic activity (Dupuis et al. 1993, Deborde, 2003). High lipolytic activity is an important selection criteria for an adjunct culture (Chamba & Irlinger, 2004).

Comparing with its lipolytic activity, the proteolytic activity of PAB is rather weak (Dupuis, Corre & Boyaval, 1995). PABs contain many active peptidases such as proline aminopeptidase, X-prolyl-dipeptidyl aminopeptidase, prolinease and prolidase (Gagnaire et al. 1998, Deborde 2003, Chamba & Irlinger 2004). However, their autolysis rates are very slow and limited during cheese ripening (Valence et al., 1998, Gagnaire et al., 2001, Sheehan et al., 2008). This limits the use of PABs as adjunct culture in cheese manufacturing (Chamba & Irlinger, 2004). Prolin, which is an important aroma component in Swiss-Type cheese and gives cheese the sweet aroma, is mainly produced by propionibacteria peptidases up to levels of 1500-6000 mg kg^{-1}. The interference of lactic acid bacteria and external factors in the production of prolin should not also be underestimated since PAB has a weak caseionolytic capacity (Langsrud et al., 1977, Rattray, 2003).

The interactions between PAB and other bacteria are important during cheese ripening. It was shown that *L. lactis* subsp *lactis* exhibited a remarkable inhibition against PAB but *L. lactis* subsp. *cremoris*, *Str*

Table 1. Propionibacterium species with milk-origin and isolation sources

Species	Isolated From
P. freudenreichii subsp. freudenreichii	Raw milk
P. freudenreichii subsp. shermanii	Swiss-type cheeses
P. jenseni	Dairy products
P. acidopropionici	Dairy products
P. thoneii	Emmental cheese
P. cyclohexanicum	Spoiled orange juice
P. microaerophilum	Olive mill wastewater

thermophilus and *Lb. helveticus* were compatible (Beresford et al., 2001). Jimeno et al. (1995) demonstrated that *Lb. rhamnosus* JCL 1211, *Lb casei* JCL 1227 strains that isolated from hard cheese showed antagonistic activity against *P. freudenreichii*. In another study, the interaction between *Lb. helveticus, Lb. acidophilus, Lb. lactis, Str. thermophilus* and *L. lactis* strains and *P. freudenreichii* and *P. acidipropionici* strains were examined and no inhibition was observed on PAB strains, on the contrary *Lb. helveticus* and *Str. thermophilus* promoted the growth of PAB (Liu & Noon, 1982, Piveteau et al., 1995).

PAB are also able to produce bacteriocins. Propionicin PLG-1, for example, is produced by *Propionibacterium thoenii* and has an inhibitory activity against some Gr- bacteria such as, *Escherichia coli, Vibrio parahaemolyticus, Pseudomonas fluorescensis*. Jensenin G produced by *Propionibacterium jensenii* can remain stable at 100°C and especially inhibits the development of clostridial spores (Frank & Hassan 1998, Holo et al. 2002, Ekinci & Barefoot, 2006).

Apart from the dairy industry, PAB is also used in biotechnology, bakery, fermentation of some vegetables due to its ability to produce vitamin B12 and propionic acid. In addition, it can be used in food industry as anti-micoitic agent (Deborde 2003).

Heterofermentative Lactobacilli

Lactic flora of many cheese varieties is classified into two main groups: starter lactic acid bacteria (SLAB) and non-starter lactic acid bacteria (NSLAB). They constitute an important source for microbial population of all ripened cheeses. The major sources of NSLAB in the cheeses are indigeneous flora of milk remained alive after inadequate heat treatment, contamination after heat treatment, biofilms which can grow in plant environment and processing equipment (Cogan 2003, Beresford & William 2004). While NSLAB are quite complex, they can be categorized as non-starter lactobacilli, leuconostocs, pediococci and enterococci (Cogan, 2003, Beresford et al. 2001, Beresford 2007).

Among NSLAB, facultative heterofermentative mesophilic lactobacilli have been widely studied. Many species were isolated from different types of cheeses but the most common ones can be listed as: *Lb. casei, Lb. paracasei, Lb. plantarum, Lb. rhamnosus, L. brevis* and *Lb. curvatus*.

They can grow easily and quickly in cheese and reach high numbers (i.e. 10^8 cfu g^{-1}) in hard and semi-hard cheeses such as Cheddar and Emmental during ripening (Chamba & Irlinger 2004, Beresford 2007). Heterofermentative lactobacilli (FHL) have many beneficial effects on the cheese. These include; reaching and maintaining high cell count from the beginning until the end of ripening, not causing any defects in cheese and enhancing organoleptic features and textural characteristics especially in reduced fat cheeses depending on the enzyme activity of strain employed (Drake et al.,1997, Crow et al., 2001). Lactobacilli can grow in cheese conditions easily and reach high numbers during the entire ripening period. Therefore, their inoculation rate to cheese milk should be low (Beresford, 2007). Lactobacilli are widely used in cheese-making as adjunct culture. Apart from their technological advantageous, in some cases they may cause aroma/flavour defects in cheese (Chamba & Irlinger 2004, Beresford 2007). On the other hand, adjunct non-starter Lactobacillihave been reported to contribute to the development of aroma in cheese and help to control of some defects like gas production in cheese during storage (Beresford 2007, Settanni & Moschetti, 2010).

Overall, the proteolytic activity of FHL used in cheese-making as adjunct are low (Lane & Fox., 1996, Chamba & Irlinger, 2004). Lynch et al., (1997), for example, found that the degree of proteolysis in cheeses in which lactic culture and lactic culture / lactobacillus mixture were used did not differ much. Gagnaire et al. (2001) stated that certain strains of *Lb. casei* are able to hydrolyse bitter peptides to non-bitter peptides of which are further hydrolysed into free amino acids by aminopeptidases of these strains. Glutamate dehydrogenase activity of FHLs is an important property as adjunct culture. Transamination of amino acids for the formation of aroma substances with the mentioned activity is carried out with α-ketoglutarate (Chamba & Irlinger, 2004). Compared to PAB, molds, yeast and corynebacteria, FHL shows rather weak lipolytic activity (Fox et al., 1993).

Among all the positive characteristics of FHL in cheese manufacturing, the production of biogenic amines by some species/strains such as *Lb. buchnerii* is noteworthy. In recent years, biogenic amines are considered as one of the major criteria of selection of starter and/or adjunct cultures for food production (Crow et al. 2001). Another improtant feature of lactobacilli is their anti-microbial acvtivity and it is well-known that majority of heterofermentative lactobacilli (*Lb. brevis, Lb. fermentum, Lb. curvatus, Lb. plantarum, Lb. rhamnosus*) have such an activity especially on Gr(+) bacteria including enterococci, clostridia, *S. aureus* and *Listeria* spp. Lactobacillus GG, for example, is known to have very effective anti-microbial activity against Salmonella spp., owing to low-molecular and heat-stable non-proteinaceus compounds produced by this species (Keersmaecker et al., 2006). Similarly, *Lb. rhamnosus* LC705 (DSM7051) produces a component showing inhibitory activity against *Clostridium tyrobutyricum*. It has been patented and is commercially produced.

Moulds

Molds are a group of microorganisms that are important for the ripening of many cheese varieties. Molds are widely used in a) surfa-mould ripened cheeses and b) blue vein/blue cheeses. The surface mould-ripened cheeses are characterized with white crumbs of *P. camemberti* covering the surface of the cheese that create the unique appearanace and taste. On contrary, in Blue veined cheeses, *P. roqueforti* spreads into the cheese block and creates blue-green veins in the cheese. Each country has established its own standard in the case of moldy cheeses and almost all varieties are rated by Protected Designation of Origin / Protected Geographical Indication (PDO / PGI) status. Camembert, Brie, and Coulommier (France) cheeses are the examples of white moldy cheeses. Roquefort, Bleu des Causses, Bleu d'Auvergne (France), Bavarian Blue (Germany), Stilton (Great Britain), Cabrales (Spain), Danablu (Denmark) Gorgonzola (Italy) and Kopanisti (Greece) are the examples of Blue-veined cheeses (Spinnier & Gripon, 2004, Cantor et al., 2004). Other mold species; *Chrysosporium sulfureum* (forms a yellow spot on the surface of St Nectaire cheese), *P. album* and *P. casualvum* are also employed as adjunct cultures in various cheeses (Chamba & Irlinger, 2004). Many *Penicillium* spp. such as *P. commune, P. nalgiovensis, P. veericosum* and *P. gladioli* can be found in hard, semi-hard or semi-soft cheeses such as Cheddar, Danbo, Saint Paulin (Chamba & Irlinger, 2004, Banjara et al., 2015).

Complex flora including *Penicillium, Mucor, Cladosporium, Geotrichum, Epicoccum, Sporotrichum* is present on the surface of surface ripened cheeses. The surface flora of cheese takes active place in the development of structure, flavour/aroma and texture in cheese (Gripon, 1993, Spinner & Gripon, 2004). *P. camemberti* is able to metabolize lactose and to reduce the acidity of the cheese surface during ripening (e.g. within 3 weeks time). The cheese surface, initially hard and brittle, softens as the ripening progresses. Due to the usage of lactic acid and formation of ammonia on the surface of cheese, the pH decreases towards the inside. At the end of ripening period, a pH-gradient occurs between inside and surface of the cheese (i.e. the pH value at the surface and inside is about 7.0 and 6.0, respectively). High pH on the surface and the fragmentation of α_{s1}-casein lead to a softening towards the center of the cheese (Rattray, 2003, Beresford et al., 2007). Although the ability of *P. roqueforti* to use lactose is known, the exact mechanism of deacidification by P. roqueforti is yet to be laid open. P. roqueforti used in association with LAB, yeasts and NSLAB in blue-veined cheese, contribute to the ripening of cheese. Mold spores may be added either to milk or sprinkled on the curd during the blu cheese production. In order to stimulate the growth of molds, the pressed and salted cheese blocks are pierced and the then left for ripening. The yeasts found in complex flora, citrate using *L. lactis* and *Leuconostoc* species produce CO_2 from citrate and form cracks and clefts in the cheese. Since molds are obligate aerobeic microorganisms, they need oxygen for development. However, *P. roqueforti* may continue to develop at a lower oxygen rate than other molds require. For this reason, very little oxygen, which is diffused to the clefts and cracks in the inner part of the cheese after the brining step, is sufficient for its development. The formation of mold veins occurs within 2-3 weeks. The methyl ketones produced by *P. roqueforti* prevent excessive mold formation (Cogan, 2003, Cantor et al., 2004).

P. camemberti and *P. roqueforti* have endopeptidase and exopeptidase activities which regulate proteolysis in cheese. Therefore, ripening starts from the surface. Both mold species have similar extracellular proteolytic systems. Both hydrolyse α_{s1}- β- and κ-caseins, and produce extracellular proteinases, namely aspartyl proteinase and metalloproteinase, which are particularly active against casein. *P. camemberti* also has intracellular acid proteinase activity, and *P. roqueforti* is able to secrete a wide range of oro-active carboxy peptidase, extracellular metallo-aminopeptidase and amino and carboxypeptidase. The major function of these peptidases are the formation of ammonia and other volatile compounds from free amino acids. These peptidases are also able to de-bitter the bitter peptides (McSeweeney & Saouza, 2000, Rattray, 2003, Chamba & Irlinger, 2004).

Lipolysis occurs more intensely in mold ripened cheeses than the others and is particularly important for the development of peppery or piquant flavor which is special to this type of cheeses. Comparing with their proteolytic activities, the lipolytic activities of *P. camemberti* and *P. roqueforti* are quite high. All *Penicillium* spp. secrete extracellular alkaline lipase (pH 7.5-9.0) and extracellular acid lipase (pH 6.0-6.5) enzymes which are the primarily responsible for intense lipolysis. The concentration of free fatty acids in mould-ripened cheeses can reach very high levels as a result of lipase activity (e.g. 30.000 mg kg^{-1} in Roqueforti cheese). (McSweeney & Souza, 2000, Rattray, 2003, Cantor et al., 2004). Free fatty acids may undergo β-oxidation reactions which yield methyl ketones, primary and secondary alcohols and lactones. In addition, esters, aldehydes and volatile amines are also produced. All these compounds constitute the typical aroma of mold-ripened cheeses (Molimard & Spinnler, 1996, Cantor et al., 2004). Therefore, the type and amount of lipolytic and proteolytic aroma substances are of critical importance in the selection of mold cultures as primary or adjunct cultures. In addition, mycelium density, length and color-forming ability are the other important parameters. *P. roqueforti* strains that develop light blue and yellowish color, are used in Gorgonzola cheese and those who develop dark green color are also used in Roquefort, Stilton cheeses (Chamba & Irlinger, 2004).

Penicillium spp. may stimulate or suppress the growth of other microorganisms in mixed cheese flora. *P. roqueforti*, for example, stimulates the development of *Leuconostoc*, *Lactobacillus*, *Lactococcus* and some *Streptococci* depending on the strain. The "cat hair" defect caused by *Rhizomucor* spp. is a common defect in surface ripened cheeses and is effectively prevented by *P. camemberti*. *P. camemberti* also supports the development of *B. linens* by increasing the pH value of the environment. *B. linens* is able to metabolize the biogenic amines produced by molds through its deaminase activity (Chamba & Irlinger, 2004). *P. camemberti* also produces compounds such as acetaldehyde, benzaldehyde, 3-methylbutanal and 1-octen-3-ol which show inhibitory effects on pathogens including *L. monocytogenes* (Brennan et al., 2004). *P. camemberti* produces chloplazonic acid (Bars, 1979) and *P. requeforti* produces mycotoxins such as PR-toxin, roquefortine C and isofumigaclavinee (Lund et al., 1995) at low levels that do not cause any health problem for human body. Since production of mycotoxin and biogenic amines is strain-dependent, strain specificity should be considered during preparing starters.

Yeasts

Yeasts are microorganisms that are found in many cheese varieties and can develop under low temperature, pH and water activity and high salt concentration conditions. Some species of yeasts may cause quality deteriorations and color changes in cheeses. The adverse effects of yeasts can be eliminated by employing effective sanitation programs (Jakobsen & Narvhus, 1996). Yeasts may contribute ripening of cheese and may well be used as adjunct culture in some cheese varieties. Many yeast species take part in ripening process of especially mold and surface ripening cheeses together *with P. camemberti, P. roqueforti* and *B. linens*. *G.candidum, D.hansenii, Yarrowia lipoytica, Kluyveromyces marxianus, Kluyveromyces lactis, Saccharomyces cerevisiae, Candida parapsilosis, Galactomyces candidus, Candida sake, Rhodotorula mucilaginosa, Torulaspora delbrueckii, Trichosporon cutaneum, Pichia fermentans* are the yeast species, isolated from cheeeses. Among the mentioned species, the most common ones present in almost all cheeses types are *D. hansenii, G.candidum Yarrowia lipoytica, Kluyveromyces marxianus* (Jakobsen & Narvhus, 1996, Corsetti et al. 2001, Chamba & Irlinger, 2004, Bockelman, 2010, Banjara et al., 2015).

The yeasts begin to develop at the early stages of cheese production and reach colony counts of 10^6-10^8 cfu cm^{-2} within the first week of ripening. No significant change is observed after this step until the end of ripening (Chamba & Irlinger, 2004). On the other hand, brine is also a potential source of yeasts. $1.9x10^4$ to $2.3x10^6$ cfu g^{-1} yeast were determined in Danish Blue depending on the manufacturing plant (Tempel & Jakopsen, 1998).

Yeasts contribute to cheese quality in many way and their characteristics are often considered for their selection as adjunct culture. Since they are primarily located on the surface of cheese, yeasts are effective on the structure and appearance of the cheese surface. For example, *G. candidum* forms slimy cream to velvet mold-like structure depending on the strain. Therefore, the appropriate strain is added to the starter culture composition according to the type of cheese to be produced (Chamba & Irlinger, 2004).

The surface-growing yeasts are capable of metabolizing the sugars in different forms. *D. hansenii* and *Kluyveromyces marxianus* used as adjunct culture are able to ferment lactose. *G. candidum*, which is found in the surface of all surface-ripened cheeses, breaks down galactose and lactose and causes pH to increase. This stimulates the development of corynebacteria and molds and initiates the second phase of ripening (Fleet, 1990, Mc Sweeney & Souza, 2000).

The proteolytic activity (caseinolytic, aminopeptidase and carboxypeptidase activities) of the yeasts show a species *Yarrowia lipolytica, Kluyveromyces marxianus, Kluyveromyces lactis, Saccharomyces cerevisiae* are known to have higher proteolyitc capabilities than the other yeasts. Some strains of *Yarrowia lipolytica* cause the brown color formation in cheese as a defect. Brown colour formation is linked to the formation of pyomelanin pigments from extracellular auto-oxidation of tyrosine. This has been reported to have no adverse effect on the nutritional value and taste of cheese (Groenewald et al., 2014). Owing to their high lipolytic characterisitcs, yeasts take active part in the formation of aroma components such as carbonyl compounds, fatty acids, amino acids, esters and methyl ketones (Ferreira & Viljoen, 2003). *G. candidum* is best known as the species producing high level of aromatic compounds in cheese (Chamba & Irlinger, 2004). Yeasts also release growth factors such as B- vitamins, pantothenic acid, niacin, riboflavin and biotin.

In cheese environment, there are interactions between yeasts and other microorganisms. Microbial interactions occur in three ways: inhibiting the pathogenic microorganisms causing deterioration, preventing the development of starter cultures or contributing to ripening by promoting the development of starter bacteria (Jakobsen & Narvhus, 1996). The development of *G. candidum* is supported by the yeasts and this actions prevents the development of undesired molds such as *Aspergillus* spp., *Mucor* spp. so on. *D. hansenii* inhibits the proliferation of *C. butyricum* and *C. tyrobutyricum* in cheese and brine (Ferreira & Viljoen, 2003).

Coryneform Bacteria and Staphylococci

Both coryneform bacteria and *Streptococci* spp. can grow on the surface of many cheese varieties. They are used as adjunct for bacterial surface ripened, soft and semi-hard cheese types such as Tilsit, Tallegio, Brick, monterey, Munster, Limburger. They are generally isolated from soft cheese types such as Camembert, Blue-veined cheese and even from hard cheese types such as Gruyere and Comte (Chamba & Irlinger, 2004).

The bacterial surface ripened cheeses have fairly comples microflora. The surface flora consists mainly of yeasts, micrococci, staphylococci and coryneform bacteria (Bockelmann et al., 1997, Bockelmann & Seyler, 2001, Bockelmann, 2002, Bockelmann et al., 2005, Bockelmann, 2010, Brennan et al., 2002, Brennan et al., 2004).

Coryneform bacteria includes the species of *Arthrobacter, Brevibacterium, Corynebacterium, Brachybacterium, Microbacterium* and *Rhodococcus* spp. Identification of coryneform bacteria is rather difficult. In order to a proper identification of these species it is necessary to determine if they undergo rod/cocci change during exponential growth by analyzing cell wall and performing 16S rRNA sequence analyses (Brennan et al., 2004, Beresford et al., 2001).

Surface color of smeared cheeses are as important feature as cheese aroma. Smeared cheeses such as Saint Pulin, St. St. Nectaire, Limburger, Tilsiter and Münster are often characterised with red-orange surface color. The colour of the cheese is regulated by variations in numbers of moulds and yeast at the early stages and in the numbers of bacteria at the later stages of ripening. (Rattray & Fox, 1999). Coryneform bacteria are classifed into three groups according to the color formation feature; orange coryneform (*Brevibacterium linens, Arthrobacter casei, Brevibacterium aurantiacum*), yellow coryneform (*Arthrobacter nicotianae, Microbacterium barkeri*) and beige coryneform (*Corynebacterium ammoniagenes, Corynebacterium variabile*). However, there are also gray-white coryneforms and colorless strains (Eliskases-Lechner & Ginzinger, 1995, Chamba & Irlinger, 2004, Bockelmann, 2010). *B. linens* can produce bright cell-bound orange pigments on the surface of cheese (Reps, 1993). Bockelmann & Hope-Seyler (2001) demonstrated that in the absence of *B. linens*, yellow-pigmented *Arthrobacter* strains could develop typical red-brown colour in model Tilsit cheese.

The most studied microorganism from smear microorganisms is *Brevibacterium linens*. The main features of this bacteria are: obligatory aerobic, rod-coccus in shape, optimum growth temperature and pH ranges, 20-30 °C and 6.5-8.5, respectively. They can withstand NaCl level of 15%. Yeast and molds from other surface microorganisms support the development of other coryneform bacteria along with *Brevibacterium linens* by raising the medium pH by deacidification (Rattray, 2003, Brennan et al., 2004). Surface bacteria have generally high proteolytic activity Extracellular amino-peptidases and proteinases produced by *B. linens* are highly active on α_{s1}-casein and β-casein (Rattray & Fox, 1999, Rattray, 2003).

Proteolysis affects the typical surface color of the cheese, and tyrosine, phenylalanine and methionine must be present in the medium for the development of orange pigments (Bockelmann et al., 1997).

B. linens produces methanethiol, a basic sulphur compound causing undesired cabbagy aroma in cheese, through its L-methionine-γ-demethylase activity. It has been stated in recent years that other coryneforms such as *B. acetobutylicum, Stahylococcus equorum Arthrobacter* spp are also effective in the formation of methanethiol (Brennan et al., 2004, Rattray, 2003).

Coryneform group bacteria including mainly *B. linens*, can produce bacteriocins and antimicrobial components. Valdes-Stauber & Sherer (1994) stated that bacteriocin produced by *B. linens* M18 (Linocin M18) isolated from red smear cheese inhibited the *Listeria* spp., some coryneforms and gram-positive bacteria. Similarly, antagonisitc effects of *Brevibacterium linens, Arthrobacter nicotianae* and *Arthrobacter nucleogenes* which were isolated from red smear cheese, on *Listeria innocua, Listeria ivanovii* and *Listyeria monocytogenes* were reported by Valdes-Stauber et al., (1991) and Moaz et al., (2003). *Koocuria varians* NCC 1482 produces variacin which is an antimicrobial component similar to nisin. Variacin has an inhibitory effect on *Enterococcus* spp., *Staph. Aureus, Bacillus cereus* and *Clostridium botulinum* (Chamba & Irlinger 2004, Rattray & Fox, 1999, Motta & Brandelli, 2002). Staphylococci spp. were also reported to produce anti-bacterial components (Brennan et al., 2002).

Novobiocin-resistant coagulase-negative staphylococci are found in especially hard cheeses produced from sheep and goat milk. They grow very rapidly in the early stages of ripening and constitute 5-25% of total surface bacteria and may become more dominant than coryneform bacteria in later stages (Brennan et al., 2002). *Staph. equorum, Staph. vitulinus, Staph. xylosus, Staph. fleuretti, Staph. succinus spp. casei, Staph. caseolyticus* are some species isolated from French and German cheeses (Bockelmann et al., 1997, Brennan et al, 2002, Brennan et al, 2004). Among the above mentioned bacteria, *Staph. xylosus* is commercially available as an adjunct culture (Chamba & Irlinger, 2004).

COMMERCIAL ADJUNCT STARTER CULTURES

Adjunct cultures are commercially produced and sold in different forms. Information about usage conditions and quantities can be obtained from the manufacturers. There are cultures belong to, Chr. Hansen, DSM Food Specialists, DuPont™Danisco®, Degussa, Clerici-Sacco, Standa Industries, Centro Sperimentale del Latte (CSL), CSK Food, Standa Industrie and Wiesby at the present time. Adjunct cultures are produced in three different forms: frozen cell concentrates, freeze-dried preparations or liquid concentrates which is no longer preferred due to handling and storage difficulties. Secondary cultures can either be added directly to the cheese milk, or by spraying onto the cheese surface or by inoculating to salt or salty whey solution. Shelf life for yeast and bacterial cultures is 2 years at -18 °C and max 2 months at 4 °C (Bockelmann, 2010).

The selection criteria of adjunct cultures show some differences from primary cultures. Adjunct cultures are expected to meet one or more of the following features: rapid development, pigment production, halotolerance, lipolytic activity, proteolytic activity, amino acid catabolic activity, de-acidification, interactions between bacterial yeast and molds, specific aroma production, bacteriocin production, micellar density length and micellar color for molds, and pigment production for smear bacteria. (Chamba & Irlinger 2004, Rattray, 2003, Bockelmann, 2010).

D. hansenii, K. marxianus and *S. cerevisiae* are the most widely comemrcialized among the adjunct yeasts. In addition to the properties mentioned above, especially appearance, color and surface appearance characteristics in the production of *G.candidum* strains and inhibition characteristics on *Mucor* spp. are taken into consideration. It is added to the cheese milk at the level of 10^5 cfu ml^{-1}. *S cerevisiae* is used in blue-veined cheese due to its ablity to produce CO_2 at high levels. *S cerevisiae* is usually added to cheese milk or to smear solution at a level of *ca.* 10^5-10^6 cfu ml^{-1}.

P. camemberti and *P. roqueforti* cultures are sold in liquid or freeze-dried form. *P. camemberti* is added directly to the milk or sprayed to the surface after brine application. *P. roqueforti* is inoculated directly into the cheese milk. Recommended inoculation rates of both moulds are around 10^6-10^7 spores L^{-1} (Chamba & Irlinger, 2004).

Although smear ripened cheeses have highly complex microbial flora, only a limited number of species and strains are commercially available. It is obvious that commercial culture combinations do not reflect natural cheese flora. In any case, microflora of smeared soft, semi-soft and acid-curd cheeses should contain *Debaryomyces hansenii, Staphylococcus aureus, Staphylococcus equorum, Cotynebacterium casei, Arthrobacter nicotianae, Brevibacterium linens* and *Geotrichum candidum* (Bockelmann, 2002, Bockelman et al., 2005). Some strains of *Brevibacterium linens, Staphylococcus xyloasus* and *Staphylococcus carnosus* are commercially available.

PROBIOTIC CHARACTERISTICS OF ADJUNCT CULTURES

It is known that yoghurt and fermented dairy products are proper vehicles for probiotic uptake to human body. However, when compared to fermented milk, it was stated that cheese may be an alternative food carrier for probiotic microorganisms because of some of its characteristics (Stanton et al., 1998, Shah, 2007, da Cruz et al., 2009, Özer & Kırmacı, 2011). Particularly, low acidity, high fat content, reduced oxygen level and more closed matrix of the cheese are thought to provide high level of protection in the transition of probiotic microorganisms to the digestive system. At the early stages of ripening, cheese microflora provides an anaerobic medium for the rapid growth of probiotics. (Boylston et al., 2004). On the other hand, although the production conditions of cheese and longer ripening periods which might extend to 1-2 years in some hard cheeses are considered to be limiting factors for the growth and viability of probiotics. Neverthless, by the selection of appropriate strains and by applying the controlled conditions, the vitality can be increased and the anticipated health effects can be performed (Ross et al., 2005, Phillips et al., 2006; da Cruz et al., 2009, Özer & Kırmacı, 2011).

Bifidobacterium spp. (*B. adolescentis, B. bifidum, B. animalis, B. breve, B. longum), Lactobacillus* spp. *(Lb. acidophilus, Lb. fermentum, Lb. casei, Lb. rhamnosus, Lb. reutei, Lb. lactis, Lb. paracasei, Lb. casei, Lb. plantarum*) are mostly used cultures in the production of probiotic cheese. Despite they exhibit probiotic characteristics, less studies were carried out on enterococci and propionibacterium (Table 2).

Probiotic bacteria are not expected to affect the overall composition of the cheese (Bergamini et al., 2005, Ong et al., 2006, Özer et al., 2008). On the othert hand, development of proteolysis and lipolysis and formation of aroma compounds take place at different levels depending on the strain used in the cheese production. By using different strains of *Bifidobacterium* spp and *Lactobacillus* spp in Cheddar cheese, Ong & Shah (2009) reported that level of proteolysis increased when the ripening temperature was raised from 4°C to 8°C. On the contrary, Bergamini et al. (2006) stated a limited level of proteolysis in Cheddar cheese which are produced by *Lactobacillus acidophilus* and *Lactobacillus* subsp *paracasei*. The lipolytic and esterolytic activities of probiotics are limited (Özer et al, 2008). However, researchers

Table 2. Some examples of probiotic cheeses and bacteria

Cheese Type	Bacteria	Reference
Prato cheese	*Lb. casei* 01	Silva et al., 2017
Italico cheese	*Lb. rhamnosus* LbGG, *Lb. rhamnosus* SP1	Blaiotta et al. 2017
Fior di Latte cheese	*Lb. rhamnosus* GG	Cuffia et al., 2017
Coalho cheese	*Lb. acidophilus* (LA-5), *Lb. casei* subsp. *paracasei, B. lactis* (BB12) *Lb. rhamnosus* EM1107 *Lb. acidophilus, Lb. paracasei, B. lactis*	de Oliveira et al., 2014 Rolim et al., 2015 Bezerra et al., 2017
Minas Frescal cheese	*Lactobacillus acidophilus* *Lactobacillus acidophilus* LA 14, *Bifidobacterium longum* BL 05 *Bifidobacterium* BB-12 *Lactobacillus acidophilus* La5 *L. casei* Zhang	Gomes et al., 2011 Lollo et al., 2015 Verruck et al., 2015 Felicio et al., 2016 Dantas et al., 2016
Feta cheese	*Propionibacterium freudenreichii* subsp. *shermanii*	Angelopoulou et al., 2017
Pecorino Siciliano cheese	*Lactobacillus rhamnosus* H25, *Lactobacillus paracasei* N24	Pino et al., 2017
Caciotta cheese	*Lactobacillus paracasei* A13	Burns et al., 2015
low-fat akawi cheese	*Lactobacillus plantarum*	Al-Dhaheri et al., 2017
Petit suisse cheese	*Lactobacillus acidophilus* LA C4, *Bifidobacterium animalis* ssp. *lactis* DN 173–010	Pereira et al., 2016
Goat ricotta cheese	*Bifidobacterium animalis subsp. lactis* Bb-12, *Lactobacillus acidophilus* La-05	Meira et al., 2015
Kalari cheese	*Lb. casei, Lb. plantarum, Lb. brevis*	Mushtaq et al., 2016
Cottage cheese	*Lb. casei* ATCC 373, *Lb. rhamnosus* GG ATCC 53103	Abadia-García et al., 2013
Panela cheese	*B. breve* ATCC 15700, *Lb. rhamnosus* GG ATCC 53103	Escobar et al., 2012
Cheddar cheese	*B. longum* 1941, *B. animalis* subsp. *lactis* B94, *Lb. casei* 279, *Lb. casei* L26, *Lb. acidophilus* 4962, *Lb. acidophilus* L10 *Lb. acidophilus, Lb. casei, Lb. paracasei, Bifidobacterium*	Ong & Shah, 2009 Oberg et al. 2011
White brined cheese	*Lb. rhamnosus* 6134	Liu et al., 2017
Ovine cheese	*Lb. acidophilus* LA-5, *B. lactis* BB-12, *B. longum* BB-46	Albenzio et al. 2010
Scamorza cheese	*B. longum, B.lactis, Lb. acidophilus*	Albenzio et al., 2013
Pikantne cheese	*Lb. fermentum* ME-3	Songisepp et al., 2004
Argentinian probiotic fresh cheese	*Lb. paracasei* A13, *B. bifidum* A1, *Lb. acidophilus* A3	Vinderola et al., 2009
Cremoso cheese	*Lb. plantarum* 191, *Lb. paracasei* 190	Burns et al., 2012
Turkish Beyaz cheese	*Lb. acidophilus* 593 N *Lb. fermentum* (AB5-18 and AK4-120), *Lb. plantarum* (AB16-65 and AC18-82)	Kasımoglu et al. 2004 Basyigit Kılıc et al., 2009

also reported a rapid lipolysis development in goat cheese (Kefalotyri) produced by using lactobacilli (Kalavrouzioti et al., 2005).

Some strains of *Lb. casei, Lb. plantarum* and *Lb. rhamnosus* have health effects such as prevention of diarrhea, intestinal inflammation and permeability disorder, as well as allegedly tumor prevention and immunomodulation (Chamba & Irlinger, 2004). Some strains of *Lb rhamnosus* can prevent the development of bacteria causing dental caries. Another feature of probiotics is that they can produce conjugated linoleic acid (CLA). This increases the health effects of the cheese. *Lb. paracasei* and *Lb. plantarum* which are isolated from traditional Italian cheese has been reported to produce CLA (Abd El Salam et al., 2010). Another important characteristic of cheese in terms of health is its high content of ACE-I inhibitor peptides. Bioactive peptides related to the breakdown of casein fractions exhibit antihypertensive action by regulating blood pressure. It has been reported that Cheddar cheese produced with *Lb casei* Zang shows ACE-I exhibits high level (85-100%) of inhibitory activity (Özer & Kırmacı, 2011).

Probiotic properties of PAB is very limited compare to lactobacilli and bifidobacteria. Some characteristics of dairy propinibacteria such as; safety, gastrointestinal transit survival and adherence to intestinal mucosa, offer good prospects for their use as human probiotics (Mantere-Alhonen, 1995, Ouwehand, 2004 & Cousin et al., 2011). It is stated that the short chain fatty acids such as acetic, propionic and butyric acids create feeling of fullness. Particularly, propionates will be a new tool to prevent type 2 diabetes and obesity (Al- Laham et al., 2010, Arora et al. 2011) Ruijchop et al. (2008) reported that fermented beverages produced by using PAB and lactic acid bacteria reduced the desire to eat by creating more sense of fullness. Apart from the dairy industry, PAB is mainly used in biotechnology, bakery, fermentation of some vegetables due to its production of vitamin B12 and propionic acid. In addition, it can be added to foods as a preservative to prevent yeast and mold growth and to extend the shelf life (Deborde 2003).

Enterococci which are especially present in the traditional microflora of Mediterranean cheese types rank as secondary starter in some cheeses. They are used in some cheese varieties because of their bacteriocin production, lipolytic and proteolytic activities, contribution to aroma formation and diacetyl production. *Enterococcus faecium* and *E. faecalis* are the most studied strains (Garde et al., 1997, Giraffa, 2003, Moreno et al., 2006, Bhardwaj et al., 2008). *E. faecium* PR88 is used in the treatment of antibiotic-associated diarrhea and irritable bowel syndrome (Gardiner et al., 1999). It has been stated that *E. faecium* FAIR-E 198 (Göncüoğlu et al., 2009) and *E. faecium* ATCC-407 (Gürsoy & Kınık, 2010) strains could be used as probiotics in Turkish White cheese. Savedra et al., (2003) have determined that a large number of *E. faecium* strains that were isolated from traditional Argentine cheese produce bacteriocin active against Listeria.

CONCLUSION

Adjunct starters are bacterial strains that are added into many varieties of cheese, together with acid-developing cultures to provide aroma, texture and color development or to accelerate the ripening. Nowadays, the source of indigenous secondary flora in raw milk have reduced due to the reasons such as the enhancement of microbiological quality of raw milk, increased use of applications such as milk thermization and bactofugation. For this reason, more adjunct cultures are needed than before. This type of cultures used in cheese production can be grouped as heterofermentative lactobacilli, propionic acid bacteria, yeasts, molds and surface bacteria. Factors such as interactions between microorganisms, proteolytic and lipolytic activity, aroma formation, rapid development, color formation and antimicrobial agent production are taken into consideration in the selection of starter strains. Secondary cultures are produced by many commercial companies at the present time, either singly or as a mixture.

REFERENCES

Abadía-García, L., Cardador, A., del Campo, S. T. M., Arvízu, S. M., Castaño-Tostado, E., Regalado González, C., ... Amaya-Llano, S. L. (2013). Influence of probiotic strains added to cottage cheese on generation of potentially antioxidant peptides, anti-listerial activity, and survival of probiotic microorganisms in simulated gastrointestinal conditions. *International Dairy Journal*, *33*(2), 191–197. doi:10.1016/j.idairyj.2013.04.005

Al-Dhaheri, A. S., Al-Hemeiri, R., Kizhakkayil, J., Al-Nabulsi, A., Abushelaibi, A., Shah, N. P., & Ayyash, M. (2017). Health-promoting benefits of low-fat akawi cheese made by exopolysaccharide-producing probiotic Lactobacillus plantarum isolated from camel milk. *Journal of Dairy Science*, *100*(10), 1–9. doi:10.3168/jds.2017-12761 PMID:28755944

Al-Lahham, S. H., Peppelenbosch, M. P., Roelofsen, H., Vonk, R. J., & Venema, K. (2010). Biological effects of propionic acid in human; metabolism, potential applications and underlying mechanisms. *Biochimica et Biophysica Acta*, *1801*(11), 1175–1183. doi:10.1016/j.bbalip.2010.07.007 PMID:20691280

Albenzio, M., Santillo, A., Caroprese, M., Marino, R., Trani, A., & Faccia, M. (2010). Biochemical patterns in ovine cheese: Influence of probiotic strains. *Journal of Dairy Science*, *93*(8), 3487–3496. doi:10.3168/jds.2009-3036 PMID:20655416

Albenzio, M., Santillo, A., Caroprese, M., Ruggieri, D., Napolitano, F., & Sevi, A. (2013). Physicochemical properties of Scamorza ewe milk cheese manufactured with different probiotic cultures. *Journal of Dairy Science*, *96*(5), 2781–2791. doi:10.3168/jds.2012-6218 PMID:23522678

Angelopoulou, A., Alexandraki, V., Georgalaki, M., Anastasiou, R., Manolopoulou, E., Tsakalidou, E., & Papadimitriou, K. (2017). Production of probiotic Feta cheese using *Propionibacterium freudenreichii* subsp. *shermanii* as adjunct. *International Dairy Journal*, *66*, 135–139. doi:10.1016/j.idairyj.2016.11.011

Arora, T., Sharma, R., & Frost, G. (2011). Propionate. Anti-obesity and satiety enhancing factor? *Appetite*, *56*(2), 511–515. doi:10.1016/j.appet.2011.01.016 PMID:21255628

Banjara, N., Suhr, M. J., & Hallen-Adams, H. E. (2015). Diversity of yeast and mold species from a variety of cheese types. *Current Microbiology*, 1–9. PMID:25694357

Bars, J. L. E. (1979). Cyclopiazonic acid production by Penicillium camemberti Thom and natural occurence of this mycotoxin in cheese. *Applied and Environmental Microbiology*, *38*(6), 1052–1055. PMID:526014

Basyigit Kılıc, G., Kuleasan, H., Eralp, I., & Karahan, A. G. (2009). Manufacture of Turkish Beyaz cheese added with probiotic strains. *Lebensmittel-Wissenschaft + Technologie*, *42*(5), 1003–1008. doi:10.1016/j.lwt.2008.12.015

Beresford, T. (2007). What are non-starter lactic acid bacteria and how do they affect cheese quality? In P. L. H. McSweeney (Ed.), *Cheese Problems Solved* (pp. 124–130). Boca Raton, FL: CRC Press.

Beresford, T., & Williams, A. (2004). The microbiology of cheese. In P. F. Fox, P. L. H. McSweeney, T. M. Cogan & T. P. Guinee (Eds.), Cheese: Chemistry, Physics and Microbiology: Vol. 1. General Aspects (3rd ed.; pp. 287-317). London, UK: Academic Press.

Beresford, T. P., Fitzsimons, N. A., Brennan, N. L., & Cogan, T. M. (2001). Recent advances in cheese microbiology. *International Dairy Journal, 11*(4-7), 259–274. doi:10.1016/S0958-6946(01)00056-5

Bergamini, C. V., Haynes, E. R., & Zalazar, C. A. (2006). Influence of probiotic bacteria on the pretolysis profile of a semi-hard cheese. *International Dairy Journal, 8*(16), 856–866. doi:10.1016/j.idairyj.2005.09.004

Bezerra, T. K. A., Arcanjo, N. M. O., de Araujo, A. R. R., de Queiroz, A. L. M., de Oliveira, M. E. G., Gomes, A. M. P., & Madruga, M. S. (2017). Volatile profile in goat coalho cheese supplemented with probiotic lactic acid bacteria. *Lebensmittel-Wissenschaft + Technologie, 76*, 209–215. doi:10.1016/j.lwt.2016.03.041

Bhardwaj, A., Malik, R. K., & Chauhan, P. (2008). Functional and safety aspects of enterococci in dairy foods. *Indian Journal of Microbiology, 48*(3), 317–325. doi:10.100712088-008-0041-2 PMID:23100728

Blaiotta, G., Murru, N., Di Cerbo, A., Succi, M., Coppola, R., & Aponte, M. (2017). Commercially standardized process for probiotic "Italico" cheese production. *Lebensmittel-Wissenschaft + Technologie, 79*, 601–608. doi:10.1016/j.lwt.2016.11.008

Bockelman, W., Krusch, U., Engel, G., Klijn, N., Smit, G., & Heller, K. J. (1997). The microflora of Tilsit cheese. Part 1. Variability of the smear flora. *Die Nahrung, 41*(4), 208–212. doi:10.1002/food.19970410405

Bockelman, W., Willems, K. P., Neve, H., & Heller, K. H. (2005). Cultures fort he ripening of the smear cheeses. *International Dairy Journal, 15*(6-9), 719–732. doi:10.1016/j.idairyj.2004.08.022

Bockelmann, W. (2002). Development of defined surface starter cultures fort he ripening of smear cheeses. *International Dairy Journal, 12*(2-3), 123–131. doi:10.1016/S0958-6946(01)00152-2

Bockelmann, W. (2010). Secondary cheese starter culture. In B. A. Law & A. Y. Tamime (Eds.), *Technology of Cheese Making* (2nd ed.; pp. 193–230). London, UK: Blackwell Pub. Ltd. doi:10.1002/9781444323740.ch6

Bockelmann, W., & Hoppe-Seyler, T. (2001). The surface flora of bacterial smear-ripened cheese from cow's and goat's milk. *International Dairy Journal, 11*(4-7), 307–314. doi:10.1016/S0958-6946(01)00060-7

Boylston, T. D., Vinderola, C. G., Ghoddusi, H. B., & Reinheimer, J. A. (2004). Incorporation of bifidobacteria into cheeses: Challenges and rewards. *International Dairy Journal, 14*(5), 375–387. doi:10.1016/j.idairyj.2003.08.008

Brennan, N. M., Cogan, T. M., Loessner, M., & Scherer, S. (2004). Bacterial surface-ripened cheeses. In P. F. Fox, P. L. H. McSweeney, T. M. Cogan & T. P. Guinee (Eds.), Cheese: Chemistry, Physics and Microbiology: Vol. 2. Major Cheese Groups (3rd ed.; pp. 200-225). London, UK: Academic Press. doi:10.1016/S1874-558X(04)80045-9

Brennan, N. M., Ward, A. C., Beresford, T. P., Fox, P. F., Goodfellow, M., & Cogan, T. M. (2002). Biodiversity of the bacterial flora on the surface of a smear cheese. *Applied and Environmental Microbiology, 68*(2), 820–830. doi:10.1128/AEM.68.2.820-830.2002 PMID:11823224

Broome, M. C., Powell, I. B., & Limsowtin, G. K. Y. (2003). Starter cultures: Spesific properties. In H. Roginski, J. W. Fuquay, & P. F. Fox (Eds.), *Encyclopedia of Dairy Sciences* (pp. 269–275). London, UK: Academic Press.

Burns, P. G., Patrignani, F., Tabanelli, G., Vinderola, G. C., Siroli, L., Reinheimer, J. A., ... Lanciotti, R. (2015). Potential of high pressure homogenisation on probiotic Caciotta cheese quality and functionality. *Journal of Functional Foods*, *13*, 126–136. doi:10.1016/j.jff.2014.12.037

Cağlar, E., Kargül, B., & Tanboğa, T. (2005). Bacteriotherapy and probiotics'role on the oral health. *Oral Diseases*, *3*(11), 131–137. doi:10.1111/j.1601-0825.2005.01109.x PMID:15888102

Cantor, M. D., den Tempel, T., Hansen, T. K., & Ardö, Y. (2004). Blue cheese. In P. F. Fox, P. L. H. McSweeney, T. M. Cogan & T. P. Guinee (Eds.), Cheese: Chemistry, Physics and Microbiology: Vol. 2. Major Cheese Groups (3rd ed.; pp. 175-198). London, UK: Academic Press. doi:10.1016/S1874-558X(04)80044-7

Chamba, J. F., & Irlinger, F. (2004). Secondary and adjunct cultures. In P. F. Fox, P. L. H. McSweeney, T. M. Cogan & T. P. Guinee (Eds.), Cheese: Chemistry, Physics and Microbiology: Vol. 1. General Aspects (3rd ed.; pp. 191-206). London, UK: Academic Press. doi:10.1016/S1874-558X(04)80068-X

Cogan, T. M. (2003). Microbiology of cheese. In H. Roginski, J. W. Fuquay, & P. F. Fox (Eds.), *Encyclopedia of Dairy Sciences* (pp. 306–314). London, UK: Academic Press.

Corsetti, A., Rossi, J., & Gobbetti, M. (2001). Interactions between yeast and bacteria in the smear surface-ripened cheeses. *International Journal of Food Microbiology*, *69*(1-2), 1–10. doi:10.1016/S0168-1605(01)00567-0 PMID:11589547

Cousin, F. J., Mater, D. D. G., Foline, B., & Jan, G. (2011). Dairy propionibacteria as human probiotics: A review of recent evidence. *Dairy Science & Technology*, *91*, 1–26.

Crow, V., Curry, B., & Hayes, M. (2001). The ecology of non-starter lactic acid bacteria (NSLAB) and their use as adjuncts in New Zealand Cheddar. *International Dairy Journal*, *11*(4-7), 275–283. doi:10.1016/S0958-6946(01)00057-7

Cuffia, F., George, G., Renzulli, P., Reinheimer, J., Meinardi, C., & Burns, P. (2017). Technological challenges in the production of a probiotic pasta filata soft cheese. *Lebensmittel-Wissenschaft + Technologie*, *81*, 111–117. doi:10.1016/j.lwt.2017.03.039

Da Cruz, A. G., Buriti, F. C. A., de Souza, C. H. B., Faria, J. A. F., & Saad, S. M. I. (2009). Probiotic cheese: Health benefits, technological and stability aspect. *Trends in Food Science & Technology*, *20*(8), 344–354. doi:10.1016/j.tifs.2009.05.001

Dantas, A. B., Jesus, V. F., Silva, R., Almada, C. N., Esmerino, E. A., Cappato, L. P., ... Cruz, A. G. (2016). Manufacture of probiotic Minas Frescal cheese with *Lactobacillus casei* Zhang. *Journal of Dairy Science*, *99*(1), 18–30. doi:10.3168/jds.2015-9880 PMID:26519974 de Oliveira, M. E. G., Garcia, E. F., de Oliveira, C. E. V., Gomes, A. M. P., Pintado, M. M. E., Madureira, A. R. M. F., & da Conceicao, M. L. (2014). Addition of probiotic bacteria in a semi-hard goat cheese (coalho): Survival to simulated gastrointestinal conditions and inhibitory effect against pathogenic bacteria. *Food Research International*, *64*, 241–247. doi:10.1016/j.foodres.2014.06.032

Deborde, C. (2003). *Propionibacterium* spp. In H. Roginski, J. W. Fuquay, & P. F. Fox (Eds.), *Encyclopedia of Dairy Sciences* (pp. 2330–2339). London, UK: Academic Press.

Downes, J., & Wade, W. G. (2009). *Propionibacterium acidifaciens* sp. nov., isolated from the human mouth. *International Journal of Systematic and Evolutionary Microbiology*, *59*(11), 2778–2781. doi:10.1099/ijs.0.010470-0 PMID:19625418

Dupuis, C., Corre, C., & Boyaval, P. (1993). Lipase and esterase activity of Propionibacterium freudenreichii subsp. Freudenreichii. *Applied and Environmental Microbiology*, *59*(12), 4004–4009. PMID:16349102

Dupuis, C., Corre, C., & Boyaval, P. (1995). Proteinase activity of dairy *Propionibacterium*. *Applied Microbiology and Biotechnology*, *42*(5), 750–755. doi:10.1007/BF00171957

Ekinci, F. Y., & Barefoot, S. F. (2006). Fed-batch enhancement of jenseniin G, a bacteriocin produced by *Propionibacterium thoenii (jensenii)* P126. *Food Microbiology*, *23*(4), 325–330. doi:10.1016/j.fm.2005.05.012 PMID:16943021

El Soda, M., Madkor, S. A., & Tong, P. S. (2000). Adjunct cultures: Recent developments and potential significance to the cheese industry. *Journal of Dairy Science*, *83*(4), 609–619. doi:10.3168/jds.S0022-0302(00)74920-4 PMID:10791774

Eliskases-Lechner, F., & Ginzinger, W. (1995). The bacterial flora of surface-ripened cheeses with special regard to coryneform. *Le Lait*, *75*(6), 571–584. doi:10.1051/lait:1995644

Escobar, M. C., Van Tassell, M. L., Martínez-Bustos, F., Singh, M., Castaño-Tostado, E., Amaya Llano, S. L., & Miller, M. J. (2012). Characterization of a Panela cheese with added probiotics and fava bean starch. *Journal of Dairy Science*, *95*(6), 2779–2787. doi:10.3168/jds.2011-4655 PMID:22612915

Felicio, T. L., Esmerino, E. A., Vidal, V. A. S., Cappato, L. P., Garcia, R. K. A., Cavalcanti, R. N., ... Cruz, A. G. (2016). Physico-chemical changes during storage and sensory acceptance of low sodium probiotic Minas cheese added with arginine. *Food Chemistry*, *196*, 628–637. doi:10.1016/j.foodchem.2015.09.102 PMID:26593536

Ferreira, A. D., & Viljoen, B. C. (2003). Yeasts as adjunct starters in matured Cheddar cheese. *International Journal of Food Microbiology*, *86*(1-2), 131–140. doi:10.1016/S0168-1605(03)00252-6 PMID:12892928

Fleet, G. (1990). Yeast in dairy products-a review. *The Journal of Applied Bacteriology*, *68*(3), 199–211. doi:10.1111/j.1365-2672.1990.tb02566.x PMID:2187843

Fox, P. F., Guinee, T. P., Cogan, T. M., & McSweeney, P. L. H. (Eds.). (2000). *Fundamentals of cheese science*. Gaithersburg, Maryland: Aspen Publishers, Inc.

Fox, P. F., Law, J., McSweeney, P. L. H., & Wallace, J. (1993). Biochemistry of cheese ripening. In P. F. Fox, (Eds.), Cheese: Chemistry, Physics and Microbiology: Vol. 1. General Aspects (2nd ed.; pp. 389-438). London, UK: Chapman & Hall. doi:10.1007/978-1-4615-2650-6_10

Frank, J. F., & Hassan, A. N. (1998). Starter cultures and their use. In E. H. Marth & J. L. Steele (Eds.), *Applied Dairy Microbiology* (pp. 131–172). Marcel and Dekker Inc.

Fröhlich-Wyder, M. T., & Bachmann, H. P. (2004). Cheeses with propionic acid fermentation. In P. F. Fox, P. L. H. McSweeney, T. M. Cogan & T. P. Guinee (Eds.), Cheese: Chemistry, Physics and Microbiology: Vol. 2. Major Cheese Groups (3rd ed.; pp. 141-156). London, UK: Academic Press. doi:10.1016/S1874-558X(04)80042-3

Fröhlich-Wyder, M. T., Bachmann, H. P., & Casey, M. G. (2002). Interaction between propionibacteria and starter/ non-starter lactic acid bacteria in Swiss- type cheeses. *Le Lait*, *82*(1), 1–15. doi:10.1051/lait:2001001

Gagnaire, V., Lortal, S., & Leonil, J. (1998). Free active peptidases are detected in Emmental juice extracted before ripening in the warm room. *The Journal of Dairy Research*, *65*(1), 119–128. doi:10.1017/S0022029997002525

Gagnaire, V., Thierry, A., & Leonil, J. (2001). Propionibacteria aand facultatively heterofermentative lactobacilli weakly contribute to secondary proteolysis of Emmental cheese. *Le Lait*, *81*(3), 339–353. doi:10.1051/lait:2001136

Garde, S., Gaya, P., Medina, M., & Nunez, M. (1997). Acceleration of flovur formation in cheese by bacteriocin-producing adjunct lactic culture. *Biotechnology Letters*, *10*(9), 1011–1014. doi:10.1023/A:1018451401925

Gardiner, G. E., Ross, R. P., Wallace, J. M., Scanlan, F. P., Jägers, P. P. J. M., Fitzgerald, G. F., ... Stanton, C. (1999). Influence of a probiotic adjunct culture of *Enterococcus faecium* on the quality of Cheddar cheese. *Journal of Agricultural and Food Chemistry*, *47*(12), 4907–4916. doi:10.1021/jf990277m PMID:10606551

Giraffa, G. (2003). Functionality of enterococci in dairy products. *International Journal of Food Microbiology*, *2-3*(88), 215–222. doi:10.1016/S0168-1605(03)00183-1 PMID:14596993

Gomes, A. A., Braga, S. P., Cruz, A. G., Cadena, R. S., Lollo, P. C. B., Carvalho, C., ... Bolini, H. M. A. (2011). Effect of the inoculation level of Lactobacillus acidophilus in probiotic cheese on the physicochemical features and sensory performance compared with commercial cheeses. *Journal of Dairy Science*, *94*(10), 4777–4786. doi:10.3168/jds.2011-4175 PMID:21943729

Göncüoğlu, M., Ormancı, F. S. B., & Doğru, A. K. (2009). Beyaz peynir üretiminde Enterococcus faecium un starter kültür olarak kullanılması. *Ankara Üniversitesi Veteriner Fakültesi Dergisi*, *56*(1), 249–254. doi:10.1501/Vetfak_0000002290

Gripon, J. C. (1993). Mould-ripened cheeses. In. F. Fox, (Eds.), Cheese: Chemistry, Physics and Microbiology: Vol. 2. Major Cheese Groups (2nd ed.; pp. 111-136). London, UK: Chapman & Hall. doi:10.1007/978-1-4615-2648-3_4

Groenewald, M., Boekhout, T., Neuveglise, C., Caillardin, C., van Dijck, P. W. M., & Wyss, M. (2014). Yarrowia lipolytica: Safety assestment of an oleaginaus yeast with a great industrial potential. *Critical Reviews in Microbiology*, *40*(3), 187–206. doi:10.3109/1040841X.2013.770386 PMID:23488872

Gursoy, O., & Kinik, O. (2010). Incorporation of adjunct cultures of *Enterococcus faecium, Lactobacillus paracasei* subsp. *paracasei* and *Bifidobacterium bifidum* into White cheese. *Journal of Food Agriculture and Environment*, *2*(8), 107–112.

Hatakka, K., Ahola, A. J., Yli-Knutila, H., Richardson, M., Poussa, T., Meurman, J. H., & Karpela, R. (2007). Probiotics reduce the prevalence, of oral Candida in the elderly-a randomized controlled Trial. *Journal of Dental Research*, *86*(2), 125–130. doi:10.1177/154405910708600204 PMID:17251510

Holo, H., Faye, T., Brede, D. A., Nilsen, T., Odegard, I., Langsrud, T., ... Nes, I. F. (2002). Bacteriocins of propionic acid bacteria. *Le Lait*, *82*(1), 59–68. doi:10.1051/lait:2001005

Jakobsen, M., & Narvhus, J. (1996). Yeast and their possible beneficial and negativeeffects on the quality of dairy products. *International Dairy Journal*, *6*(8-9), 755–768. doi:10.1016/0958-6946(95)00071-2

Jimeno, J., Lazaro, M. J., & Sollberger, H. (1995). Antagonistic interactions between propionic acid bacteria and non-starter lactic acid bacteria. *Le Lait*, *75*(4-5), 401–413. doi:10.1051/lait:19954-530

Kalvrouzioti, I., Hatzıkamari, M., Litopoulou-Tzanetaki, E., & Tzanetakis, N. (2005). Production of hard cheese from caprine milk by the use of two types of probiotic cultures as adjunct. *International Journal of Dairy Technology*, *1*(58), 30–38. doi:10.1111/j.1471-0307.2005.00179.x

Kasimoglu, A., Goncuoglu, M., & Akgun, S. (2004). Probiotic white cheese with *Lactobacillus acidophilus. International Dairy Journal*, *14*(12), 1067–1073. doi:10.1016/j.idairyj.2004.04.006

Keersmaecker, S. C. J., Verhoeven, T. L. A., Desair, J., Marchal, K., Vanderleyden, J., & Nagy, I. (2006). Strong antimicrobial activity of *Lactobacillus rhamnosus* GG against *Salmonella typhimurium* is due to accumulation of lactic acid. *FEMS Microbiology Letters*, *259*(1), 89–96. doi:10.1111/j.1574-6968.2006.00250.x PMID:16684107

Lane, C. N., & Fox, P. F. (1996). Contribution of starter and adjunct Lactobacilli to proteolysis in cheddar cheese during ripening. *International Dairy Journal*, *6*(7), 715–728. doi:10.1016/0958-6946(95)00067-4

Langsrud, T., Reinbold, G. W., & Hammond, E. G. (1977). Proline production by Propionibacterium shermanii P59. *Journal of Dairy Science*, *1*(60), 16–23. doi:10.3168/jds.S0022-0302(77)83822-8

Liu, J. A. P., & Moon, N. J. (1982). CommensalisticInteractionBetweenLactobacillusacidophilus and Propionibacterium shermanii. *Applied and Environmental Microbiology*, 715–722. PMID:16346098

Liu, L., Chen, P., Zhao, W., Li, X., Wang, H., & Qu, X. (2017). Effect of microencapsulation with the Maillard reaction products of whey proteins and isomaltooligosaccharide on the survival rate of Lactobacillus rhamnosus in white brined cheese. *Food Control*, *79*, 44–49. doi:10.1016/j.foodcont.2017.03.016

Lollo, P. C. B., Morato, P. N., Moura, C. S., Almada, C. N., Felicio, T. L., Esmerino, E. A., ... Cruz, A. G. (2015). Hypertension parameters are attenuated by the continuous consumption of probiotic Minas cheese. *Food Research International*, *76*, 611–617. doi:10.1016/j.foodres.2015.07.015 PMID:28455044

Lund, F., Filtenborg, O., & Frisvad, J. C. (1995). Associated mycoflora of cheese. *Food Microbiology*, *12*, 173–180. doi:10.1016/S0740-0020(95)80094-8

Lynch, C. M., McSweeney, P. L. H., Fox, P. F., Cogad, T. M., & Drinan, F. D. (1996). Manufacture of cheddar cheese with and without adjunct Lactobacilli under controlled microbiological conditions. *International Dairy Journal*, *6*(8-9), 851–867. doi:10.1016/0958-6946(96)00019-2

Lynch, C. M., McSweeney, P. L. H., Fox, P. F., Cogan, T. M., & Drinan, F. D. (1997). Contribution of starter lactococci and non-starter lactobacilli to proteolysis in Cheddar cheese with a controlled microflora. *Le Lait*, *77*(4), 441–459. doi:10.1051/lait:1997431

Madkor, S. A., Tong, P. S., & El Soda, M. (2000). Ripening of Cheddar cheese with added attenuated adjunct cultures of Lactobacilli. *Journal of Dairy Science*, *83*(8), 1684–1691. doi:10.3168/jds.S0022-0302(00)75037-5 PMID:10984143

Mantere-Alhonen, S. (1995). Propiobacteria used as probiotics-A review. *Le Lait*, *75*(4-5), 447–452. doi:10.1051/lait:19954-534

Maoz, A., Mayr, R., & Scherer, S. (2003). Temporal stability and Biodiversity of two complex antiristerial cheese-ripening microbial consortia. *Applied and Environmental Microbiology*, *69*(7), 4012–4018. doi:10.1128/AEM.69.7.4012-4018.2003 PMID:12839776

McSweeney, L. H., & Sousa, M. J. (2000). Biochemical pathways for the production of flavour compounds in cheeses during ripening: A review. *Le Lait*, *80*(3), 293–324. doi:10.1051/lait:2000127

Meira, Q. G. S., Magnani, M., Júnior, F. C. M., Queiroga, R. C. R. E., Madruga, M. S., Gullóne, B., ... de Souza, E. L. (2015). Effects of added Lactobacillus acidophilus and Bifidobacterium lactis probiotics on the quality characteristics of goat ricotta and their survival under simulated gastrointestinal conditions. *Food Research International*, *76*, 828–838. doi:10.1016/j.foodres.2015.08.002 PMID:28455069

Molimard, P., & Spinnler, H. E. (1996). Review: Compounds involvd in the flavor of surface mold ripened Cheeses: Origin and properties. *Journal of Dairy Science*, *79*(2), 169–184. doi:10.3168/jds.S0022-0302(96)76348-8

Moreno, M. R. F., Sarantinopoulos, P., Tsakalidou, E., & De Vuyst, L. (2006). The role and application of enterococci in food and health. *International Journal of Food Microbiology*, *106*(1), 1–24. doi:10.1016/j.ijfoodmicro.2005.06.026 PMID:16216368

Motta, A. S., & Brandelli, A. (2002). Characterization of an antibacterial peptideproduced by *Brevibacterium linens*. *Journal of Applied Microbiology*, *92*(1), 63–70. doi:10.1046/j.1365-2672.2002.01490.x PMID:11849329

Mushtaq, M., Gani, A., Masoodi, F. A., & Ahmad, M. (2016). Himalayan cheese (Kalari/Kradi) - Effect of different probiotic strains on oxidative stability, microbiological, sensory and nutraceutical properties during storage. *Lebensmittel-Wissenschaft + Technologie*, *67*, 74–81. doi:10.1016/j.lwt.2015.11.039

Oberg, C. J., Moyes, L. V., Domek, M. J., Brothersen, C., & McMahon, D. J. (2011). Survival of probiotic adjunct cultures in cheese and challenges in their enumeration using selective media. *Journal of Dairy Science*, *94*(5), 2220–2230. doi:10.3168/jds.2010-3934 PMID:21524512

Ong, L., Henriksson, S., & Shah, N. P. (2006). Development of probiotic Cheddar cheese containing *Lactobacillus acidophilus, Lb. casei, Lb. paracasei* and *Bifidobacterium* spp. and the influence of these bacteria on proteolytic patterns and production of organic acid. *International Dairy Journal, 16*(5), 446–456. doi:10.1016/j.idairyj.2005.05.008

Ong, L., Henriksson, S., & Shah, N. P. (2007). Proteolytic pattern and organic acid profiles of probiotic Cheddar cheese as influenced by probiotic strains of *Lactobacillus acidophilus, Lb. paracasei, Lb. casei* or *Bifidobacterium* sp. *International Dairy Journal, 17*(1), 67–78. doi:10.1016/j.idairyj.2005.12.009

Ong, L., & Shah, N. P. (2009). Probiotic Cheddar cheese: Influence of ripening temperatures on survival of probiotic microorganisms, cheese composition and organic acid profiles. *Lebensmittel-Wissenschaft + Technologie, 42*(7), 1260–1268. doi:10.1016/j.lwt.2009.01.011

Ouwehand, A. (2004). The probiotic potential of Propionibacteria. In S. Salminen, A. Wright, & A. Ouwehand (Eds.), *Lactic Acid Bacteria: Microbiological and Functional Aspects* (pp. 125–159). London: Maecel and Dekker Inc. doi:10.1201/9780824752033.ch4

Özer, B., & Kırmacı, H. A. (2011). Technological and health aspects of probiotic cheese. In R. D. Foster (Ed.), *Cheese: Types, Nutrition and Consumption* (pp. 1–42). Nova Science Publishers, Inc.

Parente, E., & Cogan, T. M. (2004). Starter cultures: General aspects. In P. F. Fox, P. L. H. McSweeney, T. M. Cogan & T. P. Guinee (Eds.), Cheese: Chemistry, Physics and Microbiology: Vol. 1. General Aspects (3rd ed.; pp. 123-147). London, UK: Academic Press.

Pereira, E. P. R., Cavalcanti, R. N., Esmerino, E. A., Silva, R., Guerreiro, L. R. M., Cunha, R. L., ... Cruz, A. G. (2016). Effect of incorporation of antioxidants on the chemical, rheological, and sensory properties of probiotic petit suisse cheese. *Journal of Dairy Science, 99*(3), 1762–1772. doi:10.3168/jds.2015-9701 PMID:26805976

Pino, A., Hoorde, K. V., Pitino, I., Russo, N., Carpino, S., Caggia, C., & Randazzo, C. L. (2017). Survival of potential probiotic lactobacilli used as adjunct cultures on Pecorino Siciliano cheese ripening and passage through the gastrointestinal tract of healthy volunteers. *International Journal of Food Microbiology, 252*, 42–52. doi:10.1016/j.ijfoodmicro.2017.04.012 PMID:28458191

Piveteau, P. G., Condon, S., & Cogan, T. M. (1995). Interactions between lactic and propionic acid bacteria. *Le Lait, 75*(4-5), 331–343. doi:10.1051/lait:19954-525

Powell, I. B., Broome, M. C., & Limsowtin, G. K. Y. (2003). Starter cultures: General aspects. In H. Roginski, J. W. Fuquay, & P. F. Fox (Eds.), *Encyclopedia of Dairy Sciences* (pp. 261–268). London, UK: Academic Press.

Rattray, F. P. (2003). Secondary cultures. In H. Roginski, J. W. Fuquay, & P. F. Fox (Eds.), *Encyclopedia of Dairy Sciences* (pp. 275–281). London, UK: Academic Press.

Rattray, F. P., & Fox, P. F. (1999). Aspect of enzymology and biochemical properties of Brevibacterium linens relevant to cheese ripening: A review. *Journal of Dairy Science, 82*(5), 891–909. doi:10.3168/jds.S0022-0302(99)75308-7 PMID:10342227

Reps, A. (1993). Bacterial surface-ripened cheeses. In P. F. Fox (Ed.), *Cheese: Chemistry, Physics and Microbiology* (2nd ed.; Vol. 2, pp. 137–172). Major Cheese Groups. doi:10.1007/978-1-4615-2648-3_5

Rolim, F. R. L., dos Santos, K. M. O., de Barcelos, S. C., do Egito, A. S., Ribeiro, T. S., da Conceiçao, M. L., ... Queiroga, R. C. R. E. (2015). Survival of *Lactobacillus rhamnosus* EM1107 in simulated gastrointestinal conditions and its inhibitory effect against pathogenic bacteria in semi-hard goat cheese. *Lebensmittel-Wissenschaft + Technologie*, *63*(2), 807–813. doi:10.1016/j.lwt.2015.05.004

Ross, R. P., Desmond, C., Fitzgerald, G. F., & Stanton, C. (2005). Overcoming the technological hurdles in the development of probiotic foods. *Journal of Applied Microbiology*, *98*(6), 1410–1417. doi:10.1111/j.1365-2672.2005.02654.x PMID:15916653

Ruijschop, R. M. A. J., Boelrijk, A. E. M., & te Giffel, M. C. (2008). Satiety effects of a dairy beverage fermented with propionic acid bacteria. *International Dairy Journal*, *18*(9), 945–950. doi:10.1016/j.idairyj.2008.01.004

Saavedra, L., & Taranto, M. P., Sesma, F., & Front de Valdez, G. (2003). Homade traditional cheeses fort he isolation of probiotic Enterococcus faecium strains. *International Journal of Food Microbiology*, *2-3*(88), 242–245. PMID:14596996

Settani, L., & Moschetti, G. (2010). Non-starter lactic acid bacteria used to improve cheese quality and provide health benefits. *Food Microbiology*, *27*(6), 691–697. doi:10.1016/j.fm.2010.05.023 PMID:20630311

Shah, N. P. (2007). Functional cultures and health benefits. *International Dairy Journal*, *11*(17), 1262–1277. doi:10.1016/j.idairyj.2007.01.014

Sheehan, J. J., Wilkinson, M. G., & McSweeney, P. L. H. (2008). Influence of processing and ripening parameters on starter, non-starter and propionic acid bacteria and on the ripening characteristics of semi-hard cheeses. *International Dairy Journal*, *18*(9), 905–917. doi:10.1016/j.idairyj.2007.11.024

Silva, H. L. A., Balthazar, C. F., Esmerino, E. A., Vieira, A. H., Cappato, L. P., Neto, R. P. C., ... Cruz, A. G. (2017). Effect of sodium reduction and flavor enhancer addition on probiotic Prato cheese processing. *Food Research International*, *99*, 247–255. doi:10.1016/j.foodres.2017.05.018 PMID:28784481

Songisepp, E., Kullisaar, T., Hütt, P., Elias, P., Brilene, T., Zilmer, M., & Mikelsaar, M. (2004). A new probiotic cheese with antioxidative and antimicrobial activity. *Journal of Dairy Science*, *87*(7), 2017–2023. doi:10.3168/jds.S0022-0302(04)70019-3 PMID:15328213

Spinnier, H. E., & Gripon, J. C. (2004). Surface mould-ripened cheeses. In P. F. Fox, P. L. H. McSweeney, T. M. Cogan, & T. P. Guinee (Eds.), *Cheese: Chemistry, Physics and Microbiology* (3rd ed.; Vol. 2, pp. 157–174). Major Cheese Groups.

Stanton, C., Gardiner, G., Lynch, P. B., Collins, J. K., Fitzgerald, G., & Ross, R. P. (1998). Probiotic cheese. *International Dairy Journal*, *8*(5-6), 491–496. doi:10.1016/S0958-6946(98)00080-6

Steffen, C., Eberhard, P., Bosset, J. O., & Rüegg, M. (1993). Swiss-Type Variedties. In P. F. Fox (Ed.), *Cheese: Chemistry, Physics and Microbiology* (2nd ed.; Vol. 2, pp. 83–110). Major Cheese Groups. doi:10.1007/978-1-4615-2648-3_3

Thierry, A., Deutsch, S., Falentin, H., Dalmasso, M., Cousin, F. J., & Jan, G. (2011). New insight into physiology and metabolisms of *Propionibacterium freudenreichii. International Journal of Food Microbiology, 149*(1), 19–27. doi:10.1016/j.ijfoodmicro.2011.04.026 PMID:21620505

Valdes-Stauber, N., & Götz, H., & Busse, M. (1991). Antagonistic effect of coryneform bacteria from red smear cheese against Listeria species. *International Journal of Food Microbiology, 2*(13), 131–130. PMID:1909545

Valdes-Stauber, N., & Scherer, S. (1994). Isolation and characterization of Linocin M18, A bacteriosin produced by *Brevibacterium linens. Applied and Environmental Microbiology, 60*(10), 3809–3814. PMID:7986050

Valence, F., Richoux, R., Thierry, A., Palva, A., & Lortal, S. (1998). Autolysis of *Lactobacillus helveticus* and *Propionibacterium freudenreichii* in Swiss cheeses: First evidence by using species-specific lysis markers. *The Journal of Dairy Research, 65*(4), 609–620. doi:10.1017/S0022029998003021

van den Tempel, T. (1998). Yeast associated with Danablu. *International Dairy Journal, 8*(1), 25–31. doi:10.1016/S0958-6946(98)00013-2

Verruck, S., Prudencio, E. S., Vieira, C. R. W., Amante, E. R., & Amboni, R. D. M. C. (2015). The buffalo Minas Frescal cheese as a protective matrix of *Bifidobacterium* BB-12 under in vitro simulated gastrointestinal conditions. *Lebensmittel-Wissenschaft + Technologie, 63*(2), 1179–1183. doi:10.1016/j.lwt.2015.04.014

Vinderola, C. G., Prosello, W., Ghiberto, D., & Reinheimer, J. A. (2000). Viability of probiotic (*Bifidobacterium, Lactobacillus acidophilus* and *Lactobacillus casei*) and nonprobiotic microflora in Argentinian Fresco Cheese. *Journal of Dairy Science, 83*(9), 1905–1911. doi:10.3168/jds.S0022-0302(00)75065-X PMID:11003217

Vinderola, G., Prosello, W., Molinari, F., Ghiberto, D., & Reinheimer, J. (2009). Growth of Lactobacillus paracasei A13 in Argentinian probiotic cheese and its impact on the characteristics of the product. *International Journal of Food Microbiology, 135*(2), 171–174. doi:10.1016/j.ijfoodmicro.2009.08.021 PMID:19751954

Chapter 14
Filamentous Fungi in Cheese Production

Banu Metin
Istanbul Sabahattin Zaim University, Turkey

ABSTRACT

Filamentous fungi play important roles in the production of a variety of cheeses. The most famous are the blue cheeses, such as Roquefort or Gorgonzola, in which Penicillium roqueforti *is the principal mold, and the moldy soft cheeses, such as Camembert or Brie, in which production involves* Penicillium camemberti. *There are also other filamentous fungi associated with certain types of cheeses, such as* Mucor *spp.,* Trichothecium roseum, *and* Fusarium domesticum *in Saint Nectaire,* Sporendonema casei *in Cantal, Salers and Rodez cheeses,* Scopulariopsis *species in various French and Austrian cheeses, and* Mucor mucedo *and* Mucor racemosus *in the traditional Norwegian cheese Gamalost. These fungi are either inoculated on the cheese as a starter culture or stand out in mixed cultures during spontaneous fermentation. This chapter reviews the filamentous fungi used to produce different kinds of cheeses in terms of taxonomy, physiology, ecology, and mycotoxins, and the microbiological or biochemical effects of these fungi on cheese production.*

INTRODUCTION

Filamentous fungi are used in the manufacture of different kinds of cheeses. The mold-ripened cheeses can be divided into two broad categories: internally ripened and surface-ripened (Figure 1). In internally ripened cheeses, mold growth is visible both on the surface and in the interior part of the cheese. The most well-known of these cheeses are the blue cheeses, which are produced in different countries under different names. For the production of blue cheeses, the main ripening agent is *Penicillium roqueforti*, which creates blue veins in the cheese as it grows. Another lesser-known, internally ripened variety is Norwegian Gamalost, which involves *Mucor* as the ripening mold that gives the cheese its characteristic flavor, texture and yellow-brown color. The second category of mold-ripened cheeses is the surface-ripened cheeses. The most famous of these are the Camembert-type soft cheeses, in which *Penicillium camemberti* is involved in the production as the ripening mold; the mold covers the surface and forms

DOI: 10.4018/978-1-5225-5363-2.ch014

Figure 1. Classification of mold-ripened cheeses

Mold-ripened cheeses

Internally mold-ripened cheeses

- Blue cheeses
  - ✓ *Penicillium roqueforti*
- Gamalost
  - ✓ *Mucor* spp.

Surface mold-ripened cheeses

- Camembert-type soft cheeses
  - ✓ *Penicillium camemberti*
- Semi-hard cheeses (e.g. Saint-Nectaire and Tomme de Savoie)
  - ✓ *Mucor* spp.
  - ✓ *Trichothecium roseum*
  - ✓ *Fusarium domesticum*
- Hard cheeses (e.g. Cantal, Salers, Rodez)
  - ✓ S*copulariopsis* spp.
  - ✓ *Sporendonema casei*

a white coat. There are other mold-ripened cheeses such as the semi-hard cheese Saint Nectaire, which uses *Mucor* spp., *Trichothecium roseum* and *Fusarium domesticum* as the ripening molds, and the hard surface-ripened cheeses such as French Cantal, Salers and Rodez, which involve *Scopulariopsis* species or *Sporendonema casei*. The sources of filamentous fungi used for the production of these cheeses can be milk when raw milk is used, or the fungi may be derived from the environment in which ripening occurs or from starter cultures directly added into the milk or curd. This chapter summarizes information on the filamentous fungi involved in the production of mold-ripened cheeses in terms of taxonomy, physiological characteristics, ecological habitats, and mycotoxins and the importance of these fungi in cheese production.

INTERNALLY MOLD-RIPENED CHEESES

Blue Cheeses and *Penicillium Roqueforti*

Blue-veined cheeses are produced in most countries but under different names, including Roquefort in France, Gorgonzola in Italy, Stilton in UK, and Danablu in Denmark (Figure 2), among others, and these cheeses are given a Protected Designation of Origin/Protected Geographical Indication (PDO/PGI) status (Cantor, van den Tempel, Hansen, & Ardö, 2004). In these kinds of cheeses, lactic acid bacteria are used as the primary starter culture and the ripening culture or the secondary starter culture is *Penicillium roqueforti*. Other yeasts and non-starter lactic acid bacteria are also involved but are not intentionally added (Martín & Coton, 2017). Although mostly starter cultures are added during blue cheese manufacture, some production disciplines do not involve starter cultures and instead are preferentially manufactured under spontaneous or natural conditions involving microorganisms from milk or the environment (Cantor et al., 2004). Specific manufacturing processes, in addition to the origin of the

Figure 2. Blue-veined cheeses in various countries (a) Roquefort in France ("Du Roquefort, fromage de l'Aveyron", 2015), (b) Gorgonzola in Italy ("Consorzio per la tutela del formaggio gorgonzola DOP", 2017), (c) Stilton in UK ("Stilton cheese made by Cropwell Bishop sitting on waxed paper", 2013), and (d) Danablu in Denmark ("Danish Blue cheese - picture by RN Marshman", 2005)

milk and the different strains of starter cultures, lead to the production of distinctive products (Martín & Coton, 2017).

For the blue cheeses, *P. roqueforti* conidia are either added to the milk or sprayed on the curd or may begin growing on the cheese naturally (Cantor et al., 2004). Piercing might be applied at the beginning of maturation to form channels allowing air passage, which promotes *P. roqueforti* growth (Gripon, 1993). The use of heterofermentative lactic acid bacteria as a primary starter also helps to create openings in the curd by means of CO_2 formation (Cantor et al., 2004; Martín & Coton, 2017). Ripening occurs at 8-15°C with a relative humidity of 85%-95% for 1.5-9 months either in classical ripening cellars or traditional caves according to the cheese type (Cantor et al., 2004; Martín & Coton, 2017). The blue veins of *Penicillium* become visible after 2-3 weeks (Gripon, 1993). During ripening, the proteins and lipids are degraded, and aroma compounds are formed (Martín & Coton, 2017).

Although all microorganisms present in cheese contribute to the biochemical reactions occurring, a very important role is played by *P. roqueforti* (Martín & Coton, 2017). Compared to other cheeses, extensive proteolysis occurs in blue cheeses. This proteolysis might occur due to proteases originating from the milk, the rennet or the microorganisms in the cheese; however, the highest proteolytic activity is attributed to *P. roqueforti* (Gripon, 1993). *P. roqueforti* has two extracellular proteases: a metalloprotease, which is active at a pH of 4.5-8.5 (Gripon & Hermier, 1975), and an aspartic protease, which is stable at a pH of 3.5-6.0 (Modler, Brunner, & Stine, 1974a, 1974b); both show activity at the cheese pH (Ardö, 2011). The peptides produced by these enzymes are further broken down by several extracellular exopeptidases of *P. roqueforti*, including an acid carboxypeptidase that is stable at pH 5.0-5.5 (Gripon & Nardi, 1977) and an alkaline metalloaminopeptidase with an optimum pH of 8.0 (Gripon, 1977) for which activity might increase as the pH rises during ripening (Ardö, 2011). *P. roqueforti* also contains several other intracellular peptidases (Ardö, 2011; Ichishima, Takeuchi, Yamamoto, Sano, & Kikuchi, 1978). The extensive proteolysis that takes place in blue cheeses is very important for both the texture and flavor. The proteolytic potential of *P. roqueforti* is strain-dependent (Farahat, Rabie, & Farag, 1990; Larsen, Kristiansen, & Hansen, 1998); therefore, specific strains used for the production of different types of blue cheeses have specific proteolytic properties that contribute to the final cheese characteristics. Another important biochemical reaction occurring in blue cheeses is lipolysis, which occurs more extensively than in other cheeses, including Camembert (Ardö, 2011; Martín & Coton, 2017). The most important lipolytic activity is due to *P. roqueforti*, although milk lipase and bacterial or yeast lipases also contribute (Martín & Coton, 2017). *P. roqueforti* participates in the lipolysis of blue cheeses through two extracellular lipases: an acidic lipase with an optimum pH of 6.0 at 30°C (Lamberet & Menassa, 1983), and an alkaline lipase with an optimum pH of 9.0 that is also partially active under acidic pH values (Menassa & Lamberet, 1982). *P. roqueforti* also shows intracellular lipase activity (Stepaniak, Kornacki, Grabska, Rymaszewski, & Cichosz, 1980). The products of this extensive proteolytic and lipolytic activity are directly involved in the formation of the typical blue cheese flavor (Cantor et al., 2004; Gripon, 1993; Martín & Coton, 2017).

Two other similar blue cheese varieties are Turkish mold-ripened Civil and Konya Kuflu (moldy) or Konya Yesil (green) cheese. Civil cheese that has been given a Protected Geographical Indication status has a fibrous nature (Figure 3(a)) obtained by hanging the curd allowing to stretch by its own weight (Cakmakci et al., 2013). The cheese is ripened in goat skins or plastic bags for at least 3 months without the addition of starter cultures. Majority of the molds growing on or within the fibers of the cheese were determined to be *P. roqueforti* (Cakmakci, Cetin, Gurses, Dagdemir, & Hayaloglu, 2012; Erdogan, Gurses, & Sert, 2003). Another Turkish moldy cheese variety is Konya Kuflu cheese that is a semi-hard cheese ripened in goat skin bags in caves with 80-90% relative humidity at 6-12°C for 3-4 months (Figure 3(b); Hayaloglu, Brechany, Deegan, & McSweeney, 2008; Hayaloglu & Kirbag, 2007). After this first ripening period, the cheese is cut into pieces to allow spontaneous mold growth. The second ripening phase continues until the blue-green mold growth is visible. *Penicillium* was found to be the predominant genus with *Penicillium commune*, *P. roqueforti* and *Penicillium verrucosum* the most abundant species (Hayaloglu & Kirbag, 2007).

Figure 3. Blue cheese varieties from Turkey (a) Mold-ripened Civil cheese ("Yeşil Peynirin Faydaları-Erzurum", 2016), (b) Konya Kuflu cheese (Kuru, 2011)

(a) Mold-ripened Civil cheese

(b) Kuflu cheese

P. roqueforti, in addition to being the principal filamentous fungal species of the blue cheeses, is also a common contaminant of other food substrates, such as rye bread and dairy products (Frisvad & Samson, 2004), and it can also be found in other environments (e.g., silage and occasionally wood) (Sumarah, Miller, & Blackwell, 2005). In fact, the traditional mold cultures of Roquefort were obtained by leaving barley or rye bread in caves for a period of 3-6 months and making a powder of the molded bread (Stark, 2007). The tolerance of *P. roqueforti* to weak acid preservatives and high levels of CO_2 combined with low oxygen requirements for growth make cheese the perfect substrate for this mold (Frisvad & Samson, 2004; Pitt & Hocking, 2009a). In addition, this species is a psychrophile that grows well at refrigeration temperatures but cannot grow above 35°C and has a wide pH tolerance (pH 3-10) (Moreau, 1980; Pitt & Hocking, 2009a).

Phylogenetic analyses indicate that *P. roqueforti* belongs to the family *Aspergillaceae* within the order *Eurotiales* in the *Eurotiomycetes* class and *Pezizomycotina* subphylum of *Ascomycota* (Figure 4). Recently, DNA sequencing and secondary metabolite profiling have demonstrated that the "*P. roqueforti*" is actually composed of four different species: *P. roqueforti* sensu stricto, *Penicillium carneum*, *Penicillium paneum* and *Penicillium psychrosexualis* (Boysen, Skouboe, Frisvad, & Rossen, 1996; Houbraken & Samson, 2011; Pitt & Hocking, 2009a). Reports of *P. roqueforti* in meat and meat products were later speculated to be *P. carneum* (Pitt & Hocking, 2009a).

P. roqueforti has been shown to produce a number of secondary metabolites that have bioactive properties involving andrastins with antitumor activity (Nielsen, Dalsgaard, Smedsgaard, & Larsen, 2005) and mycophenolic acid, which has antimicrobial and immunosuppressive activity (Bentley, 2000; Engel, Milczewski, Prokopek, & Teuber, 1982). Moreover, it also synthesizes several mycotoxins including roquefortines C and D, ergot alkaloids such as isofumigaclavine A and B and festuclavine, and PR-toxin and its derivatives (García-Estrada & Martín, 2016). An important mycotoxin, patulin, has been shown to be produced by three other members of the *P. roqueforti* species group, *P. carneum*, *P. paneum* and *P. psychrosexualis*, but not by *P. roqueforti* (Houbraken, Frisvad, & Samson, 2010; Martín & Liras,

Figure 4. Phylogenetic classification of filamentous fungi used in cheese production. Phylogenetic tree constructed using neighbor-joining analysis with 100 bootstrap replicates by CLC Main Workbench 7.6.2 software. Batrachochytrium dendrobatidis was selected as outgroup. The isolates and Genbank accession numbers used to construct the tree are as follows: P. roqueforti *CBS221.30 (KF465778), P.* camemberti *CBS 299.48 (NR_119396),* M. racemosus *CBS 260.68 (DQ118996),* T. roseum *CBS 113334 (EU552162),* F. domesticum *CBS 434.34 (JQ434585),* S. brevicaulis *CBS 108960 (LN850804),* S. casei *MUCL 40630 (JQ434573) and* B. dendrobatidis *KR11 (JX983051).*

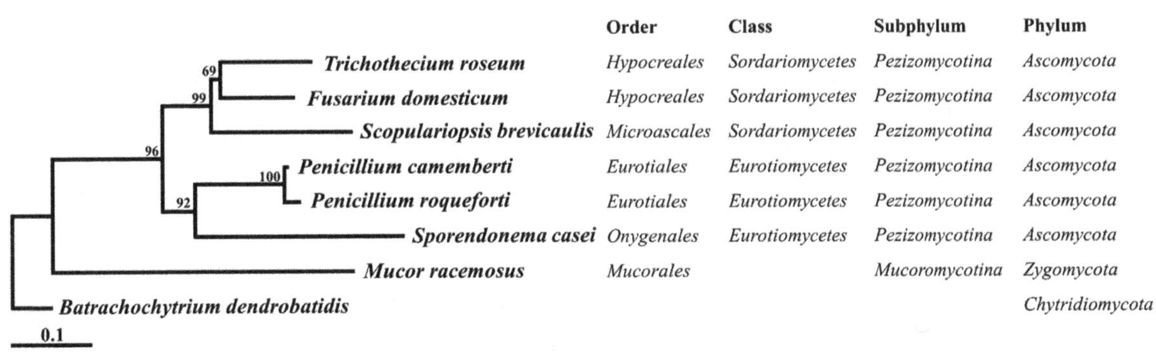

2017). Fortunately, one of the toxins produced by *P. roqueforti*, the PR-toxin, is not stable in the cheese environment and appears to be converted into less toxic metabolites (Chang, Lu, & Yeh, 1993; Scott & Kanhere, 1979). Similarly, the levels of ergot alkaloids and roquefortines detected in blue cheeses are not significant (Finoli, Vecchio, Galli, & Dragoni, 2001; Fontaine et al., 2015; Ohmomo, Oguma, Ohashi, & Abe, 1978; Ohmomo, Sato, Utagawa, & Abe, 1975; Scott & Kennedy, 1976). Considering the toxicological data and the historical safe consumption, it is generally accepted that the mycotoxins in blue cheese do not pose a toxicity risk for humans (O'Callaghan, O'Connor, & O'Brien, 2017).

For different types of blue cheeses, populations of *P. roqueforti* that were highly differentiated in terms of genetic, morphological and functional diversity were observed, which shows that either different cheese-making procedures were effective in shaping the cheese-associated *P. roqueforti* population or that different populations were selected for different kinds of cheeses (Gillot et al., 2015, 2017). The differences in *P. roqueforti* isolates together with different cheese production processes lead to the production of various types of blue cheeses around the world with distinctive characteristics.

Growth in the cheese environment influenced the evolution of not only *P. roqueforti* but also *P. camemberti* and some other contaminating *Penicillium* species of cheese because they share horizontally transferred genomic regions with nearly 100% identity that might have important functions in cheese adaptation (Ropars et al., 2015). One of these regions, named *Wallaby*, is over 500 kb in size and bears approximately 250 genes, including some that are predicted to have antimicrobial functions and was predicted to confer competitive advantage to these fungi (Cheeseman et al., 2014). Another horizontally transferred region is called *CheesyTer*, which is an 80-kb region carrying approximately 40 genes, some of which function in lactose metabolism such as lactose permease and beta-galactosidase (Ropars et al., 2015). Both *Wallaby* and *CheesyTer* were shown experimentally to be associated with faster growth and higher competitiveness on cheese (Ropars et al., 2015). Given that some fungi have only been isolated from cheese, such as *P. camemberti*, horizontal gene transfer seems to have taken place in cheese (Ropars et al., 2015). Therefore, cheese-making appears to be an important process in shaping the characteristics of cheese-associated fungi.

Sexual reproduction has recently been described in *P. roqueforti*, in which two isolates with compatible mating types mate to form recombinant progeny (Ropars et al., 2014). This finding might have important implications for the cheese industry since sexual reproduction is an efficient and easy method for strain improvement (Böhm et al., 2013).

Traditional Norwegian Cheese Gamalost and *Mucor* spp.

Gamalost is a traditional Norwegian moldy cheese not only on the surface but also internally ripened (Figure 5; Gripon, 1993). The name means "old cheese", and this variety can be traced back to the Viking times (Werner, Nielsen, Ardö, Rage, & Antila, 1999). In the production of Gamalost, *Mucor* species, namely, *Mucor mucedo* and *Mucor racemosus*, are used (Werner, Nielsen, Ardö, Rage, & Antila, 1999). The mold is sprayed on the curd, and during ripening, as the mold grows, a yellow-brown color develops both on the surface and inside of the cheese (Gripon, 1993). Ripening occurs over five weeks, and at the end of ripening, a granular-textured product with a pungent flavor is obtained (Gripon, 1993; Werner, Nielsen, Ardö, Rage, & Antila, 1999).

Mucor is a very common genus in nature, present in soil, decaying vegetation and manure (Pitt & Hocking, 2009). *Mucor* spp. belong to the *Mucoraceae* family in the order *Mucorales* and subphylum *Mucoromycotina* in *Zygomycota* (Figure 4; James et al., 2006). Current knowledge of the taxonomy of

Figure 5. Traditional Norwegion Cheese, Gamalost ("Gamalost (also Gammelost, Gammalost), which translates as "old cheese", is a pungent traditional Norwegian cheese", 2013)

Mucor is rather limited and largely based on morphology (Hermet, Méheust, Mounier, Barbier, & Jany, 2012). However, recent studies using multi-locus sequences have been able to resolve the phylogeny of *Mucor*, including strains isolated from cheeses (de Souza, Pires-Zottarelli, dos Santos, Costa, & Harakava, 2012; Hermet et al., 2012; Jacops & Botha, 2008). In a recent study, at least six species of *Mucor* isolated from cheeses have been identified (Hermet et al., 2012). *Mucor* species have been reported as good producers of proteolytic and lipolytic enzymes (Takó et al., 2015; Yegin et al., 2011). One *Mucor* spp. with significant proteolytic activity isolated from a Chinese fermented soybean product was successfully used in the ripening of a smeared semi-hard cheese (Zhang & Zhao, 2010). *Mucor* species have not been reported to produce mycotoxins (Pitt & Hocking, 2009).

SURFACE MOLD-RIPENED CHEESES

Camembert-Type Soft Cheeses and *Penicillium Camemberti*

Other well-known moldy cheese types include the surface mold-ripened soft cheeses such as Camembert or Brie (Figure 6). The original Camembert is from the Normandy region of France and has a protected designation of origin (PDO) as Camembert de Normandie; however, the name Camembert alone is not protected and can be used to name these kinds of cheeses when produced elsewhere (Spinnler & Gripon, 2004). Another similar type of cheese is Brie, but the diameter is larger than that of Camembert. Brie also has a PDO, as Brie from Meaux or from Melun (Spinnler & Gripon, 2004).

Figure 6. Moldy soft cheeses such as (a) Camembert ("Camembert of Normandy - French cheese", 2006) and (b) Brie ("Le brie de Melun", 2015)

(a) Camembert de Normandie (b) Brie de Melun

The production of Camembert-type cheeses involves complex microbial interactions, especially when raw milk is used. Since the starter cultures used are generally *Lactococcus lactis* subsp. *lactis* and *Lactococcus lactis* subsp. *cremoris*, the curd initially contains predominantly lactococci. The maturation period lasts for at least 3 weeks in a cellar at 11-13°C with a relative humidity of 90% (Spinnler & Gripon, 2004). Slight acidification of the curd by lactococci generates an environment that selects for the growth of yeasts and filamentous fungi. Therefore, just after manufacture, yeasts such as *Kluyveromyces lactis*, *Saccharomyces cerevisiae* and *Debaryomyces hansenii* start to grow on the surface (Baroiller & Schmidt, 1990). The yeast *Geotrichum candidum* that has filamentous characteristics grows at the same time as yeasts; however, growth of this species might be affected by salting. After one week when some of the fats and proteins are hydrolyzed by yeasts, *P. camemberti* starts to grow on the surface; in 2-3 days, white felt covers the entire surface (Spinnler & Gripon, 2004). After 2-3 weeks, the surface is de-acidified by the action of the lactic acid-catabolizing enzymes of *Penicillium*, which then permits acid-sensitive aerophilic bacteria such as staphylococci and coryneform bacteria including *Brevibacterium linens* to grow on the surface (Gripon, 1993; Spinnler & Gripon, 2004). As a starter, commercial *P. camemberti* strains are added either to the milk or to the surface of the curd. Currently, *G. candidum*, yeasts and coryneform bacteria are also added as ripening starter cultures in addition to *P. camemberti* to achieve a product similar to traditional Camembert (Spinnler & Gripon, 2004). These microorganisms and their enzymes are involved in the characteristic flavor and soft texture of the cheese.

The proteinases of *P. camemberti* are important for the high-level proteolysis observed in Camembert-type cheeses (Spinnler & Gripon, 2004). *P. camemberti* has a metalloprotease with an optimum pH of 5.5-6.0 and an aspartic protease with maximum activity at pH 4.0 (Lenoir, 1984). Both enzymes are stable in the cheese environment. *P. camemberti* also has extracellular exopeptidases such as acid carboxypeptidase with an optimum pH of 3.5 (Ahiko, Iwasawa, Ulda, & Nigata, 1981) and an alkaline aminopeptidase with an optimum pH of 8.5 (Matsuoka, Fuke, Kaminogawa, & Yamauchi, 1991). These enzymes play roles in the release of amino acids from the casein hydrolysate (Spinnler & Gripon, 2004). *G. candidum* also helps proteolysis by breaking down the small peptides produced by *P. camemberti*, which reduces the bitterness of the cheese (Boutrou, Aziza, & Amrane, 2006; Marcellino & Benson, 2013). Lipolysis is another key phenomenon which mainly occurs just under the rind (Hassouna & Guizani, 1995). An extensively produced alkaline lipase of *P. camemberti* is the main factor in lipolysis (Spinnler & Gripon,

2004), and although its pH optimum is approximately 9.0, it retains half of its maximum activity at the cheese pH and ripening temperatures (Lamberet & Lenoir, 1976). Free fatty acids and their metabolic products together with the products of proteolysis play roles in the formation of the typical flavor and texture of the Camembert-type cheeses.

Just like *P. roqueforti*, *P. camemberti* belongs to the *Aspergillaceae* family of the order *Eurotiales* in the class *Eurotiomycetes* and subphylum *Pezizomycotina* in *Ascomycota* (Figure 4; Houbraken & Samson, 2011). Within this family, *P. camemberti* is found in the section *Fasciculata* together with *Penicillium caseifulvum* and *Penicillium commune*, among others. The species in this section are tolerant to refrigeration temperatures and low water activity (Houbraken & Samson, 2011). *P. camemberti* is described as the domesticated species derived from *P. commune* (Pitt, Cruickshank, & Leistner, 1986; Pitt & Hocking, 2009a). *P. caseifulvum* and *P. camemberti* can only be differentiated by conidial color; while *P. caseifulvum* remains white, *P. camemberti* turns gray-green over time (Gripon, 1993; Pitt & Hocking, 2009a). Therefore, *P. caseifulvum* is thought to be a white mutant of *P. camemberti* that is not a separate species, and only the name *P. camemberti* is used (Gripon, 1993; Samson, Eckardt, & Orth, 1977; Spinnler & Gripon, 2004). *P. camemberti* and its white mutants have been isolated almost exclusively from cheeses, and therefore cheese has been described as the main ecological habitat of *P. camemberti* (Frisvad & Samson, 2004; Pitt & Hocking, 2009a).

P. camemberti is known to produce the mycotoxin cyclopiazonic acid in addition to a number of other secondary metabolites such as hadacidin, which is an aspereynone-like terpene compound, and 2-methyl-isoborneol (Frisvad, Smedsgaard, Larsen, & Samson, 2004). Cyclopiazonic acid has been detected in low quantities in cheese (Bars, 1979), but the presence of the other compounds in cheese or their toxic effects are unknown (Martín & Liras, 2017).

Semi-Hard Cheeses and *Mucor* spp., *Trichothecium Roseum* and *Fusarium Domesticum*

The uncooked, semi-hard cheese Saint Nectaire (PDO) is a traditional surface-ripened cheese produced in the Auvergne region of France using unpasteurized bovine milk without the addition of a starter culture (Figure 7(a); Harbutt, 2009; Marcellino & Benson, 1992; Marcellino & Benson, 2013). Ripening occurs for 60 days in cellars or caves (Marcellino & Benson, 2013). Scanning electron microscopy and light microscopy images indicate the start of the growth of *Mucor* spp. on day 6 with a hairy rind and the appearance of conidiophores of *Trichothecium roseum* in the late ripening phase at day 31 as a pink powder on the gray-brown *Mucor* coating (Marcellino & Benson, 1992). However, this morphological identification has not been confirmed by molecular methods. Recent molecular data indicated that a newly described species, *Fusarium domesticum*, is inoculated in the manufacture of Saint Nectaire (Ropars, Cruaud, Lacoste, & DuPont, 2012).

T. roseum, which is characterized by pinkish colonies, is a ubiquitous saprophyte of decaying matter and has been isolated from a wide variety of foods (Pitt & Hocking, 2009b). Phylogenetic analysis places *T. roseum* in the order *Hypocreales* in the class *Sordariomycetes* of the subphylum *Pezizomycotina* in *Ascomycota* (Figure 4; Summerbell et al., 2011). The optimum growth temperature of this species is 25°C, but it can also grow at 5°C (Pitt & Hocking, 2009b) and at a water activity value of 0.9 (Snow, 1949). *T. roseum* has been reported to have lipolytic activity (Sipahioglu & Heperkan, 2000), which might play an important role in the production of flavor compounds in cheese and has been reported to produce trichothecene-type mycotoxins (Konishi et al., 2003; Tang et al., 2015).

F. domesticum is a recently described species that was formerly called *Trichothecium domesticum* or *Cylindrocarpon* spp. but was later shown to be more closely related to *Fusarium dimerum* than *T. roseum* (Bachmann et al., 2005; Ropars et al., 2012). In addition to having been identified in Saint Nectaire (Ropars et al., 2012), it was also reported to be present in the smear bacteria-ripened Swiss cheeses such as Gruyère or Tilsit, in which it prevents stickiness of the smear (Bachmann et al., 2005). In fact, before characterizing it as *F. domesticum*, it was called "anticollanti", meaning "against stickiness" (Bachmann et al., 2005). Phylogenetically, *F. domesticum* belongs to the order *Hypocreales* of the class *Sordariomycetes* under *Ascomycota* (Figure 4; Bachmann et al., 2005; Summerbell et al., 2011). Its main habitat is cheese, and it has not been isolated from any other source (Ropars et al., 2012).

Other filamentous fungi have also been isolated from Saint Nectaire and a similar semi-hard French cheese, Tomme de Savoie (Figure 7(b)), in different studies and include *Chrysosporium sulfureum* (Ratomahenina, Chabalier, Galzy, & Dieu, 1995), *Cylindrocarpon* sp. (Ratomahenina, van den Boom, Galzy, & Dieu, 1995), which was later shown to be *Fusarium domesticum* (Ropars et al., 2012), *Penicillium* spp., *Epicoccum purpurascens* and *Cladosporium cladosporioides* in addition to *Mucor* spp. (Bachmann et al., 2005; Gripon, 1993).

Hard Cheeses and *Scopulariopsis* spp. and *Sporendonema Casei*

Cantal and Salers cheeses are uncooked French hard cheeses with PDO that are ripened at least for 3 months (Figure 8 (a) and (b); Cornu et al., 2009; Didienne et al., 2012; Montel et al., 2014). Few studies have determined the filamentous fungal species involved in the production of these cheeses. In one study, *Scopulariopsis brevicaulis* was isolated from Cantal cheese (Chabalier, Ratomahenina, Galzy, & Dieu, 1995), and in a recent study, *Sporendonema casei* was identified from both Cantal and Salers cheeses (Ropars et al., 2012). A similar hard cheese is Rodez (Figure 8(c), in which *S. casei* is responsible for the characteristic sunburnt color of the cheese (Ratomahenina, Chabalier, & Galzy, 1994).

Scopulariopsis is a genus related to *Penicillium*. The best known species, *S. brevicaulis*, has distinctive brown-colored conidia (Pitt & Hocking, 2009a). *S. brevicaulis* can grow at water activity values of 0.90 and grows best under NaCl concentrations of 0.1-1 M (Galloway, 1935; Yoder, Benoit, & Zettler, 2003). This species favors alkaline conditions and high nitrogen substrates with an optimum pH of 9.0

Figure 7. Semi-hard mold-ripened cheeses such as (a) Saint Nectaire ("Fromage AOP de Saint-Nectaire, France", 2013) and (b) Tomme de Savoie ("Tomme de Savoie", 2015)

(a) Saint Nectaire cheese

(b) Tomme de Savoie

Figure 8. Hard mold-ripened cheeses such as (a) Cantal ("Cantal", 2014), (b) Salers ("Salers", 2015) and (c) Rodez ("Le Rodez est un fromage produit dans la ville du même nom", 2017)

(a) Cantal (b) Salers

(c) Rodez

(Bothast, Lancaster, & Hesseltine, 1975). It has been isolated from several food sources including rice, barley, wheat, meat products, soybean, pepper and cheese (Pitt & Hocking, 2009a). In addition to having been isolated from Cantal cheese (Chabalier et al., 1995), *S. brevicaulis* was also found to be the most abundant mold species in one Austrian and one Danish bacterial smear ripened cheese variety (Gori, Ryssel, Arneborg, & Jespersen, 2013; Schornsteiner, Mann, Bereuter, Wagner, & Schmitz-Esser, 2014). *Scopulariopsis* species are taxonomically placed in the *Microascaceae* family in the order *Microascales* and class *Sordariomycetes* in the *Pezizomycotina* subphylum of *Ascomycota* (Figure 4; Jagielski et al., 2016; Ropars et al., 2012). No mycotoxin produced by *S. brevicaulis* has been reported (Pitt & Hocking, 2009a).

S. casei is an uncommon fungus isolated only from cheeses and is closely related to another cheese fungus, *Sphaerosporium equinum* (Ropars et al., 2012). Phylogenetic analysis indicates that these two species belong to the *Onygenales* order under the *Eurotiomycetes* class of the subphylum *Pezizomycotina* in *Ascomycota* (Figure 4; Ropars et al., 2012).

CONCLUSION

Filamentous fungal species are used as ripening starter cultures in the production of both surface and internally mold-ripened cheeses. The *Penicillium* species *P. roqueforti* and *P. camemberti*, which belong to the *Eurotiomycetes* class of *Ascomycota*, are used in the production of Roquefort- and Camembert-type cheeses, respectively. Similarly, *S. casei*, which belongs to another order of *Eurotiomycetes*, is isolated from surface mold-ripened cheeses such as Cantal, Salers and Rodez. Another class of *Ascomycota*, *Sordariomycetes*, contains *T. roseum* and *F. domesticum*, which are involved in the production of Saint Nectaire and *Scopulariopsis* spp. isolated from Cantal in addition to Austrian and Danish smear-ripened cheeses. *Mucor* spp. from the phylum *Zygomycota* are used to manufacture many mold-ripened cheeses such as Saint Nectaire, Tomme de Savoie and Gamalost. Apparently, molds from different branches of the kingdom Fungi have evolved to grow in cheese as an environmental niche. Some of these fungi, specifically *P. camemberti*, *F. domesticum* and *S. casei*, are almost exclusively isolated from cheeses, indicating that their main ecological habitat is cheese, whereas the others are ubiquitous saprophytes that can be isolated from foods or from nature. In cheeses, fungi play roles in the establishment of the microbiological profile by participating in complex microbial interactions and carrying out biochemical reactions; consequently, these activities form the characteristic flavor and texture of the final product.

REFERENCES

Ahiko, K., Iwasawa, S., Ulda, M., & Nigata, N. (1981). Studies on acid carboxypeptidase from *Penicillium caseicolum*. II. Hydrolysis of bitter peptides by acid carbozypeptidases and large scale preparation of the enzyme. *Report of Research of Laboratory. Snow Brand Milk Products Co., 77*, 135–140.

Ardö, Y. (2011). Cheese | Blue Mold Cheese. In J. W. Fuquay, P. F. Fox, & P. L. H. McSweeney (Eds.), *Encyclopedia of Dairy Sciences* (pp. 767–772). San Diego, CA: Academic Press. doi:10.1016/B978-0-12-374407-4.00090-X

Bachmann, H. P., Bobst, C., Bütikofer, U., Casey, M. G., Dalla Torre, M., Fröhlich-Wyder, M. T., & Fürst, M. (2005). Occurrence and significance of *Fusarium domesticum* alias Anticollanti on smear-ripened cheeses. *Lebensmittel-Wissenschaft + Technologie, 38*(4), 399–407. doi:10.1016/j.lwt.2004.05.018

Baroiller, C., Schmidt, J. L., & Lapadu-Hargues, M. (1990). Contribution `a l'´etude de l'origine des levures du fromage de Camembert. *Le Lait, 70*(1), 67–84. doi:10.1051/lait:199017

Bars, J. L. (1979). Cyclopiazonic acid production by *Penicillium camemberti* Thom and natural occurrence of this mycotoxin in cheese. *Applied and Environmental Microbiology, 38*(6), 1052–1055. PMID:526014

Bentley, R. (2000). Mycophenolic acid: A one hundred year odyssey from antibiotic to immunosuppressant. *Chemical Reviews, 100*(10), 3801–3826. doi:10.1021/cr990097b PMID:11749328

Böhm, J., Hoff, B., O'Gorman, C. M., Wolfers, S., Klix, V., & Binger, D., … Kück, U. (2013). Sexual reproduction and mating-type–mediated strain development in the penicillin-producing fungus *Penicillium chrysogenum. Proceedings of the National Academy of Sciences, 110*(4), 1476–1481.

Bothast, R. J., Lancaster, E. B., & Hesseltine, C. W. (1975). *Scopulariopsis brevicaulis*: Effect of pH and substrate on growth. *European Journal of Applied Microbiology and Biotechnology, 1*(1), 55–66. doi:10.1007/BF01880620

Boutrou, R., Aziza, M., & Amrane, A. (2006). Enhanced proteolytic activities of *Geotrichum candidum* and *Penicillium camembertii* in mixed culture. *Enzyme and Microbial Technology, 39*(2), 325–331. doi:10.1016/j.enzmictec.2005.11.003

Boysen, M., Skouboe, P., Frisvad, J., & Rossen, L. (1996). Reclassification of the *Penicillium roqueforti* group into three species on the basis of molecular genetic and biochemical profiles. *Microbiology, 142*(3), 541–549. doi:10.1099/13500872-142-3-541 PMID:8868429

Cakmakci, S., Cetin, B., Gurses, M., Dagdemir, E., & Hayaloglu, A. A. (2012). Morphological, molecular, and mycotoxigenic identification of dominant filamentous fungi from moldy civil cheese. *Journal of Food Protection, 75*(11), 2045–2049. doi:10.4315/0362-028X.JFP-12-107 PMID:23127715

Cakmakci, S., Dagdemir, E., Hayaloglu, A. A., Gurses, M., Cetin, B., & Tahmas-Kahyaoglu, D. (2013). Effect of *Penicillium roqueforti* and incorporation of whey cheese on volatile profiles and sensory characteristics of mould-ripened Civil cheese. *International Journal of Dairy Technology, 66*(4), 512–526.

Camembert of Normandy - French cheese. (2006, November 12). Retrieved from https://commons.wikimedia.org/wiki/File:Camembert.JPG

Cantal. (2014, September 25). Retrieved from https://commons.wikimedia.org/wiki/File:Cantal_06.jpg

Cantor, M. D., van den Tempel, T., Hansen, T. K., & Ardö, Y. (2004). Blue cheese. In P. F. Fox, P. L. H. McSweeney, T. M. Cogan, & T. P. Guinee (Eds.), *Cheese: Chemistry, Physics and Microbiology* (Vol. 2, pp. 175–198). Amsterdam: Elsevier Academic Press.

Chabalier, C., Ratomahenina, R., Galzy, P., & Dieu, B. (1995). Remarks on the appearance of "violine" on Cantal cheese. *Acta Microbiologica et Immunologica Hungarica, 42*(2), 199–202. PMID:7551713

Chang, S. C., Lu, K. L., & Yeh, S. F. (1993). Secondary metabolites resulting from degradation of PR toxin by *Penicillium roqueforti. Applied and Environmental Microbiology, 59*(4), 981–986. PMID:8476299

Cheeseman, K., Ropars, J., Renault, P., Dupont, J., Gouzy, J., Branca, A., … Brygoo, Y. (2014). Multiple recent horizontal transfers of a large genomic region in cheese making fungi. *Nature Communications, 5*(3876).

Consorzio per la tutela del formaggio gorgonzola DOP. (2017, April 4). Retrieved from https://commons.wikimedia.org/wiki/File:Gorgonzoladop.jpg

Cornu, A., Rabiau, N., Kondjoyan, N., Verdier-Metz, I., Pradel, P., Tournayre, P., ... Martin, B. (2009). Odour-active compound profiles in Cantal-type cheese: Effect of cow diet, milk pasteurization and cheese ripening. *International Dairy Journal, 19*(10), 588–594. doi:10.1016/j.idairyj.2009.04.008

Danish Blue cheese - picture by RN Marshman. (2005, November 4). Retrieved from https://commons.wikimedia.org/wiki/File:Danish_Blue_cheese.jpg

de Souza, J. I., Pires-Zottarelli, C. L. A., dos Santos, J. F., Costa, J. P., & Harakava, R. (2012). *Isomucor* (*Mucoromycotina*): A new genus from a Cerrado reserve in state of São Paulo, Brazil. *Mycologia, 104*(1), 232–241. doi:10.3852/11-133 PMID:21933930

Didienne, R., Defargues, C., Callon, C., Meylheuc, T., Hulin, S., & Montel, M.-C. (2012). Characteristics of microbial biofilm on wooden vats ("gerles") in PDO Salers cheese. *International Journal of Food Microbiology, 156*(2), 91–101. doi:10.1016/j.ijfoodmicro.2012.03.007 PMID:22483401

Du Roquefort, fromage de l'Aveyron. (2015, April 17). Retrieved from https://commons.wikimedia.org/wiki/File:Wikicheese_-_Roquefort_-_20150417_-_003.jpg

Engel, G., von Milczewski, K. E., Prokopek, D., & Teuber, M. (1982). Strain-specific synthesis of mycophenolic acid by *Penicillium roqueforti* in blue-veined cheese. *Applied and Environmental Microbiology, 43*(5), 1034–1040. PMID:16346004

Erdogan, A., Gurses, M., & Sert, S. (2003). Isolation of moulds capable of producing mycotoxins from blue mouldy Tulum cheeses produced in Turkey. *International Journal of Food Microbiology, 85*(1), 83–85. doi:10.1016/S0168-1605(02)00485-3 PMID:12810273

Farahat, S. M., Rabie, A. M., & Farag, A. A. (1990). Evaluation of the proteolytic and lipolytic activity of different *Penicillium roqueforti* strains. *Food Chemistry, 36*(3), 169–180. doi:10.1016/0308-8146(90)90052-6

Finoli, C., Vecchio, A., Galli, A., & Dragoni, I. (2001). Roquefortine C occurrence in blue cheese. *Journal of Food Protection, 64*(2), 246–251. doi:10.4315/0362-028X-64.2.246 PMID:11271775

Fontaine, K., Passeró, E., Vallone, L., Hymery, N., Coton, M., Jany, J.-L., ... Coton, E. (2015). Occurrence of roquefortine C, mycophenolic acid and aflatoxin M1 mycotoxins in blue-veined cheeses. *Food Control, 47*, 634–640. doi:10.1016/j.foodcont.2014.07.046

Frisvad, Jens C, & Samson, R. A. (2004). Polyphasic taxonomy of *Penicillium* subgenus *Penicillium*. A guide to identification of food and air-borne terverticillate *Penicillia* and their mycotoxins. *Studies in Mycology, 49*(1), 1–174.

Frisvad, J. C., Smedsgaard, J., Larsen, T. O., & Samson, R. A. (2004). Mycotoxins, drugs and other extrolites produced by species in *Penicillium* subgenus *Penicillium. Studies in Mycology, 49*, 201–241.

Fromage AOP de Saint-Nectaire, France. (2013, August 26). Retrieved from https://commons.wikimedia.org/wiki/File:Saint-Nectaire.jpg

Galloway, L. D. (1935). The moisture requirements of mould fungi, with special reference to mildew in textiles. *Journal of the Textile Institute Transactions, 26*(4), T123–T129. doi:10.1080/19447023508661646

Gamalost (also Gammelost, Gammalost), which translates as "old cheese", is a pungent traditional Norwegian cheese. (2013, January 24). Retrieved from https://commons.wikimedia.org/wiki/File:Gamalost-NorwegianOldCheese.jpg

García-Estrada, C., & Martín, J.-F. (2016). Biosynthetic gene clusters for relevant secondary metabolites produced by *Penicillium roqueforti* in blue cheeses. *Applied Microbiology and Biotechnology, 100*(19), 8303–8313. doi:10.100700253-016-7788-x PMID:27554495

Gillot, G., Jany, J.-L., Coton, M., Floch, G. L., Debaets, S., Ropars, J., ... Coton, E. (2015). Insights into *Penicillium roqueforti* morphological and genetic diversity. *PLoS One, 10*(6), e0129849. doi:10.1371/journal.pone.0129849 PMID:26091176

Gillot, G., Jany, J.-L., Poirier, E., Maillard, M.-B., Debaets, S., Thierry, A., ... Coton, M. (2017). Functional diversity within the *Penicillium roqueforti* species. *International Journal of Food Microbiology, 241*, 141–150. doi:10.1016/j.ijfoodmicro.2016.10.001 PMID:27771579

Gori, K., Ryssel, M., Arneborg, N., & Jespersen, L. (2013). Isolation and identification of the microbiota of Danish farmhouse and industrially produced surface-ripened cheeses. *Microbial Ecology*, *65*(3), 602–615. doi:10.100700248-012-0138-3 PMID:23224222

Gripon, J.-C. (1977). The proteolytic system of *Penicillium roqueforti*. *Biochimie*, *59*(8), 679–686. doi:10.1016/S0300-9084(77)80246-0 PMID:23180

Gripon, J. C. (1993). Mould-ripened cheeses. In P. F. Fox (Ed.), *Cheese: Chemistry, Physics and Microbiology* (pp. 111–136). Boston, MA: Springer. doi:10.1007/978-1-4615-2648-3_4

Gripon, J.-C., & Hermier, J. (1975). Le système protéolytique de *Penicillium roqueforti*. *Biochimie*, *56*(10), 1323–1332. doi:10.1016/S0300-9084(75)80017-4 PMID:4219726

Gripon, J.-C., & Nardi, M. (1977). Le système protéolytique de *Penicillium roqueforti*. IV. Propriétés d'une carboxypeptidase acide. *Annales de Biologie Animale, Biochimie, Biophysique*, *17*(3A), 283–298. doi:10.1051/rnd:19770304

Harbutt, J. (2009). *World Cheese Book*. London: Dorling Kindersley Limited.

Hassouna, M., & Guizani, N. (1995). Evolution de la flore microbienne et des caractéristiques physico-chimiques au cours de la maturation du fromage Tunisien de type camembert fabriqué avec du lait pasteurisé. *Microbiolgy et Hygiene Alimentaire.*, *7*(18), 14–23.

Hayaloglu, A. A., Brechany, E. Y., Deegan, K. C., & McSweeney, P. L. H. (2008). Characterization of the chemistry, biochemistry and volatile profile of Kuflu cheese, a mould-ripened variety. *Lebensmittel-Wissenschaft + Technologie*, *41*(7), 1323–1334. doi:10.1016/j.lwt.2007.08.020

Hayaloglu, A. A., & Kirbag, S. (2007). Microbial quality and presence of moulds in Kuflu cheese. *International Journal of Food Microbiology*, *115*(3), 376–380. doi:10.1016/j.ijfoodmicro.2006.12.002 PMID:17258341

Hermet, A., Méheust, D., Mounier, J., Barbier, G., & Jany, J.-L. (2012). Molecular systematics in the genus Mucor with special regards to species encountered in cheese. *Fungal Biology*, *116*(6), 692–705. doi:10.1016/j.funbio.2012.04.002 PMID:22658314

Houbraken, J., Frisvad, J. C., & Samson, R. A. (2010). Sex in *Penicillium* series *Roqueforti*. *IMA Fungus: The Global Mycological Journal*, *1*(2), 171–180.

Houbraken, J., & Samson, R. (2011). Phylogeny of *Penicillium* and the segregation of *Trichocomaceae* into three families. *Studies in Mycology*, *70*(1), 1–51. doi:10.3114im.2011.70.01 PMID:22308045

Ichishima, E., Takeuchi, M., Yamamoto, K., Sano, Y., & Kikuchi, T. (1978). Intracellular acid carboxypeptidase of *Penicillium roqueforti* isolated from blue cheese. *Current Microbiology*, *1*(2), 95–98. doi:10.1007/BF02605424

Jacops, K., & Botha, A. (2008). *Mucor renispora* sp. nov., a new coprophilous species from Southern Africa. *Fungal Diversity*, *29*, 27–35.

Jagielski, T., Sandoval-Denis, M., Yu, J., Yao, L., Bakuła, Z., Kalita, J., ... Gené, J. (2016). Molecular taxonomy of *Scopulariopsis*-like fungi with description of new clinical and environmental species. *Fungal Biology, 120*(4), 586–602. doi:10.1016/j.funbio.2016.01.014 PMID:27020159

James, T. Y., Kauff, F., Schoch, C. L., Matheny, P. B., Hofstetter, V., Cox, C. J., ... Vilgalys, R. (2006). Reconstructing the early evolution of Fungi using a six-gene phylogeny. *Nature, 443*(7113), 818–822. doi:10.1038/nature05110 PMID:17051209

Konishi, K., Iida, A., Kaneko, M., Tomioka, K., Tokuda, H., Nishino, H., & Kumeda, Y. (2003). Cancer preventive potential of trichothecenes from *Trichothecium roseum. Bioorganic & Medicinal Chemistry, 11*(12), 2511–2518. doi:10.1016/S0968-0896(03)00215-3 PMID:12757719

Kuru, A. (Photographer). (2011, February 13). *Yillar sonra Konya'da "kuflu peynir"in pazarda hala serbestce satilabildigini gormek sasirtici* [digital image]. Retrieved from https://www.flickr.com/photos/alikuru/5440805421

Lamberet, G., & Lenoir, J. (1976). Les caractères du système lipolytique de l'espèce *Penicillium caseicolum*. Purification et propriétes de la lipase majeure. *Le Lait, 56*(559–560), 622–644. doi:10.1051/lait:1976559-56033

Lamberet, G., & Menassa, A. (1983). Purification and properties of an acid lipase from *Penicillium roqueforti. The Journal of Dairy Research, 50*(4), 459–468. doi:10.1017/S0022029900032696

Larsen, M. D., Kristiansen, K. R., & Hansen, T. K. (1998). Characterization of the proteolytic activity of starter cultures of *Penicillium roqueforti* for production of blue veined cheeses. *International Journal of Food Microbiology, 43*(3), 215–221. doi:10.1016/S0168-1605(98)00114-7 PMID:9801197

Le brie de Melun est une appellation d'origine désignant un fromage au lait cru de la région de la Brie et de Melun. (2015, May 15). Retrieved from https://commons.wikimedia.org/wiki/File:Wikicheese_-_Brie_de_Melun_-_20150515_-_015.jpg

Le Rodez est un fromage produit dans la ville du même nom. (2017, February 5). Retrieved from https://commons.wikimedia.org/wiki/File:Wikicheese-XX_-_Rodez_-_20170905_-_011.jpg

Lenoir, J. (1984). The surface flora and its role in the ripening of cheese. In *Federation Internationale de Laiterie Bulletin Annuel* (Vol. 171, pp. 3–20). Brussels: International Dairy Federation.

Marcellino, N., & Benson, D. R. (1992). Scanning electron and light microscopic study of microbial succession on Bethlehem St. Nectaire cheese. *Applied and Environmental Microbiology, 58*(11), 3448–3454. PMID:16348797

Marcellino, S. N. M., & Benson, D. R. (2013). The good, the bad, and the ugly: Tales of mold-ripened cheese. *Microbiology Spectrum, 1*(1), CM-0005–CM-0012. PMID:26184810

Martín, J. F., & Liras, P. (2017). Secondary Metabolites in Cheese Fungi. In J. M. Mérillon & K. G. Ramawat (Eds.), Fungal Metabolites (pp. 293–315). Cham: Springer International Publishing.

Martín, J. F., & Coton, M. (2017). Blue Cheese: Microbiota and Fungal Metabolites. In J. Frias, C. Martinez-Villaluenga, & E. Peñas (Eds.), *Fermented Foods in Health and Disease Prevention* (pp. 275–303). Boston: Academic Press. doi:10.1016/B978-0-12-802309-9.00012-1

Matsuoka, H., Fuke, Y., Kaminogawa, S., & Yamauchi, K. (1991). Purification and debittering effect of aminopeptidase II from *Penicillium caseicolum. Journal of Agricultural and Food Chemistry*, *39*(8), 1392–1395. doi:10.1021/jf00008a007

Menassa, A., & Lamberet, G. (1982). Contribution à l'étude du système lipolytique de *Penicillium roqueforti*. Caractères comparés de deux activités exocellulaires. *Le Lait*, *62*(611–612), 32–43. doi:10.1051/lait:1982611-6122

Modler, H. W., Brunner, J. R., & Stine, C. M. (1974a). Extracellular protease of *Penicillium roqueforti*. I. Production and characteristics of crude enzyme preparation. *Journal of Dairy Science*, *57*(5), 523–527. doi:10.3168/jds.S0022-0302(74)84927-1 PMID:4858149

Modler, H. W., Brunner, J. R., & Stine, C. M. (1974b). Extracellular protease of *Penicillium roqueforti*. II. Characterization of a purified enzyme preparation. *Journal of Dairy Science*, *57*(5), 528–534. doi:10.3168/jds.S0022-0302(74)84928-3 PMID:4835412

Montel, M.-C., Buchin, S., Mallet, A., Delbes-Paus, C., Vuitton, D. A., Desmasures, N., & Berthier, F. (2014). Traditional cheeses: Rich and diverse microbiota with associated benefits. *International Journal of Food Microbiology*, *177*, 136–154. doi:10.1016/j.ijfoodmicro.2014.02.019 PMID:24642348

Moreau, C. (1980). Le Penicillium roqueforti, morphologie, physiologie, intérêt en industrie fromagère, mycotoxines. (Révision bibliographique). *Le Lait*, *60*(595–596), 254–271. doi:10.1051/lait:1980595-59615

Nielsen, K. F., Dalsgaard, P. W., Smedsgaard, J., & Larsen, T. O. (2005). Andrastins A–D, P*enicillium roqueforti* metabolites consistently produced in blue-mold-ripened cheese. *Journal of Agricultural and Food Chemistry*, *53*(8), 2908–2913. doi:10.1021/jf047983u PMID:15826038

O'Callaghan, Y. C., O'Connor, T. P., & O'Brien, N. M. (2017). Nutritional Aspects of Cheese. In P. F. Fox, T. P. Guinee, T. M. Cogan, & P. L. H. McSweeney (Eds.), *Fundamentals of Cheese Science* (pp. 715–730). Boston, MA: Springer. doi:10.1007/978-1-4899-7681-9_20

Ohmomo, S., Oguma, K., Ohashi, T., & Abe, M. (1978). Isolation of a new indole alkaloid, roquefortine D, from the cultures of *Penicillium roqueforti. Agricultural and Biological Chemistry*, *42*(12), 2387–2389.

Ohmomo, S., Sato, T., Utagawa, T., & Abe, M. (1975). Isolation of festuclavine and three new indole alkaloids, roquefortine A, B and C from the cultures of *Penicillium roqueforti. Agricultural and Biological Chemistry*, *39*(6), 1333–1334.

Pitt, J. I., Cruickshank, R. H., & Leistner, L. (1986). *Penicillium commune, P. camembertii*, the origin of white cheese moulds, and the production of cyclopiazonic acid. *Food Microbiology*, *3*(4), 363–371. doi:10.1016/0740-0020(86)90022-5

Pitt, J. I., & Hocking, A. D. (2009). Zygomycetes. In *Fungi and Food Spoilage* (pp. 145–168). Boston, MA: Springer. doi:10.1007/978-0-387-92207-2_6

Pitt, J. I., & Hocking, A. D. (2009a). *Penicillium* and Related Genera. In J. I. Pitt & A. D. Hocking (Eds.), *Fungi and Food Spoilage* (pp. 169–273). Boston, MA: Springer. doi:10.1007/978-0-387-92207-2_7

Pitt, J. I., & Hocking, A. D. (2009b). Primary Keys and Miscellaneous Fungi. In J. I. Pitt & A. D. Hocking (Eds.), *Fungi and Food Spoilage* (pp. 53–143). Boston, MA: Springer. doi:10.1007/978-0-387-92207-2_5

Ratomahenina, R., Chabalier, C., & Galzy, P. (1994). Concerning *Sporendonema casei* Desmazieres. *Latte*, *19*(6), 616–617.

Ratomahenina, R., Chabalier, C., Galzy, P., & Dieu, B. (1995). Study of *Chrysosporium sulfureum*, the mould responsible for Fleur jaune on Saint-Nectaire cheese. *Milchwissenschaft. Milk Science International*, *50*(5), 266–267.

Ratomahenina, R., van den Boom, S., Galzy, P., & Dieu, B. (1995). Study of growth parameters of *Cylindrocarpon* sp., a mould isolated from Saint-Nectaire cheese. *Chemie, Mikrobiologie, Technologie. Der Lebensmittel*, *17*, 169–171.

Ropars, J., Cruaud, C., Lacoste, S., & Dupont, J. (2012). A taxonomic and ecological overview of cheese fungi. *International Journal of Food Microbiology*, *155*(3), 199–210. doi:10.1016/j.ijfoodmicro.2012.02.005 PMID:22381457

Ropars, J., López-Villavicencio, M., Dupont, J., Snirc, A., Gillot, G., Coton, M., ... Giraud, T. (2014). Induction of sexual reproduction and genetic diversity in the cheese fungus *Penicillium roqueforti*. *Evolutionary Applications*, *7*(4), 433–441. doi:10.1111/eva.12140 PMID:24822078

Ropars, J., Rodríguez de la Vega, R. C., López-Villavicencio, M., Gouzy, J., Sallet, E., Dumas, É., ... Giraud, T. (2015). Adaptive horizontal gene transfers between multiple cheese-associated fungi. *Current Biology*, *25*(19), 2562–2569. doi:10.1016/j.cub.2015.08.025 PMID:26412136

Salers. (2015, February 20). Retrieved from https://commons.wikimedia.org/wiki/File:Salers_(fromage)_01.jpg

Samson, R. A., Eckardt, C., & Orth, R. (1977). The taxonomy of *Penicillium* species from fermented cheeses. *Antonie van Leeuwenhoek*, *43*(3), 341–350. doi:10.1007/BF02313761 PMID:603235

Schornsteiner, E., Mann, E., Bereuter, O., Wagner, M., & Schmitz-Esser, S. (2014). Cultivation-independent analysis of microbial communities on Austrian raw milk hard cheese rinds. *International Journal of Food Microbiology*, *180*, 88–97. doi:10.1016/j.ijfoodmicro.2014.04.010 PMID:24794620

Scott, P., & Kanhere, S. (1979). Instability of PR toxin in blue cheese. *Journal - Association of Official Analytical Chemists*, *62*(1), 141–147. PMID:422496

Scott, P. M., & Kennedy, B. P. C. (1976). Analysis of blue cheese for roquefortine and other alkaloids from *Penicillium roqueforti*. *Journal of Agricultural and Food Chemistry*, *24*(4), 865–868. doi:10.1021/jf60206a028 PMID:956545

Sipahioglu, H. N., & Heperkan, D. (2000). Lipolytic activity of *Trichothecium roseum* on hazelnut. *Food Microbiology*, *17*(4), 401–405. doi:10.1006/fmic.2000.0328

Snow, D. (1949). The Germination of mould spores at controlled humidities. *Annals of Applied Biology*, *36*(1), 1–13. doi:10.1111/j.1744-7348.1949.tb06395.x PMID:18116920

Spinnler, H. E., & Gripon, J. C. (2004). Surface mould-ripened cheeses. In P. F. Fox, P. L. H. McSweeney, T. M. Cogan, & T. P. Guinee (Eds.), *Cheese: Chemistry, Physics and Microbiology* (Vol. 2, pp. 157–174). Amsterdam: Elsevier Academic Press.

Stark, J. (2007). Cheese and fermented sausages. In J. Dijksterhuis & R. A. Samson (Eds.), *Food Mycology* (pp. 319–331). Boca Raton, FL: CRC Press.

Stepaniak, L., Kornacki, K., Grabska, J., Rymaszewski, J., & Cichosz, G. (1980). Lipolytic and proteolytic activity of *Penicillium roqueforti, Penicillium candidum* and *Penicillium camemberti* strains. *Acta Alimentaria Polonica, 6*(3), 155–165.

Stilton cheese made by Cropwell Bishop sitting on waxed paper. (2013, October 11). Retrieved from https://commons.wikimedia.org/wiki/File:Stilton_Cheese_02.png

Sumarah, M. W., Miller, J. D., & Blackwell, B. A. (2005). Isolation and metabolite production by *Penicillium roqueforti, P. paneum* and *P. crustosum* isolated in Canada. *Mycopathologia, 159*(4), 571–577. doi:10.100711046-005-5257-7 PMID:15983744

Summerbell, R., Gueidan, C., Schroers, H.-J., de Hoog, G., Starink, M., Rosete, Y. A., ... Scott, J. (2011). *Acremonium* phylogenetic overview and revision of *Gliomastix, Sarocladium*, and *Trichothecium*. *Studies in Mycology, 68*, 139–162. doi:10.3114im.2011.68.06 PMID:21523192

Takó, M., Kotogán, A., Krisch, J., Vágvölgyi, C., Mondal, K. C., & Papp, T. (2015). Enhanced production of industrial enzymes in *Mucoromycotina* fungi during solid-state fermentation of agricultural wastes/by-products. *Acta Biologica Hungarica, 66*(3), 348–360. doi:10.1556/018.66.2015.3.10 PMID:26344030

Tang, Y., Xue, H., Bi, Y., Li, Y., Wang, Y., Zhao, Y., & Shen, K. (2015). A method of analysis for T-2 toxin and neosolaniol by UPLC-MS/MS in apple fruit inoculated with *Trichothecium roseum*. *Food Additives & Contaminants: Part A, 32*(4), 480–487. doi:10.1080/19440049.2014.968884 PMID:25254921

Tomme de Savoie. (2015, June 19). Retrieved from https://commons.wikimedia.org/wiki/File:WikiCheese_-_Tomme_de_Savoie_-_20150619_-_002.jpg

Werner, H., Nielsen, E. W., Ardö, Y., Rage, A., & Antila, V. (1999). North European varieties of cheese. In P. F. Fox (Ed.), *Cheese: Chemistry, Physics and Microbiology* (pp. 245–262). Boston, MA: Springer. doi:10.1007/978-1-4615-2800-5_8

Yegin, S., Fernandez-Lahore, M., Jose Gama Salgado, A., Guvenc, U., Goksungur, Y., & Tari, C. (2011). Aspartic proteinases from *Mucor* spp. in cheese manufacturing. *Applied Microbiology and Biotechnology, 89*(4), 949–960. doi:10.100700253-010-3020-6 PMID:21127856

Yeşil Peynirin Faydaları- Erzurum. (2016, October 23). Retrieved from http://www.erzurumsayfasi.com/haber/yesil-peynirin-faydalari-erzurum-h1089.html

Yoder, J. A., Benoit, J. B., & Zettler, L. W. (2003). Effects of salt and temperature on the growth rate of a tick-associated fungus, *Scopulariopsis brevicaulis* Bainier (Deuteromycota). *International Journal of Acarology, 29*(3), 265–269. doi:10.1080/01647950308684338

Zhang, N., & Zhao, X.-H. (2010). Study of *Mucor* spp. in semi-hard cheese ripening. *Journal of Food Science and Technology, 47*(6), 613–619. doi:10.100713197-010-0108-z PMID:23572694

Chapter 15
Potential Probiotic Microorganisms in Kefir

H. Ceren Akal
Ankara University, Turkey

Şebnem Öztürkoğlu Budak
Ankara University, Turkey

Atila Yetisemiyen
Ankara University, Turkey

ABSTRACT

Probiotic microorganisms are defined as living microorganisms that provide health benefits on the host when administered in adequate amounts. The benefits include improvement of microbial balance immune system and oral health, provision of cholesterol-lowering effect, and antimicrobial activity against a wide variety of bacteria and some fungi. Kefir microbiota contains active living microorganisms. Many researches were carried out that potential probiotic bacteria such as Lactobacillus acidophilus, Lactobacillus plantarum, Lactobacillus kefir, Lactobacillus kefiranofaciens, Leuconostoc mesenteroides, *or yeasts like microorganisms such as* Saccharomyces cerevisiae, Kluyveromyces lactis, *and* Kluyveromyces marxianus *were isolated from kefir grains. This chapter presents the data both on the probiotic bacteria isolated from kefir grains or kefir and the probiotic properties of kefir produced with these microorganisms.*

INTRODUCTION

Kefir is a traditional fermented dairy beverage which was originated in the Caucasus (Kim, Chon, Kim, & Seo, 2015). It has a slightly acidic flavor and a creamy consistency that consumed all over the world but most commonly in Turkey. The world "kefir" originates from a Turkish world "keyif" which means good feeling and pleasure (Kabak & Dobson, 2011). Kefir can be produced in two ways; by using kefir grains or Direct Vat Set (DVS) cultures. Traditional kefir product is produced by kefir grains. Kefir grains are a cluster of microorganisms with yellowish-white colour and cauliflower-like shape. Microbiota of kefir

DOI: 10.4018/978-1-5225-5363-2.ch015

grains is very complex which composed of lactic acid bacteria, acetic acid bacteria and yeasts (Pintado, Da Silva, Fernandes, Malcata, & Hogg, 1996). The microbiota of kefir grains strongly depends on the origin of the grains but not on the microbial composition of grains (Witthuhn, Schoema, & Britz, 2005).

Kefir is thought as a probiotic product because kefir grains provide potential probiotic microorganisms to kefir. Qualification of a microorganism as a probiotic mainly depends on determination of some properties they should possess. Some of these properties are; resistance to bile and gastric acids, ability to survive under the gastrointestinal conditions, exhibition of proven beneficial effects on human health such as showing antipathogenic or antiviral effects and enhancement of immune system.

In this chapter, the definition of probiotics that isolated from kefir and kefir grains were given and their properties were presented in detail.

PROBIOTIC MICROORGANISMS

Probiotic microorganisms are live microorganisms that provide a benefit on the human health. The term "probiotic" means "for life" in Greek and first used in 1965 to explain secretions of a microorganism that stimulate the growth of other microorganism by Lilly and Stillwell (Lee, 2009). In 1974, it was understood that there was a relationship between probiotics and host system. In addition, probiotics are described as "organisms and substances with beneficial effects for animals by influencing the intestinal microbiota". Finally, probiotics are defined as "living microorganisms which upon ingested in adequate amounts provide health benefits beyond inherent general nutrition" by FAO/WHO (FAO/WHO, 2006).

Fermented dairy products are the most common foods containing probiotics. Among them yoghurt is the leading (Brown & Valiere, 2004). In 1907, it was first stated that bacteria in the intestines had a positive effect on human health. *Lactobacillus delbrueckii* subsp. *bulgaricus* which is a yoghurt bacteria, was determined to be responsible for a longer life and better health conditions (Brown & Valiere, 2004). In 1935, positive effect of *Lactobacillus acidophilus* on human health was shown by Retteger, Levy, Weinstein, & Weiss (1935). After that date, many microorganisms such as *Bifidobacterium*, *Streptococcus*, *Enterococcus* and *Lactococcus* have been identified as probiotic by different researches (Goldin, 2011).

In order for a microorganism to be evaluated as a probiotic, it must have certain criteria:

- Having a proven beneficial effect on human health,
- Being non-pathogenic and non-toxic,
- Maintaining the stability in the product,
- Containing an appropriate number of viable cells,
- Surviving in the gastrointestinal system,
- Having acceptable sensory properties.

HEALTH EFFECTS OF PROBIOTIC MICROORGANISMS

Many studies were performed about the health effects of probiotic microorganisms. Some of these effects are listed below:

Prevention of Diarrhea

Diarrhea is defined as an increase in liquidity or decrease in consistency of stools. This disease causes excessive fluid and electrolyte losses and results approximately 4% of all deaths and 5% of days lost in disability all over the world (Lee & Salminen, 2009). Thus the prevention of this disease is significant with regard to the maintaining of public health.

Several studies focused on prevention of different types of diarrhea by means of consuming foods that containing probiotics which are detailed below.

Antibiotic-Associated Diarrhea

Antimicrobial treatment could adversely affect the microbiota of gastrointestinal system which may cause some disorders particularly diarrhea in human body. The effects of *Lactobacillus rhamnosus, Bifidobacterium lactis, Lactobacillus reuterii, Saccharomyces boulardii* and *Lactobacillus acidophilus* or mixture of some probiotics on prevention of antibiotic-associated diarrhea and shortening the diarrhea time are reported in different studies (Arvola et al., 1999; Pant et al., 1996; de Vrese & Marteau, 2007; Hempel et al., 2012).

Acute Diarrhea

Acute diarrhea is a pediatric disease. Rotaviruses are known as the most common infectious agents associated with acute diarrhea. Rotavirus is responsible for approximately 440 000 diarrheal deaths every year (Teran, Teran-Escalera, & Villarroel, 2009). Probiotic microorganisms such as *Lactabacillus reuteri, Saccharomyces boulardii, Lactobacillus rhamnosus* were shown to have beneficial effects on acute diarrhea (Oberhelman et al., 1999; Allen, Okoko, Martinez, Gregorio, & Dans, 2003; Goldin, 2011).

Traveler's Diarrhea

International travelers could be affected by a gastrointestinal disease called "Traveler's Diarrhea". *Escherichia coli, Campylobacter, Salmonella, Bacillus cereus, Shigella* and *Aeromonas* are well known bacteria which cause traveler's diarrhea. It was reported that rotaviruses and parasites like *Entamoeba histolytica, Giardia lamblia* could also cause traveler's diarrhea (de Bruyn, Hahn, & Borwick, 2000). Traveler's diarrhea was shown that it could be prevented by some probiotic bacteria such as *Lactobacillus acidophilus, Lactobacillus fermentum, Lactobacillus delbrueckii* subsp. *bulgaricus* (Briand et al., 2006; Samaržija et al., 2009). In addition, according to the results of meta-analysis that performed during the 12 random studies, the mixture of *Lactobacillus acidophilus, Bifidobacterium bifidum* and *Saccharomyces boulardii* has a high effect on prevention of traveler's diarrhea (McFarland, 2007).

Effect on Lactose Intolerance

Millions of people have lactose malabsorption problem all over the world. This health problem originated from inadequate activity of the β-galactosidase enzyme and the dis-assimilation of lactose cause abdominal cramps, nausea, bloating and diarrheas. Consumption of yoghurt is known as not have a negative effect on lactose intolerants because of the activity of *Lactobacillus delbrueckii* subsp. *bulgaricus* and *Streptococcus thermophilus*. The positive effect of probiotics can be explained by the production of lactase by these bacteria or penetration of active lactase into the intestine with probiotics (Yavuzdurmaz, 2007; Seth & Maulik, 2011). Other probiotic microorganisms could also limited the negative effects of consumption of dairy products depending on their strains or species (Goldin, 2011).

Cholesterol Lowering Effect

Reducing low density lipoprotein (LDL) cholesterol is very important for human health due to decreasing coronary artery disease. There are many studies on cholesterol lowering effect of probiotics. These studies demonstrated that consumption of fermented dairy products or yoghurt with *Lactobacillus acidophilus* and *Bifidobacterium animalis* subsp. *lactis* reduce the LDL cholesterol (Liong & Shah, 2005; Ataie-Jafari, Larijani, Majd, & Tahbaz, 2009). Binding of cholesterol into the cell membrane or increase the breakdown of cholesterol by probiotics could be the explanations of reduction of cholesterol with these microorganisms (Yavuzdurmaz, 2007).

Improvement of Immune System

The microbiota of intestines is normally stable but when a negative effect occurs the composition could be change. Some physiological stressors have an adverse effect on gut microbiota and make it susceptible to pathogens (Manuel, Elena, Carolina, & Gabriela, 2017). Probiotic microorganisms support immune system against pathogens. Perdigon, Alvarez, de Macias, Roux, & de Ruiz Holgado (1990) demonstrated that mice which were fed with *Lactobacillus casei* and *Streptococcus thermophilus* had shown a resistance to the growth of *Salmonella thyphimurium*. However, feeding mice with *Lactobacillus acidophilus* had not defended the mice against *Salmonella thyphimurium* infection. In a similar study, it was shown that feeding mice with *Lactobacilli* sp. or yoghurt as itself increased the concentration of Immunoglobulin A in the body (Scheinbach, 1998).

Manuel et al. (2017) demonstrated that usage of probiotic supplements increased the systematic immune response and provided protection against *Salmonella* infection. Due to the improvement of immune system probiotic microorganisms mainly *Lactobacillus acidophilus* might cause milk allergy in children (Özdemir, 2010).

Anticarcinogenic Effect

Cancer is described as uncontrolled proliferation of cells. Probiotics may decrease the risk of many cancer types particularly colon cancer by preventing the intestinal activity of bacterial enzymes and thus by producing short-chain fatty acids which can be indicated as protective agents (Goldin, 2011).

Lactobacillus acidophilus was reported to decrease the urinary mutagenic activity significantly (Lidbeck, Övervik, Rafter, Nord, & Gustafsson, 1992). Similarly, fermented milk or colostrum with *Lactobacillus acidophilus* resulted in reduction in proliferation of tumor (Shahani & Ayebo, 1980). *Lactobacillus* species effect the prevention of cancer by the production of antioxidants and anti-angiogenic factors; in addition have a positive role in decreasing the DNA damage, tumor size and pro-carcinogenic enzymes (Dasari, Kathera, Janardhan, Kumar, & Viswanath, 2016). Many other studies also indicated that consumption of probiotic bacteria could change the gut microbiota and decrease the development of tumors (Hirayama & Rafter, 2000; dos Reis et al., 2017).

Other Effects

There are some effects of probiotics on human health other than the mentioned ones before. These are;

- Prevention of gastritis and ulcer through the inhibition effect on the activity of *Helicobacter pylori* (Çakır, 2003),
- Prevention of uregenital infection by inhibition effect on the of activity of *Candida albicans* (Juarez-Tomas, Duhart, De Gregorio, Pingitore, & Nader-Macias, 2011; Tachedjian, Aldunate, Bradshaw, & Cone, 2017),
- Increase the synthesis of vitamin B (Shahani & Ayebo, 1980),
- Prevention of dental caries by the way of antimicrobial activity against *Streptococcus* spp. (Goldin, 2011),
- Reducing allergic reactions such as atopic eczema (Kalliomäki, Salminen, Poussa, Arvilommi, & Isolauri, 2003; Kukkonen et al., 2007).

PROBIOTIC MICROORGANISMS IN KEFIR

Kefir is produced by the biochemical activities of microorganisms present in kefir grains (Leite et al., 2012; Dertli & Çon, 2017). Kefir grains are composed of lactic acid bacteria (LAB) (10^8 cfu/g), acetic acid bacteria(10^5 cfu/g) and some yeast species such as *Torula* and *Saccharomyces* (10^6-10^7 cfu/g) (Prado et al., 2015). *Lactobacillus* species are dominant in the count of total LAB and *Lactobacillus kefiri* represent 80% of the *Lactobacillus* species. Other species are composed of *Lactobacillus acidophilus*, *Lactobacillus delbrueckii* subsp. *bulgaricus*, *Lactobacillus plantarum* and *Lactobacillus kefiranofaciens*. These species are in a small ratio but they are counted as pre-dominant species (Zanirati et al., 2015; Prado et al., 2015).

Lactobacillus Species

In many studies that are carried out up to nowadays reported there are a wide variety of *Lactobacillus* species isolated from different kefir grains. These species are given in Table 1.

Lactobacillus Kefiri

L. kefiri that isolated from kefir grains is a dominant LAB in kefir microbiota. Likotrafiti, Valavani, Argiriou, & Rhoades (2015) found that *L. kefiri* has a potential to be a probiotic because of its resistance to bile and anti-pathogenic activity. Golowczyc, Mobili, Garrote, Abraham, & de Antoni (2007) reported that *L. kefiri* showed a protective action against *Salmonella enteritica*. Similarly, Carasi et al. (2014) demonstrated an inhibition effect of *L. kefiri* on Gram positive bacteria such as *Listeria monocytogenes, Bacillus cereus, Staphylococcus aureus* and Gram negative pathogens such as *Eschericia coli, Pseudomonas aeroginosa, Salmonella enteridis, Shigella aflexneri*.

Bacterial surface layers are crystalline arrays of proteinaceous subunits. *L. kefiri* has surface layer (S-layer) proteins which are composed of protein subunits that are non-covalently linked to each other to encourage the cell wall (Golowczyc et al., 2007). It is also thought that these proteins play an important

Table 1. Lactobacillus species isolated from kefir grain

Lactobacillus **Species**	**References**
Lactobacillus kefiri	Chen, Wang, & Chen, 2008
	Vardjan, Lorbeg, Rogelj, & Majheniz, 2013
	Zanirati et al., 2015
	Zhou, Xiaoli, Hanhu, & Mingsheng, 2009
Lactobacillus kefiranofaciens subsp. *kefiranofaciens*	Chen et al., 2008
Lactobacillus kefiranofaciens ssp. *kefirgranum*	Vardjan et al.,2013
Lactobacillus acidophilus	Chen et al., 2008
Lactobacillus plantarum	Chen et al., 2008
Lactobacillus delbrueckii subsp. *bulgaricus*	Chen et al., 2008
Lactobacillus delbrueckii subsp. *delbrueckii*	Chen et al., 2008
Lactobacillus delbrueckii subsp. *lactis*	Chen et al., 2008
Lactobacillus rhamnosus	Chen et al., 2008
Lactobacillus brevis	Chen et al., 2008
	Zanirati et al., 2015
Lactobacillus casei	Chen et al., 2008
Lactobacillus sake	Chen et al., 2008
Lactobacillus helveticus	Chen et al., 2008
Lactobacillus paracasei subsp. *paracasei*	Chen et al., 2008
Lactobacillus parakefiri	Vardjan et al.,2013
Lactobacillus satsumensis	Zanirati et al.,2015
Lactobacillus mali	Zanirati et al.,2015

role in the probiotic activity. In a different study performed by Carasi, Trejo, Perez, de Antoni, & Serradell (2012), antagonistic effect of S-layer proteins that were separated from *L. kefiri* on *Clostiridium difficile* toxins (TcdA and TcdB) was investigated. *L. kefiri* cells did not show any protective effect against these toxins. Authors determined it could be due to the difference between the interaction of S-layer proteins and clostridial toxins and the interaction of S-layer proteins and bacteria.

All studies about *L. kefiri* defined that it has a potential to be used as probiotic for human consumption and also can be evaluated as an ingredient during the production of functional products.

Lactobacillus Kefiranofaciens

L. kefiranofaciens is responsible for the production of kefiran. Kefiran is a water soluble polysaccharide which present on the surface of kefir grains. It is composed of 24% of the solid content of the kefir grains (Fujisawa, Adachi, Toba, Arihara, & Mitsuoka, 1988). This exopolysaccharide (EPS) is known as having bioactive functions as immunostimulatory, anti-tumor and cholesterol lowering effect (Vinderola, Perdigon, Duarte, Farnworth, & Matar, 2006; Kim et al., 2015; Jeong et al., 2017). EPS produced by *L. kefiranofaciens* ZW3 isolated from Tibet kefir was reported to have a good emulsifying capability (Wang, Ahmed, Feng, Li, & Song, 2008). In a different study, it was found that the same EPS type had a good solubility and binding capacity in water and oil (Ahmed, Wang, Anjum, Ahmad, & Khan, 2013). Moreover, EPS could play a role in prevention of the dehydration and adhesion to surface (Chersilp, Shoji, Shimizu, & Shioya, 2003).

Jeong et al. (2017) studied the antibacterial effect of *L. kefiranofaciens* on *Listeria monocytogenes* and *Salmonella enteridis*. *L. kefiranofaciens* EPS-DN1 was found to exhibit a bactericidal effect on these pathogens. Other than its antibacterial effect, anti-colitis effect of this strain was also determined (Chen, Hsiao, Hong, Dai, & Chen, 2012). Chen, Lee, Hong, Hsieh, & Chen (2013) also demonstrated that *L. kefiranofaciens* had an antimicrobial effect on *Escherichia coli* as one of a crucial pathogen.

Lactobacillus Acidophilus

L. acidophilus is a Gram positive and rod shaped bacteria with rounded ends. It can synthesize antimicrobial compounds such as bacteriocin, hydrogen peroxide and lactic acid. It was reported to show anti-colitis effect *in vivo* models (Selle, Klaenhammer, & Russell, 2014). This microorganism was first isolated from infant feces and approved as therapeutic for a long time. It has a potential to prevent vaginal disorders, some enteric infections, dermatitis and diarrhea (Neish, 2017).

Probiotic properties of *L. acidophilus* such as bile tolerance, acid resistance, cholesterol assimilation activities were studied on mice by Song et al. (2015). It was determined that this bacterium could be accepted as probiotic based on its LDL cholesterol lowering effect. In another study, Lee et al. (2010) investigated the factors affecting the serum cholesterol reduction *by L. acidophilus*. With this aim, wild type and BA9 mutant type of *L. acidophilus* were analyzed and it was found that reduction rate of serum cholesterol by BA9 mutant type of *L. acidophilus* was lower than the wild type strain.

The effect of *L. acidophilus* and other three probiotic microorganisms (*Bifidobacterium bifidum*, *Lactobacillus casei* and *Lactabacillus salivarius*) on atopic dermatitis of pediatric patients was studied by Yeşilova, Çalka, Akdeniz, & Berktaş (2012). It was determined that probiotic treatment caused a decrease in clinical symptoms and a reduction of immunoglobulin E in serum. In addition, the anti-pathogenic effect of *L. acidophilus* on *Clostridium difficile* which causes antibiotic related diarrhea and colitis was also reported by Yun, Oh, & Griffiths (2014).

EPS produced by different strains of *L. acidophilus* was also determined to have beneficial effects as anti-tumor and hypocholesterolemia (London et al., 2016). In addition, released EPS (r-EPS) produced by *L. acidophilus* A4 contributed to the prevention of biofilm formation of *E. coli* O157:H7 (Kim, Oh, & Kim, 2009). Kim, Oh, Yun, Oh, & Kim (2010) also showed EPS separated from *L. acidophilus* 606 promoted the death of colon cancer cells *in vitro*.

Similar to *L. kefiri*, *L. acidophilus* strains have surface layer associated protein that demonstrates immunomodulatory effect (Johnson, Selle, O'Flaherty, Goh, & Klaenhammer, 2013). Meng, Gao, Zhang, & Lu (2015) determined that the murein hydrolase activity of surface layer proteins from different subspecies of *L. acidophilus* inhibited the *E. coli*.

Lactobacillus Plantarum

L. plantarum is a typical LAB to be used in silage and play an important role on the fermentation of many plant products (Ashenafi & Busse, 1991). Besides, L. *plantarum* has known as probiotic for decades.

Cebeci & Gürakan (2003) investigated the probiotic properties of thirteen different *L. plantarum* strains. In this scope, tolerance of the strains to bile salts and acids, susceptibility to antibiotics, and β-galactosidase activities of these strains were determined. As a result, *L. plantarum* 37, *L. plantarum* 80, *L. plantarum* NCIMB 1193 and *L. plantarum* HU were decided as the most promising strains.

In a similar research, Ramos, Thorsen, Schwan, & Jespersen (2013) studied on different *L. plantarum* strains (additionally *L. brevis* and *L. fermentum*) to determine their probiotic properties. Their ability to grow at low pH values, tolerance to bile, co-aggregation capability with *E. coli*, and adhesion ability to CaCo-2 cells were analyzed. In conclusion, *L. plantarum* CH3 and *L. plantarum* CH41 were described as the potential probiotic strains. Zago et al. (2011) also investigated the probiotic strains of *L. plantarum* isolated from cheese and found that three strains of *L. plantarum* showed the best probiotic properties.

The protective effect of *L. plantarum* species was proven by many authors. Sunanliganon, Thong-Ngam, Tumwasorn, & Klaikeaw (2012) showed the protective effect of *L. plantarum* B7 against *Helicobacter pylori* in rats. Zhao et al. (2017) also defined the protective effect of *L. plantarum* 88 against alcoholic liver injury. The protective effect of *L. plantarum* might derived from "plantaricin" which is a bacteriocin produced by this strain. Plantaricin is resistant to environmental conditions and prevent the growth of some food spoilage bacteria such as *E. faecalis* and *C. tyrobutyricum* (Gonzalez, Arca, Mayo, & Suarez, 1994). Additionally, prevention of Th2-releated immune disorders with *L. plantarum* was investigated by Song, Lee, Park, Oh, & Lim (2016).

Lactobacillus Delbrueckii and Streptococcus Thermophilus

L. delbrueckii subsp. *bulgaricus*, *L. delbrueckii* subsp. *delbrueckii* and *L. delbrueckii* subsp. *lactis* were the other LAB species frequently isolated from kefir. *L. delbrueckii* subsp. *bulgaricus* cells are rod shaped with rounded ends. These microorganisms play an important role in the fermentation of foods mainly in dairy products. *L. delbrueckii* subsp. *bulgaricus* has some critical properties such as proteolytic activity, EPS and lactic acid production (Teixeira, 2014).

There are a few studies on survival of *L. delbrueckii* subsp. *bulgaricus* during gastrointestinal transit in humans (Marteu, Minekus, Havenaar, & Veld, 1997). However, in recent years the probiotic properties of this bacterium were decided as sensible to bile and gastric acids (Teixeira, 2014).

Contrary to *L. delbrueckii, Streptococcus thermophilus* is a cocci shaped bacteria and classified as thermophile that is growing at 45°C. This species is sensitive to salts and high osmotic strength media. It is commonly used as starter in production of yoghurt and different cheese types (Mullan, 2014).

Kanamarlapudi and Muddada (2017) investigated the characterization of EPS produced by *Streptococcus thermophilus* CC30 and Tang et al. (2017) researched on EPS produced by *L. delbrueckii* subsp. *bulgaricus*. Both studies indicated that EPS isolated from yoghurt bacteria presented antioxidant activities.

Other Lactobacilli

- **Lactobacillus Rhamnosus:** A lot of studies were performed on probiotic properties of L. rhamnosus. One of the properties that have to be presented by probiotic bacteria is its survival in the gastrointestinal tract. Pitino et al. (2010) examined six different strains of L. rhamnosus for this aim and determined a sufficient survival rate during gastric and duodenal digestion. Similarly, Rolim et al. (2015) showed that L. rhamnosus EM1107 could survive during the artificial digestion and have an inhibition effect on Listeria monocytogenes and Staphylococcus aureus. Nase et al. (2001) also determined that L. rhamnosus GG have a beneficial effect on children's dental health. The progress of immunity by L. rhamnosus HN001 was proved by Gill, Rutherfurd, Prasad, & Gopal (2000) and the antimicrobial effect of L. rhamnosus DR20 on E. coli strains by Gopal, Prasad, Smart, & Gill (2001).

- **Lactobacillus Brevis:** This species is Gram positive and a heterofermentative bacterium. It is approved to have probiotic properties and accepted as a potential probiotic microorganism due to the effects such as enhancement of immune functions (Rönka et al. 2003).
- **Lactobacillus Casei:** L. casei is a Gram positive, catalase negative, and a rod shaped bacteria. This species supports the enhancement of immune system by stimulating the growth of beneficial bacteria in the intestinal tract. Mishra and Prasad (2005) indicated that L. casei showed the ability to survive at low pH values, to tolerate bile salts and to reduce the cholesterol level depending on the strains. The antifungal effect of L. casei AST18 (Li et al. 2014) and EPS production of L. casei LC2W (Li et al. 2015) were determined in different studies.
- **Lactobacillus Sake:** Importance of this bacterium stands on the production of "lactocin", which is a bacteriocin that effect on L. monocytogenes, as a pathogen bacteria (Martinis & Franco, 1998).

Leuconostoc Mesenteroides Species

The main *Leuconostoc* species isolated from kefir grains up to now are *Leuconostoc mesenteroides* subsp. *mesenteroides*, *Leuconostoc mesenteroides* subsp. *dextranicum* and *Leuconostoc mesenteroides* subsp. *cremoris* (Chen et al., 2008). Some isolated species in literature are given in detail in Table 2.

Leuconostoc mesenteroides species are Gram positive and non-spore forming bacteria. They can be grown under aerobic or anaerobic conditions. *Leuconostoc mesenteroides* subsp. *mesenteroides* is the most frequently isolated bacterium on plants and fruits. *Leuconostoc mesenteroides* subsp. *mesenteroides* and *Leuconostoc mesenteroides* subsp. *dextranicum* could synthesize the dextran from sucrose.

D'Angelo et al. (2017) searched the behavior of 29 different strains of *Leuconostoc* under food stress conditions and determined that all strains survive under these conditions. EPS produced by *Leuconostoc mesenteroides* subsp. *mesenteroides* NTM048 has immunostimulant effect in regards to improvement of immunoglobulin A production (Matsuzaki et al., 2017). In addition, *Leuconostoc mesenteroides* was known to produce a bacteriocin referred as "mesentericin". Mesentericin produced by this bacterium was found adversely effect on the growth of *L. monocytogenes* (Hechard, Derijard, Letellier, & Cenatiempo, 1992). Moreover, an antiviral effect of *Leuconostoc mesenteroides* YML003 strain on H9N2 viruses of chickens was also identified.

Table 2. Leuconostoc mesenteroides species

Species	References
Leuconostoc mesenteroides subsp. *mesenteroides*	Chen et al., 2008
Leuconostoc mesenteroides subsp. *dextranicum*	Chen et al., 2008
Leuconostoc mesenteroides subsp. *cremoris*	Chen et al., 2008

Yeasts

The term "yeast" was derived from a Dutch word "gist" that means foam formed during fermentation of beer. Yeasts are single celled microorganisms which are important for human and animal nutrition.

Kefir grains mostly contain lactose fermenting yeasts such as *Kluyveromyces lactis*, *Kluyveromyces marxianus*, *Torula kefir* and non-lactose fermenting yeasts such as *Saccharomyces cerevisiae*.

Competitiveness for nutrient, ability to cell adhesion, production of antimicrobial components, cholesterol reduction and enhancement of immune system can be counted as some of the beneficial effects of probiotic yeasts (Moslehi-Jenabian, Pedersen, & Jespersen, 2010; de Lima et al., 2017).

The yeasts species reported to have isolated from kefir and kefir grains in literature are given in Table 3.

Saccharomyces Cerevisiae

S. cerevisiae is a well-known and the oldest industrial microorganism used as yeast culture during food productions. In a study performed on identifying the probiotic properties of *S. cerevisiae*, 12 strains of *S. cerevisiae* were tested on their survival under stimulated gastrointestinal conditions and the antimicrobial effects on *E. coli* and *C. difficile* of *S. cerevisiae* strains (Martins et al., 2005).

In another study, 18 strains of *S. cerevisiae* also determined to show probiotic properties such as resistance to low pH values and prevention of the activity of *E. coli* (van der AA Kühle, Skovgaard, & Jespersen, 2005). Similarly, 28 strains of *S. cerevisiae* exhibited antimicrobial effect on some pathogen microorganisms (*Klebsiella pneumoniae, Pseudomonas aeruginosa, Staphylococcus aureus, Bacillus subtilis)*, antioxidant activity and ability to produce β-galactosidase (de Lima et al., 2017). Etienne-Mesmin et al. (2011) also demonstrated the antibacterial effect of *S. cerevisiae* CNCMI 3856 strain on *E. coli* O:157: H7. In addition, Chiquette (2009) confirmed that *S. cerevisiae* stimulate the vitamin B and organic acid productions.

Table 3. Yeasts isolated from kefir and kefir grains

Species	References
Saccharomyces cerevisiae	Vardjan et al., 2013
	Zhou et al., 2009
Kluyveromyces lactis	Zhou et al., 2009
Kluyveromyces marxianus	Vardjan et al., 2013
	Zhou et al., 2009
Kluyveromyces siamensis	Zhou et al., 2009
Torulaspora delbrueckii	Vardjan et al., 2013
Kazachstania exigua	Vardjan et al., 2013
	Zhou et al., 2009
Kazachstania unispora	Vardjan et al., 2013
	Zhou et al., 2009

Kluyveromyces Species

Kluyveromyces genus belongs to Ascomycetes. They form endospore and mycelia without clamp connection. *K. lactis* and *K. marxianus* are the most well-known species of *Kluyveromyces* genus (Batt, 2014). *K. marxianus* is known as homothallic, hemiascomycetes yeast and related to *S. cerevisiae* (Llorente et al., 2000). *K. lactis* was reported to be a sister species of *K. marxianus* by Lane and Morrissy (2010).

K. lactis has informed to have many beneficial effects on human health (van Ooyen et al., 2006; Spohner, Schaum, Quitmann, & Czernak, 2016). Some of these effects are;

- Production of human interleukin 1-β that helps the treatment of immune system disorders (Maedler, Dharmadhikari, Schumann, & Størling, 2009),
- Production of recombined interferon-a that have anti-carcinogenic effect (Kumar et al., 2005),
- Production of β-lactoglobulin, some antibodies and insulin precursors (Robin et al., 2003),
- Efficient secretion of lysozyme that provides anti-inflammation and bactericide effect (Iwata et al., 2004).

To determine the probiotic potential of some *K. marxianus* and *K. lactis* strains, the antibacterial effects of these strains were studied on *Salmonella enterica* subsp. *enterica*. In addition, their ability to survive under the gastrointestinal environment were also studied. Results showed that both strains could live in vitro gastrointestinal system and inhibit *Salmonella* by downregulation of the sop*D* gene (Ceugniez et al., 2017).

Other than these yeasts, *Candida inconspicua*, *Candida maris* are responsible for the production of ethanol and carbon dioxide from lactose during fermentation (Simova et al., 2002).

FUTURE RESEARCH DIRECTIONS

Although there are many studies about probiotic bacteria and their properties, data on probiotic properties of microorganisms that are available in Kefir is inadequate. Probiotic specifications of special strains obtained from kefir should be examined.

CONCLUSION

Kefir, a traditional dairy beverage, is very important for human health due to the probiotic microorganisms it includes. It is recognized as a potential source of probiotics however the characterization of microbial species of kefir is thought lacking. Microbiota of each kefir varies according to its origin and more studies required for the evaluation the kefir microbiota.

REFERENCES

Ahmed, Z., Wang, Y., Anjum, N., Ahmad, A., & Khan, S. T. (2013). Characterization of exopolysaccharide produced by *Lactobacillus kefiranofaciens* ZW3 isolated from Tibet kefir e Part II. *Food Hydrocolloids, 30*(1), 343–350. doi:10.1016/j.foodhyd.2012.06.009

Allen, S. J., Okoko, B., Martinez, E. G., Gregorio, G. V., & Dans, L. F. (2003). *Probiotics for treating infectious diarrhoea (Review)*. Wiley & Sons.

Arvola, T., Laiho, K., Torkkeli, S., Mykkänen, H., Salminen, S., Maunula, L., & Isolauri, E. (1999). Prophylactic *Lactobacillus* GG Reduces Antibiotic-Associated Diarrhea in Children With Respiratory Infections: A Randomized Study. *Pediatrics, 104*(5), 64–68. doi:10.1542/peds.104.5.e64 PMID:10545590

Ashenafi, M., & Busse, M. (1991). Growth Potential of *Salmonella infantis* and *Escherichia coli* in Fermenting Tempeh Made from Horsebean, Pea and Chickpea and their Inhibition by *Lactobacillus plantarum*. *Journal of the Science of Food and Agriculture, 55*(4), 607–615. doi:10.1002/jsfa.2740550412

Ataie-Jafari, A., Larijani, B., Majd, H. A., & Tahbaz, F. (2009). Cholesterol-lowering effect of probiotic yogurt in comparison with ordinary yogurt in mildly to moderately hypercholesterolemic subjects. *Annals of Nutrition & Metabolism, 54*(1), 22–27. doi:10.1159/000203284 PMID:19229114

Batt, C. A. (2014). Kluyveromyces. In C. A. Batt & M. L. Tortorello (Eds.), *Encyclopedia of Food Microbiology* (vol. 2, pp. 389–391). Elsevier. doi:10.1016/B978-0-12-384730-0.00173-7

Briand, V., Buffet, P., Genty, S., Lacombe, K., Godineau, N., Salomon, J., ... Bouchaud, O. (2006). Absence of Efficacy of Nonviable *Lactobacillus acidophilus* for the Prevention of Traveler's Diarrhea: A Randomized, Double-Blind, Controlled Study. *Clinical Infectious Diseases, 43*(9), 1170–1175. doi:10.1086/508178 PMID:17029137

Brown, A. C., & Valiere, A. (2004). Probiotics and medical nutrition therapy. *Nutrition in Clinical Care*, 56–68. PMID:15481739

Çakır, İ. (2003). Laktobasillus ve Bifidobakterlerde Bazı Probiyotik Özelliklerin Belirlenmesi [Determination of Some Probiotic Properties on *Lactobacilli* And *Bifidobacteria*] (PhD Thesis). Ankara University, Graduate School of Natural and Applied Sciences, Department of Food Engineering, Ankara, Turkey.

Carasi, P., Diaz, M., Racedo, S. M., De Antoni, G., Urdaci, M. C., & de los Angeles Serradell, M. (2014). Safety Characterization and Antimicrobial Properties of Kefir-Isolated *Lactobacillus kefiri*. *BioMed Research International, 2014*, 1–8. doi:10.1155/2014/208974 PMID:24955346

Carasi, P., Trejo, F. M., Perez, P. F., de Antoni, G. L., & Serradell, M. A. (2012). Surface proteins from *Lactobacillus kefir* antagonize in vitro cytotoxic effect of *Clostridium difficile* toxins. *Anaerobe, 18*(1), 135–142. doi:10.1016/j.anaerobe.2011.11.002 PMID:22126976

Cebeci, A., & Gürakan, C. (2003). Properties of potential probiotic *Lactobacillus plantarum* strains. *Food Microbiology, 20*(5), 511–518. doi:10.1016/S0740-0020(02)00174-0

Ceugniez, A., Coucheney, F., Jacques, P., Daube, G., Delcenserie, V., & Drider, D. (2017). Anti-Salmonella activity and probiotic trends of *Kluyveromyces marxianus* S-2-05 and *Kluyveromyces lactis* S-3-05 isolated from a French cheese, Tomme d'Orchies. *Research in Microbiology, 168*(6), 575–582. doi:10.1016/j.resmic.2017.03.004 PMID:28366839

Chen, H. C., Wang, S. Y., & Chen, M. J. (2008). Microbiological study of lactic acid bacteria in kefir grains by culture-dependent and culture-independent methods. *Food Microbiology, 25*(3), 492–501. doi:10.1016/j.fm.2008.01.003 PMID:18355674

Chen, Y. P., Hsiao, P. J., Hong, W. S., Dai, T. Y., & Chen, M. J. (2012). *Lactobacillus kefiranofaciens* M1 isolated from milk kefir grains ameliorates experimental colitis in vitro and in vivo. *Journal of Dairy Science, 95*(1), 63–74. doi:10.3168/jds.2011-4696 PMID:22192184

Chen, Y. P., Lee, T. Y., Hong, W. S., Hsieh, H. H., & Chen, M. J. (2013). Effects of *Lactobacillus kefiranofaciens* M1 isolated from kefir grains on enterohemorrhagic *Escherichia coli* infection using mouse and intestinal cell models. *Journal of Dairy Science, 96*(12), 7467–7477. doi:10.3168/jds.2013-7015 PMID:24140320

Chersilp, B., Shoji, H., Shimizu, H., & Shioya, S. (2003). Interactions between *Lactobacillus kefiranofaciens* and *Saccharomyces cerevisiae* in Mixed Culture for Kefiran Production. *Journal of Bioscience and Bioengineering, 96*(3), 279–284. doi:10.1016/S1389-1723(03)80194-9 PMID:16233522

Chiquette, J. (2009). Evaluation of the protective effect of probiotics fed to dairy cows during a subacute ruminal acidosis challenge. *Animal Feed Science and Technology, 153*(3-4), 278–291. doi:10.1016/j.anifeedsci.2009.07.001

D'Angelo, L., Cicotello, J., Zago, M., Guglielmotti, D., Quiberoni, A., & Suarez, V. (2017, September). Leuconostoc strains isolated from dairy products: Response against food stress conditions. *Food Microbiology, 66*, 28–29. doi:10.1016/j.fm.2017.04.001 PMID:28576370

Dasari, S., Kathera, C., Janardhan, A., Kumar, A. P., & Viswanath, B. (2016). Surfacing role of probiotics in cancer prophylaxis and therapy: A systematic review. *Clinical Nutrition (Edinburgh, Lothian)*, 1–8. PMID:27923508

de Bruyn, G., Hahn, S., & Borwick, A. (2012). Antibiotic treatment for travellers' diarrhoea [Review]. *Cochrane Database of Systematic Reviews, 3*, CD002242. PMID:10908534

de Lima, M. D. S. F., de Souza, K. M. S., Albuquerque, W. W. C., Teixeira, J. A. C., Cavalcanti, M. T. H., & Porto, A. L. F. (2017). *Saccharomyces cerevisiae* from Brazilian kefir-fermented milk: An in vitro evaluation of probiotic properties. *Microbial Pathogenesis*, 1–8. PMID:28478200

de Vrese, M., & Marteu, P. R. (2007). Probiotics and Prebiotics: Effects on Diarrhea. *The Journal of Nutrition, 137*(3), 803S–811S. doi:10.1093/jn/137.3.803S PMID:17311979

Dertli, E., & Çon, A. H. (2017). Microbial diversity of traditional kefir grains and their role on kefir aroma. *Lebensmittel-Wissenschaft + Technologie, 85*, 151–157. doi:10.1016/j.lwt.2017.07.017

dos Reis, S. A., da Conceição, L. L., Siqueira, N. P., Rosa, D. D., da Silva, L. L., Carmo, M., & Peluzio, G. (2017). Review of the mechanisms of probiotic actions in the prevention of colorectal cancer. *Nutrition Research (New York, N.Y.)*, *37*, 1–19. doi:10.1016/j.nutres.2016.11.009 PMID:28215310

Etienne-Mesmin, L., Livrelli, V., Privat, M., Denis, S., Cardot, J. M., Alric, M., & Blanquet-Diot, S. (2011). Effect of a New Probiotic Saccharomyces cerevisiae Strain on Survival of *Escherichia coli* O157:H7 in a Dynamic Gastrointestinal Model. *Applied and Environmental Microbiology*, *77*(3), 1127–1131. doi:10.1128/AEM.02130-10 PMID:21131521

FAO/WHO. (2006). *Probiotics in Food: Health and nutritional properties and guidelines for evaluation, FAO food and nutrition paper no. 85*. Rome, Italy: FAO/WHO.

Fujisawa, T., Adachi, S., Toba, T., Arihara, K., & Mitsuoka, T. (1988). *Lactobacillus kefiranofaciens* sp. nov. Isolated from Kefir Grains. *International Journal of Systematic Bacteriology*, *38*(1), 12–14. doi:10.1099/00207713-38-1-12

Gill, H. S., Rutherfurd, K. J., Prasad, J., & Gopal, P. K. (2000). Enhancement of natural and acquired immunity by *Lactobacillus rhamnosus* (HN001), *Lactobacillus acidophilus* (HN017) and *Bifidobacterium lactis* (HN019). *British Journal of Nutrition*, *83*(02), 167–176. doi:10.1017/S0007114500000210 PMID:10743496

Goldin, B. R. (2011). Probiotics and Health: From History to Future. In W. Kneifel & S. Salminen (Eds.), *Probiotics and Health Claims* (pp. 1–16). Blackwell Publishing. doi:10.1002/9781444329384.ch1

Golowczyc, M. A., Mobili, P., Garrote, G. L., Abraham, A. G., & de Antoni, G. L. (2007). Protective action of *Lactobacillus kefir* carrying S-layer protein against *Salmonella enterica* serovar *enteritidis*. *International Journal of Food Microbiology*, *118*(3), 264–273. doi:10.1016/j.ijfoodmicro.2007.07.042 PMID:17719671

Gonzalez, B., Arca, P., Mayo, B., & Suarez, J. E. (1994). Detection, purification, and partial characterization of plantaricin C, a bacteriocin produced by a *Lactobacillus plantarum* strain of dairy origin. *Applied and Environmental Microbiology*, *60*(6), 2158–2163. PMID:8080262

Gopal, P. K., Prasad, J., Smart, J., & Gill, H. S. (2001). In vitro adherence properties of *Lactobacillus rhamnosus* DR20 and *Bifidobacterium lactis* DR10 strains and their antagonistic activity against an enterotoxigenic *Escherichia coli*. *International Journal of Food Microbiology*, *67*(3), 207–216. doi:10.1016/S0168-1605(01)00440-8 PMID:11518430

Hechard, Y., Derijard, B., Letellier, F., & Cenatiempo, Y. (1992). Characterization and purification of mesentericin Y 105, an anti-listeria bacteriocin from *Leuconostoc mesenteroides*. *Journal of General Microbiology*, *138*(12), 2725–2731. doi:10.1099/00221287-138-12-2725 PMID:1487737

Hempel, S., Newberry, A. J., Maher, A. R., Wang, A., Miles, J. N. V., Shanman, R., ... Shekelle, P. G. (2012). Probiotics for the Prevention and Treatment of Antibiotic-Associated Diarrhea. *Journal of the American Medical Association*, *307*(18), 1959–1969. doi:10.1001/jama.2012.3507 PMID:22570464

Hirayama, K., & Rafter, J. (2000). The role of probiotic bacteria in cancer prevention. *Microbes and Infection*, *2*(6), 681–686. doi:10.1016/S1286-4579(00)00357-9 PMID:10884619

Iwata, T., Tanaka, R., Suetsugu, M., Ishibashi, M., Tokunaga, H., Kikuchi, M., & Tokunaga, M. (2004). Efficient secretion of human lysozyme from the yeast, *Kluyveromyces lactis*. *Biotechnology Letters*, *26*(23), 1803–1808. doi:10.100710529-004-4614-9 PMID:15672218

Jeong, D., Kim, D. H., Kang, B., Kim, H., Song, K. Y., Kim, H. S., & Seo, K. H. (2017). Characterization and antibacterial activity of a novel exopolysaccharide produced by *Lactobacillus kefiranofaciens* DN1 isolated from kefir. *Food Control*, *78*, 436–442. doi:10.1016/j.foodcont.2017.02.033

Johnson, B., Selle, K., O'Flaherty, S., Goh, Y. J., & Klaenhammer, T. (2013). Identification of extracellular surface-layer associated proteins in *Lactobacillus acidophilus* NCFM. *Microbiology*, *159*(Pt_11), 2269–2282. doi:10.1099/mic.0.070755-0 PMID:24002751

Juarez-Tomas, M. S., Duhart, C. I. S., De Gregorio, P. R., Pingitore, E. V., & Nader-Macias, M. E. (2011). Urogenital pathogen inhibition and compatibility between vaginal *Lactobacillus* strains to be considered as probiotic candidates. *European Journal of Obstetrics, Gynecology, and Reproductive Biology*, *159*(2), 399–406. doi:10.1016/j.ejogrb.2011.07.010 PMID:21862199

Kabak, B., & Dobson, D. W. (2011). An Introduction to the traditional fermented foods and beverages of Turkey. *Critical Reviews in Food Science and Nutrition*, *51*(3), 248–260. doi:10.1080/10408390903569640 PMID:21390945

Kalliomäki, M., Salminen, S., Poussa, T., Arvilommi, H., & Isolauri, E. (2003). Probiotics and prevention of atopic disease: 4-year follow-up of a randomised placebo-controlled trial. *Lancet*, *361*(9372), 1869–1871. doi:10.1016/S0140-6736(03)13490-3 PMID:12788576

Kanamarlapudi, S. L. R., & Muddada, S. (2017). Characterization of Exopolysaccharide Produced by *Streptococcus thermophilus* CC30. *BioMed Research International*, *2017*, 1–12. doi:10.1155/2017/4201809 PMID:28815181

Kim, D. H., Chon, J. W., Kim, H., & Seo, K. H. (2015). Modulation of intestinal microbiota in mice by kefir administration. *Food Science and Biotechnology*, *24*(4), 1397–1403. doi:10.100710068-015-0179-8

Kim, Y., Oh, S., & Kim, S. H. (2009). Released exopolysaccharide (r-EPS) produced from probiotic bacteria reduce biofilm formation of enterohemorrhagic Escherichia coli O157:H7. *Biochemical and Biophysical Research Communications*, *379*(2), 324–329. doi:10.1016/j.bbrc.2008.12.053 PMID:19103165

Kim, Y., Oh, S., Yun, H. S., Oh, S., & Kim, S. H. (2010). Cell-bound exopolysaccharide from probiotic bacteria induces autophagic cell death of tumour cells. *Letters in Applied Microbiology*, *51*, 123–130. PMID:20536712

Kukkonen, K., Savilahti, E., Haahtela, T., Juntunen-Backman, K., Korpela, R., Poussa, T., ... Kuitunen, M. (2007). Probiotics and prebiotic galactooligosaccharides in the prevention of allergic diseases: A randomized, double-blind, placebo-controlled trial. *The Journal of Allergy and Clinical Immunology*, *119*(1), 192–198. doi:10.1016/j.jaci.2006.09.009 PMID:17208601

Kumar, L., Gangadharan, V. P., Rao, D. R., Saikia, T., Shah, S., Malhotra, H., ... Rao, R. (2005). Safety and efficacy of an indigenous recombinant interferonalpha-2b in patients with chronic myelogenous leukaemia: Results of a multicentre trial from India. *The National Medical Journal of India*, *18*(2), 66–70. PMID:15981440

Lane, M. M., & Morrissey, J. P. (2010). *Kluyveromyces marxianus*: A yeast emerging from its sister's shadow. *Fungal Biology Reviews, 24*(1-2), 17–26. doi:10.1016/j.fbr.2010.01.001

Lee, J., Kim, Y., Yun, H. S., Kim, J. G., Oh, S., & Kim, S. H. (2010). Genetic and Proteomic Analysis of Factors Affecting Serum Cholesterol Reduction by *Lactobacillus acidophilus* A4. *Applied and Environmental Microbiology, 76*(14), 4829–4835. doi:10.1128/AEM.02892-09 PMID:20495044

Lee, Y. K. (2009). Probiotic Microorganisms. In Y. K. Lee & S. Salminen (Eds.), *Handbook of Probiotics and Prebiotics* (2nd ed.; pp. 3–176). Wiley & Sons.

Lee, Y. K., & Salminen, S. (2009). Effects on Human Health and Diseases. In Y. K. Lee & S. Salminen (Eds.), *Handbook of Probiotics and Prebiotics* (2nd ed.; pp. 267–321). Wiley & Sons.

Leite, A. M. O., Mayo, B., Rachid, C. T. C. C., Peixoto, R. S., Silva, J. T., Paschoalin, V. M. F., & Delgado, S. (2012). Assessment of the microbial diversity of Brazilian kefir grains by PCR-DGGE and pyrosequencing analysis. *Food Microbiology, 31*(2), 215–221. doi:10.1016/j.fm.2012.03.011 PMID:22608226

Li, H., Zhang, S., Lu, J., Liu, L., Uluko, H., Pang, X., ... Lv, J. (2014). Antifungal activities and effect of *Lactobacillus casei* AST18 on the mycelia morphology and ultrastructure of *Penicillium chrysogenum. Food Control, 43*, 57–64. doi:10.1016/j.foodcont.2014.02.045

Li, N., Wang, Y., Zhu, P., Liu, Z., Guo, B., & Ren, J. (2015). Improvement of exopolysaccharide production in *Lactobacillus casei* LC2W by overexpression of NADH oxidase gene. *Microbiological Research, 171*, 73–77. doi:10.1016/j.micres.2014.12.006 PMID:25644955

Lidbeck, A., Övervik, E., Rafter, J., Nord, C. E., & Gustafsson, J. A. (1992). Effect of *Lactobacillus acidophilus* Supplements on Mutagen Excretion in Faeces and Urine in Humans. *Microbial Ecology in Health and Disease, 5*(1), 59–67. doi:10.3109/08910609209141305

Likotrafiti, E., Valavani, P., Argiriou, A., & Rhoades, J. (2015). In vitro evaluation of potential antimicrobial synbiotics using *Lactobacillus kefiri* isolated from kefir grains. *International Dairy Journal, 45*, 23–30. doi:10.1016/j.idairyj.2015.01.013

Liong, M. T., & Shah, N. P. (2005). Optimization of cholesterol removal by probiotics in prebiotics by using a response surface method. *Applied and Environmental Microbiology, 71*(4), 1745–1753. doi:10.1128/AEM.71.4.1745-1753.2005 PMID:15811997

Llorente, B., Malpertuy, A., Blandin, G., Artiguenave, F., Wincker, P., & Dujon, B. (2000). Genomic Exploration of the Hemiascomycetous Yeasts: 12. *Kluyveromyces marxianus* var. *marxianus. FEBS Letters, 487*(1), 71–75. doi:10.1016/S0014-5793(00)02283-3 PMID:11152887

London, L. E. E., Ross, R. P., Fitzgerald, G. F., Shanahan, F., Caplice, N. M., & Stanton, C. (2016). Probiotics as Cell Factories for Bioactive Ingredients: Focus on Microbial Polysaccharides and Health Beneficial Effects. In P. Foerst & C. Santivarangkna (Eds.), *Advances in Probiotic Technology* (pp. 77–102). New York: CRC Press.

Maedler, K., Dharmadhikari, G., Schumann, D. M., & Størling, J. (2009). Interleukin-1 beta targeted therapy for type 2 diabetes. *Expert Opinion on Biological Therapy, 9*(9), 1177–1188. doi:10.1517/14712590903136688 PMID:19604125

Manuel, P. M., Elena, B., Carolina, M. G., & Gabriela, P. (2017). Oral probiotics supplementation can stimulate the immune system in a stress process. *Journal of Nutrition & Intermediary Metabolism*, *8*, 29–40. doi:10.1016/j.jnim.2017.06.001

Marteu, P., Minekus, M., Havenaar, R., & Veld, J. H. J. (1997). Survival of Lactic Acid Bacteria in a Dynamic Model of the Stomach and Small Intestine: Validation and the Effects of Bile. *Journal of Dairy Science*, *80*(6), 1031–1037. doi:10.3168/jds.S0022-0302(97)76027-2 PMID:9201571

Martinis, E. C. P., & Franco, B. D. G. M. (1998). Inhibition of *Listeria monocytogenes* in a pork product by a *Lactobacillus sake* strain. *International Journal of Food Microbiology*, *42*(1-2), 119–126. doi:10.1016/S0168-1605(98)00059-2 PMID:9706804

Martins, F. S., Nardi, R. M. D., Arantes, R. M. E., Rosa, C. A., Neves, M. J., & Nicoli, J. R. (2005). Screening of yeasts as probiotic based on capacities to colonize the gastrointestinal tract and to protect against enteropathogen challenge in mice. *The Journal of General and Applied Microbiology*, *51*(2), 83–92. doi:10.2323/jgam.51.83 PMID:15942869

Matsuzaki, C., Takagaki, C., Tomabechi, Y., Forsbeg, L. S., Heiss, C., Azadi, P., ... Hisa, K. (2017). Structural characterization of the immunostimulatory exopolysaccharide produced by *Leuconostoc mesenteroides* strain NTM048. *Carbohydrate Research*, *448*, 95–102. doi:10.1016/j.carres.2017.06.004 PMID:28633071

McFarland, L. V. (2007). Meta-analysis of probiotics for the prevention of traveler's diarrhea. *Travel Medicine and Infectious Disease*, *5*(2), 97–105. doi:10.1016/j.tmaid.2005.10.003 PMID:17298915

Meng, J., Gao, S. M., Zhang, Q. X., & Lu, R. R. (2015). Murein hydrolase activity of surface layer proteins from *Lactobacillus acidophilus* against *Escherichia coli*. *International Journal of Biological Macromolecules*, *79*, 527–532. doi:10.1016/j.ijbiomac.2015.03.057 PMID:25841372

Mishra, V., & Prasad, D. N. (2005). Application of in vitro methods for selection of *Lactobacillus casei* strains as potential probiotics. *International Journal of Food Microbiology*, *103*, 109–115. doi:10.1016/j.ijfoodmicro.2004.10.047 PMID:16040148

Moslehi-Jenabian, S., Pedersen, L. L., & Jespersen, L. (2010). Beneficial Effects of Probiotic and Food Borne Yeasts on Human Health. *Nutrients*, *2*(4), 449–473. doi:10.3390/nu2040449 PMID:22254033

Mullan, W. M. A. (2014). Importance of Selected Genera. In C. A. Batt & M. L. Tortorello (Eds.), *Encyclopedia of Food Microbiology, Volume:2* (pp. 515–521). USA: Elsevier. doi:10.1016/B978-0-12-384730-0.00321-9

Nase, L., Hatakka, K., Savilahti, E., Saxelin, M., Ponka, A., Poussa, T., ... Meurman, J. H. (2001). Effect of Long-Term Consumption of a Probiotic Bacterium, *Lactobacillus rhamnosus* GG, in Milk on Dental Caries and Caries Risk in Children. *Caries Research*, *35*(6), 412–420. doi:10.1159/000047484 PMID:11799281

Neish, A. S. (2017). Probiotics of the Acidophilus Group: *Lactobacillus acidophilus, delbrueckii* subsp. *bulgaricus* and *johnsonii*. In M. H. Floch, Y. Ringel, & A. Walker (Eds.), *The Microbiota in Gastrointestinal Pathophysiology* (pp. 71–78). Elsevier. doi:10.1016/B978-0-12-804024-9.00006-9

Oberhelman, R. A., Gilman, R. H., Sheen, P., Taylor, D., Black, R. E., Cabrera, L., ... Madico, G. (1999). A placebo-controlled trial of Lactobacillus GG to prevent diarrhea in undernourished Peruvian children. *The Journal of Pediatrics*, *134*(1), 15–20. doi:10.1016/S0022-3476(99)70366-5 PMID:9880443

Özdemir, Ö. (2010). Various effects of different probiotic strains in allergic disorders: An update from laboratory and clinical data. *Clinical and Experimental Immunology*, *160*(3), 295–304. doi:10.1111/j.1365-2249.2010.04109.x PMID:20345982

Pant, A. R., Graham, S. M., Allen, S. J., Harikul, S., Sabchareon, A., Cuevas, L., ... Hart, C. A. (1996). *Lactobacillus* GG and Acute Diarrhoea in Young Children in the Tropics. *Journal of Tropical Pediatrics*, *42*(3), 162–165. doi:10.1093/tropej/42.3.162 PMID:8699584

Perdigon, G., Alvarez, S., de Macias, M. E., Roux, M., & de Ruiz Holgado, A. P. (1990). The Oral Administration of Lactic Acid Bacteria Increase the Mucosal Intestinal Immunity in Response to Enteropathogens. *Journal of Food Protection*, *53*(5), 404–410. doi:10.4315/0362-028X-53.5.404

Pintado, M. E., Da Silva, L., Fernandes, P B., Malcata, F. X. & Hogg, T. A. (1996). Microbiological and rheological studies on Portuguese kefir grains. *International of Food Science + Technology, 31*(1), 15-26.

Pitino, I., Randazzo, C. L., Mandalari, G., Curto, A. L., Faulks, R. M., Marc, Y. L., ... Wickham, M. S. J. (2010). Survival of *Lactobacillus rhamnosus* strains in the upper gastrointestinal tract. *Food Microbiology*, *27*(8), 1121–1127. doi:10.1016/j.fm.2010.07.019 PMID:20832693

Prado, M. R., Blandon, L. M., Vandenberghe, P. S., Rodrigues, C., Castro, G. R., Thomaz-Soccol, V., & Soccol, C. R. (2015). Milk kefir: Composition, microbial cultures, biological activities, and related products. *Frontiers in Microbiology*, *6*, 1–10. doi:10.3389/fmicb.2015.01177 PMID:26579086

Ramos, C. L., Thorsen, L., Schwan, R. F., & Jespersen, L. (2013). Strain-specific probiotics properties of *Lactobacillus fermentum, Lactobacillus plantarum* and *Lactobacillus brevis* isolates from Brazilian food products. *Food Microbiology*, *36*(1), 22–29. doi:10.1016/j.fm.2013.03.010 PMID:23764216

Retteger, L. F., Levy, M. N., Weinstein, L., & Weiss, J. E. (1935). *Lactobacillus acidophilus and its therapeutic Application*. New Haven, CT: Yale University Press.

Robin, S., Petrov, K., Dintinger, T., Kujumdzvieva, A., Tellier, C., & Dion, M. (2003). Comparison of three microbial hosts for the expression of an active catalytic scFv. *Molecular Immunology*, *39*(12), 729–738. doi:10.1016/S0161-5890(02)00253-5 PMID:12531284

Rolim, F. R. L., dos Santos, K. M. O., de Barcelos, S. C., do Egito, A. S., Ribeiro, T. S., da Conceiçao, M. L., ... do Egypto Queiroga, R. C. R. (2015). Survival of *Lactobacillus rhamnosus* EM1107 in simulated gastrointestinal conditions and its inhibitory effect against pathogenic bacteria in semi-hard goat cheese. *Lebensmittel-Wissenschaft + Technologie*, *63*(2), 807–813. doi:10.1016/j.lwt.2015.05.004

Rönka, E., Malinen, E., Saarela, M., Rinta-Koski, M., Aarnikunnas, J., & Palva, A. (2003). Probiotic and milk technological properties of *Lactobacillus brevis*. *International Journal of Food Microbiology*, *83*(1), 63–74. doi:10.1016/S0168-1605(02)00315-X PMID:12672593

Samaržija, D., Tudor, M., Prtilo, T., Špehar, I. D., Zamberlin, S., & Havranek, J. (2009). Probiotic bacteria in prevention and treatment of diarrhea. *Mljekarstvo*, *59*(1), 28–32.

Scheinbach, S. (1998). Probiotics: Functionality and Commercial Status. *Biotechnology Advances, 16*(3), 581–608. doi:10.1016/S0734-9750(98)00002-0 PMID:14538145

Selle, K. M., Klaenhammer, T. R., & Russell, W. M. (2014). Lactobacillus acidophilus. In C. A. Batt & M. L. Tortorello (Eds.), *Encyclopedia of Food Microbiology* (vol. 2, pp. 412–417). Elsevier. doi:10.1016/B978-0-12-384730-0.00179-8

Seo, B. J., Rather, I. A., Kumar, V. J. R., Choi, U. H., Moon, M. R., Lim, J. H., & Park, Y. H. (2012). Evaluation of *Leuconostoc mesenteroides* YML003 as a probiotic against low-pathogenic avian influenza (H9N2) virus in chickens. *Journal of Applied Microbiology, 113*(1), 163–171. doi:10.1111/j.1365-2672.2012.05326.x PMID:22548634

Seth, S. D., & Maulik, M. (2011). Probiotics: A Pharmacologist's Perspective. In G. B. Nair & Y. Takeda (Eds.), *Probiotic Foods In Health And Disease* (pp. 42–48). CRC Press. doi:10.1201/b10770-6

Shahani, K. M., & Ayebo, A. D. (1980). Role of dietary lactobacilli in gastrointestinal microecology. *The American Journal of Clinical Nutrition, 33*(11), 2448–2457. doi:10.1093/ajcn/33.11.2448 PMID:6449143

Simova, E., Beshkova, D., Angelov, A., Hristozov, T., Frengova, G., & Spasov, Z. (2002). Lactic acid bacteria and yeasts in kefir grains and kefir made from them. *Journal of Industrial Microbiology & Biotechnology, 28*(1), 1–6. doi:10.1038j/jim/7000186 PMID:11938463

Song, M., Park, S., Lee, H., Min, B., Jung, S., Park, S., ... Oh, S. (2015). Effect of *Lactobacillus acidophilus* NS1 on plasma cholesterol levels in diet-induced obese mice. *Journal of Dairy Science, 98*(3), 1492–1501. doi:10.3168/jds.2014-8586 PMID:25468703

Song, S., Lee, S. J., Park, D. J., Oh, S., & Lim, K. T. (2016). The anti-allergic activity of *Lactobacillus plantarum* L67 and its application to yogurt. *Journal of Dairy Science, 99*(12), 9372–9382. doi:10.3168/jds.2016-11809 PMID:27743673

Spohner, S. C., Schaum, V., Quitmann, H., & Czernak, P. (2016). Kluyveromyces lactis: An emerging tool in biotechnology. *Journal of Biotechnology, 222*, 104–116. doi:10.1016/j.jbiotec.2016.02.023 PMID:26912289

Sunanliganon, C., Thong-Ngam, D., Tumwasorn, S., & Klaikeaw, N. (2012). *Lactobacillus plantarum* B7 inhibits *Helicobacter pylori* growth and attenuates gastric inflammation. *World Journal of Gastroenterology, 18*(20), 2472–2480. doi:10.3748/wjg.v18.i20.2472 PMID:22654444

Tachedjian, G., Aldunate, M., Bradshaw, C. S., & Cone, R. A. (2017). The role of lactic acid production by probiotic *Lactobacillus* species in vaginal health. *Research in Microbiology*, 1–11. PMID:28435139

Tang, W., Dong, M., Wang, W., Han, S., Rui, X., Chen, X., ... Li, W. (2017). Structural characterization and antioxidant property of released exopolysaccharides from *Lactobacillus delbrueckii* ssp. *bulgaricus* SRFM-1. *Carbohydrate Polymers, 173*, 654–664. doi:10.1016/j.carbpol.2017.06.039 PMID:28732910

Teixeira, P. (2014). *Lactobacillus delbrueckii* ssp. *bulgaricus*. In C. A. Batt & M. L. Tortorello (Eds.), *Encyclopedia of Food Microbiology* (vol. 2, pp. 425–431). Elsevier. doi:10.1016/B978-0-12-384730-0.00177-4

Teran, C. G., Teran-Escalera, C. N., & Villarroel, P. (2009). Nitazoxanide vs. probiotics for the treatment of acute rotavirus diarrhea in children: A randomized, single-blind, controlled trial in Bolivian children. *International Journal of Infectious Diseases*, *13*(4), 518–523. doi:10.1016/j.ijid.2008.09.014 PMID:19070525

van der Aa Kühle, A., Skovgaard, K., & Jespersen, L. (2005). In vitro screening of probiotic properties of *Saccharomyces cerevisiae* var. *boulardii* and food-borne *Saccharomyces cerevisiae* strains. *International Journal of Food Microbiology*, *101*(1), 29–39. doi:10.1016/j.ijfoodmicro.2004.10.039 PMID:15878404

van Ooyen, A. J. J., Dekker, P., Huang, M., Olsthoorn, M. M. A., Jacobs, D. I., Colussi, P. A., & Taron, C. H. (2006). Heterologous protein production in the yeast *Kluyveromyces lactis*. *FEMS Yeast Research*, *6*(3), 381–392. doi:10.1111/j.1567-1364.2006.00049.x PMID:16630278

Vardjan, T., Lorbeg, P. M., Rogelj, I., & Majheniz, A. C. (2013). Characterization and stability of lactobacilli and yeast microbiota in kefir grains. *Journal of Dairy Science*, *96*(5), 2729–2736. doi:10.3168/jds.2012-5829 PMID:23497996

Vinderola, G., Perdigon, G., Duarte, J., Farnworth, E., & Matar, C. (2006). Effects of the oral administration of the exopolysaccharide produced by *Lactobacillus kefiranofaciens* on the gut mucosal immunity. *Cytokine*, *36*(5-6), 254–260. doi:10.1016/j.cyto.2007.01.003 PMID:17363262

Wang, Y., Ahmed, Z., Feng, W., Li, C., & Song, S. (2008). Physicochemical properties of exopolysaccharide produced by *Lactobacillus kefiranofaciens* ZW3 isolated from Tibet kefir. *International Journal of Biological Macromolecules*, *43*(3), 283–288. doi:10.1016/j.ijbiomac.2008.06.011 PMID:18662712

Witthuhn, R. C., Schoema, T., & Britz, T. J. (2005). Characterisation of the microbial population at different stages of Kefir production and Kefir grain mass cultivation. *International Dairy Journal*, *15*(4), 383–389. doi:10.1016/j.idairyj.2004.07.016

Yavuzdurmaz, H. (2007). *Isolation*. İzmir: Characterization, Determination of Probiotic Properties of Lactic Acid Bacteria From Human Milk. Master of Science, Food Engineering, the Graduate School of Engineering and Sciences of Izmir Institute of Technology.

Yeşilova, Y., Çalka, Ö., Akdeniz, N., & Berktaş, M. (2012). Effect of Probiotics on the Treatment of Children with Atopic Dermatitis. *Annals of Dermatology*, *24*(2), 189–193. doi:10.5021/ad.2012.24.2.189 PMID:22577270

Yun, B., Oh, S., & Griffiths, M. W. (2014). Lactobacillus acidophilus modulates the virulence of *Clostridium difficile*. *Journal of Dairy Science*, *97*(8), 4745–4758. doi:10.3168/jds.2014-7921 PMID:24856984

Zago, M., Fornasari, M. E., Carminati, D., Burns, P., Suarez, V., Vinderola, G., & Giraffa, G. (2011). Characterization and probiotic potential of *Lactobacillus plantarum* strains isolated from cheeses. *Food Microbiology*, *28*(5), 1033–1040. doi:10.1016/j.fm.2011.02.009 PMID:21569949

Zanirati, D. F., Abatemarco Jr, M., de Cicco Sandes, S. H., Nicoli, J. R., Nunes, A. C., & Neumann, E. (2015). Selection of lactic acid bacteria from Brazilian kefir grains for potential use as starter or probiotic cultures. *Anaerobe*, *32*, 70–76. doi:10.1016/j.anaerobe.2014.12.007 PMID:25542841

Zhao, L., Jiang, Y., Zhang, T., Duan, C., Huang, C., Zhao, Y., ... Li, S. (2017). Protective effects of *Lactobacillus plantarum* C88 on chronic ethanolinduced liver injury in mice. *Journal of Functional Foods*, *35*, 97–104. doi:10.1016/j.jff.2017.05.017

Zhou, J., Xiaoli, L., Hanhu, J., & Mingsheng, D. (2009). Analysis of the microflora in Tibetan kefir grains using denaturing gradient gel electrophoresis. *Food Microbiology*, *26*(8), 770–775. doi:10.1016/j.fm.2009.04.009 PMID:19835760

Chapter 16
The Effects of Probiotic Cultures on Quality Characteristics of Ice Cream

Nihat Akın
Selçuk University, Turkey

Hale İnci Öztürk
Selçuk University, Turkey

ABSTRACT

The manufacturing design and probiotic strain that are used in ice cream production affect sensorial, nutritional, and physicochemical properties of ice cream. Therefore, it is possible to produce better quality products or to develop existing methods by considering the effects of probiotic cultures on these properties of ice cream. The International Dairy Federation has recommended that probiotic products contain at least 10^7 CFU/g of viable probiotic strain before consumption, so that the therapeutic value of a probiotic product can be associated with the viability of these bacteria. The nutritional value of probiotic ice cream is related not only to the dietary components it contains, but also to the maintaining of the viability and the activity of the probiotic bacteria. This chapter provides information on the viability of probiotic strains during ice cream processing and storage. Furthermore, the effects of probiotic cultures on sensory and physicochemical properties of ice cream are also discussed.

INTRODUCTION

Ice cream is a dairy product with a complex physicochemical system which is obtained by freezing a pasteurized mix prepared with fat, milk solid nonfat, sugar, stabilizers and emulsifiers, sometimes aroma and color substances, and fresh or dried egg yolk. Ice cream definition contains plain ice cream, nonfat, reduced-fat, low-fat, fruit added, and probiotic ice creams. As a result of its beneficial composition, ice cream can be described as a dairy dessert which is fondly consumed and which is used by manufacturers to deliver probiotics. Ice cream is an ideal vehicle for probiotic bacteria thanks to milk proteins, lactose, lipids, and other contents in its composition (Cruz, Antunes, Sousa, Faria, & Saad, 2009). Also,

DOI: 10.4018/978-1-5225-5363-2.ch016

it has a higher pH value, compared to other fermented foods, which is another important advantage for probiotic bacteria.

The World Health Organization has defined probiotics as viable microorganisms which provide health benefits when they are consumed in sufficient amount (FAO/WHO, 2001). The International Dairy Federation has reported that probiotics products should have at least 10^7 CFU/g of viable cell in order to be beneficial to consumer health (Champagne, Gardner, & Roy, 2005). Viability and metabolic activity of probiotics are associated with food composition, processing, and storage conditions. In addition, probiotics influence the sensory and physicochemical properties of foods. Thus, it is necessary to assess of the effects of probiotics on the quality characteristics of ice cream and determine the factors that affect the viability of probiotics from ice cream production to consumption. The aim of this chapter is to give an overview of how probiotics have an impact on the characteristics of ice cream and of which factors affect the survivability of probiotics in the ice cream matrix.

BACKGROUND

The demand of functional foods has increased rapidly, as consumers have more awareness of the health effects of foods. Probiotic dairy products are an important part of the functional food industry. Among the dairy products containing live culture, probiotic ice creams have attracted great interest in recent studies (Arslan et al., 2016; Cruxen et al., 2017; Yilmaz-Ersan, Ozcan, Akpinar-Bayizit, Turan, & Taskin, 2017). Ice cream ensures suitable conditions for the growth of probiotic bacteria in large counts and their viability over storage time. Several studies have informed that ice cream is appropriate to deliver probiotics in the diet (Da Silva, De Fátima Bezerra, Dos Santos, & Correia, 2015; Hekmat & McMahon, 1992). An ice cream matrix is a complex system consisting of fat globules, nutritional compounds, ice crystals, and air cells. However, probiotics, depending on their characteristics, cannot maintain viability in the ice cream due to oxidative, mechanical, cold, and osmotic stresses. In this context, studies focused on the viability of probiotic bacteria during ice cream production and storage (Arslan et al., 2016; Magarinos, Selaive, Costa, Flores, & Pizarro, 2007; Mohammadi, Mortazavian, Khosrokhavar, & Da Cruz, 2011; Nousia, Androulakis, & Fletouris, 2011). In addition, another focus of studies has been how probiotic cultures which are used in production affect the characteristics of ice creams.

Lactobacillus and *Bifidobacterium* species are the most commonly used probiotic bacteria in the manufacturing of probiotic ice creams. In probiotic ice cream formulations, appropriate cultures should be selected for maintain their viability in freezing and storage conditions. Probiotic strain characteristics, freezing, storage temperature and time, composition of ice cream, and packaging material have significant effects on the viability of cultures in the ice cream matrix. These factors will be discussed in detail in the following section. Homayouni, Azizi, Ehsani, Yarmand, and Razavi (2008) mentioned that *Bifidobacterium lactis* (Bb-12) and *Lactobacillus casei* (Lc01) showed better resistance to ice cream conditions compared to other probiotics. However, the viability and growth of probiotic bacteria can be optimized with different production techniques and formulations. The probiotic cultures which are utilized in ice cream production are shown in Table 1.

Overall, the production of probiotic ice cream is carried out with four different techniques. Probiotics are inoculated at a concentration of about 10^8-10^9 CFU/g of bacterial cells for each technique (Abghari, Sheikh-Zeinoddin, & Soleimanian-Zad, 2011; Arslan et al., 2016; Mohammadi et al., 2011). The common steps of these techniques consist of preparation of ingredients (e.g., milk, sweeteners, milk powder,

Table 1. Probiotic cultures used in ice cream production

Probiotic Cultures	References
Lactobacillus acidophilus	Abghari et al. (2011); Akalın and Erişir (2008).
Lactobacillus rhamnosus	Abghari et al. (2011); Salem et al. (2005).
Bifidobacterium lactis	Kailasapathy and Sultana (2003).
Lactobacillus johnsonii	Alamprese et al. (2002).
Bifidobacterium bifidum,	Salem et al. (2005); El-Sayed, Salama, and El-Sayed (2014).
Lactobacillus reuteri	Salem et al. (2005).
Lactobacillus gasseri	Salem et al. (2005).
Lactobacillus casei	El-Sayed et al. (2014); Karthikeyan, Elango, Kumaresan, Golapakrishnamurty, and Raghunath (2014).
Bifidobacterium longum BL-04	Favaro-Trindade et al. (2006).
Bifidobacterium lactis BL-01	Favaro-Trindade et al. (2006).
Lactobacillus paracasei CRL-431	Haynes and Playne (2002).
Lactobacillus plantarum	El-Sayed et al. (2014).
Bifidobacterium animalis subsp. *lactis* Bb-12.	Magarinos et al. (2007); Da Silva et al. (2015).
Lactobacillus delbrueckii ssp *bulgaricus*	Vijayageetha, Begum, and Reddy (2011).
Streptococcus salivarius ssp *thermophilus*	Vijayageetha et al. (2011).
Propionibacterium jensenii	Ranadheera, Evans, Adams, and Baines (2013).

cream, emulsifiers, stabilizers, prebiotics, flavoring, and colorant), mixing, and pasteurization. In the first production technique, the pasteurized mix is cooled to 4 °C, and then allowed to be ageing by gently mixing for 1-2 minutes at one-hour intervals at 4 °C for 4-24 hours. After the ageing step, the mix is warmed up to fermentation temperature (30-40 °C), afterwards probiotic strains (free or encapsulated) are inoculated to the mixture. This mix is incubated to reach pH 5.5 at 30-40 °C, thereafter, cooled to 4 °C before the freezing process. Finally, the cooled mix is frozen, packaged, hardened at -28 °C, and stored between -16 and -28 °C. In the latter production technique, probiotic culture is inoculated after the pasteurized mix is cooled to 4 °C. The inoculated mix is left for ageing at 4 °C for 4-24 hours, and subsequently frozen. Processing steps following the freezing are carried out as in the first technique. In another production technique, milk fermented (pH ~ 5.5) by probiotic strains is added to the mix after the ageing step. The mix supplemented with fermented milk, is cooled to 4 °C, afterwards frozen, packaged, hardened, and stored as in the previous production techniques. The product which is obtained by these three techniques is called fermented probiotic ice cream. However, unfermented probiotic ice cream is produced in another production method. In this technique, free or encapsulated probiotic culture is added into the mix after the ageing step, and the mix is immediately frozen. The steps after the freezing process are applied as in other techniques.

In probiotic ice cream production, process steps can be optimized to enhance the viability of probiotic cultures. Various studies were carried out to determine the effect of different production techniques on probiotic survival in the ice cream (Alamprese, Foschino, Rossi, Pompei, & Savani, 2002; Arslan et al., 2016; Favaro -Trindade, Bernardi, Bodini, De Carvalho Balieiro, & De Almeida, 2006; Hagen & Narvhus, 1999). Arslan et al. (2016) reported that the viability of probiotics is higher in fermented ice cream, compared to unfermented ice cream. In addition, Hekmat and McMahon (1992) investigated the effect of production technique, in which the mix fermented with probiotic strains was frozen, on viable cell counts. They determined higher probiotic viability after freezing and throughout the storage time, compared to the outcomes of Salem, Fathi, and Awad (2005), who produced ice cream from the mix supplemented with fermented milk. As a result, studies have shown that the fermentation process which is used in ice cream production has positive effects on the viability of probiotics.

THE FACTORS THAT EFFECT ON VIABILITY OF PROBIOTICS IN ICE CREAM PRODUCTION

Probiotic Strain Characteristics

The selection of proper strains with survival ability on the conditions of high oxygen content, freezing, and storage temperatures is substantial. *L. casei*, *L. paracasei*, *L. plantarum*, *L. rhamnosus*, *L. johnsonii*, *B. adolescentis*, *B. animalis*, *B. bifidum*, *B. infantis*, *B. lactis*, *B. longum,* and *B. breve* are mesophilic probiotic bacteria. Nevertheless, probiotic bacteria with thermophilic characteristics also exist, such as *Lactobacillus acidophilus*, *L. delbrueckii* subsp. *bulgaricus*, *L. helveticus*, *L. lactis*, *L. reuteri*, and *L. fermentum* (Lee, 2009). Considering that thermophiles can be more susceptible to cold shock than mesophiles (Russell, 2002), the use of mesophilic probiotic bacteria in ice cream formulations may be more appropriate to obtain high viable counts during manufacturing and storage.

Bacterial cells with larger microscopic size are more sensitive to mechanical damage (King & Alexander, 1948). Hence, the mechanical destruction caused by ice crystals in the ice cream matrix can be more effective in large bacterial cells. Abghari et al. (2011) stated that *L. rhamnosus*, which has a smaller cell size compared to *L. acidophilus*, presented better viability than *L. acidophilus* in the ice cream.

Lactic acid bacteria (LAB) which are used as probiotics in ice cream production generally consist of bifidobacteria and lactobacilli strains. These strains show different oxygen tolerance, as bifidobacteria cells are more susceptible to oxygen than lactobacilli cells. In ice cream production, an increase in volume occurs, due to air incorporation by the vigorously agitating process during the freezing step, and is called "overrun". Given that overrun is substantial for manufacturers in terms of providing suitable texture, body, and taste to ice cream, it cannot be removed from the ice cream processing. In this context, the selection of oxygen-resistant strains is essential to maintain the viability of probiotics in the ice cream matrix.

Molecular studies can be carried out on increasing expression of stress proteins to improve the viability of probiotics in ice cream. Considering cold shock and oxidative stress, two proteins attract attention: the betaine transporter protein (BetL) and the superoxide dismutase (SodA). The expression of cold shock protein BetL improves the viability of cultures at -20 °C (Sheehan, Sleator, Fitzgerald, & Hill, 2006). SodA is an oxidative stress protein and preserves *Lactobacillus gasseri* and *Lactobacillus acidophilus* against oxygen damage (Bruno-Bárcena, Andrus, Libby, Klaenhammer, & Hassan, 2004). Hence, increased expression of these genes in probiotic bacteria assists to maintain their survivability against adverse conditions in ice cream, such as cold shock and oxidative stress.

Composition of Ice Cream

The composition of standard ice cream consists of 10-12% milk fat, 9-11% milk solids nonfat (MSNF), 10-16% sweeteners, and 0.2-0.4% stabilizers/emulsifiers, and its total solid content is between 35% and 38% (Akın, 2009). The MSNF in the ice cream composition is made of 34-36% milk proteins rich in essential amino acids, such as tryptophan and lysine. Generally, the mean nutrient content of ice cream comprises 4% of protein, 12% of fat, and 20% of carbohydrate. The 20% of carbohydrate content is lactose and the other part is composed of sucrose, glucose or fructose depending on the sweetener which is used. Ice cream is a good source for fat soluble vitamins (e.g., A, D, E, and K) and riboflavin, as well as calcium and phosphor. Consequently, the compositional properties of ice cream present an excellent media for the growth and viability of probiotics. Casein, milk fat, lactose, and sucrose, which were found in the composition of ice cream, demonstrate cryoprotective properties (Godward & Kailasapathy, 2003; Moss & Speck, 1963). These components with cryoprotective activities protect probiotic bacteria from freezing damage which affects their viability.

Studies showed that the low fat content of ice cream, compared to regular fat content, offers more favorable conditions for the viability of probiotics during storage (Favaro-Trindade, De Carvalho Balieiro, Dias, Amaral Sanino, & Boschini, 2007; Haynes & Playne, 2002). Favaro-Trindade et al. (2007) reported that the viable counts of *L. acidophilus* LAC4 and 74-2 were higher in ice creams with 5% milk fat than in those with 10% milk fat. In another study, the numbers of viable *B. bifidum* DSMZ 200456 were higher in ice cream with 10% milk fat, while the counts of *L. acidophilus* DSMZ 20079 were higher in ice cream containing 5% milk fat (Turgut & Cakmakci, 2009). It appears that the effect of fat content of ice cream on the survival of probiotics is dependent on strains.

The sugars which are used in ice cream formulations have complicated effects on probiotics. They can negatively affect the viability of probiotics by causing osmotic stress (Jay, Loessner, & Golden, 2008) or they can increase probiotic survival by showing cryoprotective properties. However, these effects depend on the sugar type and concentration which is used in the formulation. Akın, Akın, and Kırmacı (2007) determined that *L. acidophilus* La-14 and *B. lactis* BL-01 showed higher viability at a sugar concentration of 18%, in comparison with 15% and 20%. Short chain polysaccharides, such as fructooligosaccharides, exhibit higher cryoprotective activity than long chain polysaccharides (Champagne, 2009) and are considered prebiotic components that have beneficial effects on probiotics.

Viability of probiotics can be improved by altering the ice cream formulation. Whey protein and casein hydrolysates, L-cysteine, and prebiotics encourage the growth of bacteria as a result of their usage by bacterial cells as nutritional sources. Moreover, protein derivatives both advance the buffer capacity of the medium and decline the redox potential of the medium. Therefore, the use of these components in ice cream mix formulations can support the viability of probiotics. Dave and Shah (1998) reported that L-cysteine, acid casein hydrolysate, and whey protein concentrate increased the survival of bifidobacteria and *L. acidophilus*, which do not have proteolytic activity, in yoghurt during cold storage.

Prebiotic Ingredients

Prebiotics are defined as nutrient source that stimulate the growth and activity of bacteria (Soukoulis, Fisk, & Bohn, 2014). Prebiotics are nondigestible or very low digestible ingredients that are selectively metabolized by probiotics and enhanced their activity (Mohammadi et al., 2011). These compounds can be added to probiotic products as growth factors, and can substantially support the maintenance of probiotic viability. A synergistic effect occurs when properly combining probiotics and prebiotics, since prebiotics improve bacterial growth and activity (Ranadheera, Baines, & Adams, 2010). Recent studies have focused on the utilization of prebiotic ingredients to enhance bacterial growth and survivability. In this sense, prebiotic effects of various fruits, dietary fibers, oligosaccharides, and seed extracts on probiotic cultures which are used in symbiotic ice cream formulations were examined (Akalın & Erişir, 2008; Di Criscio et al., 2010; Favaro-Trindade et al., 2007; Sagdic, Ozturk, Cankurt, & Tornuk, 2012; Senanayake, Fernando, Bamunuarachchi, & Arsekularatne, 2013). Studies have been conducted on which food materials, such as fruits, and dietary fibers, have prebiotic effect on which probiotic strains.

Prebiotic ingredients which are used in ice cream production and probiotic strains in which these ingredients are effective are shown in Table 2. Fructooligosaccharide, especially oligofructose, are preferred by probiotics as a source of carbon and energy. Akalın and Erişir (2008) notified that the addition of 4% oligofructose to ice cream mix provided higher probiotic counts compared to 4% inulin. They determined that the supplementation of 4% oligofructose increased the counts of *L. acidophilus* La-5 and *B. animalis* Bb-12 by about 12% and 7%, respectively, in fermented mix. In addition to this, Akın et al. (2007) reported that *L. acidophilus* and *B. lactis* counts increased with increasing inulin concentration in ice cream. Likewise, Parussolo et al. (2017) stated that the viable counts of *L. acidophilus* NCFM in ice cream improved depending on increasing concentration of yacon flour. Studies have showed that the prebiotic potential changes depending on the prebiotic concentration and the probiotic culture. However, further researches are necessary on the development of new symbiotic ice cream formulations by using different probiotic strains and prebiotics.

Acidity

The optimum pH values for the growth of probiotic lactobacilli and bifidobacteria are 5.5-6.0 and 6.0-7.0, respectively (De Vuyst, 2000). A relatively high pH value (6.0-6.5) of ice cream offers an advantage for the viability of probiotic cells. Compared to frozen yoghurt, ice cream provides more favorable conditions

Table 2. Prebiotics used in ice cream formulations for certain probiotic strains

Prebiotics	Probiotic Strains	References
Pomegranate peel extract	*L. casei* Shirota	Sagdic et al. (2012).
Oligofructose	*L. plantarum, L. casei, B. bifidum, L. acidophilus* La-5 and *B. animalis* Bb-12	Akalın and Erişir (2008); El-Sayed et al. (2014).
Inulin	*B. bifidum, L. acidophilus, B. lactis*	Akın et al. (2007); El-Sayed et al. (2014).
Acerola	*B. lactis, B. bifidum*	Favaro-Trindade et al. (2006).
Yellow mombin	*L. acidophilus* 74-2, *L. acidophilus* LAC 4	Favaro-Trindade et al. (2007).
Soy milk	*B. bifidum* Bb-12, *L. acidophilus* La-05	Aboulfazli, Shori, and Baba (2016).
Coconut milk	*B. bifidum* Bb-12, *L. acidophilus* La-05	Aboulfazli et al. (2016).
Yacon flour	*L. acidophilus* NCFM	Parussolo et al. (2017).
Strawberry	*L. acidophilus*	Vardar and Öksüz (2007).
Peach	*B. bifidum* Bb-12	Villalva et al. (2017).
Dark blue and white myrtle fruits	*L. casei* 431	Öztürk, Demirci, and Akın (2018)

for probiotic bacteria, as it has a higher pH value. In ice creams which are fermented by probiotics, the pH of the mix should be regularly controlled throughout the fermentation, because the pH values below approximately 5.0-5.5 cause considerable loss of viable probiotic cells. In addition, the use of fruit juice or pulp in the ice cream formulation can lead to high acidity, which adversely affects the viability of probiotic bacteria, in the final product. Hence, fruit juices or pulps with low acidity should be used in the production of probiotic ice cream. Additionally, the utilization of *Lactobacillus* strains, that are more resistant to acidic conditions than *Bifidobacterium* strains thanks to their high cytoplasmic buffering capacity (Godward et al., 2000), can be preferred in acidic product formulations.

Overrun

Overrun is defined as volume increase and occurs when oxygen is incorporated in the ice cream mix as a result of the agitation performed in the freezing step. It ensures textural properties to ice cream and has an impact on the physical properties of ice cream, such as melting rate and firmness. On the other hand, oxygen shows a toxic effect on *Lactobacillus* and *Bifidobacterium* strains, since they do not contain the enzymes which are required to break down the toxic oxygen reduction products, such as hydrogen peroxide, superoxide, and hydroxyl radicals. Thereby, the accumulation of toxic metabolites in the bacterial cells causes their death. Based on this, oxygen-resistant strains should be selected in the production of ice cream. As the authors previously mentioned, the use of *Lactobacillus* strains, which show higher resistance to oxygen than *Bifidobacterium* strains do, is suggested for ice cream production. In addition, oxygen scavenging agents, such as L-ascorbic acid, can be utilized in ice cream formulations, in order to protect them against oxygen damage. Ferraz et al. (2012), who examined the effects of different overrun levels on probiotic viability, reported that high overrun levels adversely affect cell survival. *L. acidophilus* counts decreased by 2 log units for the 90% overrun application. The researchers recommended that low overrun levels should be applied in ice cream production, in order to ensure high probiotic survival throughout the storage time.

Freezing

In studies which were conducted on probiotic ice cream, the number of live cultures decreased by 0.2-1.0 log units as a result of the effect of freezing (Abghari et al., 2011; Alamprese et al., 2002; Arslan et al., 2016). The effect of freezing process on the viability of probiotics shows differences, depending on the strain, production techniques, and ice cream formulation which are used. Ranadheera et al. (2013) informed that *P. jensenii* 702, *B. animalis* subsp. *lactis* BB-12, and *L. acidophilus* LA-5 showed 88.72%, 66.46%, and 56.14% survival rates, respectively, against the freezing process. The authors suggested that the resistance differences which they observed against the freezing process might be due to some mechanisms of bacterial cells. Some bacterial cells, such as *L. acidophilus* LA-05, can have mechanisms which allow them to be dehydrated and thereby reduce the occurrence of intracellular ice crystals (Magarinos et al., 2007; Ranadheera et al., 2013).

Abghari et al. (2011) reported that probiotic cultures which they added after the ice cream mix ageing step lost their viability with a high rate by freezing. Thus, it can be recommended that probiotics are added prior to the mix ageing step or incubated before the freezing process, in order to minimize the freezing effect. The ageing process is performed by keeping the mix at 4 °C for 4-24 hours. Probiotic cultures in the mix can develop adaptation to freezing temperature, called cryotolerance, as a result of holding them at 4 °C for a long time. In another study, Arslan et al. (2016) indicated that probiotics sustain their viability at higher levels against freezing in fermented ice cream, compared to unfermented one. This can be explained by the acid adaption that occurs during fermentation, since the adaption which is developed by bacterial cells against acid stress enhances their resistance to cold stress, based on the cross-protection phenomenon (Streit, Delettre, Corrieu, & Béal, 2008).

Ice crystals which form by the freezing process in the ice cream matrix also influence the viability of probiotic microorganisms. The size of ice crystals diminishes with escalation of freezing velocity. Large ice crystals further damage microorganism cells by their mechanical action. Therefore, a quick freezing of the mix is significant to maintain probiotic viability. Moreover, agitation operation applied during the freezing shows a similar action on microorganisms, and the incorporation of oxygen into the mix by agitation has a negative effect on the viability of the culture (Hagen & Narvhus, 1999).

Storage Temperature and Time

The growth and survivability of probiotics are affected by the environmental temperature. Hence, the storage conditions of ice cream are effective on probiotics. Ice creams are stored between -16 and -28 °C in freezers until consumption. Considering their sensory properties, the shelf life of ice creams stored at -20 and -25 °C is around 10 and 40 weeks, respectively (Fu & Labuza, 1997). Studies were conducted on storage temperatures practiced between -16 and -28 °C to assess the survival of probiotics in ice creams over the storage time (Alamprese, Foschino, Rossi, Pompei, & Corti, 2005; Favaro-Trindade et al., 2006; Magarinos et al., 2007). Abghari et al. (2011) reported that *L. acidophilus* counts decreased by 0.82 log units and *L. rhamnosus* counts remained constant in probiotic ice creams which were stored at -19 °C for 12 weeks. However, the final cell counts of these probiotic bacteria were found to be higher than 7 log CFU/g, which is the lowest recommended therapeutic level. In another study, it was determined that viable *L. acidophilus* ATCC 4356 counts decreased by 1 log unit in fermented ice cream which were stored at -20 °C for 12 weeks (Arslan et al., 2016). This study suggested using the fermentation process in the production of probiotic ice cream in order to maintain required bacterial counts during the frozen storage time. Consequently, studies have shown that ice cream which is stored for a long time is a suitable media to transport *Lactobacillus* and *Bifidobacterium* species to consumers. Most of these investigations have demonstrated that probiotics maintain the required cell counts for their therapeutic effects during the frozen storage (-16 to -28 °C) for up to 6 months.

Temperature fluctuations should be avoided to maintain the viability of probiotics during the storage time, because the growth of ice crystals, as a consequence of temperature fluctuations, impresses the viability of microorganisms by damaging their cell structure. Moreover, it can be suggested that the inoculum concentration can be increased in order to provide high levels of probiotic cells throughout the storage time. After the determination of how many log units of probiotic cells decreased during ice cream production and storage, the concentration of inoculum can be increased by as much as the decreasing log units.

Microencapsulation

The application of encapsulation protects bacteria from adverse environmental conditions and provides bacterial stability. Microencapsulation is defined as a technology of tiny liquid or solid particles packaging (Chandramouli, Kailasapathy, Peiris, & Jones, 2004). Probiotic bacteria can be microencapsulated to provide physical protection and enhance their survival ability during the frozen storage time. In recent years, microencapsulation technique has attracted considerable interest to improve the viability of probiotics in ice cream (Homayouni et al., 2008; Kailasapathy & Sultana, 2003). Spray drying, spray congealing, fluidized bed coating/air suspension, extrusion, coacervation/phase separation technique, and electrostatic method are used for microencapsulation of probiotic organisms. For this purpose, water soluble polymers, water insoluble polymers, lipids, fatty acids, waxes, and negatively charged polymers or components are used as coating materials in microencapsulation techniques (Anal & Singh, 2007).

Cui, Goh, Kim, Choi, and Lee (2000) informed that two-stage microencapsulation with alginate and poly-l-lisin materials protects bifidobacteria from cold shock. Distinct coating materials can be selected against different stress conditions. The substances which are used as coating materials in microencapsulation of lactobacilli and bifidobacteria have the following properties:

- Alginate protects against oxygen toxicity (Talwalkar & Kailasapathy, 2003).
- The two-stage coating system containing alginate and chitosan preserves from acidic and alkali conditions (Krasaekoopt, Bhandari, & Deeth, 2004).
- Skim milk protects against the effect of heat treatment (Simpson, Stanton, Fitzgerald, & Ross, 2005).

Homayouni et al. (2008) reported that microencapsulated *L. lactis* (Lc01) and *B. lactis* (Bb12) cells, which were obtained by using alginate as a covering material, survived 30% more than uncapsulated cells in the ice cream matrix which was stored at -20 °C for 180 days. In another study, viable counts of microencapsulated *B. bifidum* and *L. plantarum* were determined above 8 log CFU/g in ice cream samples stored at -20 °C for 90 days (El-Sayed et al., 2014). These results highlighted that the required cell numbers for therapeutic effect in probiotic ice cream during storage time can be provided by microencapsulation of probiotics.

Packaging Material

Air incorporation during the freezing process is substantial to provide overrun to ice cream. Oxygen, that constitutes 21% of air, decreases probiotic counts in ice cream by influencing the viability of anaerobic *Bifidobacterium* and microaerophilic *Lactobacillus*. Oxygen passing through the package can also have a negative impact on probiotic survival. Therefore, the level of oxygen permeation through package should be kept as low as possible to prevent damage. In accordance with this purpose, various package or packaging material designs can be realized. Miller, Nguyen, Rooney, and Kailasapathy (2003) and Talwalkar, Miller, Kailasapathy, and Nguyen (2004) reported that the usage of oxygen impermeable packing materials, thick packaging materials, and packaging system containing oxygen scavenging agents reduce the oxygen level in probiotic products. A limited number of studies have been conducted to determine the effect of packaging materials on the survival of probiotics in ice cream (Ranadheera et al., 2013). The oxygen permeability of glass containers is very low, compared to plastic packages (Da Cruz, Faria, & Van Dender, 2007). However, the effects of glass, polypropylene, and polyethylene which are used as packaging materials are the same on the viability of probiotics in ice creams (Ranadheera et al., 2013). The effect of the packaging material can depend on the storage temperature, because low storage temperatures, at which the ice creams are stored, limit the biochemical reactions of probiotics. Therefore, polypropylene and polyethylene materials, instead of glass materials, can be used in the packaging of ice creams to reduce production cost.

THE EFFECTS OF PROBIOTICS ON THE PROPERTIES OF ICE CREAM

Sensorial Properties

Probiotic ice cream presents different sensory profiles if it is compared to regular ice cream. Favaro-Trindade et al. (2007) found that the sensorial acceptability of the ice cream which is fermented by probiotic cultures was lower than the sensorial acceptability of unfermented probiotic ice cream. Since ice cream is not generally described as an acidic food, low pH values between 4.0 and 4.5 have an adverse effect on sensory acceptability. Therefore, in the production of fermented ice cream, incubation should be terminated when the pH reaches about 5.0-5.5 in the fermentation step. Hekmat and McMahon (1992), who investigated the effect of different acidity values on the overall acceptability of ice creams, stated that the ice cream at pH 5.5 was more preferred in terms of overall acceptability in comparison to those at pH 5.0 and 6.0. Also, Vardar and Öksüz (2007) asserted that the completion of the incubation at pH 5.6 provided better results in terms of taste and flavor. Contrary to fermented ice creams, no flavor problems are caused by bacterial activity in unfermented ice creams, since their freezing and storage at very low temperatures limits microbial reactions. Furthermore, relevant studies showed that other sensory attributes, such as color, appearance, and consistency, do not alter with the addition of probiotic cultures in both fermented and unfermented ice creams (Da Silva et al., 2015; Vijayageetha et al., 2011). At the same time, Nousia et al. (2011) indicated that the storage at -25 °C for 45 weeks was not effective on the sensory characteristics of probiotic ice creams.

Metabolic products which are produced by probiotic bacteria can affect sensory properties, especially flavor. Homofermentative probiotic LAB, such as *L. acidophilus*, *L. helveticus*, *L. plantarum*, and *L. johnsonii*, produce lactic acid from lactose, while probiotic *Bifidobacterium* species produce acetic and lactic acids as degradation products of glucose metabolism. Acetic acid gives an off-flavor, which is called probiotic flavor, to dairy products. Flavoring agents can be used to camouflage this defect in probiotic ice cream containing *Bifidobacterium* strains.

Various studies showed that the probiotic addition is not effective on the firmness of ice cream (Abghari et al., 2011; Akalın & Erişir, 2008; Di Criscio et al., 2010). The firmness is increased with the progression of the storage time in both conventional and probiotic ice creams (Akalın & Erişir, 2008; Arslan et al., 2016). Besides, different formulations can be developed to improve the firmness of probiotic ice creams. Inulin and oligofructose ensure favorable sensory attributes to ice cream, such as slight sweetness, creaminess, improved mouthfeel, and preservation of flavor after consumption. Akalın and Erişir (2008) showed that the addition of inulin gave firmness to probiotic ice cream. In addition to this, fruits which are used in ice cream formulations mask the acidic taste and give aroma. Thus, symbiotic ice creams have higher flavor and overall acceptability scores, compared to plain probiotic ice creams. In studies which were conducted on probiotic ice cream formulations, fermented ice creams containing fruits demonstrated good sensory properties in terms of appearance, flavor, and taste (Vardar & Öksüz, 2007; Villalva et al., 2017). In order to optimize sensory characteristics of ice cream, preexperiments should be performed using different inoculation concentrations and probiotic strains.

Nutritional Properties

Deosarkar, Kalyankar, Pawshe, and Khedkar (2016) and Goff and Hartel (2013) reviewed the nutritional effects of ice cream. In addition to this, the nutritional advantages of probiotic bacteria have been studied in dairy products fermented with lactobacilli and bifidobacteria. These fermented products contain a great deal of chemical compounds depending on probiotic strain, biochemical activities of probiotics, and manufacturing process (Gomes & Malcata, 1999). During fermentation, the probiotic bacteria produce vitamin K, folic acid, biotin, thiamine, riboflavin, cobalamin (B12), niacin, and pyridoxine. Furthermore, Alm (1982) informed that fermented dairy products offer higher folic acid content, compared to raw milk. Therefore, fermented probiotic ice cream can contain high amounts of B-group vitamins in contrast with unfermented probiotic ice cream and plain ice cream. Another important nutritional component produced by probiotics during fermentation is L(+)-lactic acid. Bifidobacteria produce L(+)-lactic acid, which is more easily metabolized by humans, compared to D(-)-lactic acid. Also, L(+)-lactic acid is used as an energy source in human metabolism and it provides 15 kJ/g energy (Gurr, 1987). That's why, the use of bifidobacteria strains can be preferred to increase nutritional value of fermented ice cream.

Senanayake et al. (2013) reported that protein and total solid contents of probiotic ice cream were higher than those of regular ice cream. These differences can be attributed to the activity of probiotic cultures in the fermentation stage. As a result of fermentation, lactose content of fermented dairy products decreases, while free amino acids and certain vitamin, especially B group vitamins, occur by metabolic activity of lactobacilli and bifidobacteria (Gomes & Malcata, 1999). Besides, Aboulfazli et al. (2016) reported that sugar content in fermented ice cream decreased because of the utilizations of lactose and sucrose by probiotic bacteria as energy sources during fermentation. Moreover, LAB and bifidobacteria synthesize exopolysaccharides from glucose, galactose or other monosaccharides by the action of glycosyl transferases (Salazar et al., 2009). Hence, the nutritional composition of the fermented probiotic ice cream differs from regular ice cream.

Physicochemical Properties

The physicochemical properties of ice cream, such as acidity, overrun, viscosity, specific gravity, and melting rate, have an important role on the quality of ice cream. These properties depend significantly on the formulation and manufacturing technique of ice cream. In probiotic ice creams, both formulations and production methods show variability by comparison with conventional ice cream. In addition to culture inoculation, prebiotic components that enhance the growth and viability of probiotic bacteria can be used in probiotic ice cream formulations. Moreover, different production techniques can be applied to further increase the viability of probiotic culture. In order to develop new probiotic ice cream with high quality or improve the quality of current product, it is substantial to determine the effect of addition of probiotic cultures on physicochemical characteristics of ice cream. Some physicochemical properties of probiotic ice creams are summarized below.

Viscosity is defined as the resistance of a liquid material to flow. The viscosity of the ice cream mix is influenced by protein, fat, total solids, and stabilizer in its composition. Furthermore, pasteurization, ageing, and low storage temperature improve ice cream mix viscosity. The smoothness of the ice cream texture and the resistance to melting improve viscosity (Goff & Hartel, 2013). Studies showed that fermented probiotic ice creams have higher viscosity than unfermented ice creams (Arslan et al., 2016; Salem et al., 2005). Moreover, the viscosities of unfermented probiotic and plain ice creams have been found to be similar. The high viscosity in fermented probiotic ice cream can be caused by metabolic products generated by probiotics during fermentation and protein precipitation, due to increased acidity. Along the fermentation, probiotic cultures can produce exopolysaccharides that cause viscosity increase (Patel, Michaud, Singhania, Soccol, & Pandey, 2010). In addition, the inulin and oligosaccharides which are used as prebiotic components in the formulations increase the mix viscosity thanks to their high hygroscopic properties (Akalın & Erişir, 2008).

Salem et al. (2005) pointed out that the freezing point of probiotic ice cream mix is relatively low by comparison with that of the conventional ice cream mix. This difference is related to the acidity value of the probiotic ice cream mix. In the production of probiotic ice cream, except for unfermented ones, either the mix is fermented or fermented milk is added into the mix, and thus the acidity of the mix enhances. The improved acidity alters the molecular weight and solubility of some mix components, such as lactose and milk salts. The quantity and molecular weight of solutes in the mix affects the freezing point.

Specific gravity of ice cream is affected by the addition of probiotic culture. El-Sayed et al. (2014) and Salem et al. (2005) showed that the supplementation of probiotics into the mix increased the specific gravity by about 3-5%. The specific gravity of ice cream hinges on the components which are used in formulation and the overrun level in the ice cream. High overrun percentage reduces the specific gravity of ice cream, since air cells have low specific gravity. Consequently, the changes in specific gravity are related to the ability of the ice cream mix to hold air bubbles. Hence, the high specific gravity of probiotic ice cream may be due to the reducing effect of probiotic addition on overrun percentage of ice cream.

Overrun is an important application in the ice cream production to obtain proper sensory and textural properties. The maximum overrun level is specified as 100-120% for standard ice creams (Goff & Hartel, 2013). When the effects of oxygen on probiotic bacteria are taken into account, the overrun level can be reduced. Notwithstanding, overrun levels which will not adversely affect product quality should be selected. Various studies have been carried out to determine the relationship between the probiotic culture addition and the overrun level. Vijayageetha et al. (2011) reported that the supplementation of probiotic cultures to the ice cream mix reduced the overrun level by about 10%. In another study, Salem et al. (2005), found out that probiotic addition affected the overrun percentage. The researchers have suggested that the overrun percentage varies according to strains. This difference has been associated with different acidity values of mixes containing fermented milk because the acidity of the mix has an impact on the freezing point and protein structure. The fact that the overrun percentage of unfermented ice cream (Abghari et al., 2011) is higher than fermented ice cream (Da Silva et al., 2015) supports this situation. However, some studies reported that the addition of probiotics to the mix did not significantly affect overrun levels of ice creams (Akalın & Erişir, 2008; Alamprese et al., 2005). Additionally, Akalın and Erişir (2008) determined that the use of inulin and oligofructose as prebiotics in the probiotic ice cream formulation increases the overrun, and inulin provides higher overrun capacity than oligofructose does.

Beside the overrun, the melting rate attribute is a significant quality parameter of ice cream. The melting behaviour is affected by overrun and fat content. High fat content and overrun percentage result in slow melting of ice cream, since air cells in ice cream function as an insulator against ambient temperature and the fat protects the foam structure of ice cream. A low freezing point leads to rapid melting, thereby the components in the ice cream formulation have an impact on melting characteristic. Some studies indicated that the inoculation of probiotic culture into the mix enhanced the melting time, due to increases in the freezing point and in the viscosity of the mix (El-Sayed et al., 2014; Salem et al., 2005). On the other hand, Abghari et al. (2011) reported that the addition of probiotics into the ice cream mix did not affect the melting behaviour of ice cream. This difference arises from the production technique of probiotic ice creams. Experiments of El-Sayed et al. (2014) and Salem et al. (2005) were conducted on fermented probiotic ice creams, while Abghari et al. (2011) evaluated unfermented ice creams in their studies. It seems that the fermentation application reduces the melting rate, which is the quality parameter of ice cream. A high melting rate creates problems, especially in countries with tropical climates. In this regard, appropriate production techniques can be selected and new probiotic ice cream formulations can be developed to increase the melting time. Akalın and Erişir (2008), who used the inulin in probiotic ice cream formulation, notified that the increase in the amount of inulin enhances the melting time by decreasing the melting rate, thanks to its water binding capacity.

The pH and titratable acidity of probiotic ice cream can show alteration depending on the production technique (fermented or unfermented), fermentation time, and probiotic strains which are used in production. In fermented ice creams, the titratable acidity of the mix shows an increase up to about 0.45% lactic acid, depending on the strain which is used after incubation with probiotic culture, and the pH of the mix decreases in parallel with this. Vijayageetha et al. (2011) determined that the titratable acidities of ice creams containing *L. acidophilus+B. bifidum* and only *L. acidophilus* were 0.27 and 0.43% lactic acid, respectively. Also, the time required to achieve the desired acidity in the production of probiotic ice cream varies according to the used strain. Magarinos et al. (2007) notified that ice cream mixes containing *L. acidophilus* reached pH 5.0 in a shorter time than those containing *B. lactis*. *Bifidobacterium* species can need more special nutrients than *Lactobacillus* species for their growth, and thus the completion of fermentation at desired acidity level takes long times. Low storage temperatures limit or terminate the bacterial activity. Abghari et al. (2011) and Favaro-Trindade et al. (2007) reported that the pH and titratable acidity of fermented ice creams did not change statistically during storage at -19 °C for 12 weeks.

FUTURE RESEARCH DIRECTIONS

Microencapsulation applications can further preserve probiotic bacteria from mechanical and freezing conditions. In this context, the effect of more different coating materials on probiotic viability can be investigated. Nevertheless, further studies should be conducted to determine the effect of microcapsules on the textural and sensory properties of ice cream. In addition to microencapsulation studies, new probiotic cultures with enhanced expression of stress proteins can be developed by molecular studies. Different probiotic strains and formulations can be utilized to offer a broader range of probiotic ice creams to consumers. Further researches are required on new ice cream formulations that support probiotic growth and survival. Especially in these studies, natural ingredients that can be used as prebiotics should be focused.

CONCLUSION

Ice cream is a potentially convenient media to deliver probiotics to the consumer, since it provides high viability of the probiotics as well as acceptable sensory and physicochemical attributes. Nevertheless, viable probiotic counts in the ice cream matrix can be lost depending on the exposure of cultures to stress factors in the formulation, freezing, overrun, and storage processes. In this context, the design of the formulation with prebiotic compounds, the election of oxygen-resistant strains, the application of microencapsulation techniques, the utilization of packaging materials with low or no oxygen permeability, and the implementation of the fermentation process in the production can improve the survival of probiotics in ice cream during storage time. Further studies should be conducted on the selection of a more appropriate strain and technique in ice cream production, in order to increase the resistance of industrially used strains and to preserve their viability. Finally, probiotics do not adversely affect the sensory attributes of ice creams which are produced with appropriate techniques and formulations. Furthermore, probiotic cultures show good survivability during frozen storage time. However, further research should be carried out to investigate whether health benefits of probiotics in ice creams are still maintained after frozen storage for a long time.

REFERENCES

Abghari, A., Sheikh-Zeinoddin, M., & Soleimanian-Zad, S. (2011). Nonfermented ice cream as a carrier for *Lactobacillus acidophilus* and *Lactobacillus rhamnosus*. *International Journal of Food Science & Technology*, *46*(1), 84–92. doi:10.1111/j.1365-2621.2010.02453.x

Aboulfazli, F., Shori, A. B., & Baba, A. S. (2016). Effects of the replacement of cow milk with vegetable milk on probiotics and nutritional profile of fermented ice cream. *Lebensmittel-Wissenschaft + Technologie*, *70*, 261–270. doi:10.1016/j.lwt.2016.02.056

Akalın, A., & Erişir, D. (2008). Effects of inulin and oligofructose on the rheological characteristics and probiotic culture survival in low-fat probiotic ice cream. *Journal of Food Science*, *73*(4), 184–188. doi:10.1111/j.1750-3841.2008.00728.x PMID:18460135

Akın, M. B., Akın, M. S., & Kırmacı, Z. (2007). Effects of inulin and sugar levels on the viability of yogurt and probiotic bacteria and the physical and sensory characteristics in probiotic ice-cream. *Food Chemistry*, *104*(1), 93–99. doi:10.1016/j.foodchem.2006.11.030

Akın, N. (2009). *Ice cream science and technology*. Konya, Turkey: Damla Press.

Alamprese, C., Foschino, R., Rossi, M., Pompei, C., & Corti, S. (2005). Effects of *Lactobacillus rhamnosus* GG addition in ice cream. *International Journal of Dairy Technology*, *58*(4), 200–206. doi:10.1111/j.1471-0307.2005.00214.x

Alamprese, C., Foschino, R., Rossi, M., Pompei, C., & Savani, L. (2002). Survival of *Lactobacillus johnsonii* La1 and influence of its addition in retail-manufactured ice cream produced with different sugar and fat concentrations. *International Dairy Journal*, *12*(2), 201–208. doi:10.1016/S0958-6946(01)00159-5

Alm, L. (1982). Effect of fermentation on B-vitamin content of milk in Sweden. *Journal of Dairy Science*, *65*(3), 353–359. doi:10.3168/jds.S0022-0302(82)82199-1 PMID:7076958

Anal, A. K., & Singh, H. (2007). Recent advances in microencapsulation of probiotics for industrial applications and targeted delivery. *Trends in Food Science & Technology*, *18*(5), 240–251. doi:10.1016/j.tifs.2007.01.004

Arslan, A. A., Gocer, E. M. C., Demir, M., Atamer, Z., Hinrichs, J., & Kücükcetin, A. (2016). Viability of *Lactobacillus acidophilus* ATCC 4356 incorporated into ice cream using three different methods. *Dairy Science & Technology*, *96*(4), 477–487. doi:10.100713594-016-0282-5

Bruno-Bárcena, J. M., Andrus, J. M., Libby, S. L., Klaenhammer, T. R., & Hassan, H. M. (2004). Expression of a heterologous manganese superoxide dismutase gene in intestinal lactobacilli provides protection against hydrogen peroxide toxicity. *Applied and Environmental Microbiology*, *70*(8), 4702–4710. doi:10.1128/AEM.70.8.4702-4710.2004 PMID:15294805

Champagne, C. P. (2009). Some technological challenges in the addition of probiotic bacteria to foods. In D. Charalampopoulos & R. A. Rastall (Eds.), *Prebiotics and probiotics science and technology* (pp. 761–804). New York, NY: Springer. doi:10.1007/978-0-387-79058-9_19

Champagne, C. P., Gardner, N. J., & Roy, D. (2005). Challenges in the addition of probiotic cultures to foods. *Critical Reviews in Food Science and Nutrition*, *45*(1), 61–84. doi:10.1080/10408690590900144 PMID:15730189

Chandramouli, V., Kailasapathy, K., Peiris, P., & Jones, M. (2004). An improved method of microencapsulation and its evaluation to protect *Lactobacillus* spp. in simulated gastric conditions. *Journal of Microbiological Methods*, *56*(1), 27–35. doi:10.1016/j.mimet.2003.09.002 PMID:14706748

Cruxen, C. D. S., Hoffmann, J. F., Zandoná, G. P., Fiorentini, Â. M., Rombaldi, C. V., & Chaves, F. C. (2017). Probiotic butiá (*Butia odorata*) ice cream: Development, characterization, stability of bioactive compounds, and viability of *Bifidobacterium lactis* during storage. *Lebensmittel-Wissenschaft + Technologie*, *75*, 379–385. doi:10.1016/j.lwt.2016.09.011

Cruz, A. G., Antunes, A. E., Sousa, A. L. O., Faria, J. A., & Saad, S. M. (2009). Ice-cream as a probiotic food carrier. *Food Research International*, *42*(9), 1233–1239. doi:10.1016/j.foodres.2009.03.020

Cui, J. H., Goh, J. S., Kim, P. H., Choi, S. H., & Lee, B. J. (2000). Survival and stability of bifidobacteria loaded in alginate poly-l-lysine microparticles. *International Journal of Pharmaceutics*, *210*(1), 51–59. doi:10.1016/S0378-5173(00)00560-3 PMID:11163987

Da Cruz, A. G., Faria, J. D. A. F., & Van Dender, A. G. (2007). Packaging system and probiotic dairy foods. *Food Research International*, *40*(8), 951–956. doi:10.1016/j.foodres.2007.05.003

Da Silva, P. D. L., De Fátima Bezerra, M., Dos Santos, K. M. O., & Correia, R. T. P. (2015). Potentially probiotic ice cream from goat's milk: Characterization and cell viability during processing, storage and simulated gastrointestinal conditions. *Lebensmittel-Wissenschaft + Technologie*, *62*(1), 452–457. doi:10.1016/j.lwt.2014.02.055

Dave, R., & Shah, N. (1998). Ingredient supplementation effects on viability of probiotic bacteria in yogurt. *Journal of Dairy Science*, *81*(11), 2804–2816. doi:10.3168/jds.S0022-0302(98)75839-4 PMID:9839222

De Vuyst, L. (2000). Technology aspects related to the application of functional starter cultures. *Food Technology and Biotechnology*, *38*(2), 105–112.

Deosarkar, S. S., Kalyankar, S. D., Pawshe, R. D., & Khedkar, C. D. (2016). Ice cream: Composition and health effects. In B. Caballero, P. M. Finglas, & F. Toldrá (Eds.), *Encyclopedia of food and health* (pp. 385–390). Waltham, MA: Academic Press. doi:10.1016/B978-0-12-384947-2.00385-8

Di Criscio, T., Fratianni, A., Mignogna, R., Cinquanta, L., Coppola, R., Sorrentino, E., & Panfili, G. (2010). Production of functional probiotic, prebiotic, and synbiotic ice creams. *Journal of Dairy Science*, *93*(10), 4555–4564. doi:10.3168/jds.2010-3355 PMID:20854989

El-Sayed, H. S., Salama, H. H., & El-Sayed, S. M. (2014). Production of synbiotic ice cream. *International Journal of Chemical Technology Research*, *7*(1), 138–147.

FAO/WHO. (2001). Health and nutritional properties of probiotics in food including powder milk with live lactic acid bacteria. Córdoba, Argentina: Academic Press.

Favaro-Trindade, C. S., Bernardi, S., Bodini, R. B., De Carvalho Balieiro, J. C., & De Almeida, E. (2006). Sensory acceptability and stability of probiotic microorganisms and vitamin C in fermented acerola (*Malpighia emarginata* DC.) ice cream. *Journal of Food Science*, *71*(6), 491–495. doi:10.1111/j.1750-3841.2006.00100.x

Favaro-Trindade, C. S., De Carvalho Balieiro, J. C., Dias, P. F., Amaral Sanino, F., & Boschini, C. (2007). Effects of culture, pH and fat concentration on melting rate and sensory characteristics of probiotic fermented yellow mombin (*Spondias mombin* L) ice creams. *Food Science & Technology International*, *13*(4), 285–291. doi:10.1177/1082013207082387

Ferraz, J. L., Cruz, A. G., Cadena, R. S., Freitas, M. Q., Pinto, U. M., Carvalho, C. C., ... Bolini, H. (2012). Sensory acceptance and survival of probiotic bacteria in ice cream produced with different overrun levels. *Journal of Food Science*, *77*(1), S24–S28. doi:10.1111/j.1750-3841.2011.02508.x PMID:22260128

Fu, B., & Labuza, T. P. (1997). Shelf-life testing: procedures and prediction methods. In M. C. Erickson & Y. Hung (Eds.), *Quality in frozen food* (pp. 377–415). New York, NY: Springer. doi:10.1007/978-1-4615-5975-7_19

Godward, G., & Kailasapathy, K. (2003). Viability and survival of free, encapsulated and co-encapsulated probiotic bacteria in ice cream. *Milchwissenschaft. Milk Science International*, *58*(3-4), 161–164.

Godward, G., Sultana, K., Kailasapathy, K., Peiris, P., Arumugaswamy, R., & Reynolds, N. (2000). The importance of strain selection on the viability and survival of probiotic bacteria in dairy foods. *Milchwissenschaft. Milk Science International*, *55*(8), 441–445.

Goff, H. D., & Hartel, R. W. (2013). *Ice cream*. New York, NY: Springer. doi:10.1007/978-1-4614-6096-1

Gomes, A. M., & Malcata, F. X. (1999). Bifidobacterium spp. and *Lactobacillus acidophilus*: Biological, biochemical, technological and therapeutical properties relevant for use as probiotics. *Trends in Food Science & Technology*, *10*(4), 139–157. doi:10.1016/S0924-2244(99)00033-3

Gurr, M. I. (1987). Nutritional aspects of fermented milk products. *FEMS Microbiology Reviews*, *3*(3), 337–342. doi:10.1111/j.1574-6968.1987.tb02470.x

Hagen, M., & Narvhus, J. (1999). Production of ice cream containing probiotic bacteria. *Milchwissenschaft. Milk Science International*, *54*(5), 265–268.

Haynes, I., & Playne, M. (2002). Survival of probiotic cultures in low-fat ice-cream. *Australian Journal of Dairy Technology*, *57*(1), 10–14.

Hekmat, S., & McMahon, D. J. (1992). Survival of *Lactobacillus acidophilus* and *Bifidobacterium bifidum* in ice cream for use as a probiotic food. *Journal of Dairy Science*, *75*(6), 1415–1422. doi:10.3168/jds.S0022-0302(92)77895-3 PMID:1500547

Homayouni, A., Azizi, A., Ehsani, M., Yarmand, M., & Razavi, S. (2008). Effect of microencapsulation and resistant starch on the probiotic survival and sensory properties of synbiotic ice cream. *Food Chemistry*, *111*(1), 50–55. doi:10.1016/j.foodchem.2008.03.036

Jay, J. M., Loessner, M. J., & Golden, D. A. (2008). *Modern food microbiology*. New York, NY: Springer.

Kailasapathy, K., & Sultana, K. (2003). Survival and [beta]-D-galactosidase activity of encapsulated and free *Lactobacillus acidophilus* and *Bifidobacterium lactis* in ice-cream. *Australian Journal of Dairy Technology*, *58*(3), 223–227.

Karthikeyan, N., Elango, A., Kumaresan, G., Golapakrishnamurty, T., & Raghunath, B. (2014). Enhancement of probiotic viability in ice cream by microencapsulation. *International Journal of Science. Environmental Technology*, *3*(1), 339–347.

King, H., & Alexander, H. (1948). The mechanical destruction of bacteria. *Microbiology*, *2*(3), 315–324.

Krasaekoopt, W., Bhandari, B., & Deeth, H. (2004). The influence of coating materials on some properties of alginate beads and survivability of microencapsulated probiotic bacteria. *International Dairy Journal*, *14*(8), 737–743. doi:10.1016/j.idairyj.2004.01.004

Lee, Y. K. (2009). Probiotic microorganisms. In Y. K. Lee & S. Salminen (Eds.), *Handbook of probiotics and prebiotics* (pp. 1–139). Hoboken, NJ: John Wiley & Sons.

Magarinos, H., Selaive, S., Costa, M., Flores, M., & Pizarro, O. (2007). Viability of probiotic micro-organisms (*Lactobacillus acidophilus* la-5 and *Bifidobacterium animalis* subsp. *lactis* bb-12) in ice cream. *International Journal of Dairy Technology*, *60*(2), 128–134. doi:10.1111/j.1471-0307.2007.00307.x

Miller, C. W., Nguyen, M. H., Rooney, M., & Kailasapathy, K. (2003). The control of dissolved oxygen content in probiotic yoghurts by alternative packaging materials. *Packaging Technology & Science*, *16*(2), 61–67. doi:10.1002/pts.612

Mohammadi, R., Mortazavian, A. M., Khosrokhavar, R., & Da Cruz, A. G. (2011). Probiotic ice cream: Viability of probiotic bacteria and sensory properties. *Annals of Microbiology*, *61*(3), 411–424. doi:10.100713213-010-0188-z

Moss, C. W., & Speck, M. L. (1963). Injury and death of *Streptococcus lactis* due to freezing and frozen storage. *Applied Microbiology*, *11*(4), 326–329. PMID:13936199

Nousia, F. G., Androulakis, P. I., & Fletouris, D. J. (2011). Survival of *Lactobacillus acidophilus* LMGP-21381 in probiotic ice cream and its influence on sensory acceptability. *International Journal of Dairy Technology*, *64*(1), 130–136. doi:10.1111/j.1471-0307.2010.00645.x

Öztürk, H. İ., Demirci, T., & Akın, N. (2017). Production of functional probiotic ice creams with white and dark blue fruits of Myrtus communis: The comparison of the prebiotic potentials on Lactobacillus casei 431 and functional characteristics. *Lebensmittel-Wissenschaft + Technologie*, *90*, 339–345.

Parussolo, G., Busatto, R. T., Schmitt, J., Pauletto, R., Schons, P. F., & Ries, E. F. (2017). Synbiotic ice cream containing yacon flour and *Lactobacillus acidophylus* NCFM. *Lebensmittel-Wissenschaft + Technologie*, *82*, 192–198. doi:10.1016/j.lwt.2017.04.049

Patel, A. K., Michaud, P., Singhania, R. R., Soccol, C. R., & Pandey, A. (2010). Polysaccharides from probiotics: New developments as food additives. *Food Technology and Biotechnology*, *48*(4), 451–463.

Ranadheera, C. S., Evans, C., Adams, M., & Baines, S. (2013). Production of probiotic ice cream from goat's milk and effect of packaging materials on product quality. *Small Ruminant Research*, *112*(1), 174–180. doi:10.1016/j.smallrumres.2012.12.020

Ranadheera, R. D. C. S., Baines, S. K., & Adams, M. C. (2010). Importance of food in probiotic efficacy. *Food Research International*, *43*(1), 1–7. doi:10.1016/j.foodres.2009.09.009

Russell, N. J. (2002). Bacterial membranes: The effects of chill storage and food processing. An overview. *International Journal of Food Microbiology*, *79*(1), 27–34. doi:10.1016/S0168-1605(02)00176-9 PMID:12382682

Sagdic, O., Ozturk, I., Cankurt, H., & Tornuk, F. (2012). Interaction between some phenolic compounds and probiotic bacterium in functional ice cream production. *Food and Bioprocess Technology*, *5*(8), 2964–2971. doi:10.100711947-011-0611-x

Salazar, N., Ruas-Madiedo, P., Kolida, S., Collins, M., Rastall, R., Gibson, G., & Clara, G. (2009). Exopolysaccharides produced by *Bifidobacterium longum* IPLA E44 and *Bifidobacterium animalis* subsp. *lactis* IPLA R1 modify the composition and metabolic activity of human faecal microbiota in pH-controlled batch cultures. *International Journal of Food Microbiology*, *135*(3), 260–267. doi:10.1016/j.ijfoodmicro.2009.08.017 PMID:19735956

Salem, M. M., Fathi, F. A., & Awad, R. (2005). Production of probiotic ice cream. *Polish Journal of Food and Nutrition Sciences*, *14*(3), 267–271.

Senanayake, S. A., Fernando, S., Bamunuarachchi, A., & Arsekularatne, M. (2013). Application of *Lactobacillus acidophilus* (LA 5) strain in fruit-based ice cream. *Food Science & Nutrition*, *1*(6), 428–431. doi:10.1002/fsn3.66 PMID:24804052

Sheehan, V. M., Sleator, R. D., Fitzgerald, G. F., & Hill, C. (2006). Heterologous expression of BetL, a betaine uptake system, enhances the stress tolerance of *Lactobacillus salivarius* UCC118. *Applied and Environmental Microbiology*, *72*(3), 2170–2177. doi:10.1128/AEM.72.3.2170-2177.2006 PMID:16517668

Simpson, P., Stanton, C., Fitzgerald, G., & Ross, R. (2005). Intrinsic tolerance of *Bifidobacterium* species to heat and oxygen and survival following spray drying and storage. *Journal of Applied Microbiology*, *99*(3), 493–501. doi:10.1111/j.1365-2672.2005.02648.x PMID:16108790

Soukoulis, C., Fisk, I. D., & Bohn, T. (2014). Ice cream as a vehicle for incorporating health-promoting ingredients: Conceptualization and overview of quality and storage stability. *Comprehensive Reviews in Food Science and Food Safety*, *13*(4), 627–655. doi:10.1111/1541-4337.12083

Streit, F., Delettre, J., Corrieu, G., & Béal, C. (2008). Acid adaptation of *Lactobacillus delbrueckii* subsp. *bulgaricus* induces physiological responses at membrane and cytosolic levels that improves cryotolerance. *Journal of Applied Microbiology*, *105*(4), 1071–1080. doi:10.1111/j.1365-2672.2008.03848.x PMID:18498349

Talwalkar, A., & Kailasapathy, K. (2003). Effect of microencapsulation on oxygen toxicity in probiotic bacteria. *Australian Journal of Dairy Technology*, *58*(1), 36–39.

Talwalkar, A., Miller, C. W., Kailasapathy, K., & Nguyen, M. H. (2004). Effect of packaging materials and dissolved oxygen on the survival of probiotic bacteria in yoghurt. *International Journal of Food Science & Technology*, *39*(6), 605–611. doi:10.1111/j.1365-2621.2004.00820.x

Turgut, T., & Cakmakci, S. (2009). Investigation of the possible use of probiotics in ice cream manufacture. *International Journal of Dairy Technology*, *62*(3), 444–451. doi:10.1111/j.1471-0307.2009.00494.x

Vardar, N. B., & Öksüz, Ö. (2007). Artisan strawberry ice cream made with supplementation of Lactococci or *Lactobacillus acidophilus*. *Italian Journal of Food Science*, *19*(4), 403–412.

Vijayageetha, V., Begum, S. K., & Reddy, Y. K. (2011). Technology and quality attributes of probiotic ice cream. *Tamilnadu. Journal of Veterinary & Animal Sciences (Lahore)*, *7*(6), 299–302.

Villalva, F. J., Cravero Bruneri, A. P., Vinderola, G., Gonçalvez de Oliveira, E., Paz, N. F., & Ramón, A. N. (2017). Formulation of a peach ice cream as potential symbiotic food. *Food Science and Technology (Campinas)*, *37*(3), 456–461. doi:10.1590/1678-457x.19716

Yilmaz-Ersan, L., Ozcan, T., Akpinar-Bayizit, A., Turan, M. A., & Taskin, M. B. (2017). Probiotic cream: Viability of probiotic bacteria and chemical characterization. *Journal of Food Processing and Preservation*, *41*(1), 1–9. doi:10.1111/jfpp.12797

KEY TERMS AND DEFINITIONS

Cryotolerance: An adaption that is developed by bacteria against freezing conditions. It supports the maintenance of viability in freezing conditions.

Ice Cream: A sweetened dairy product containing milk, milk solid nonfat, sugar, emulsifiers, stabilizers, cream, flavoring agents, and colorant. It is widely consumed around the world.

Microencapsulation: A process that surrounds active tiny substances by one or more coating materials.

Overrun: An incorporation of air that occurs by vigorously agitating the mix during freezing, and is defined as an increase in volume.

Prebiotics: Nutrition sources that promote the growth and viability of target microorganisms, for example, dietary fibers.

Probiotics: Viable bacterial cells that ensure health benefits to consumers who use them in adequate amounts.

Symbiotics: Foods or nutrients containing probiotics and prebiotics together. These foods are synergic, so they are called symbiotics. Not only do symbiotics provide transport of probiotic bacteria, but they also improve the viability and growth of probiotic bacteria.

Viable Bacterial Counts: This term provides quantitative information about the presence of bacteria in the media or sample. Viable bacterial counts are expressed in terms of colony forming units per g or mL.

Compilation of References

Aaltonen, T., Huumonen, I., & Myllärinen, P. (2014). Controlled transglutaminase treatment in Edam cheese-making. *International Dairy Journal*, *38*(2), 179–182. doi:10.1016/j.idairyj.2013.12.004

Abadía-García, L., Cardador, A., del Campo, S. T. M., Arvízu, S. M., Castaño-Tostado, E., Regalado González, C., ... Amaya-Llano, S. L. (2013). Influence of probiotic strains added to cottage cheese on generation of potentially antioxidant peptides, anti-listerial activity, and survival of probiotic microorganisms in simulated gastrointestinal conditions. *International Dairy Journal*, *33*(2), 191–197. doi:10.1016/j.idairyj.2013.04.005

Abbasi, M., Amiri, R., Bordbar, A.-K., Ranjbakhsh, E., & Khosropour, A. R. (2016). Improvement of the stability and activity of immobilized glucose oxidase on modified iron oxide magnetic nanoparticles. *Applied Surface Science*, *364*, 752–757. doi:10.1016/j.apsusc.2015.12.120

Abd El-Salam, M. H., Hippen, A. R., El-Shafie, K., Assem, F. M., Abbas, H., Abd El-Aziz, M., ... El-Aassar, M. (2011). Preparation and properties of probiotic concentrated yoghurt (labneh) fortified with conjugated linoleic acid. *International Journal of Food Science & Technology*, *46*(10), 2103–2110. doi:10.1111/j.1365-2621.2011.02722.x

Abghari, A., Sheikh-Zeinoddin, M., & Soleimanian-Zad, S. (2011). Nonfermented ice cream as a carrier for *Lactobacillus acidophilus* and *Lactobacillus rhamnosus*. *International Journal of Food Science & Technology*, *46*(1), 84–92. doi:10.1111/j.1365-2621.2010.02453.x

Aboulfazli, F., Shori, A. B., & Baba, A. S. (2016). Effects of the replacement of cow milk with vegetable milk on probiotics and nutritional profile of fermented ice cream. *Lebensmittel-Wissenschaft + Technologie*, *70*, 261–270. doi:10.1016/j.lwt.2016.02.056

Abou-Soliman, N. H. I., Sakr, S. S., & Awad, S. (2017). Physico-chemical, microstructural and rheological properties of camel-milk yogurt as enhanced by microbial transglutamianse. *Journal of Food Science and Technology*, *54*(6), 1616–1627. doi:10.100713197-017-2593-9 PMID:28559621

Abushelaibi, A., Al-Mahadin, S., El-Tarabily, K., Shah, N. P., & Ayyash, M. (2017). Characterization of potential probiotic lactic acid bacteria isolated from camel milk. *Lebensmittel-Wissenschaft + Technologie*, *79*, 316–325. doi:10.1016/j.lwt.2017.01.041

Aeschlimann, D., & Paulsson, M. (1994). Transglutaminases: Protein cross-linking enzymes in tissues and body fluids. *Thrombosis and Haemostasis*, *73*, 256–263. PMID:7914385

Afroz, Q. M., Khan, K. A., & Ahmed, P. & S, U. (2015). Enzymes used in dairy industries. *International Journal of Applied Research*, *1*(10), 523–527.

Agostoni, C., Axelsson, I., Goulet, O., Koletzko, B., Michaelsen, K. F., & Puntis, J. W. L. ESPGHAN Committee on Nutrition. (2004). Prebiotic oligosaccharides in dietetic products for infants: a commentary by the ESPGHAN Committee on Nutrition. *Journal of Pediatric Gastroenterology and Nutrition, 39*(5), 465–73. Retrieved from http://www.ncbi.nlm.nih.gov/pubmed/15572882

Aguilar, C. N., Gutiérrez-Sánchez, G., Rado-Barragán, P. A., Rodríguez-Herrera, R., Martínez-Hernandez, J. L., & Contreras-Esquivel, J. C. (2008). Perspectives of solid-state fermentation for production of food enzymes. *American Journal of Biochemistry and Biotechnology*, *4*(4), 354–366. doi:10.3844/ajbbsp.2008.354.366

Agyare, K. K., Addo, K., & Xiong, Y. L. L. (2009). Emulsifying and foaming properites of transglutaminase-treated wheat gluten hydrolysateas influenced by pH, temperature and salt. *Food Hydrocolloids*, *23*(1), 72–81. doi:10.1016/j.foodhyd.2007.11.012

Agyare, K. K., & Damodaran, S. (2010). pH-stability and thermal properties of microbial transglutaminase-treated whey protein isolate. *Journal of Agricultural and Food Chemistry*, *58*(3), 1946–1953. doi:10.1021/jf903530d PMID:20088508

Ahiko, K., Iwasawa, S., Ulda, M., & Nigata, N. (1981). Studies on acid carboxypeptidase from *Penicillium caseicolum.* II. Hydrolysis of bitter peptides by acid carbozypeptidases and large scale preparation of the enzyme. *Report of Research of Laboratory. Snow Brand Milk Products Co.*, *77*, 135–140.

Ahmed, Z., Wang, Y., Anjum, N., Ahmad, A., & Khan, S. T. (2013). Characterization of exopolysaccharide produced by *Lactobacillus kefiranofaciens* ZW3 isolated from Tibet kefir e Part II. *Food Hydrocolloids*, *30*(1), 343–350. doi:10.1016/j.foodhyd.2012.06.009

Akalın, A., & Erişir, D. (2008). Effects of inulin and oligofructose on the rheological characteristics and probiotic culture survival in low-fat probiotic ice cream. *Journal of Food Science*, *73*(4), 184–188. doi:10.1111/j.1750-3841.2008.00728.x PMID:18460135

Akın, M. B., Akın, M. S., & Kırmacı, Z. (2007). Effects of inulin and sugar levels on the viability of yogurt and probiotic bacteria and the physical and sensory characteristics in probiotic ice-cream. *Food Chemistry*, *104*(1), 93–99. doi:10.1016/j.foodchem.2006.11.030

Akin, M. S., & Güler-Akin, M. B. (2017). Effects of capsule materials on the textural and sensory characteristics of Kashar cheese ripened with encapsulated lipase and protease. *Bulgarian Journal of Agricultural Science*, *23*(2), 330–337.

Akin, M. S., Güler-Akin, M. B., Kirmaci, H. A., Atasoy, A. F., & Türkoglu, H. (2012). The effects of lipase-encapsulating carriers on the accelerated ripening of Kashar cheese. *International Journal of Dairy Technology*, *65*(2), 243–249. doi:10.1111/j.1471-0307.2012.00821.x

Akın, N. (2009). *Ice cream science and technology*. Konya, Turkey: Damla Press.

Alamprese, C., Foschino, R., Rossi, M., Pompei, C., & Corti, S. (2005). Effects of *Lactobacillus rhamnosus* GG addition in ice cream. *International Journal of Dairy Technology*, *58*(4), 200–206. doi:10.1111/j.1471-0307.2005.00214.x

Alamprese, C., Foschino, R., Rossi, M., Pompei, C., & Savani, L. (2002). Survival of *Lactobacillus johnsonii* La1 and influence of its addition in retail-manufactured ice cream produced with different sugar and fat concentrations. *International Dairy Journal*, *12*(2), 201–208. doi:10.1016/S0958-6946(01)00159-5

Albano, H., Pinho, C., Leite, D., Barbosa, J., Silva, J., Carneiro, L., ... Teixeira, P. (2009). Evaluation of a bacteriocin-producing strain of *Pediococcus acidilactici* as a biopreservative for "Alheira", a fermented meat sausage. *Food Control*, *20*(8), 764–770. doi:10.1016/j.foodcont.2008.09.021

Albenzio, M., Santillo, A., Caroprese, M., Marino, R., Trani, A., & Faccia, M. (2010). Biochemical patterns in ovine cheese: Influence of probiotic strains. *Journal of Dairy Science*, *93*(8), 3487–3496. doi:10.3168/jds.2009-3036 PMID:20655416

Albenzio, M., Santillo, A., Caroprese, M., Ruggieri, D., Napolitano, F., & Sevi, A. (2013). Physicochemical properties of Scamorza ewe milk cheese manufactured with different probiotic cultures. *Journal of Dairy Science*, *96*(5), 2781–2791. doi:10.3168/jds.2012-6218 PMID:23522678

Al-Dhaheri, A. S., Al-Hemeiri, R., Kizhakkayil, J., Al-Nabulsi, A., Abushelaibi, A., Shah, N. P., & Ayyash, M. (2017). Health-promoting benefits of low-fat akawi cheese made by exopolysaccharide-producing probiotic Lactobacillus plantarum isolated from camel milk. *Journal of Dairy Science*, *100*(10), 1–9. doi:10.3168/jds.2017-12761 PMID:28755944

Alhelli, A. M., Manap, M. Y. A., Mohammed, A. S., Mirhosseini, H., Suliman, E., Shad, Z., ... Hussin, A. S. M. (2016). Response surface methodology modelling of an aqueous two-phase system for purification of protease from *Penicillium candidum (PCA 1/TT031)* under solid-state fermentation and its biochemical characterization. *International Journal of Molecular Sciences*, *17*(11), 1872. doi:10.3390/ijms17111872 PMID:27845736

Ali, N. A., Ahmed, S. H., Mohamed, E. A., Ahmed, I. A. M., & Babiker, E. E. (2010). Effect of transglutaminase cross linking on the functional properties as a function of NaCl concentration of legumes protein isolate. *International Journal of Biological and Life Sciences*, *6*(1), 8–13.

Al-Lahham, S. H., Peppelenbosch, M. P., Roelofsen, H., Vonk, R. J., & Venema, K. (2010). Biological effects of propionic acid in human; metabolism, potential applications and underlying mechanisms. *Biochimica et Biophysica Acta*, *1801*(11), 1175–1183. doi:10.1016/j.bbalip.2010.07.007 PMID:20691280

Allen, S. J., Okoko, B., Martinez, E. G., Gregorio, G. V., & Dans, L. F. (2003). *Probiotics for treating infectious diarrhoea (Review)*. Wiley & Sons.

Almeida, C. M., Gomes, D., Faro, C., & Simoes, I. (2015). Engineering a cardosin B-derived rennet for sheep and goat cheese manufacture. *Applied Microbiology and Biotechnology*, *99*(1), 269–281. doi:10.100700253-014-5902-5 PMID:24986551

Alm, L. (1982). Effect of fermentation on B-vitamin content of milk in Sweden. *Journal of Dairy Science*, *65*(3), 353–359. doi:10.3168/jds.S0022-0302(82)82199-1 PMID:7076958

Altschul, S. F., Gish, W., Miller, W., Myers, E. W., & Lipman, D. J. (1990). Basic local alignment search tool. *Journal of Molecular Biology*, *215*(3), 403–410. doi:10.1016/S0022-2836(05)80360-2 PMID:2231712

al-Zoreky, N., Ayres, J. W., & Sandine, W. E. (1991). Antimicrobial activity of microgard against food spoilage and pathogenic microorganisms. *Journal of Dairy Science*, *74*(3), 758–763. doi:10.3168/jds.S0022-0302(91)78222-2 PMID:1906486

Aminifar, M., & Emam-Djomeh, Z. (2014). Changes of texture, microstructure and free fatty acid contents of Lighvan cheese during accelerated ripening with lipase. *Journal of Agricultural Science and Technology*, *16*(1), 113–123.

Amiri, Z. R., Khandelwal, P., & Aruna, B. R. (2010). Development of acidophilus milk via selected probiotics and prebiotics using artificial neural network. *Advances in Bioscience and Biotechnology*, *1*(03), 224–231. doi:10.4236/abb.2010.13031

Anal, A. K., & Singh, H. (2007). Recent advances in microencapsulation of probiotics for industrial applications and targeted delivery. *Trends in Food Science & Technology*, *18*(5), 240–251. doi:10.1016/j.tifs.2007.01.004

Anbukkarasi, K., UmaMaheswari, T., Hemalatha, T., Nanda, D. K., Singh, P., & Singh, R. (2014). Preparation of low galactose yogurt using cultures of Gal+ *Streptococcus thermophilus* in combination with *Lactobacillus delbrueckii* ssp. *bulgaricus*. *Journal of Food Science and Technology*, *51*(9), 2183–2189. doi:10.100713197-014-1262-5 PMID:25190881

Ando, H., Adachi, M., Umeda, K., Matsuura, A., Nonaka, M., Uchio, R., ... Motoki, M. (1989). Purification and characteristics of a novel transglutaminase derived from microorganisms. *Agricultural and Biological Chemistry*, *53*, 2613–2617.

Anema, S. G., Lauber, S., Lee, S. K., Henle, T., & Klostermeyer, H. (2005). Rheological propertis of acid gels prepared from pressure- and transglutaminase-treated skim milk. *Food Hydrocolloids*, *19*(5), 879–887. doi:10.1016/j.foodhyd.2004.12.001

Angelopoulou, A., Alexandraki, V., Georgalaki, M., Anastasiou, R., Manolopoulou, E., Tsakalidou, E., & Papadimitriou, K. (2017). Production of probiotic Feta cheese using *Propionibacterium freudenreichii* subsp. *shermanii* as adjunct. *International Dairy Journal*, *66*, 135–139. doi:10.1016/j.idairyj.2016.11.011

Anjani, K., Kailasapathy, K., & Phillips, M. (2007). Microencapsulation of enzymes for potential application in acceleration of cheese ripening. *International Dairy Journal*, *17*(1), 79–86. doi:10.1016/j.idairyj.2006.01.005

Ansari, S. A., Satar, R., Zaidi, S. K., & Ahmad, A. (2014). Immobilization of *Aspergillus oryzae* β-Galactosidase on cellulose acetate-polymethylmethacrylate membrane and its application in hydrolysis of lactose from milk and whey. *International Scholarly Research Notices*, *163987*, 1–6. PMID:27350979

Anusha, R., & Bindhu, O. S. (2016). *Bioactive Peptides from Milk. In Milk Proteins-From Structure to Biological Properties and Health Aspects*. InTech.

Apajalahti, J. H., Särkilahti, L. K., Mäki, B. R., Heikkinen, J. P., Nurminen, P. H., & Holben, W. E. (1998). Effective recovery of bacterial DNA and percent-guanine-plus-cytosine-based analysis of community structure in the gastrointestinal tract of broiler chickens. *Applied and Environmental Microbiology*, *64*(10), 4084–4088. PMID:9758849

Ardebili, H., Pecht, M. G., Ardebili, H., & Pecht, M. G. (2009). Encapsulation Process Technology. In Encapsulation Technologies for Electronic Applications (pp. 129–179). Academic Press. doi:10.1016/B978-0-8155-1576-0.50007-3

Ardelean, A. I., Jaros, D., & Rohm, H. (2013). Influence of microbial transglutaminase cross-linking on gelation kinetics and texture of acid gels made from whole goats and cows milk. *Dairy Science & Technology*, *93*(1), 63–71. doi:10.100713594-012-0092-3

Ardelean, A. I., Otto, C., Jaros, D., & Rohm, H. (2012). Transglutaminase treatment to improve physical properties of gels made from enriched goat milk. *Small Ruminant Research*, *106*(1), 47–53. doi:10.1016/j.smallrumres.2012.05.002

Ardö, Y. (2006). Flavor formation by amino acid catabolism. *Biotechnology Advances*, *24*(2), 238–242. doi:10.1016/j.biotechadv.2005.11.005 PMID:16406465

Ardö, Y. (2011). Cheese | Blue Mold Cheese. In J. W. Fuquay, P. F. Fox, & P. L. H. McSweeney (Eds.), *Encyclopedia of Dairy Sciences* (pp. 767–772). San Diego, CA: Academic Press. doi:10.1016/B978-0-12-374407-4.00090-X

Argyri, A. A., Zoumpopoulou, G., Karatzas, K. A., Tsakalidou, E., Nychas, G. J., Panagou, E. Z., & Tassou, C. C. (2013). Selection of potential probiotic lactic acid bacteria from fermented olives by *in vitro* tests. *Food Microbiology*, *33*(2), 282–291. doi:10.1016/j.fm.2012.10.005 PMID:23200662

Arici, M., & Coskun, F. (2001). Hardaliye: Fermented grape juice as a traditional Turkish beverage. *Food Microbiology*, *18*(4), 417–421. doi:10.1006/fmic.2001.0413

Arihara, K. (2006). Strategies for designing novel functional meat products. *Meat Science*, *74*(1), 219–229. doi:10.1016/j.meatsci.2006.04.028 PMID:22062731

Arioli, S., Zambelli, D., Guglielmetti, S., De Noni, I., Pedersen, M. B., Pedersen, P. D., ... Mora, D. (2013). Increasing the heme-dependent respiratory efficiency of *Lactococcus lactis* by inhibition of lactate dehydrogenase. *Applied and Environmental Microbiology*, *79*(1), 376–380. doi:10.1128/AEM.02734-12 PMID:23064338

Arnau, J., Lauritzen, C., Petersen, G. E., & Pedersen, J. (2006). Current strategies for the use of affinity tags and tag removal for the purification of recombinant proteins. *Protein Expression and Purification*, *48*(1), 1–13. doi:10.1016/j.pep.2005.12.002 PMID:16427311

Arora, T., Sharma, R., & Frost, G. (2011). Propionate. Anti-obesity and satiety enhancing factor? *Appetite*, *56*(2), 511–515. doi:10.1016/j.appet.2011.01.016 PMID:21255628

Arslan, A. A., Gocer, E. M. C., Demir, M., Atamer, Z., Hinrichs, J., & Kücükcetin, A. (2016). Viability of *Lactobacillus acidophilus* ATCC 4356 incorporated into ice cream using three different methods. *Dairy Science & Technology*, *96*(4), 477–487. doi:10.100713594-016-0282-5

Arvola, T., Laiho, K., Torkkeli, S., Mykkänen, H., Salminen, S., Maunula, L., & Isolauri, E. (1999). Prophylactic *Lactobacillus* GG Reduces Antibiotic-Associated Diarrhea in Children With Respiratory Infections: A Randomized Study. *Pediatrics*, *104*(5), 64–68. doi:10.1542/peds.104.5.e64 PMID:10545590

Ashenafi, M., & Busse, M. (1991). Growth Potential of *Salmonella infantis* and *Escherichia coli* in Fermenting Tempeh Made from Horsebean, Pea and Chickpea and their Inhibition by *Lactobacillus plantarum*. *Journal of the Science of Food and Agriculture*, *55*(4), 607–615. doi:10.1002/jsfa.2740550412

Ashour, M. M., Abdel Baky, A. A., El Neshawy, A. A., & Salem, O. M. (1986). Improving the quality of Domiati cheese made from recombined milk. *Food Chemistry*, *20*(2), 85–96. doi:10.1016/0308-8146(86)90145-7

Ataie-Jafari, A., Larijani, B., Majd, H. A., & Tahbaz, F. (2009). Cholesterol-lowering effect of probiotic yogurt in comparison with ordinary yogurt in mildly to moderately hypercholesterolemic subjects. *Annals of Nutrition & Metabolism*, *54*(1), 22–27. doi:10.1159/000203284 PMID:19229114

Auch, A. F., von Jan, M., Klenk, H.-P., & Göker, M. (2010). Digital DNA-DNA hybridization for microbial species delineation by means of genome-to-genome sequence comparison. *Standards in Genomic Sciences*, *2*(1), 117–134. doi:10.4056igs.531120 PMID:21304684

Auty, M. A. E., Gardiner, G. E., McBrearty, S. J., O' Sullivan, E. O., Mulvihill, D. M., Collins, J. K., ... Ross, R. P. (2001). Direct In Situ Viability Assessment of Bacteria in Probiotic Dairy Products Using Viability Staining in Conjunction with Confocal Scanning Laser Microscopy. *Applied and Environmental Microbiology*, *67*(1), 420–425. doi:10.1128/AEM.67.1.420-425.2001 PMID:11133474

Awad, M., Ouda, O., El-Refy, A., El-Feky, F. A., Mosa, K. A., & Helmy, M. (2015). FN-identify: Novel restriction enzymes-based method for bacterial identification in absence of genome sequencing. *Advances in Bioinformatics*, *2015*, 1–14. doi:10.1155/2015/303605 PMID:26880910

Awaisheh, S. S. (2011). Development of probiotic soft cheese manufactured using goat's milk with the addition of thyme. *Milchwissenschaft. Milk Science International*, *66*(1), 51–54.

Awaisheh, S. S. (2012). Probiotic food products classes, types, and processing. In C. R. Everlon (Ed.), *Probiotics* (pp. 551–582). Croatia: Intech Publisher.

Awaisheh, S. S., Al-Dmoor, H. M., Omar, S. S., Hawari, A., & Al-Rwaily, M. M. (2012). Impact of Selected nutraceuticals on viability of probiotic strains in milk during refrigerated storage at 4°C for 15 days. *International Journal of Dairy Technology*, *65*(2), 268–273. doi:10.1111/j.1471-0307.2011.00817.x

Ayad, E. H. E., Verheul, A., de Jong, C., Wouters, J. T. M., & Smit, G. (1999). Flavour forming abilities and amino acid requirements of *Lactococcus lactis* strains isolated from artisanal and non-dairy origin. *International Dairy Journal*, *9*(10), 725–735. doi:10.1016/S0958-6946(99)00140-5

Azarnia, S., Robert, N., & Lee, B. H. (2006). Biotechnological methods to accelerate Cheddar cheese ripening. *Critical Reviews in Biotechnology*, *26*(3), 121–143. doi:10.1080/07388550600840525 PMID:16923531

Bachmann, H. P., Bobst, C., Bütikofer, U., Casey, M. G., Dalla Torre, M., Fröhlich-Wyder, M. T., & Fürst, M. (2005). Occurrence and significance of *Fusarium domesticum* alias Anticollanti on smear-ripened cheeses. *Lebensmittel-Wissenschaft + Technologie*, *38*(4), 399–407. doi:10.1016/j.lwt.2004.05.018

Bachmann, H., Fischlechner, M., Rabbers, I., Barfa, N., Branco dos Santos, F., Molenaar, D., & Teusink, B. (2013). Availability of public goods shapes the evolution of competing metabolic strategies. *Proceedings of the National Academy of Sciences of the United States of America*, *110*(35), 14302–14307. doi:10.1073/pnas.1308523110 PMID:23940318

Bachmann, H., Pronk, J. T., Kleerebezem, M., & Teusink, B. (2014). Evolutionary engineering to enhance starter culture performance in food fermentations. *Current Opinion in Biotechnology*, *32C*, 1–7. PMID:25269887

Bachmann, H., Starrenburg, M. J., Dijkstra, A., Molenaar, D., Kleerebezem, M., Rademaker, J. L., & van Hylckama Vlieg, J. E. (2009). Regulatory phenotyping reveals important diversity within the species *Lactococcus lactis*. *Applied and Environmental Microbiology*, *75*(17), 5687–5694. doi:10.1128/AEM.00919-09 PMID:19592532

Bachmann, H., Starrenburg, M. J., Molenaar, D., Kleerebezem, M., & van Hylckama Vlieg, J. E. (2012). Microbial domestication signatures of *Lactococcus lactis* can be reproduced by experimental evolution. *Genome Research*, *22*(1), 115–124. doi:10.1101/gr.121285.111 PMID:22080491

Banat, I. M., Satpute, S. K., Cameotra, S. S., Patil, R., & Nyayanit, N. V. (2014). Cost effective technologies and renewable substrates for biosurfactants' production. *Frontiers in Microbiology*, *5*(697), 1–18. PMID:25566213

Banjara, N., Suhr, M. J., & Hallen-Adams, H. E. (2015). Diversity of yeast and mold species from a variety of cheese types. *Current Microbiology*, 1–9. PMID:25694357

Bansal, S., Mangal, M., Sharma, S. K., & Gupta, R. K. (2016). Non-dairy based probiotics: A healthy treat for intestine. *Critical Reviews in Food Science and Nutrition*, *56*(11), 1856–1867. doi:10.1080/10408398.2013.790780 PMID:25747894

Barbano, B. M., & Rasmussen, R. R. (1992). Cheese yield performance of fermentation-produced chymosin and other milk coagulants. *Journal of Dairy Science*, *75*(1), 1–12. doi:10.3168/jds.S0022-0302(92)77731-5 PMID:1592942

Barbosa-Cánovas, G. V. (2005). *Food powders : physical properties, processing, and functionality*. Kluwer Academic/Plenum Publishers.

Barenholz, Y. (2001). Liposome application: Problems and prospects. *Current Opinion in Colloid & Interface Science*, *6*(1), 66–77. doi:10.1016/S1359-0294(00)00090-X

Barholt-Perdersen, V., Asbaek-Christensen, K., & Foltmann, B. (1979). Investigations on the activation of bovine prochymosin. *European Journal of Biochemistry*, *94*(2), 573–580. doi:10.1111/j.1432-1033.1979.tb12927.x PMID:34519

Baroiller, C., Schmidt, J. L., & Lapadu-Hargues, M. (1990). Contribution `a l'´etude de l'origine des levures du fromage de Camembert. *Le Lait*, *70*(1), 67–84. doi:10.1051/lait:199017

Barron, L. J. R., Hernández, I., Bilbao, A., Flanagan, C. E., Nájera, A. I., Virto, M., ... De Renobales, M. (2004). Changes in lipid fractions and sensory properties of Idiazabal cheese induced by lipase addition. *The Journal of Dairy Research*, *71*(3), 372–379. doi:10.1017/S0022029904000135 PMID:15354585

Bars, J. L. (1979). Cyclopiazonic acid production by *Penicillium camemberti* Thom and natural occurrence of this mycotoxin in cheese. *Applied and Environmental Microbiology*, *38*(6), 1052–1055. PMID:526014

Bars, J. L. E. (1979). Cyclopiazonic acid production by Penicillium camemberti Thom and natural occurence of this mycotoxin in cheese. *Applied and Environmental Microbiology*, *38*(6), 1052–1055. PMID:526014

Bastian, E. D., & Brown, R. J. (1996). Plasmin in Milk and Dairy Products: An Update. *International Dairy Journal*, *6*(5), 435–457. doi:10.1016/0958-6946(95)00021-6

Başyiğit Kılıç, G., & Akpınar, D. (2013). The effects of different levels of β-glucan on yoghurt manufactured with *Lactobacillus plantarum* strains as adjunct culture. *Journal of Food Agriculture and Environment*, *11*(1), 281–287.

Başyiğit Kılıç, G., Kuleaşan, H., Eralp, I., & Karahan, A. G. (2009). Manufacture of Turkish Beyaz cheese added with probiotic strains. *Lebensmittel-Wissenschaft + Technologie*, *42*(5), 1003–1008. doi:10.1016/j.lwt.2008.12.015

Başyiğit, Kılıç, G., Ağdaş, K., Karahan, A.G., & Çakmakçı, M. L. (2016). Effect of *Lactobacillus plantarum* AK4-11 and different grape varieties on the properties of Hardaliye. *Journal of Agricultural Sciences*, *22*(4), 512–521.

Başyiğit, G., Kuleaşan, H., & Karahan, A. G. (2006). Viability of human derived probiotic lactobacilli in ice-cream produced with sucrose and aspartame. *Journal of Industrial Microbiology & Biotechnology*, *33*(9), 796–800. doi:10.100710295-006-0128-x PMID:16639576

Batt, C. A. (2014). Kluyveromyces. In C. A. Batt & M. L. Tortorello (Eds.), *Encyclopedia of Food Microbiology* (vol. 2, pp. 389–391). Elsevier. doi:10.1016/B978-0-12-384730-0.00173-7

Bautista, L., & Kroll, R. (1988). Survival of some non-starter bacteria in naturally ripened and enzyme-accelerated Cheddar cheese. *The Journal of Dairy Research*, *55*(4), 597–602. doi:10.1017/S0022029900033380 PMID:3148643

Bayjanov, J. R., Starrenburg, M. J., van der Sijde, M. R., Siezen, R. J., & van Hijum, S. A. (2013). Genotype-phenotype matching analysis of 38 *Lactococcus lactis* strains using random forest methods. *BMC Microbiology*, *13*(1), 68. doi:10.1186/1471-2180-13-68 PMID:23530958

Bayjanov, J. R., Wels, M., Starrenburg, M., van Hylckama Vlieg, J. E., Siezen, R. J., & Molenaar, D. (2009). PanCGH: A genotype-calling algorithm for pangenome CGH data. *Bioinformatics (Oxford, England)*, *25*(3), 309–314. doi:10.1093/bioinformatics/btn632 PMID:19129208

Bedani, R., Vieira, A. D. S., Rossi, E. A., & Saad, S. M. I. (2014). Tropical fruit pulps decreased probiotic survival to in vitro gastrointestinal stress in synbiotic soy yoghurt with okara during storage. *Lebensmittel-Wissenschaft + Technologie*, *55*(2), 436–443. doi:10.1016/j.lwt.2013.10.015

Beganović, J., Pavunc, A. L., Gjuračić, K., Spoljarec, M., Šušković, J., & Kos, B. (2011). Improved sauerkraut production with probiotic strain *Lactobacillus plantarum* L4 and *Leuconostoc mesenteroides* LMG 7954. *Journal of Food Science*, *76*(2), 124–129. doi:10.1111/j.1750-3841.2010.02030.x PMID:21535775

Begley, M., Gahan, C. G., & Hill, C. (2005). The interaction between bacteria and bile. *FEMS Microbiology Reviews*, *29*(4), 625–651. doi:10.1016/j.femsre.2004.09.003 PMID:16102595

Behare, P., Singh, R., & Singh, R. P. (2009). Exopolysaccharide-producing mesophilic lactic cultures for preparation of fat-free Dahi–an Indian fermented milk. *The Journal of Dairy Research*, *76*(1), 90–97. doi:10.1017/S0022029908003865 PMID:19121242

Bekar, O., Yilmaz, Y., & Gulten, M. (2011). Kefir improves the efficacy and tolerability of triple therapy in eradicating *Helicobacter pylori*. *Journal of Medicinal Food*, *14*(4), 344–347. doi:10.1089/jmf.2010.0099 PMID:21186984

Benjakul, S., Phatcharat, S., Tammatinna, A., Visessanguan, W., & Kishimura, H. (2008). Improvement of gelling properties of lizardfish mince as influenced by microbial transglutaminase and fish freshness. *Journal of Food Science*, *73*(6), S239–S246. doi:10.1111/j.1750-3841.2008.00813.x PMID:19241566

Benkerroum, N., Ghouati, Y., Ghalfi, H., Elmejdoub, T., Roblain, D., Jacques, P., & Thonart, P. (2002). Biocontrol of *Listeria monocytogenes* in a model cultured milk (lben) by in situ bacteriocin production from *Lactococcus lactis* ssp. *lactis*. *International Journal of Dairy Technology*, *55*(3), 145–151. doi:10.1046/j.1471-0307.2002.00053.x

Benkerroum, N., & Sandine, W. E. (1988). Inhibitory action of nisin against *Listeria monocytogenes*. *Journal of Dairy Science*, *71*(12), 3237–3245. doi:10.3168/jds.S0022-0302(88)79929-4 PMID:3148644

Bentley, R. (2000). Mycophenolic acid: A one hundred year odyssey from antibiotic to immunosuppressant. *Chemical Reviews*, *100*(10), 3801–3826. doi:10.1021/cr990097b PMID:11749328

Beresford, T. P., & Williams, A. (2004). The microbiology of cheese ripening. In P. F. Fox, McSweeney, P. L. H., Cogan, T. M. & Guinee, T. P. (Ed.), Cheese: Chemistry, Physics and Microbiology: Vol 1. General Aspects (pp. 287-318). London: Elsevier. doi:10.1016/S1874-558X(04)80071-X

Beresford, T., & Williams, A. (2004). The microbiology of cheese. In P. F. Fox, P. L. H. McSweeney, T. M. Cogan & T. P. Guinee (Eds.), Cheese: Chemistry, Physics and Microbiology: Vol. 1. General Aspects (3rd ed.; pp. 287-317). London, UK: Academic Press.

Beresford, T. (2007). What are non-starter lactic acid bacteria and how do they affect cheese quality? In P. L. H. McSweeney (Ed.), *Cheese Problems Solved* (pp. 124–130). Boca Raton, FL: CRC Press.

Beresford, T. P., Fitzsimons, N. A., Brennan, N. L., & Cogan, T. M. (2001). Recent advances in cheese microbiology. *International Dairy Journal*, *11*(4-7), 259–274. doi:10.1016/S0958-6946(01)00056-5

Bergamini, C. V., Haynes, E. R., & Zalazar, C. A. (2006). Influence of probiotic bacteria on the pretolysis profile of a semi-hard cheese. *International Dairy Journal*, *8*(16), 856–866. doi:10.1016/j.idairyj.2005.09.004

Bergamini, C. V., Wolf, I. V., Perotti, M. C., & Zalazar, C. A. (2010). Characterisation of biochemical changes during ripening in Argentinean sheep cheeses. *Small Ruminant Research*, *94*(1-3), 79–89. doi:10.1016/j.smallrumres.2010.07.004

Berlec, A., Tompa, G., Slapar, N., Fonović, U. P., Rogelj, I., & Štrukelj, B. (2008). Optimization of fermentation conditions for the expression of sweet-tasting protein brazzein in *Lactococcus lactis*. *Letters in Applied Microbiology*, *46*(2), 227–231. doi:10.1111/j.1472-765X.2007.02297.x PMID:18215220

Bernet, M. F., Brassart, D., Neeser, J. R., & Servin, A. L. (1993). Adhesion of human bifidobacterial strains to cultured human intestinal epithelial cells and inhibition of enteropathogen-cell interactions. *Applied and Environmental Microbiology*, *59*(12), 4121–4128. Retrieved from http://www.ncbi.nlm.nih.gov/pubmed/8285709 PMID:8285709

Beshkova, D., & Frengova, G. (2012). Bacteriocins from lactic acid bacteria: Microorganisms of potential biotechnological importance for the dairy industry. *Engineering in Life Sciences*, *12*(4), 419–432. doi:10.1002/elsc.201100127

Beshkova, D., Simova, E., Frengova, G., & Simov, Z. (1998). Production of flavour compounds by yogurt starter cultures. *Journal of Industrial Microbiology & Biotechnology*, *20*(3-4), 180–186. doi:10.1038j.jim.2900504

Betz, M., Hörmansperger, J., Fuchs, T., & Kulozik, U. (2012). Swelling behaviour, charge and mesh size of thermal protein hydrogels as influenced by pH during gelation. *Soft Matter*, *8*(8), 2477–2485. doi:10.1039/c2sm06976h

Beuvier, E., & Buchin, S. (2004). Raw milk cheeses. In P. F. Fox, McSweeney, P. L. H., Cogan, T. M. & Guinee, T. P. (Ed.), Cheese: Chemistry, Physics and Microbiology: Vol 1. General Aspects (pp. 319-345). London: Elsevier. doi:10.1016/S1874-558X(04)80072-1

Bezerra, T. K. A., Arcanjo, N. M. O., de Araujo, A. R. R., de Queiroz, A. L. M., de Oliveira, M. E. G., Gomes, A. M. P., & Madruga, M. S. (2017). Volatile profile in goat coalho cheese supplemented with probiotic lactic acid bacteria. *Lebensmittel-Wissenschaft + Technologie*, *76*, 209–215. doi:10.1016/j.lwt.2016.03.041

Bhakta, J. N., Ohnishi, K., Munekage, Y., Iwasaki, K., & Wei, M. Q. (2012). Characterization of lactic acid bacteria-based probiotics as potential heavy metal sorbents. *Journal of Applied Microbiology*, *112*(6), 1193–1206. doi:10.1111/j.1365-2672.2012.05284.x PMID:22404232

Bhardwaj, A., Malik, R. K., & Chauhan, P. (2008). Functional and safety aspects of enterococci in dairy foods. *Indian Journal of Microbiology*, *48*(3), 317–325. doi:10.100712088-008-0041-2 PMID:23100728

Bhowmik, T., & Marth, E. H. (1990). Role of Micrococcus and Pediococcus species in cheese ripening. *Journal of Dairy Science*, *73*(4), 859–866. doi:10.3168/jds.S0022-0302(90)78740-1

Bhushan, B., Tomar, S. K., & Chauhan, A. (2017). Techno-functional differentiation of two vitamin B12 producing *Lactobacillus plantarum* strains: An elucidation for diverse future use. *Applied Microbiology and Biotechnology*, *101*(2), 697–709. doi:10.100700253-016-7903-z PMID:27752783

Bi, J., Fang, F., Lu, S., Du, G., & Chen, J. (2013). New insight into the catalytic properties of bile salt hydrolase. *Journal of Molecular Catalysis. B, Enzymatic*, *96*, 46–51. doi:10.1016/j.molcatb.2013.06.010

Bimbo, F., Bonanno, A., Nocella, G., Viscecchia, R., Nardone, G., De Devitiis, B., & Carlucci, D. (2017). Consumers' acceptance and preferences for nutrition-modified and functional dairy products: A systematic review. *Appetite*, *113*, 141–154. doi:10.1016/j.appet.2017.02.031 PMID:28235616

Bines, V. E., Young, P., & Law, B. A. (1989). Comparison of cheddar cheese made with recombinant calf chymosin and standard calf rennet. *The Journal of Dairy Research*, *56*(4), 657–664. doi:10.1017/S0022029900029186

Birschbach, P. (1994). Origins of lipases and their characteristics. *Bulletin of the International Dairy Federation*, (294): 7–10.

Bishop, P. D., Teller, D. C., Smith, R. A., Lasser, G. W., Gilbert, T., & Seale, R. L. (1990). Expression, purification and characterization of human factor XIII in *Saccharomyces cerevisiae*. *Biochemistry*, *29*(7), 1861–1869. doi:10.1021/bi00459a028 PMID:2184890

Biswas, S. R., Ray, P., Johnson, M. C., & Ray, B. (1991). Influence of growth conditions on the prod of a bacteriocin, pediocin AcH, by *Pediococcus acidilactic* H. *Applied and Environmental Microbiology*, *57*, 1265–1267. PMID:16348467

Blaiotta, G., Murru, N., Di Cerbo, A., Succi, M., Coppola, R., & Aponte, M. (2017). Commercially standardized process for probiotic "Italico" cheese production. *Lebensmittel-Wissenschaft + Technologie*, *79*, 601–608. doi:10.1016/j.lwt.2016.11.008

Blank, L. M., Koebmann, B. J., Michelsen, O., Nielsen, L. K., & Jensen, P. R. (2001). Hemin reconstitutes proton extrusion in an H+-ATPase-Negative Mutant of *Lactococcus lactis*. *Journal of Bacteriology*, *183*(22), 6707–6709. doi:10.1128/JB.183.22.6707-6709.2001 PMID:11673444

Bleve, G., Rizzotti, L., Dellaglio, F., & Torriani, S. (2003). Development of reverse transcription (RT)-PCR and real-time RT-PCR assays for rapid detection and quantification of viable yeasts and molds contaminating yogurts and pasteurized food products. *Applied and Environmental Microbiology*, *69*(7), 4116–4122. doi:10.1128/AEM.69.7.4116-4122.2003 PMID:12839789

Bockelmann, W. (2002). Development of defined surface starter cultures fort he ripening of smear cheeses. *International Dairy Journal*, *12*(2-3), 123–131. doi:10.1016/S0958-6946(01)00152-2

Bockelmann, W. (2010). Secondary cheese starter culture. In B. A. Law & A. Y. Tamime (Eds.), *Technology of Cheese Making* (2nd ed.; pp. 193–230). London, UK: Blackwell Pub. Ltd. doi:10.1002/9781444323740.ch6

Bockelmann, W., & Hoppe-Seyler, T. (2001). The surface flora of bacterial smear-ripened cheese from cow's and goat's milk. *International Dairy Journal*, *11*(4-7), 307–314. doi:10.1016/S0958-6946(01)00060-7

Bockelman, W., Krusch, U., Engel, G., Klijn, N., Smit, G., & Heller, K. J. (1997). The microflora of Tilsit cheese. Part 1. Variability of the smear flora. *Die Nahrung*, *41*(4), 208–212. doi:10.1002/food.19970410405

Bockelman, W., Willems, K. P., Neve, H., & Heller, K. H. (2005). Cultures fort he ripening of the smear cheeses. *International Dairy Journal*, *15*(6-9), 719–732. doi:10.1016/j.idairyj.2004.08.022

Böhm, J., Hoff, B., O'Gorman, C. M., Wolfers, S., Klix, V., & Binger, D., ... Kück, U. (2013). Sexual reproduction and mating-type–mediated strain development in the penicillin-producing fungus *Penicillium chrysogenum*. *Proceedings of the National Academy of Sciences*, *110*(4), 1476–1481.

Bolotin, A., Wincker, P., Mauger, S., Jaillon, O., Malarme, K., Weissenbach, J., ... Sorokin, A. (2001). The complete genome sequence of the lactic acid bacterium *Lactococcus lactis* ssp. *lactis* IL1403. *Genome Research*, *11*(5), 731–753. doi:10.1101/gr.GR-1697R PMID:11337471

Bönisch, M. P., Heidebach, T. C., & Kulozik, U. (2008). Influence of transglutaminase protein cross-linking on the rennet coagulation of casein. *Food Hydrocolloids*, *22*(2), 288–297. doi:10.1016/j.foodhyd.2006.11.015

Bönisch, M. P., Huss, M., Lauber, S., & Kulozik, U. (2007b). Yoghurt gel formation by means of enzymatic protein cross-linkling during microbial fermentation. *Food Hydrocolloids*, *21*(4), 585–595. doi:10.1016/j.foodhyd.2006.07.002

Bönisch, M. P., Huss, M., Weitl, K., & Kulozik, U. (2007c). Transglutaminase cross-linking of milk proteins and impact on yoghurt gel properties. *International Dairy Journal*, *17*(11), 1360–1371. doi:10.1016/j.idairyj.2007.01.019

Bönisch, M. P., Lauber, S., & Kulozik, U. (2004). Effect of ultra-high temperature treatment on the enzymatic cross-linking of micellar casein and sodium caseinate by transglutaminase. *Journal of Food Science*, *69*(8), E398–E404. doi:10.1111/j.1365-2621.2004.tb09902.x

Bönisch, M. P., Lauber, S., & Kulozik, U. (2007a). Improvement of enzymatic cross-linking of casein micelles with transglutaminase by glutathione addition. *International Dairy Journal*, *17*(1), 3–11. doi:10.1016/j.idairyj.2006.01.007

Bönisch, M. P., Tolkach, A., & Kulozik, U. (2006). Inactivation of an indigenous transglutaminase inhibitor in milk serum by means of UHT-treatment and membrane separation techniques. *International Dairy Journal*, *16*(6), 669–678. doi:10.1016/j.idairyj.2005.08.014

Bontinis, T. G., Mallatou, H., Pappa, E. C., Massouras, T., & Alichanidis, E. (2012). Study of proteolysis, lipolysis and volatile profile of a traditional Greek goat cheese (Xinotyri) during ripening. *Small Ruminant Research*, *105*(1-3), 193–201. doi:10.1016/j.smallrumres.2012.01.003

Bordoni, A., Amaretti, A., Leonardi, A., Boschetti, E., Danesi, F., Matteuzzi, D., ... Rossi, M. (2013). Cholesterol-lowering probiotics: *In vitro* selection and *in vivo* testing of bifidobacteria. *Applied Microbiology and Biotechnology*, *97*(18), 8273–8281. doi:10.100700253-013-5088-2 PMID:23872958

Botelho, G., Canas, S., & Lameiras, J. (2017). Development of phenolic compounds encapsulation techniques as a major challenge for food industry and for health and nutrition fields. In Nutrient Delivery (pp. 535–586). Academic Press. doi:10.1016/B978-0-12-804304-2.00014-7

Bothast, R. J., Lancaster, E. B., & Hesseltine, C. W. (1975). *Scopulariopsis brevicaulis*: Effect of pH and substrate on growth. *European Journal of Applied Microbiology and Biotechnology*, *1*(1), 55–66. doi:10.1007/BF01880620

Bottari, B., Agrimonti, C., Gatti, M., Neviani, E., & Marmiroli, N. (2013). Development of a multiplex real time PCR to detect thermophilic lactic acid bacteria in natural whey starters. *International Journal of Food Microbiology*, *160*(3), 290–297. doi:10.1016/j.ijfoodmicro.2012.10.011 PMID:23290237

Bottari, B., Santarelli, M., Neviani, E., & Gatti, M. (2010). Natural whey starter for Parmigiano Reggiano: Culture-independent approach. *Journal of Applied Microbiology*, *108*(5), 1676–1684. doi:10.1111/j.1365-2672.2009.04564.x PMID:19849773

Bourdichon, F., Casaregola, S., Farrokh, C., Frisvad, J. C., Gerds, M. L., Hammes, W. P., ... Hansen, E. B. (2012). Food fermentations: Microorganisms with technological beneficial use. *International Journal of Food Microbiology*, *154*(3), 87–97. doi:10.1016/j.ijfoodmicro.2011.12.030 PMID:22257932

Boutrou, R., Aziza, M., & Amrane, A. (2006). Enhanced proteolytic activities of *Geotrichum candidum* and *Penicillium camembertii* in mixed culture. *Enzyme and Microbial Technology*, *39*(2), 325–331. doi:10.1016/j.enzmictec.2005.11.003

Boutrou, R., Sepulchre, A., Gripon, J. C., & Monnet, V. (1998). Simple Tests for Predicting the Lytic Behavior and Proteolytic Activity of Lactococcal Strains in Cheese. *Journal of Dairy Science*, *81*(9), 2321–2328. doi:10.3168/jds.S0022-0302(98)70121-3

Boylston, T. D., Vinderola, C. G., Ghoddusi, H. B., & Reinheimer, J. A. (2004). Incorporation of bifidobacteria into cheeses: Challenges and rewards. *International Dairy Journal*, *14*(5), 375–387. doi:10.1016/j.idairyj.2003.08.008

Boysen, M., Skouboe, P., Frisvad, J., & Rossen, L. (1996). Reclassification of the *Penicillium roqueforti* group into three species on the basis of molecular genetic and biochemical profiles. *Microbiology*, *142*(3), 541–549. doi:10.1099/13500872-142-3-541 PMID:8868429

Brandsma, R. L., Mistry, V. V., Anderson, D. L., & Baldwin, K. A. (1994). Reduced Fat Cheddar Cheese from Condensed Milk. 3. Accelerated Ripening. *Journal of Dairy Science*, *77*(4), 897–906. doi:10.3168/jds.S0022-0302(94)77024-7

Brennan, N. M., Cogan, T. M., Loessner, M., & Scherer, S. (2004). Bacterial surface-ripened cheeses. In P. F. Fox, P. L. H. McSweeney, T. M. Cogan & T. P. Guinee (Eds.), Cheese: Chemistry, Physics and Microbiology: Vol. 2. Major Cheese Groups (3rd ed.; pp. 200-225). London, UK: Academic Press. doi:10.1016/S1874-558X(04)80045-9

Brennan, N. M., Ward, A. C., Beresford, T. P., Fox, P. F., Goodfellow, M., & Cogan, T. M. (2002). Biodiversity of the bacterial flora on the surface of a smear cheese. *Applied and Environmental Microbiology*, *68*(2), 820–830. doi:10.1128/AEM.68.2.820-830.2002 PMID:11823224

Breukink, E., Weidemann, I., van Kraaij, C., Kuipers, O. P., Sahl, H. G., & de Kruijff, B. (1999). Use of cell wall precursor lipid II by a poreforming peptide antibiotic. *Science*, *286*(5448), 2361–2364. doi:10.1126cience.286.5448.2361 PMID:10600751

Briand, V., Buffet, P., Genty, S., Lacombe, K., Godineau, N., Salomon, J., ... Bouchaud, O. (2006). Absence of Efficacy of Nonviable *Lactobacillus acidophilus* for the Prevention of Traveler's Diarrhea: A Randomized, Double-Blind, Controlled Study. *Clinical Infectious Diseases*, *43*(9), 1170–1175. doi:10.1086/508178 PMID:17029137

Broadbent, J. R., & Lin, C. (1999). Effect of heat shock or cold shock treatment on the resistance of Lactococcus lactis to freezing and lyophilization. *Cryobiology*, *39*(1), 88–102. doi:10.1006/cryo.1999.2190 PMID:10458904

Broadbent, J. R., McMahon, D. J., Oberg, C. J., & Welker, D. L. (2001). Use of exopolysaccharide-producing cultures to improve the functionality of low fat cheese. *International Dairy Journal*, *11*(4), 433–439. doi:10.1016/S0958-6946(01)00084-X

Bron, P. A., & Kleerebezem, M. (2011). Engineering lactic acid bacteria for increased industrial functionality. *Bioengineered Bugs*, *2*(2), 80–87. doi:10.4161/bbug.2.2.13910 PMID:21636994

Bron, P. A., Wels, M., Bongers, R. S., van Bokhorst-van de Veen, H., Wiersma, A., Overmars, L., ... Kleerebezem, M. (2012). Transcriptomes reveal genetic signatures underlying physiological variations imposed by different fermentation conditions in *Lactobacillus plantarum*. *PLoS One*, *7*(7), e38720. doi:10.1371/journal.pone.0038720 PMID:22802930

Brooijmans, R. J. W., Poolman, B., Schuurman-Wolters, G. K., De Vos, W. M., & Hugenholtz, J. (2007). Generation of a membrane potential by *Lactococcus lactis* through aerobic electron transport. *Journal of Bacteriology*, *189*(14), 5203–5209. doi:10.1128/JB.00361-07 PMID:17496098

Brooijmans, R., De Vos, W. M., & Hugenholtz, J. (2009a). *Lactobacillus plantarum* WCFS1 electron transport chains. *Applied and Environmental Microbiology*, *75*(11), 3580–3585. doi:10.1128/AEM.00147-09 PMID:19346351

Brooijmans, R., Smit, B., Santos, F., Van Riel, J., de Vos, W. M., & Hugenholtz, J. (2009b). Heme and menaquinone induced electron transport in lactic acid bacteria. *Microbial Cell Factories*, *8*(1), 28. doi:10.1186/1475-2859-8-28 PMID:19480672

Broome, M. C., Krause, D. A., & Hickey, M. W. (1990). The use of non-starter lactobacilli in Cheddar cheese manufacture. *Australian Journal of Dairy Technology*, *45*, 67–73.

Broome, M. C., Powell, I. B., & Limsowtin, G. K. Y. (2003). Starter cultures: Spesific properties. In H. Roginski, J. W. Fuquay, & P. F. Fox (Eds.), *Encyclopedia of Dairy Sciences* (pp. 269–275). London, UK: Academic Press.

Brotz, H., Bierbaum, G., Leopold, K., Reynolds, P. E., & Sahl, H. G. (1998). The lantibiotic mersacidin inhibits peptidoglycan synthesis by targeting lipid II. *Antimicrobial Agents and Chemotherapy*, *42*, 154–160. PMID:9449277

Brown, A. C., & Valiere, A. (2004). Probiotics and medical nutrition therapy. *Nutrition in Clinical Care*, 56–68. PMID:15481739

Brown, L., Pingitore, E. V., Mozzi, F., Saavedra, L., Villegas, J. M., & Hebert, E. M. (2017). Lactic Acid Bacteria as Cell Factories for the Generation of Bioactive Peptides. *Protein and Peptide Letters*, *24*(2), 146–155. doi:10.2174/0929866524666161123111333 PMID:27890004

Bruno-Bárcena, J. M., Andrus, J. M., Libby, S. L., Klaenhammer, T. R., & Hassan, H. M. (2004). Expression of a heterologous manganese superoxide dismutase gene in intestinal lactobacilli provides protection against hydrogen peroxide toxicity. *Applied and Environmental Microbiology*, *70*(8), 4702–4710. doi:10.1128/AEM.70.8.4702-4710.2004 PMID:15294805

Bruzzese, E., Volpicelli, M., Squaglia, M., Tartaglione, A., & Guarino, A. (2006). Impact of prebiotics on human health. *Digestive and Liver Disease*, *38*, 283–287. doi:10.1016/S1590-8658(07)60011-5 PMID:17259092

Buard, A., Carlton, B. D., Floch, F., & Simon, G. S. (2003). Subchronic toxicity, mutagenicity and allergenicity studies of a cultured dextrose food product. *Food and Chemical Toxicology*, *41*(5), 689–694. doi:10.1016/S0278-6915(03)00006-1 PMID:12659722

Buchert, J., Ercili-Cura, D., Ma, H., Gasparetti, C., Monogioudi, E., Faccio, G., ... Kruus, K. (2010). Crosslinking food proteins for improved functionality. *Annual Review of Food Science and Technology*, *1*(1), 113–138. doi:10.1146/annurev.food.080708.100841 PMID:22129332

Bunthof, C. J., & Abee, T. (2002). *Development of a Flow Cytometric Method To Analyze Subpopulations of Bacteria in Probiotic Products and Dairy Starters*. Academic Press.

Burns, P. G., Patrignani, F., Tabanelli, G., Vinderola, G. C., Siroli, L., Reinheimer, J. A., ... Lanciotti, R. (2015). Potential of high pressure homogenisation on probiotic Caciotta cheese quality and functionality. *Journal of Functional Foods*, *13*, 126–136. doi:10.1016/j.jff.2014.12.037

Bzducha, A., Mieczysław, W., & Obiedziński, M. W. (2007). Influence of two probiotic *Lactobacillus* strains on CLA content in model ripening cheeses. *Polish Journal of Food And Nutrition Sciences, 57*(4A), 65-69.

Cağlar, E., Kargül, B., & Tanboğa, T. (2005). Bacteriotherapy and probiotics'role on the oral health. *Oral Diseases*, *3*(11), 131–137. doi:10.1111/j.1601-0825.2005.01109.x PMID:15888102

Çakır, İ. (2003). Laktobasillus ve Bifidobakterlerde Bazı Probiyotik Özelliklerin Belirlenmesi [Determination of Some Probiotic Properties on *Lactobacilli* And *Bifidobacteria*] (PhD Thesis). Ankara University, Graduate School of Natural and Applied Sciences, Department of Food Engineering, Ankara, Turkey.

Cakmakci, S., Cetin, B., Gurses, M., Dagdemir, E., & Hayaloglu, A. A. (2012). Morphological, molecular, and mycotoxigenic identification of dominant filamentous fungi from moldy civil cheese. *Journal of Food Protection*, *75*(11), 2045–2049. doi:10.4315/0362-028X.JFP-12-107 PMID:23127715

Cakmakci, S., Dagdemir, E., Hayaloglu, A. A., Gurses, M., Cetin, B., & Tahmas-Kahyaoglu, D. (2013). Effect of *Penicillium roqueforti* and incorporation of whey cheese on volatile profiles and sensory characteristics of mould-ripened Civil cheese. *International Journal of Dairy Technology*, *66*(4), 512–526.

Calasso, M., Mancini, L., Di Cagno, R., Cardinali, G., & Gobbetti, M. (2015). Microbial cell-free extracts as sources of enzyme activities to be used for enhancement flavor development of ewe milk cheese. *Journal of Dairy Science*, *98*(9), 5874–5889. doi:10.3168/jds.2015-9362 PMID:26142846

Camembert of Normandy - French cheese. (2006, November 12). Retrieved from https://commons.wikimedia.org/wiki/File:Camembert.JPG

Cantal. (2014, September 25). Retrieved from https://commons.wikimedia.org/wiki/File:Cantal_06.jpg

Cantor, M. D., den Tempel, T., Hansen, T. K., & Ardö, Y. (2004). Blue cheese. In P. F. Fox, P. L. H. McSweeney, T. M. Cogan & T. P. Guinee (Eds.), Cheese: Chemistry, Physics and Microbiology: Vol. 2. Major Cheese Groups (3rd ed.; pp. 175-198). London, UK: Academic Press. doi:10.1016/S1874-558X(04)80044-7

Cantor, M. D., van den Tempel, T., Hansen, T. K., & Ardö, Y. (2004). Blue cheese. In P. F. Fox, P. L. H. McSweeney, T. M. Cogan, & T. P. Guinee (Eds.), *Cheese: Chemistry, Physics and Microbiology* (Vol. 2, pp. 175–198). Amsterdam: Elsevier Academic Press.

Cao, M., Lia, Z., Wanga, J., Gea, W., Yuea, T., Lib, R., ... Yuc, W. W. (2012). Food related applications of magnetic iron oxide nanoparticles: Enzyme immobilization, protein purification, and food analysis. *Trends in Food Science & Technology*, *27*(1), 47–56. doi:10.1016/j.tifs.2012.04.003

Cao, Y., Fanning, S., Proos, S., Jordan, K., & Srikumar, S. (2017). A review on the applications of next generation sequencing technologies as applied to food-related microbiome studies. *Frontiers in Microbiology*, *8*, 1–16. doi:10.3389/fmicb.2017.01829 PMID:29033905

Capela, P., Hay, T. K. C., & Shah, N. P. (2006). Effect of cryoprotectants, prebiotics and microencapsulation on survival of probiotic organisms in yoghurt and freeze-dried yoghurt. *Food Research International*, *39*(2), 203–211. doi:10.1016/j.foodres.2005.07.007

Carasi, P., Diaz, M., Racedo, S. M., De Antoni, G., Urdaci, M. C., & de los Angeles Serradell, M. (2014). Safety Characterization and Antimicrobial Properties of Kefir-Isolated *Lactobacillus kefiri. BioMed Research International, 2014*, 1–8. doi:10.1155/2014/208974 PMID:24955346

Carasi, P., Trejo, F. M., Perez, P. F., de Antoni, G. L., & Serradell, M. A. (2012). Surface proteins from *Lactobacillus kefir* antagonize in vitro cytotoxic effect of *Clostridium difficile* toxins. *Anaerobe*, *18*(1), 135–142. doi:10.1016/j.anaerobe.2011.11.002 PMID:22126976

Carbonell, M., Nunez, M., & Fernandez-Garcia, E. (2002). Evolution of the volatile components of ewe raw milk La Serena cheese during ripening. Correlation with flavour characteristics. *Le Lait*, *82*(6), 683–698. doi:10.1051/lait:2002042

Cardoza, R. E., Gutierrez, S., Ortega, N., Casqueiro, J., & Martin, J. F. (2003). Expression of a synthetic copy of the bovine chymosin gene in *Aspergillus awamori* from constitutive and pH-regulated promoters and secretion using two different pre-pro sequences. *Biotechnology and Bioengineering*, *83*(3), 249–259. doi:10.1002/bit.10666 PMID:12783481

Carlton, R. M., Noordman, W. H., Biswas, B., de Meester, E. D., & Loessner, M. J. (2005). Bacteriophage P100 for control of *Listeria monocytogenes* in foods: Genome sequence, bioinformatic analyses, oral toxicity study, and application. *Regulatory Toxicology and Pharmacology*, *43*(3), 301–312. doi:10.1016/j.yrtph.2005.08.005 PMID:16188359

Carneiro, H. C. F., Tonon, R. V., Grosso, C. R. F., & Hubinger, M. D. (2013). Encapsulation efficiency and oxidative stability of flaxseed oil microencapsulated by spray drying using different combinations of wall materials. *Journal of Food Engineering*, *115*(4), 443–451. doi:10.1016/j.jfoodeng.2012.03.033

Carvalhoa, A. S., Silvaa, J., Hob, P., Teixeiraa, P., Malcataa, F. X., & Gibbsa, P. (2004). Relevant factors for the preparation of freeze-dried lactic acid bacteria. *International Dairy Journal*, *14*(10), 835–847. doi:10.1016/j.idairyj.2004.02.001

Casarotti, S. N., Monteiro, D. A., Moretti, M. M. S., & Penna, A. L. B. (2014). Influence of the combination of probiotic cultures during fermentation and storage of fermented milk. *Food Research International*, *59*, 67–75. doi:10.1016/j.foodres.2014.01.068

Cebeci, A., & Gürakan, C. (2003). Properties of potential probiotic *Lactobacillus plantarum* strains. *Food Microbiology*, *20*(5), 511–518. doi:10.1016/S0740-0020(02)00174-0

Celligoi, M., Baldo, C., Rodrigues de Melo, M., Gasparin, F., Marques, T., & Barros, M. Lipase. Properties, functions and food applications. In R. C. Ray & C. M. Rosell (Eds.), *Microbial enzyme technology in food applications* (pp. 214–240). Boca Raton, FL: CRC Press.

Cesselin, B., Derré-Bobillot, A., Fernandez, A., Lamberet, G., Lechardeur, D., Yamamoto, Y., ... Gaudu, P. (2010). Respiration, a strategy to avoid oxidative stress in *Lactococcus lactis*, is regulated by the heme status. *Nihon Nyusankin Gakkaishi*, *21*(1), 10–15. doi:10.4109/jslab.21.10

Ceugniez, A., Coucheney, F., Jacques, P., Daube, G., Delcenserie, V., & Drider, D. (2017). Anti-Salmonella activity and probiotic trends of *Kluyveromyces marxianus* S-2-05 and *Kluyveromyces lactis* S-3-05 isolated from a French cheese, Tomme d'Orchies. *Research in Microbiology*, *168*(6), 575–582. doi:10.1016/j.resmic.2017.03.004 PMID:28366839

Chabalier, C., Ratomahenina, R., Galzy, P., & Dieu, B. (1995). Remarks on the appearance of "violine" on Cantal cheese. *Acta Microbiologica et Immunologica Hungarica*, *42*(2), 199–202. PMID:7551713

Chae, J. P., Valeriano, V. D., Kim, G. B., & Kang, D. K. (2013). Molecular cloning, characterization and comparison of bile salt hydrolases from *Lactobacillus johnsonii* PF01. *Journal of Applied Microbiology*, *114*(1), 121–133. doi:10.1111/jam.12027 PMID:23035872

Chamba, J. F., & Irlinger, F. (2004). Secondary and adjunct cultures. In M. P. L. H. Fox P. F., Cogan T. M., & Guinee T. P. (Ed.), Cheese: Chemistry, Physics and Microbiology (Vol. 1, pp. 191-206). London: Elsevier Academic Press. doi:10.1016/S1874-558X(04)80068-X

Chambi, H., & Grosso, C. (2006). Edible films produced with gelatin and casein cross-linked with transglutaminase. *Food Research International*, *39*(4), 458–466. doi:10.1016/j.foodres.2005.09.009

Champagne, C. P. (2009). Some technological challenges in the addition of probiotic bacteria to foods. In D. Charalampopoulos & R. A. Rastall (Eds.), *Prebiotics and probiotics science and technology* (pp. 761–804). New York, NY: Springer. doi:10.1007/978-0-387-79058-9_19

Champagne, C. P., & Fustier, P. (2007). Microencapsulation for the improved delivery of bioactive compounds into foods. *Current Opinion in Biotechnology*, *18*(2), 184–190. doi:10.1016/j.copbio.2007.03.001 PMID:17368017

Champagne, C. P., Gardner, N. J., & Roy, D. (2005). Challenges in the addition of probiotic cultures to foods. *Critical Reviews in Food Science and Nutrition*, *45*(1), 61–84. doi:10.1080/10408690590900144 PMID:15730189

Chand, D., Avinash, V. S., Yadav, Y., Pundle, A. V., Suresh, C. G., & Ramasamy, S. (2017). Molecular features of bile salt hydrolases and relevance in human health. *Biochimica et Biophysica Acta*, *1861*(1), 2981–2991. doi:10.1016/j.bbagen.2016.09.024 PMID:27681686

Chandramouli, V., Kailasapathy, K., Peiris, P., & Jones, M. (2004). An improved method of microencapsulation and its evaluation to protect *Lactobacillus* spp. in simulated gastric conditions. *Journal of Microbiological Methods*, *56*(1), 27–35. doi:10.1016/j.mimet.2003.09.002 PMID:14706748

Chang, S. C., Lu, K. L., & Yeh, S. F. (1993). Secondary metabolites resulting from degradation of PR toxin by *Penicillium roqueforti. Applied and Environmental Microbiology*, *59*(4), 981–986. PMID:8476299

Charteris, W. P., Kelly, P. M., Morelli, L., & Collins, J. K. (1998). Development and application of an in vitro methodology to determine the transit tolerance of potentially probiotic *Lactobacillus* and *Bifidobacterium* species in the upper human gastrointestinal tract. *Journal of Applied Microbiology*, *84*(5), 759–768. doi:10.1046/j.1365-2672.1998.00407.x PMID:9674129

Cheeseman, K., Ropars, J., Renault, P., Dupont, J., Gouzy, J., Branca, A., ... Brygoo, Y. (2014). Multiple recent horizontal transfers of a large genomic region in cheese making fungi. *Nature Communications*, *5*(3876).

Chen, H. C., Wang, S. Y., & Chen, M. J. (2008). Microbiological study of lactic acid bacteria in kefir grains by culture-dependent and culture-independent methods. *Food Microbiology*, *25*(3), 492–501. doi:10.1016/j.fm.2008.01.003 PMID:18355674

Chen, H., & Hoover, D. G. (2003). Bacteriocins and their food applications. *Comprehensive Reviews in Food Science and Food Safety*, *2*(3), 82–100. doi:10.1111/j.1541-4337.2003.tb00016.x

Chen, K.-N., Chen, M.-J., & Lin, C.-W. (2006). Optimal combination of the encapsulating materials for probiotic microcapsules and its experimental verification (R1). *Journal of Food Engineering*, *76*(3), 313–320. doi:10.1016/j.jfoodeng.2005.05.036

Chen, L., Romandetto, G. E., & Subirade, M. (2006). Food protein-based materials as nutraceutical delivery system. *Trends in Food Science & Technology*, *17*(5), 272–283. doi:10.1016/j.tifs.2005.12.011

Chen, L., & Subirade, M. (2006). Alginate-whey protein granular microspheres as oral delivery vehicles for bioactive compounds. *Biomaterials*, *27*(26), 4646–4654. doi:10.1016/j.biomaterials.2006.04.037 PMID:16714058

Chen, W., Chen, H., Xia, Y., Yang, J., Zhao, J., Tian, F., ... Zhang, H. (2009). Immobilization of recombinant thermostable β-galactosidase from *Bacillus stearothermophilus* for lactose hydrolysis in milk. *Journal of Dairy Science*, *92*(2), 491–498. doi:10.3168/jds.2008-1618 PMID:19164659

Chen, X., Pfeil, J. E., & Gal, S. (2002). The three typical aspartic proteinase genes of *Arabidopsis thaliana* are differentially expressed. *European Journal of Biochemistry*, *269*(18), 4675–4684. doi:10.1046/j.1432-1033.2002.03168.x PMID:12230581

Chen, Y. P., Hsiao, P. J., Hong, W. S., Dai, T. Y., & Chen, M. J. (2012). *Lactobacillus kefiranofaciens* M1 isolated from milk kefir grains ameliorates experimental colitis in vitro and in vivo. *Journal of Dairy Science*, *95*(1), 63–74. doi:10.3168/jds.2011-4696 PMID:22192184

Chen, Y. P., Lee, T. Y., Hong, W. S., Hsieh, H. H., & Chen, M. J. (2013). Effects of *Lactobacillus kefiranofaciens* M1 isolated from kefir grains on enterohemorrhagic *Escherichia coli* infection using mouse and intestinal cell models. *Journal of Dairy Science*, *96*(12), 7467–7477. doi:10.3168/jds.2013-7015 PMID:24140320

Chersilp, B., Shoji, H., Shimizu, H., & Shioya, S. (2003). Interactions between *Lactobacillus kefiranofaciens* and *Saccharomyces cerevisiae* in Mixed Culture for Kefiran Production. *Journal of Bioscience and Bioengineering*, *96*(3), 279–284. doi:10.1016/S1389-1723(03)80194-9 PMID:16233522

Chichlowski, M. (2006). *Effect of probiotic consortium on level and mechanism of intestine function*. Academic Press.

Chinachoti, N., Matsuaki, H., Sonomoto, K., & Ishikazi, A. (1997a). Utilization of xylose as an alternative carbon source for nisin Z production by *Lactococcus lactis* IO-1. *Fac Agric. Kyushu Univ*, *42*, 171–181.

Chinachoti, N., Zaima, T., Masuaki, H., Sonomoto, K., & Ishikazi, A. (1997b). Relationship between nisin Z fermentaire prod and aeration condition using *Lactococcus lactis* I0-1. *Journal of the Faculty of Agriculture, Kyushu University*, *43*, 437–448.

Chiquette, J. (2009). Evaluation of the protective effect of probiotics fed to dairy cows during a subacute ruminal acidosis challenge. *Animal Feed Science and Technology*, *153*(3-4), 278–291. doi:10.1016/j.anifeedsci.2009.07.001

Cho, S. K., Lee, S. J., Shin, S. Y., Moon, J. S., Li, L., Joo, W., ... Han, N. S. (2015). Development of Bile salt-resistant *Leuconostoc citreum* by expression of bile salt hydrolase gene. *Journal of Microbiology and Biotechnology*, *25*(12), 2100–2105. doi:10.4014/jmb.1505.05072 PMID:26282688

Cho, Y. H., Shim, H. K., & Park, J. (2003). Encapsulation of Fish Oil by an Enzymatic Gelation Process Using Transglutaminase Cross-linked Proteins. *Journal of Food Science*, *68*(9), 2717–2723. doi:10.1111/j.1365-2621.2003.tb05794.x

Christensen, B. M., Sørensen, E. S., Højrup, P., Petersen, T. E., & Rasmussen, L. K. (1996). Localization of potential transglutaminase cross-linking sites in bovine caseins. *Journal of Agricultural and Food Chemistry*, *44*(7), 1943–1947. doi:10.1021/jf9602131

Cieśliński, H., Kur, J., Białkowska, A., Baran, I., Makowski, K., & Turkiewicz, M. (2005). Cloning, expression, and purification of a recombinant cold-adapted β-galactosidase from antarctic bacterium *Pseudoalteromonas sp*. 22b. *Protein Expression and Purification*, *39*(1), 27–34. doi:10.1016/j.pep.2004.09.002 PMID:15596357

Claverie-Martin, F., & Vega-Hernàndez, M. C. (2007). Aspartic Proteases Used in Cheese Making. In J. Polaina & A. P. MacCabe (Eds.), *Industrial Enzymes: Structure, Function and Applications* (pp. 207–219). Dordrecht: Springer Netherlands. doi:10.1007/1-4020-5377-0_13

Coda, R., Rizzello, C. G., Pinto, D., & Gobbetti, M. (2012). Selected lactic acid bacteria synthesize antioxidant peptides during sourdough fermentation of cereal flours. *Applied and Environmental Microbiology*, *78*(4), 1087–1096. doi:10.1128/AEM.06837-11 PMID:22156436

Codex Stan., 283-1978. General standard for cheese.

Coeuret, V., Dubernet, S., Bernardeau, M., Gueguen, M., & Vernoux, J. P. (2003). Isolation, characterisation and identification of lactobacilli focusing mainly on cheeses and other dairy products. *Le Lait*, *83*(4), 269–306. doi:10.1051/lait:2003019

Coffey, A., & Ross, R. P. (2002). Bacteriophage-resistance systems in dairy starter strains: molecular analysis to application. In R. J. Siezen, J. Kok, T. Abee, & G. Schasfsma (Eds.), *Lactic Acid Bacteria: Genetics, Metabolism and Applications* (pp. 303–321). Springer. doi:10.1007/978-94-017-2029-8_20

Cogan, T. M. (2003). Microbiology of cheese. In H. Roginski, J. W. Fuquay, & P. F. Fox (Eds.), *Encyclopedia of Dairy Sciences* (pp. 306–314). London, UK: Academic Press.

Cogan, T. M. (2014). Srater cultures: Employed in Cheesemaking. In C. A. B. L. Tortorello (Ed.), *Encyclopedia of Food Microbiology* (2nd ed.; pp. 508–514). Oxford, UK: Academic Press. doi:10.1016/B978-0-12-384730-0.00322-0

Cogan, T. M., Barbosa, M., Beuvier, E., Bianchi-Salvadori, B., Cocconcelli, P. S., Fernandes, I., ... Rodriguez, E. (1997). Characterization of the lactic acid bacteria in artisanal dairy products. *The Journal of Dairy Research*, *64*(3), 409–421. doi:10.1017/S0022029997002185

Collins, Y. F., McSweeney, P. L. H., & Wilkinson, M. G. (2003). Lipolysis and free fatty acid catabolism in cheese: A review of current knowledge. *International Dairy Journal*, *13*(11), 841–866. doi:10.1016/S0958-6946(03)00109-2

Coman, M. M., Verdenelli, M. C., Cecchini, C., Silvi, S., Vasile, A., Bahrim, G. E., ... Cresci, A. (2013). Effect of buckwheat flour and oat bran on growth and cell viability of the probiotic strains *Lactobacillus rhamnosus* IMC 501®, *Lactobacillus paracasei* IMC 502® and their combination SYNBIO®, in synbiotic fermented milk. *International Journal of Food Microbiology*, *167*(2), 261–268. doi:10.1016/j.ijfoodmicro.2013.09.015 PMID:24140807

Consorzio per la tutela del formaggio gorgonzola DOP. (2017, April 4). Retrieved from https://commons.wikimedia.org/wiki/File:Gorgonzoladop.jpg

Cornu, A., Rabiau, N., Kondjoyan, N., Verdier-Metz, I., Pradel, P., Tournayre, P., ... Martin, B. (2009). Odour-active compound profiles in Cantal-type cheese: Effect of cow diet, milk pasteurization and cheese ripening. *International Dairy Journal*, *19*(10), 588–594. doi:10.1016/j.idairyj.2009.04.008

Corrales, A., Henderson, M., & Morales, I. (2007). Survival of probiotic microorganisms *Lactobacillus acidophilus* and *Bifidobacterium lactis* in whipped ice cream. *Revista Chilena de Nutrición*, *34*(2), 157–163.

Corsetti, A., Gobbetti, M., Smacchi, E., De Angelis, M., & Rossi, J. (1998). Accelerated ripening of Pecorino Umbro cheese. *The Journal of Dairy Research*, *65*(4), 631–642. doi:10.1017/S0022029998003161

Corsetti, A., Rossi, J., & Gobbetti, M. (2001). Interactions between yeast and bacteria in the smear surface-ripened cheeses. *International Journal of Food Microbiology*, *69*(1-2), 1–10. doi:10.1016/S0168-1605(01)00567-0 PMID:11589547

Coşkun, T. (2005). Fonksiyonel besinlerin sağlığımız üzerine etkileri. *Çocuk Sagligi ve Hastaliklari Dergisi*, *48*, 69–84. Retrieved from http://www.cshd.org.tr/uploads/pdf_CSH_133.pdf

Coşkun, T. (2006). Pro-, Pre- ve Sinbiyotikler. *Çocuk Sagligi ve Hastaliklari Dergisi*, *49*, 128–148. Retrieved from http://www.cshd.org.tr/uploads/pdf_CSH_204.pdf

Coultate, T. P. (2002). *Food: The chemistry of its components*. Cambridge, UK: The Royal Society of Chemistry.

Council for Agricultural Science and Technology (CAST). (2007). *Probiotics: their potential to impact human health. Issue Paper 36*. Ames, Iowa: CAST.

Cousin, F. J., Mater, D. D. G., Foline, B., & Jan, G. (2011). Dairy propionibacteria as human probiotics: A review of recent evidence. *Dairy Science & Technology*, *91*, 1–26.

Cozzolino, A., Di Pierro, P., Mariniello, L., Sorrentino, A., Masi, P., & Porta, R. (2003). Incorporation of whey proteins into cheese curd by using transglutamianse. *Biotechnology and Applied Biochemistry*, *38*(3), 289–295. doi:10.1042/BA20030102 PMID:12906711

Crowley, S., Mahony, J., Morrissey, J. P., & van Sinderen, D. (2013). Transcriptomic and morphological profiling of *Aspergillus fumigatus* Af293 in response to antifungal activity produced by *Lactobacillus plantarum* 16. *Microbiology*, *159*(Pt_10), 2014–2024. doi:10.1099/mic.0.068742-0 PMID:23876797

Crow, V., Curry, B., & Hayes, M. (2001). The ecology of non-starter lactic acid bacteria (NSLAB) and their use as adjuncts in New Zealand Cheddar. *International Dairy Journal*, *11*(4-7), 275–283. doi:10.1016/S0958-6946(01)00057-7

Cruxen, C. D. S., Hoffmann, J. F., Zandoná, G. P., Fiorentini, Â. M., Rombaldi, C. V., & Chaves, F. C. (2017). Probiotic butiá (*Butia odorata*) ice cream: Development, characterization, stability of bioactive compounds, and viability of *Bifidobacterium lactis* during storage. *Lebensmittel-Wissenschaft + Technologie*, *75*, 379–385. doi:10.1016/j.lwt.2016.09.011

Cruz, A. G., Antunes, A. E., Sousa, A. L. O. P., Faria, J. A. F., & Saad, S. M. I. (2009). Ice-cream as a probiotic food carrier. *Food Research International*, *42*(9), 1233–1239. doi:10.1016/j.foodres.2009.03.020

Cruz, A. G., Cadena, R. S., Castro, W. F., Esmerino, E. A., Rodrigues, J. B., Gaze, L., ... Bolini, H. M. A. (2013). Consumer perception of probiotic yogurt: Performance of check all that apply (CATA), projective mapping, sorting and intensity scale. *Food Research International*, *54*(1), 601–610. doi:10.1016/j.foodres.2013.07.056

Cruz, A. G., Castro, W. F., Faria, J. A. F., Bogusz, S. Jr, Granato, D., Celeguini, R. M. S., ... Godoy, H. T. (2012a). Glucose oxidase: A potential option to decrease the oxidative stress in stirred probiotic yogurt. *Lebensmittel-Wissenschaft + Technologie*, *47*(2), 512–515. doi:10.1016/j.lwt.2012.01.037

Cruz, A. G., Castro, W. F., Faria, J. A., Lollo, P. C., Amaya-Farfan, J., Freitas, M. Q., ... Godoy, H. T. (2012b). Probiotic yogurts manufactured with increased glucose oxidase levels: Postacidification, proteolytic patterns, survival of probiotic microorganisms, production of organic acid and aroma compounds. *Journal of Dairy Science*, *95*(5), 2261–2269. doi:10.3168/jds.2011-4582 PMID:22541455

Cruz, A. G., Faria, J. A. F., Walter, E. H. M., Andrade, R. R., Cavalcanti, R. N., Oliveira, C. A. F., & Granato, D. (2010). Processing optimization of probiotic yogurt containing glucose oxidase using response surface methodology. *Journal of Dairy Science*, *93*(11), 5059–5068. doi:10.3168/jds.2010-3336 PMID:20965320

Cuatrecasas, P., Wilchek, M., & Anfinsen, C. B. (1968). Selective enzyme purification by affinity chromatography. *Proceedings of the National Academy of Sciences of the United States of America*, *61*(2), 636–643. doi:10.1073/pnas.61.2.636 PMID:4971842

Cuffia, F., George, G., Renzulli, P., Reinheimer, J., Meinardi, C., & Burns, P. (2017). Technological challenges in the production of a probiotic pasta filata soft cheese. *Lebensmittel-Wissenschaft + Technologie*, *81*, 111–117. doi:10.1016/j.lwt.2017.03.039

Cui, J. H., Goh, J. S., Kim, P. H., Choi, S. H., & Lee, B. J. (2000). Survival and stability of bifidobacteria loaded in alginate poly-l-lysine microparticles. *International Journal of Pharmaceutics*, *210*(1), 51–59. doi:10.1016/S0378-5173(00)00560-3 PMID:11163987

Curioni, P. M. G., & Bosset, J. O. (2002). Key odorants in various cheese types as determined by gas chromatography-olfactometry. *International Dairy Journal*, *12*(12), 959–984. doi:10.1016/S0958-6946(02)00124-3

D'Angelo, L., Cicotello, J., Zago, M., Guglielmotti, D., Quiberoni, A., & Suarez, V. (2017, September). Leuconostoc strains isolated from dairy products: Response against food stress conditions. *Food Microbiology*, *66*, 28–29. doi:10.1016/j.fm.2017.04.001 PMID:28576370

Da Cruz, A. G., Buriti, F. C. A., de Souza, C. H. B., Faria, J. A. F., & Saad, S. M. I. (2009). Probiotic cheese: Health benefits, technological and stability aspect. *Trends in Food Science & Technology*, *20*(8), 344–354. doi:10.1016/j.tifs.2009.05.001

Da Cruz, A. G., Faria, J. D. A. F., & Van Dender, A. G. (2007). Packaging system and probiotic dairy foods. *Food Research International*, *40*(8), 951–956. doi:10.1016/j.foodres.2007.05.003

Da Silva, P. D. L., De Fátima Bezerra, M., Dos Santos, K. M. O., & Correia, R. T. P. (2015). Potentially probiotic ice cream from goat's milk: Characterization and cell viability during processing, storage and simulated gastrointestinal conditions. *Lebensmittel-Wissenschaft + Technologie*, *62*(1), 452–457. doi:10.1016/j.lwt.2014.02.055

Daigle, A., Roy, D., Belanger, D., & Vuillemard, J. (1999). Production of probiotic cheese (Cheddar like cheese) using enriched cream fermented by *Bifidobacterium infantis. Journal of Dairy Science*, *82*(6), 1081–1091. doi:10.3168/jds.S0022-0302(99)75330-0 PMID:10386294

Dalié, D. K. D., Deschamps, A. M., & Richard-Forget, F. (2010). Lactic acid bacteria–Potential for control of mould growth and mycotoxins: A review. *Food Control*, *21*(4), 370–380. doi:10.1016/j.foodcont.2009.07.011

Daly, C., Fitzgerald, G. F., O'Connor, L., & Davis, R. (1998). Technological and health benefits of dairy starter cultures. *International Dairy Journal*, *8*(3), 195–205. doi:10.1016/S0958-6946(98)00042-9

Damodaran, S., & Agyare, K. K. (2013). Effect of microbial transglutaminase treatment on thermal stability and pH-solubility of heat-shocked whey protein isolate. *Food Hydrocolloids*, *30*(1), 12–18. doi:10.1016/j.foodhyd.2012.04.012

Danish Blue cheese - picture by RN Marshman. (2005, November 4). Retrieved from https://commons.wikimedia.org/wiki/File:Danish_Blue_cheese.jpg

Dantas, A. B., Jesus, V. F., Silva, R., Almada, C. N., Esmerino, E. A., Cappato, L. P., ... Cruz, A. G. (2016). Manufacture of probiotic Minas Frescal cheese with *Lactobacillus casei* Zhang. *Journal of Dairy Science*, *99*(1), 18–30. doi:10.3168/jds.2015-9880 PMID:26519974

Darnay, L., Králik, F., Oros, G., Koncz, Á., & Firtha, F. (2017a). Monitoring the effect of transglutaminase in semi-hard cheese during ripening by hyperspectral imaging. *Journal of Food Engineering*, *196*, 123–129. doi:10.1016/j.jfoodeng.2016.10.020

Darnay, L., Len, A., Koncz, A., Friedrich, L., & Rosta, L. (2015). Small angle neutron scattering study of nanostructural changes in microbial transglutaminase-treated low-fat yogurt during fermentation. *Food Science and Biotechnology*, *24*(6), 2125–2128. doi:10.100710068-015-0282-x

Darnay, L., Tóth, A., Salamon, B., Papik, K., Oros, G., Jónás, G., ... Friedrich, L. (2017b). Texture-modifying properties of microbial transglutaminase on 2 popular hungarian products: Trappist cheese and frankfurter. *Acta Alimentaria*, *46*(1), 116–122. doi:10.1556/066.2017.46.1.15

Dasari, S., Kathera, C., Janardhan, A., Kumar, A. P., & Viswanath, B. (2016). Surfacing role of probiotics in cancer prophylaxis and therapy: A systematic review. *Clinical Nutrition (Edinburgh, Lothian)*, 1–8. PMID:27923508

Dave, R., & Shah, N. (1998). Ingredient supplementation effects on viability of probiotic bacteria in yogurt. *Journal of Dairy Science*, *81*(11), 2804–2816. doi:10.3168/jds.S0022-0302(98)75839-4 PMID:9839222

David, B., Radziejwoski, A., Toussaint, F., Fontaine, L., Henry de Frahan, M., Patout, C., van Dillen, S., Boyaval, P., Horvath, P., Fremaux, C., & Hols, P. (2017). Natural DNA transformation is functional in *Lactococcus lactis* ssp. *cremoris* KW2. *Appl. Environ. Microbiol.*

de Bruyn, G., Hahn, S., & Borwick, A. (2012). Antibiotic treatment for travellers' diarrhoea [Review]. *Cochrane Database of Systematic Reviews*, *3*, CD002242. PMID:10908534

De Filippis, F., Parente, E., & Ercolini, D. (2017). Metagenomics insights into food fermentations. *Microbial Biotechnology*, *10*(1), 91–102. doi:10.1111/1751-7915.12421 PMID:27709807

De Jong, G. A. H., Boumans, J. W. L., & Winjgaards, G. (2002). *Food grade transglutaminase inhibitor and uses thereof.* European Patent 1201136.

De Jong, G. A. H., Winjgaards, G., & Koppelman, S. J. (2003). Transglutaminase inhibitor from milk. *Journal of Food Science*, *68*(3), 820–825. doi:10.1111/j.1365-2621.2003.tb08249.x

De Kruif, C. G., Tuinier, R., Holt, C., Timmins, P. A., & Rollema, H. S. (2002). Physicochemical study of κ-casein and β-casein dispersions and the effect of cross- linking by transglutaminase. *Langmuir*, *18*(12), 4885–4891. doi:10.1021/la025543w

de Lima, M. D. S. F., de Souza, K. M. S., Albuquerque, W. W. C., Teixeira, J. A. C., Cavalcanti, M. T. H., & Porto, A. L. F. (2017). *Saccharomyces cerevisiae* from Brazilian kefir-fermented milk: An in vitro evaluation of probiotic properties. *Microbial Pathogenesis*, 1–8. PMID:28478200

de Oliveira, M. E. G., Garcia, E. F., de Oliveira, C. E. V., Gomes, A. M. P., Pintado, M. M. E., Madureira, A. R. M. F., & da Conceicao, M. L. (2014). Addition of probiotic bacteria in a semi-hard goat cheese (coalho): Survival to simulated gastrointestinal conditions and inhibitory effect against pathogenic bacteria. *Food Research International*, *64*, 241–247. doi:10.1016/j.foodres.2014.06.032

De sá, E.M.F., & Bordignon-Luiz, M.T. (2010). The effect of transglutaminase on the properties of milk gels and processed cheese. *International Journal of Dairy Technology, 63*(2), 243-251.

De Sá, E. M. F., & Bordignon-Luiz, M. (2010). The effect of transglutamianse on the properties of milk gels and processed cheese. *International Journal of Dairy Technology*, *63*(2), 243–251. doi:10.1111/j.1471-0307.2010.00568.x

de Souza, J. I., Pires-Zottarelli, C. L. A., dos Santos, J. F., Costa, J. P., & Harakava, R. (2012). *Isomucor* (*Mucoromycotina*): A new genus from a Cerrado reserve in state of São Paulo, Brazil. *Mycologia*, *104*(1), 232–241. doi:10.3852/11-133 PMID:21933930

de Vrese, M., & Marteu, P. R. (2007). Probiotics and Prebiotics: Effects on Diarrhea. *The Journal of Nutrition*, *137*(3), 803S–811S. doi:10.1093/jn/137.3.803S PMID:17311979

de Vuyst, L. (1994). Nisin production variability between natural *Lactococcus lactis* subsp. *lactis* strains. *Biotechnology Letters*, *16*(3), 287–292. doi:10.1007/BF00134627

De Vuyst, L. (2000). Technology aspects related to the application of functional starter cultures. *Food Technology and Biotechnology*, *38*(2), 105–112.

Deborde, C. (2003). *Propionibacterium* spp. In H. Roginski, J. W. Fuquay, & P. F. Fox (Eds.), *Encyclopedia of Dairy Sciences* (pp. 2330–2339). London, UK: Academic Press.

Deegan, L. H., Cotterm, P. D., Hill, C., & Ross, P. (2006). Bacteriocins: Biological tools for bio-preservation and shelf-life extension. *International Dairy Journal, 16*(9), 1058–1071. doi:10.1016/j.idairyj.2005.10.026

Deeth, H. C., & Fitz-Gerald, C. H. (1983). Lipolytic enzymes and hydrolytic rancidity in milk and milk products. In P. E. Fox (Ed.), *Developments in Dairy Chemistry-2: Lipids* (pp. 195–240). London: Applied Science Publishers. doi:10.1007/978-94-010-9231-9_6

Delavenne, E., Ismail, R., Pawtowski, A., Mounier, J., Barbier, G., & Le Blay, G. (2013). Assessment of lactobacilli strains as yogurt bioprotective cultures. *Food Control, 30*(1), 206–213. doi:10.1016/j.foodcont.2012.06.043

Delavenne, E., Mounier, J., Asmani, K., Jany, J. L., Barbier, G., & Le Blay, G. (2011). Fungal diversity in cow, goat and ewe milk. *International Journal of Food Microbiology, 151*(2), 247–251. doi:10.1016/j.ijfoodmicro.2011.08.029 PMID:21944758

Delbès, C., Ali-Mandjee, L., & Montel, M.-C. (2007). Monitoring bacterial communities in raw milk and cheese by culture-dependent and -independent 16S rRNA gene-based analyses. *Applied and Environmental Microbiology, 73*(6), 1882–1891. doi:10.1128/AEM.01716-06 PMID:17259356

Delgado, F. J., González-Crespo, J., Cava, R., García-Parra, J., & Ramírez, R. (2010). Characterisation by SPME–GC–MS of the volatile profile of a Spanish soft cheese P.D.O. Torta del Casar during ripening. *Food Chemistry, 118*(1), 182–189. doi:10.1016/j.foodchem.2009.04.081

Deosarkar, S. S., Kalyankar, S. D., Pawshe, R. D., & Khedkar, C. D. (2016). Ice cream: Composition and health effects. In B. Caballero, P. M. Finglas, & F. Toldrá (Eds.), *Encyclopedia of food and health* (pp. 385–390). Waltham, MA: Academic Press. doi:10.1016/B978-0-12-384947-2.00385-8

Derakshan, K. F., Darvishi, F., Dezfulian, M., & Madzak, C. (2017). Expression and characterization of glucose oxidase from *Aspergillus niger* in *Yarrowia lipolytica. Molecular Biotechnology, 59*(8), 307–314. doi:10.100712033-017-0017-8 PMID:28631058

Derkx, P. M., Janzen, T., Sorensen, K. I., Christensen, J. E., Stuer-Lauridsen, B., & Johansen, E. (2014). The art of strain improvement of industrial lactic acid bacteria without the use of recombinant DNA technology. *Microbial Cell Factories, 13*(Suppl 1), S5. doi:10.1186/1475-2859-13-S1-S5 PMID:25186244

Dertli, E., & Çon, A. H. (2017). Microbial diversity of traditional kefir grains and their role on kefir aroma. *Lebensmittel-Wissenschaft + Technologie, 85*, 151–157. doi:10.1016/j.lwt.2017.07.017

Desai, K. G. H., & Jin Park, H. (2005). Recent Developments in Microencapsulation of Food Ingredients. *Drying Technology, 23*(7), 1361–1394. doi:10.1081/DRT-200063478

Desmasures, N. (2014). Cheese: Mold-Ripened Varieties. In M. L. Tortorello (Ed.), *Encyclopedia of Food Microbiology* (2nd ed.; pp. 409–415). Oxford, UK: Academic Press. doi:10.1016/B978-0-12-384730-0.00060-4

Devirgiliis, C., Zinno, P., Stirpe, M., Barile, S., & Perozzi, G. (2014). Functional Screening of Antibiotic Resistance Genes from a Representative Metagenomic Library of Food Fermenting Microbiota. *BioMed Research International.* PMID:25243126

DeVuyst, L., & Vandamme, E. (1994). *Bacteriocins of Lactic acid bacteria: microbiology, genetics and application.* London, UK: Chapman & Hill Ltd. doi:10.1007/978-1-4615-2668-1

DeZarn, T. J. (1995). *Food ingredient encapsulation: an overview* (S. J. Risch & G. A. Reineccius, Eds.). Washington, DC: American Chemical Society. doi:10.1021/bk-1995-0590.ch007

Di Cagno, R., Banks, J., Sheehan, L., Fox, P. F., Brechany, E. Y., Corsetti, A., & Gobbetti, M. (2003). Comparison of the microbiological, compositional, biochemical, volatile profile and sensory characteristics of three Italian PDO ewes' milk cheeses. *International Dairy Journal*, *13*(12), 961–972. doi:10.1016/S0958-6946(03)00145-6

Di Criscio, T., Fratianni, A., Mignogna, R., Cinquanta, L., Coppola, R., Sorrentino, E., & Panfili, G. (2010). Production of functional probiotic, prebiotic, and synbiotic ice creams. *Journal of Dairy Science*, *93*(10), 4555–4564. doi:10.3168/jds.2010-3355 PMID:20854989

Di Pierro, P., Chico, B., Villalonga, R., Mariniello, L., Damiano, A. E., Masi, P., & Porta, R. (2006). Chitosan-whey protein based edible films produced in the absence or presence of transglutaminase: Analysis of their mechanical and barrier properties. *Biomacromolecules*, *7*(3), 74–749. doi:10.1021/bm050661u PMID:16529409

Di Pierro, P., Mariniello, L., Sorrentino, A., Valeria, C., Giosafatto, L., Chianese, L., & Porta, R. (2010). Transglutaminase-induced chemical and rheological properties of cheese. *Food Biotechnology*, *24*(2), 107–120. doi:10.1080/08905431003784465

Dickinson, E. (1997). Enzymatic crosslinking as a tool for food colloid rheology control and interfacial stabilization. *Trends in Food Science & Technology*, *8*(10), 334–339. doi:10.1016/S0924-2244(97)01067-4

Dickinson, E. (1999). Caseins in emulsions: Interfacial properties and interactions. *International Dairy Journal*, *9*(3-6), 305–312. doi:10.1016/S0958-6946(99)00079-5

Dickinson, E., & Yamamoto, Y. (1996). Rheology of milk protein gels and protein-stabilized emulsion gels cross-linked with transglutaminase. *Journal of Agricultural and Food Chemistry*, *44*(6), 1371–1377. doi:10.1021/jf950705y

Didienne, R., Defargues, C., Callon, C., Meylheuc, T., Hulin, S., & Montel, M.-C. (2012). Characteristics of microbial biofilm on wooden vats ("gerles") in PDO Salers cheese. *International Journal of Food Microbiology*, *156*(2), 91–101. doi:10.1016/j.ijfoodmicro.2012.03.007 PMID:22483401

Dijkstra, A. R., Starrenburg, M. J., Todt, T. J., van Hijum, S. A., Hugenholtz, J., & Bron, P. A. *Isolation of a heat-resistant subpopulation of* Lactococcus lactis *SK11 with enhanced spray drying robustness.* Submitted for publication.

Dijkstra, A. R., Alkema, W., Starrenburg, M. J., Hugenholtz, J., van Hijum, S. A., & Bron, P. A. (2016). Strain-Dependent Transcriptome Signatures for Robustness in *Lactococcus lactis. PLoS One*, *11*(12), e0167944. doi:10.1371/journal.pone.0167944 PMID:27973578

Dijkstra, A. R., Alkema, W., Starrenburg, M., Hugenholtz, J., van Hijum, S., & Bron, P. A. (2014). Fermentation-induced variation in heat and oxidative stress phenotypes of *Lactococcus lactis* MG1363 reveals transcriptome signatures for robustness. *Microbial Cell Factories*, *13*(1), 148. doi:10.118612934-014-0148-6 PMID:25366036

Dijkstra, A. R., Setyawati, M. C., Bayjanov, J. R., Alkema, W., van Hijum, S. A., Bron, P. A., & Hugenholtz, J. (2014). Diversity in robustness of Lactococcus lactis strains during heat stress, oxidative stress, and spray drying stress. *Applied and Environmental Microbiology*, *80*(2), 603–611. doi:10.1128/AEM.03434-13 PMID:24212574

Ding, W., Shi, C., Chen, M., Zhou, J., Long, R., & Guo, X. (2017). Screening for lactic acid bacteria in traditional fermented Tibetan yak milk and evaluating their probiotic and cholesterol-lowering potentials in rats fed a high-cholesterol diet. *Journal of Functional Foods*, *32*, 324–332. doi:10.1016/j.jff.2017.03.021

Dinkçi, N. (2012). The influence of transglutaminase treatment on functional properties of strained yoghurt. *Journal of Animal and Veterinary Advances*, *11*(13), 2238–2246. doi:10.3923/javaa.2012.2238.2246

Divya, J. B., Varsha, K. K., Nampoothiri, K. M., Ismail, B., & Pandey, A. (2012). Probiotic fermented foods for health benefits. *Engineering in Life Sciences*, *12*(4), 377–390. doi:10.1002/elsc.201100179

Dmytrów, I., Jasinska, M., & Dmytrów, K. (2010). Effect of microbiological transglutaminase on selected physicochemical properties of tvarog. *Italian Journal of Food Science*, *22*(4), 449–460.

do Espirito Santo, A. P., Perego, P., Converti, A., & Oliveira, M. N. (2011). Influence of food matrices on probiotic viability – a review focusing on the fruity bases. *Trends in Food Science & Technology*, *22*(7), 377–385. doi:10.1016/j.tifs.2011.04.008

Doğan, M. (2012). Probiyotik Bakterilerin Gastrointestinal Sistemdeki Etki Mekanizması. *Gıda Teknolojileri Elektronik Dergisi*, *7*(1), 20–27.

Domagała, J., Najgebauer-Lejko, D., Wieteska-Śliwa, I., Sady, M., Wszołek, M., Bonczar, G., & Filipczak-Fiutak, M. (2015). Influence of milk protein cross-linking by transglutaminase on the rennet coagulation time and the gel properties. *Journal of the Science of Food and Agriculture*, *96*(10), 3500–3507. doi:10.1002/jsfa.7534 PMID:26585165

Domagała, J., Wszołek, M., Tamime, A. Y., & Kupiec-Teahan, B. (2013). The effect of transglutamianse concentration on the texture, syneresis and microstructure of set-type goat's milk yoghurt during the storage period. *Small Ruminant Research*, *112*(1-3), 154–161. doi:10.1016/j.smallrumres.2012.12.003

Dong, Q., Yan, X., Zheng, M., & Yang, Z. (2014). Characterization of an extremely thermostable but cold-adaptive β-galactosidase from the hyperthermophilic archaeon *Pyrococcus furiosus* for use as a recombinant aggregation for batch lactose degradation at high temperature. *Journal of Bioscience and Bioengineering*, *117*(6), 706–710. doi:10.1016/j.jbiosc.2013.12.002 PMID:24462527

Doolan, I. A., Nongonierma, A. B., Kilcawley, K. N., & Wilkinson, M. G. (2014). Partitioning of starter bacteria and added exogenous enzyme activities between curd and whey during Cheddar cheese manufacture. *International Dairy Journal*, *34*(1), 159–166. doi:10.1016/j.idairyj.2013.07.005

dos Reis, S. A., da Conceição, L. L., Siqueira, N. P., Rosa, D. D., da Silva, L. L., Carmo, M., & Peluzio, G. (2017). Review of the mechanisms of probiotic actions in the prevention of colorectal cancer. *Nutrition Research (New York, N.Y.)*, *37*, 1–19. doi:10.1016/j.nutres.2016.11.009 PMID:28215310

Douillard, F. P., & de Vos, W. M. (2014). Functional genomics of lactic acid bacteria: From food to health. *Microbial Cell Factories*, *13*(Suppl 1), S8. doi:10.1186/1475-2859-13-S1-S8 PMID:25186768

Downes, J., & Wade, W. G. (2009). *Propionibacterium acidifaciens* sp. nov., isolated from the human mouth. *International Journal of Systematic and Evolutionary Microbiology*, *59*(11), 2778–2781. doi:10.1099/ijs.0.010470-0 PMID:19625418

Du Roquefort, fromage de l'Aveyron. (2015, April 17). Retrieved from https://commons.wikimedia.org/wiki/File:Wikicheese_-_Roquefort_-_20150417_-_003.jpg

Du Toit, M., Franz, C. M., Dicks, L. M., Schillinger, U., Haberer, P., Warlies, B., ... Holzapfel, W. H. (1998). Characterisation and selection of probiotic lactobacilli for a preliminary minipig feeding trial and their effect on serum cholesterol levels, faeces pH and faeces moisture content. *International Journal of Food Microbiology*, *40*(1-2), 93–104. doi:10.1016/S0168-1605(98)00024-5 PMID:9600615

Dubey, M. K., Zehra, A., Aamir, M., Meena, M., Ahirwal, L., Singh, S., ... Bajpai, V. K. (2017). Improvement strategies, cost effective production, and potential applications of fungal glucose oxidase (GOD): Current updates. *Frontiers in Microbiology*, *8*, 1032. doi:10.3389/fmicb.2017.01032 PMID:28659876

Dubey, R., & Dubey, R. (2009). Microencapsulation Technology and Applications. *Defence Science Journal*, *59*(1), 82–95. doi:10.14429/dsj.59.1489

Dugat-Bony, E., Straub, C., Teissandier, A., Onésime, D., Loux, V., Monnet, C., ... Bonnarme, P. (2015). Overview of a surface-ripened cheese community functioning by meta-omics analyses. *PLoS One*, *10*(4), 1–25. doi:10.1371/journal.pone.0124360 PMID:25867897

DuPont, H. L. (2007). The growing threat of foodborne bacterial enteropathogens of animal origin. *Clinical Infectious Diseases*, *45*(10), 1353–1361. doi:10.1086/522662 PMID:17968835

Dupuis, C., Corre, C., & Boyaval, P. (1993). Lipase and esterase activity of Propionibacterium freudenreichii subsp. Freudenreichii. *Applied and Environmental Microbiology*, *59*(12), 4004–4009. PMID:16349102

Dupuis, C., Corre, C., & Boyaval, P. (1995). Proteinase activity of dairy *Propionibacterium. Applied Microbiology and Biotechnology*, *42*(5), 750–755. doi:10.1007/BF00171957

Duran, R., Junqua, M., Schmitter, J. M., Gancet, C., & Goulas, P. (1998). Purification, characterisation, and gene cloning of transglutaminase from *Streptoverticillium cinnamoneum* CBS 683.68. *Biochimie*, *80*(4), 313–319. doi:10.1016/S0300-9084(98)80073-4 PMID:9672751

Duwat, P., Bravard, A., Sourice, S., & Gruss, A. (2004). *Process for preparing starter cultures of lactic acid bacteria*. United States Patent No. US20050032196.

Duwat, P., Cesselin, B., Sourice, S., & Gruss, A. (2000). Lactococcus lactis, a bacterial model for stress responses and survival. *International Journal of Food Microbiology*, *55*(1–3), 83–86. doi:10.1016/S0168-1605(00)00179-3 PMID:10791722

Duwat, P., Ehrlich, S. D., & Gruss, A. (1999). Effects of metabolic flux on stress response pathways in *Lactococcus lactis*. *Molecular Microbiology*, *31*(3), 845–858. doi:10.1046/j.1365-2958.1999.01222.x PMID:10048028

Duwat, P., Sourice, S., Cesselin, B., Lamberet, G., Vido, K., Gaudu, P., ... Gruss, A. (2001). Respiration capacity of the fermenting bacterium *Lactococcus lactis* and Its positive effects on growth and survival. *Journal of Bacteriology*, *183*(15), 4509–4516. doi:10.1128/JB.183.15.4509-4516.2001 PMID:11443085

Eissa, A. S., & Khan, S. A. (2006). Modulation of hydrophobic interactions in denatured whey proteins by transglutaminase enzyme. *Food Hydrocolloids*, *20*(4), 543–547. doi:10.1016/j.foodhyd.2005.07.005

Ekinci, F. Y., & Barefoot, S. F. (2006). Fed-batch enhancement of jenseniin G, a bacteriocin produced by *Propionibacterium thoenii (jensenii)* P126. *Food Microbiology*, *23*(4), 325–330. doi:10.1016/j.fm.2005.05.012 PMID:16943021

El Soda, M., Madkor, S. A., & Tong, P. S. (2000). Adjunct Cultures: Recent Developments and Potential Significance to the Cheese Industry. *Journal of Dairy Science*, *83*(4), 609–619. doi:10.3168/jds.S0022-0302(00)74920-4 PMID:10791774

El-Baradei, G., Delacroix-Buchet, A., & Ogier, J. C. (2007). Biodiversity of bacterial ecosystems in traditional Egyptian Domiati cheese. *Applied and Environmental Microbiology*, *73*(4), 1248–1255. doi:10.1128/AEM.01667-06 PMID:17189434

El-Hofi, M., El-Tanboly, E.-S., & Abd-Rabou, N. S. (2011). Industrial application of lipases in cheese making: A review. *Internet Journal of Food Safety*, *13*, 293–302.

Eliskases-Lechner, F., & Ginzinger, W. (1995). The bacterial flora of surface-ripened cheeses with special regard to coryneform. *Le Lait*, *75*(6), 571–584. doi:10.1051/lait:1995644

El-Sayed, H. S., Salama, H. H., & El-Sayed, S. M. (2014). Production of synbiotic ice cream. *International Journal of Chemical Technology Research*, *7*(1), 138–147.

El-Sayed, H. S., Salama, H. H., & El-Sayed, S. M. (2014). Production of synbiotic ice cream. *International Journal of Chemtech Research, 15*(01), 138–147.

El-Sharoud, W. M., Belloch, C., Peris, D., & Querol, A. (2009). Molecular identification of yeasts associated with traditional egyptian dairy products. *Journal of Food Science, 74*(7), M341–M346. doi:10.1111/j.1750-3841.2009.01258.x PMID:19895478

Elsser-Gravesen, D., & Elsser-Gravesen, A. (2014). Biopreservatives. *Advances in Biochemical Engineering/Biotechnology, 143*, 29–49. doi:10.1007/10_2013_234 PMID:24185748

Engel, G., von Milczewski, K. E., Prokopek, D., & Teuber, M. (1982). Strain-specific synthesis of mycophenolic acid by *Penicillium roqueforti* in blue-veined cheese. *Applied and Environmental Microbiology, 43*(5), 1034–1040. PMID:16346004

Ercili-Cura, D., Lille, M., Legland, D., Gaucel, S., Poutanen, K., Partaen, R., & Şantto, R. (2013). Structural mechanisms leading to improved water retention in acid milk gels by transglutaminase. *Food Hydrocolloids, 30*(1), 419–427. doi:10.1016/j.foodhyd.2012.07.008

Ercolini, D. (2004). PCR-DGGE fingerprinting: Novel strategies for detection of microbes in food. *Journal of Microbiological Methods, 56*(3), 297–314. doi:10.1016/j.mimet.2003.11.006 PMID:14967221

Ercolini, D., Hill, P. J., & Dodd, C. E. R. (2003). Development of a fluorescence in situ hybridization method for cheese using a 16S rRNA probe. *Journal of Microbiological Methods, 52*(2), 267–271. doi:10.1016/S0167-7012(02)00162-8 PMID:12459248

Erdogan, A., Gurses, M., & Sert, S. (2003). Isolation of moulds capable of producing mycotoxins from blue mouldy Tulum cheeses produced in Turkey. *International Journal of Food Microbiology, 85*(1), 83–85. doi:10.1016/S0168-1605(02)00485-3 PMID:12810273

Erkkilä, S. (2001). *Bioprotective and probiotic meat starter cultures for the fermentatation of dry sausages* (Dissertation). Department of Food Technology, University of Helsinki, Finland.

Erkkilä, S., Suihko, M. L., Eerola, S., Petäjä, E., & Mattila-Sandholm, T. (2001). Dry sausage fermented by *Lactobacillus rhamnosus* strains. *International Journal of Food Microbiology, 64*(1-2), 205–210. doi:10.1016/S0168-1605(00)00457-8 PMID:11252505

Erkus, O., de Jager, V. C. L., Geene, R. T. C. M., van Alen-Boerrigter, I., Hazelwood, L., van Hijum, S. A. F. T., ... Smid, E. J. (2016). Use of propidium monoazide for selective profiling of viable microbial cells during Gouda cheese ripening. *International Journal of Food Microbiology, 228*, 1–9. doi:10.1016/j.ijfoodmicro.2016.03.027 PMID:27077825

Erkuş, O., de Jager, V. C. L., Spus, M., van Alen-Boerrigter, I. J., van Rijswijck, I. M., Hazelwood, L., ... Smid, E. J. (2013). Multifactorial diversity sustains microbial community stability. *The ISME Journal, 7*(11), 2126–2136. doi:10.1038/ismej.2013.108 PMID:23823494

Ertürkmen, P., Başyiğit Kılıç, G., & Kılıç, B. (2016). Utilization of lactic acid bacteria and probiotics on meat products. *Journal of Hygienic Engineering and Design*, 78–82.

Escobar, M. C., Van Tassell, M. L., Martínez-Bustos, F., Singh, M., Castaño-Tostado, E., Amaya Llano, S. L., & Miller, M. J. (2012). Characterization of a Panela cheese with added probiotics and fava bean starch. *Journal of Dairy Science, 95*(6), 2779–2787. doi:10.3168/jds.2011-4655 PMID:22612915

Espinoza-Molina, J. A., Acosta-Muniz, C. H., Sepulveda, D. R., Zamudio-Flores, P. B., & Rios-Velasco, C. (2016). Codon optimization of the '*Bos taurus* chymosin' gene for the production of recombinant chymosin in *Pichia pastoris*. *Molecular Biotechnology, 58*(10), 657–664. doi:10.100712033-016-9965-7 PMID:27394727

Esteban-Torres, M., Mancheño, J. M., de las Rivas, B., & Muñoz, R. (2014). Production and characterization of a tributyrin esterase from *Lactobacillus plantarum* suitable for cheese lipolysis. *Journal of Dairy Science*, *97*(11), 6737–6744. doi:10.3168/jds.2014-8234 PMID:25173466

Etienne-Mesmin, L., Livrelli, V., Privat, M., Denis, S., Cardot, J. M., Alric, M., & Blanquet-Diot, S. (2011). Effect of a New Probiotic Saccharomyces cerevisiae Strain on Survival of *Escherichia coli* O157:H7 in a Dynamic Gastrointestinal Model. *Applied and Environmental Microbiology*, *77*(3), 1127–1131. doi:10.1128/AEM.02130-10 PMID:21131521

Ewaschuk, J. B., & Dieleman, L. A. (2006). Probiotics and prebiotics in chronic inflammatory bowel diseases. *World Journal of Gastroenterology*, *12*(37), 5941–5950. doi:10.3748/wjg.v12.i37.5941 PMID:17009391

Faergemand, M., & Murray, B. S. (1998). Interfacial dilatational properties of milk proteins cross-linked by transglutaminase. *Journal of Agricultural and Food Chemistry*, *46*(3), 884–890. doi:10.1021/jf970733b

Faergemand, M., Otte, J., & Qvist, K. B. (1997). Enzymatic crosslinking of whey proteins by a Ca^{+2}-independent microbial transglutaminase from *Streptomyces lydicus. Food Hydrocolloids*, *11*(1), 19–25. doi:10.1016/S0268-005X(97)80006-9

Faergemand, M., Otte, J., & Qvist, K. B. (1998). Emulsifying properties of milk proteins cross-linked with microbial transglutaminase. *International Dairy Journal*, *8*(8), 715–723. doi:10.1016/S0958-6946(98)00111-3

Faergemand, M., & Qvist, K. B. (1997). Transglutaminase: Effect on rheological properties, microstructure and permeability of set style acid skim milk gel. *Food Hydrocolloids*, *11*(3), 287–292. doi:10.1016/S0268-005X(97)80058-6

Fang, Z., & Bhandari, B. (2010). Encapsulation of polyphenols – a review. *Trends in Food Science & Technology*, *21*(10), 510–523. doi:10.1016/j.tifs.2010.08.003

FAO/WHO. (2001). Health and nutritional properties of probiotics in food including powder milk with live lactic acid bacteria. Córdoba, Argentina: Academic Press.

FAO/WHO. (2002). *Guidelines for the evaluation of probiotics in food: Report of a joint FAO/WHO working group on drafting guidelines for the evaluation of probiotics in food.* London, Canada: FAO/WHO.

FAO/WHO. (2002b). *Report of a Joint FAO/WHO Working Group on Drafting Guidelines for the Evaluation of Probiotics in Food.* Author.

FAO/WHO. (2006). *Probiotics in Food: Health and nutritional properties and guidelines for evaluation, FAO food and nutrition paper no. 85.* Rome, Italy: FAO/WHO.

Farag, A.A., Aly, M.E. & El-Alfy, M.B. (1992). Enhancement of Blue cheese flavour using sodium dodecylsulphate and lipase. *Food/Nahrung, 36*(1), 1-7.

Farahat, S. M., Rabie, A. M., & Farag, A. A. (1990). Evaluation of the proteolytic and lipolytic activity of different *Penicillium roqueforti* strains. *Food Chemistry*, *36*(3), 169–180. doi:10.1016/0308-8146(90)90052-6

Farkye, N. Y., & Imafidon, G. I. (1995). *Thermal denaturation of indigenous milk enzymes. In Heat-induced changes in milk* (pp. 331–348). Brussels, Belgium: Int. Dairy Fed.

Farnsworth, J. P., Li, J., Hendricks, G. M., & Guo, M. R. (2006). Effects of transglutaminase treatment on functional properties and probiotic culture survivability of goat milk yogurt. *Small Ruminant Research*, *65*(1-2), 113–121. doi:10.1016/j.smallrumres.2005.05.036

Favaro-Trindade, C. S., Bernardi, S., Bodini, R. B., De Carvalho Balieiro, J. C., & De Almeida, E. (2006). Sensory acceptability and stability of probiotic microorganisms and vitamin C in fermented acerola (*Malpighia emarginata* DC.) ice cream. *Journal of Food Science*, *71*(6), 491–495. doi:10.1111/j.1750-3841.2006.00100.x

Favaro-Trindade, C. S., De Carvalho Balieiro, J. C., Dias, P. F., Amaral Sanino, F., & Boschini, C. (2007). Effects of culture, pH and fat concentration on melting rate and sensory characteristics of probiotic fermented yellow mombin (*Spondias mombin* L) ice creams. *Food Science & Technology International, 13*(4), 285–291. doi:10.1177/1082013207082387

Feijoo-Siota, L., Blasco, L., Rodríguez-Rama, J. L., Barros-Velázquez, J., Miguel, T. D., Sánchez-Pérez, A., & Villa, T. G. (2014). Recent patents on microbial proteases for the dairy industry. *Recent Advances in DNA & Gene Sequences, 8*(1), 44–55. doi:10.2174/2352092208666141013231720 PMID:25564028

Felicio, T. L., Esmerino, E. A., Vidal, V. A. S., Cappato, L. P., Garcia, R. K. A., Cavalcanti, R. N., ... Cruz, A. G. (2016). Physico-chemical changes during storage and sensory acceptance of low sodium probiotic Minas cheese added with arginine. *Food Chemistry, 196*, 628–637. doi:10.1016/j.foodchem.2015.09.102 PMID:26593536

Fernández-García, E., López-Fandiño, R., & Alonso, L. (1994b). Effect of a food-grade enzyme preparation from *Aspergillus oryzae* on free fatty acid release in Manchego-type cheese from ovine and bovine milk. *Zeitschrift fur Lebensmittel-Untersuchung und -Forschung, 199*(4), 262–264. doi:10.1007/BF01193308

Fernandez-Garcia, E., Lopez-Fandiño, R., Alonso, L., & Ramos, M. (1994a). The Use of Lipolytic and Proteolytic Enzymes in the Manufacture of Manchego Type Cheese from Ovine and Bovine Milk. *Journal of Dairy Science, 77*(8), 2139–2149. doi:10.3168/jds.S0022-0302(94)77156-3

Fernandez-Garcia, E., Olano, A., Cabezudo, D., Martin-Alvarez, P. J., & Ramos, M. (1993). Accelerated ripening of Manchego type cheese by added commercial enzyme preparation from *Aspergillus oryzae. Enzyme and Microbial Technology, 15*(6), 519–524. doi:10.1016/0141-0229(93)90086-H

Fernández-García, E., Ramos, M., Polo, C., Juárez, M., & Olano, A. (1988). Enzyme accelerated ripening of Spanish hard cheese. *Food Chemistry, 28*(1), 63–80. doi:10.1016/0308-8146(88)90007-6

Ferrari, I. S., de Souza, J. V., Ramos, C. L., da Costa, M. M., Schwan, R. F., & Dias, F. S. (2016). Selection of autochthonous lactic acid bacteria from goat dairies and their addition to evaluate the inhibition of *Salmonella* typhi in artisanal cheese. *Food Microbiology, 60*, 29–38. doi:10.1016/j.fm.2016.06.014 PMID:27554143

Ferraz, J. L., Cruz, A. G., Cadena, R. S., Freitas, M. Q., Pinto, U. M., Carvalho, C. C., ... Bolini, H. (2012). Sensory acceptance and survival of probiotic bacteria in ice cream produced with different overrun levels. *Journal of Food Science, 77*(1), S24–S28. doi:10.1111/j.1750-3841.2011.02508.x PMID:22260128

Ferreira, A. D., & Viljoen, B. C. (2003). Yeasts as adjunct starters in matured Cheddar cheese. *International Journal of Food Microbiology, 86*(1-2), 131–140. doi:10.1016/S0168-1605(03)00252-6 PMID:12892928

Ferrocino, I., & Cocolin, L. (2017). Current perspectives in food-based studies exploiting multi-omics approaches. *Current Opinion in Food Science, 13*(February), 10–15. doi:10.1016/j.cofs.2017.01.002

Figueroa-González, I., Quijano, G., Ramírez, G., & Cruz-Guerrero, A. (2011). Probiotics and prebiotics. Perspectives and challenges. *Journal of the Science of Food and Agriculture, 91*(8), 1341–1348. doi:10.1002/jsfa.4367 PMID:21445871

Finoli, C., Vecchio, A., Galli, A., & Dragoni, I. (2001). Roquefortine C occurrence in blue cheese. *Journal of Food Protection, 64*(2), 246–251. doi:10.4315/0362-028X-64.2.246 PMID:11271775

Fioramonti, J., Theodorou, V., & Bueno, L. (2003). Probiotics: What are they? What are their effects on gut physiology? *Best Practice & Research. Clinical Gastroenterology, 17*(5), 711–724. doi:10.1016/S1521-6918(03)00075-1 PMID:14507583

Fitzsimons, N. A., Cogan, T. M., Condon, S., & Beresford, T. (1999). Phenotypic and genotypic characterization of non-starter lactic acid bacteria in mature cheddar cheese. *Applied and Environmental Microbiology, 65*, 3418–3426. PMID:10427029

Fleet, G. (1990). Yeast in dairy products-a review. *The Journal of Applied Bacteriology*, *68*(3), 199–211. doi:10.1111/j.1365-2672.1990.tb02566.x PMID:2187843

Flynn, S., Fitzgerald, G., Daly, C., Kiely, B., O'Sullivan, G. C., Shanahan, F., & Collins, J. K. (2001). *In vitro* selection criteria for probiotic bacteria of human origin: Correlation with in vivo findings. *The American Journal of Clinical Nutrition*, *73*, 386–392. PMID:11157346

Fontaine, K., Passeró, E., Vallone, L., Hymery, N., Coton, M., Jany, J.-L., ... Coton, E. (2015). Occurrence of roquefortine C, mycophenolic acid and aflatoxin M1 mycotoxins in blue-veined cheeses. *Food Control*, *47*, 634–640. doi:10.1016/j.foodcont.2014.07.046

Fortin, M. H., Champagne, C. P., St-Gelais, D., Britten, M., Fustier, P., & Lacroix, M. (2011). Effect of time of inoculation, starter addition, oxygen level and salting on the viability of probiotic cultures during Cheddar cheese production. *International Dairy Journal*, *21*(2), 75–82. doi:10.1016/j.idairyj.2010.09.007

Fox, P. F., Law, J., McSweeney, P. L. H., & Wallace, J. (1993). Biochemistry of cheese ripening. In P. F. Fox, (Eds.), Cheese: Chemistry, Physics and Microbiology: Vol. 1. General Aspects (2nd ed.; pp. 389-438). London, UK: Chapman & Hall. doi:10.1007/978-1-4615-2650-6_10

Fox, P. F. (1989). Proteolysis during cheese manufacture and ripening. *Journal of Dairy Science*, *72*(6), 1379–1400. doi:10.3168/jds.S0022-0302(89)79246-8

Fox, P. F., Guinee, T. P., Cogan, T. M., & McSweeney, P. L. H. (2000). *Fundamentals of Cheese Science*. Gaithersburg, MD: Aspen.

Fox, P. F., Guinee, T. P., Cogan, T. M., & McSweeney, P. L. H. (Eds.). (2000). *Fundamentals of cheese science*. Gaithersburg, Maryland: Aspen Publishers, Inc.

Fox, P. F., Law, J., McSweeney, P. L. H., & Wallace, J. (1993). *Biochemistry of Cheese Ripening. Cheese: Chemistry, Physics and Microbiology* (Vol. 1). London: Chapman & Hall.

Fox, P. F., & McSweeney, P. L. H. (1996). Proteolysis in cheese during ripening. *Food Reviews International*, *12*(4), 457–509. doi:10.1080/87559129609541091

Fox, P. F., McSweeney, P. L. H., & Lynch, C. M. (1998). Significance of non-starter lactic acid bacteria in Cheddar cheese. *Australian Journal of Dairy Technology*, *53*, 83–89.

Fox, P. F., McSweeney, P. L. H., & Lynch, C. M. (1998). Significance of non-starter lactic acid bacteria in chedder cheese. *Australian Journal of Dairy Technology*, *53*(2), 83.

Fox, P. F., & Stepaniak, L. (1993). Enzymes in cheese technology. *International Dairy Journal*, *3*(4-6), 509–530. doi:10.1016/0958-6946(93)90029-Y

Franciosi, E., Settanni, L., Carlin, S., Cavazza, A., & Poznanski, E. (2008). A Factory-Scale Application of Secondary Adjunct Cultures Selected from Lactic Acid Bacteria During Puzzone di Moena Cheese Ripening. *Journal of Dairy Science*, *91*(8), 2981–2991. doi:10.3168/jds.2007-0764 PMID:18650274

Frank, J. F., & Hassan, A. N. (1998). Starter cultures and their use. In E. H. Marth & J. L. Steele (Eds.), *Applied Dairy Microbiology* (pp. 131–172). Marcel and Dekker Inc.

Franz, C. M. A. P., van Belkum, M. J., Holzapfel, W. H., Abriouel, H., & G'alvez, A. (2007). Diversity of enterococcal bacteriocins and their grouping into a new classification scheme. *FEMS Microbiology Reviews*, *31*(3), 293–310. doi:10.1111/j.1574-6976.2007.00064.x PMID:17298586

Frece, J., Kovačević, D., Kazazić, S., Mrvčić, J., Vahčić, N., Ježek, D., ... Markov, K. (2014). Comparison of sensory properties, shelf-life and microbiological safety of industrial sausages produced with autochthonous and commercial starter cultures. *Food Technology and Biotechnology*, *52*, 307–316.

Freire, A. L., Ramos, C. L., da Costa Souza, P. N., Cardoso, M. G. B., & Schwan, R. F. (2017). Nondairy beverage produced by controlled fermentation with potential probiotic starter cultures of lactic acid bacteria and yeast. *International Journal of Food Microbiology*, *248*, 39–46. doi:10.1016/j.ijfoodmicro.2017.02.011 PMID:28242421

Frisvad, J. C., Smedsgaard, J., Larsen, T. O., & Samson, R. A. (2004). Mycotoxins, drugs and other extrolites produced by species in *Penicillium* subgenus *Penicillium*. *Studies in Mycology*, *49*, 201–241.

Frisvad, Jens C, & Samson, R. A. (2004). Polyphasic taxonomy of *Penicillium* subgenus *Penicillium*. A guide to identification of food and air-borne terverticillate *Penicillia* and their mycotoxins. *Studies in Mycology*, *49*(1), 1–174.

Fröhlich-Wyder, M. T., & Bachmann, H. P. (2004). Cheeses with propionic acid fermentation. In P. F. Fox, P. L. H. McSweeney, T. M. Cogan & T. P. Guinee (Eds.), Cheese: Chemistry, Physics and Microbiology: Vol. 2. Major Cheese Groups (3rd ed.; pp. 141-156). London, UK: Academic Press. doi:10.1016/S1874-558X(04)80042-3

Fröhlich-Wyder, M. T., Bachmann, H. P., & Casey, M. G. (2002). Interaction between propionibacteria and starter/ non-starter lactic acid bacteria in Swiss- type cheeses. *Le Lait*, *82*(1), 1–15. doi:10.1051/lait:2001001

Fromage AOP de Saint-Nectaire, France. (2013, August 26). Retrieved from https://commons.wikimedia.org/wiki/File:Saint-Nectaire.jpg

Fu, B., & Labuza, T. P. (1997). Shelf-life testing: procedures and prediction methods. In M. C. Erickson & Y. Hung (Eds.), *Quality in frozen food* (pp. 377–415). New York, NY: Springer. doi:10.1007/978-1-4615-5975-7_19

Fujisawa, T., Adachi, S., Toba, T., Arihara, K., & Mitsuoka, T. (1988). *Lactobacillus kefiranofaciens* sp. nov. Isolated from Kefir Grains. *International Journal of Systematic Bacteriology*, *38*(1), 12–14. doi:10.1099/00207713-38-1-12

G'alvez, A., Abriouel, H., Lucas L'opez, R., & Ben Omar, N. (2007). Bacteriocin based strategies for food biopreservation. *International Journal of Food Microbiology*, *120*(1-2), 51–70. doi:10.1016/j.ijfoodmicro.2007.06.001 PMID:17614151

Gagnaire, V., Lortal, S., & Leonil, J. (1998). Free active peptidases are detected in Emmental juice extracted before ripening in the warm room. *The Journal of Dairy Research*, *65*(1), 119–128. doi:10.1017/S0022029997002525

Gagnaire, V., Thierry, A., & Leonil, J. (2001). Propionibacteria aand facultatively heterofermentative lactobacilli weakly contribute to secondary proteolysis of Emmental cheese. *Le Lait*, *81*(3), 339–353. doi:10.1051/lait:2001136

Gallois, A., & Langlois, D. (1990). New results in the volatile odorous compounds of French cheeses. *Le Lait*, *70*(2), 89–106. doi:10.1051/lait:199028

Galloway, L. D. (1935). The moisture requirements of mould fungi, with special reference to mildew in textiles. *Journal of the Textile Institute Transactions*, *26*(4), T123–T129. doi:10.1080/19447023508661646

Gálvez, A., López, R. L., Pulido, R. P., & Burgos, M. J. G. (2014). *Application of Lactic Acid Bacteria and their Bacteriocins for Food Biopreservation. In Food Biopreservation* (pp. 15–22). Springer.

Gamalost (also Gammelost, Gammalost), which translates as "old cheese", is a pungent traditional Norwegian cheese. (2013, January 24). Retrieved from https://commons.wikimedia.org/wiki/File:Gamalost-NorwegianOldCheese.jpg

Gandhi, N. N. (1997). Applications of Lipase. *Journal of the American Oil Chemists' Society*, *74*(6), 621–634. doi:10.100711746-997-0194-x

Garault, P., Quere, G., Beal, C., Bomchil, N., Faurie, J.-M., Gobert, G., & Lipowski, G. (2011). *Method for obtaining variants of lactic acid bacteria usable for producing vitamin K2 and application to the preparation of food products.* United States Patent No. US7981657 B2.

Garcı'a, P., Madera, C., Martı'nez, B., & Rodrı'guez, A. (2007). Biocontrol of *Staphylococcus aureus* in curd manufacturing processes using bacteriophages. *International Dairy Journal, 17*(10), 1232–1239. doi:10.1016/j.idairyj.2007.03.014

Garcia, H. S., Lopez-Hernandez, A., & Hill, C. G. (2016). Enzyme technology - Dairy industry applications. Comprehensive Biotechnology, 4, 567-574.

García-Estrada, C., & Martín, J.-F. (2016). Biosynthetic gene clusters for relevant secondary metabolites produced by *Penicillium roqueforti* in blue cheeses. *Applied Microbiology and Biotechnology, 100*(19), 8303–8313. doi:10.100700253-016-7788-x PMID:27554495

Garde, S., Gaya, P., Medina, M., & Nunez, M. (1997). Acceleration of flovur formation in cheese by bacteriocin-producing adjunct lactic culture. *Biotechnology Letters, 10*(9), 1011–1014. doi:10.1023/A:1018451401925

Gardiner, G. E., Ross, R. P., Wallace, J. M., Scanlan, F. P., Jägers, P. P. J. M., Fitzgerald, G. F., ... Stanton, C. (1999). Influence of a probiotic adjunct culture of *Enterococcus faecium* on the quality of Cheddar cheese. *Journal of Agricultural and Food Chemistry, 47*(12), 4907–4916. doi:10.1021/jf990277m PMID:10606551

Gauche, C., Tomazi, T., Barreto, P. L. M., Ogliari, P. J., & Bordignon-Ruiz, M. T. (2009). Physical properties of yoghurt manufactured with milk whey and transglutaminase. *Lebensmittel-Wissenschaft + Technologie, 42*(1), 239–243. doi:10.1016/j.lwt.2008.05.023

Gauche, C., Vieira, J. T. C., Ogliari, P. J., & Bordignon-Luiz, M. T. (2008). Crosslinking of milk with whey proteins by transglutaminase. *Process Biochemistry, 43*(7), 788–794. doi:10.1016/j.procbio.2008.04.004

Gaudu, P., Vido, K., Cesselin, B., Kulakauskas, S., Tremblay, J., Rezaïki, L., ... Gruss, A. (2002). *Respiration capacity and consequences in Lactococcus lactis. In Lactic Acid Bacteria: Genetics, Metabolism and Applications* (pp. 263–269). Springer. doi:10.1007/978-94-017-2029-8_16

Gaynor, P. (2006). How U.S. FDA's GRAS notification program works. *Food Safety Magazine, 11*, 16–19.

Geppel, A., Kringelum, B. W., Hansen, K. F., Iversen, S. L., & Henriksen, C. M. (2016). *Porphyrin containing lactic acid bacterial cells and use therof.* United States Patent No. US 8486468 B2.

Gerard, J. A. (2002). Protein–protein crosslinking in food: Methods, consequences, applications. *Trends in Food Science & Technology, 13*(12), 391–399. doi:10.1016/S0924-2244(02)00257-1

Gharsallaoui, A., Roudaut, G., Chambin, O., Voilley, A., & Saurel, R. (2007). Applications of spray-drying in microencapsulation of food ingredients: An overview. *Food Research International, 40*(9), 1107–1121. doi:10.1016/j.foodres.2007.07.004

Gill, A. (2017). The importance of bacterial culture to food microbiology in the age of genomics. *Frontiers in Microbiology, 8*, 1–6. doi:10.3389/fmicb.2017.00777 PMID:28507541

Gill, H. S., Rutherfurd, K. J., Prasad, J., & Gopal, P. K. (2000). Enhancement of natural and acquired immunity by *Lactobacillus rhamnosus* (HN001), *Lactobacillus acidophilus* (HN017) and *Bifidobacterium lactis* (HN019). *British Journal of Nutrition, 83*(02), 167–176. doi:10.1017/S0007114500000210 PMID:10743496

Gill, J. J., Pacan, J. C., Carson, M. E., Leslie, K. E., Griffiths, M. W., & Sabour, P. M. (2006a). Efficacy and pharmacokinetics of bacteriophage therapy in treatment of subclinical *Staphylococcus aureus* mastitis in lactating dairy cattle. *Antimicrobial Agents and Chemotherapy, 50*(9), 2912–2918. doi:10.1128/AAC.01630-05 PMID:16940081

Gill, J. J., Sabour, P. M., Leslie, K. E., & Griffiths, M. W. (2006b). Bovine whey proteins inhibit the interaction of *Staphylococcus aureus* and bacteriophage K. *Journal of Applied Microbiology*, *101*(2), 377–386. doi:10.1111/j.1365-2672.2006.02918.x PMID:16882145

Gillot, G., Jany, J.-L., Coton, M., Floch, G. L., Debaets, S., Ropars, J., ... Coton, E. (2015). Insights into *Penicillium roqueforti* morphological and genetic diversity. *PLoS One*, *10*(6), e0129849. doi:10.1371/journal.pone.0129849 PMID:26091176

Gillot, G., Jany, J.-L., Poirier, E., Maillard, M.-B., Debaets, S., Thierry, A., ... Coton, M. (2017). Functional diversity within the *Penicillium roqueforti* species. *International Journal of Food Microbiology*, *241*, 141–150. doi:10.1016/j.ijfoodmicro.2016.10.001 PMID:27771579

Giosafatto, C. V. L., Rigby, N. M., Wellner, N., Ridout, M., Husband, F., & Mackie, A. R. (2012). Microbial transglutaminase-mediated modification of ovalbumin. *Food Hydrocolloids*, *26*(1), 261–267. doi:10.1016/j.foodhyd.2011.06.003

Giraffa, G. (2003). Functionality of enterococci in dairy products. *International Journal of Food Microbiology*, *2-3*(88), 215–222. doi:10.1016/S0168-1605(03)00183-1 PMID:14596993

Giridhar, A. (2013). *Increasing the Lactococcus lactis biomass through aerobic growth* (Master's thesis). Massey University, Palmerston North, New Zealand.

Gobbetti, M., di Cagno, R., & de Angelis, M. (2010). Functional microorganisms for functional food quality. *Critical Reviews in Food Science and Nutrition*, *50*(8), 716–727. doi:10.1080/10408398.2010.499770 PMID:20830633

Godward, G., & Kailasapathy, K. (2003). Viability and survival of free, encapsulated and co-encapsulated probiotic bacteria in ice cream. *Milchwissenschaft. Milk Science International*, *58*(3-4), 161–164.

Godward, G., Sultana, K., Kailasapathy, K., Peiris, P., Arumugaswamy, R., & Reynolds, N. (2000). The importance of strain selection on the viability and survival of probiotic bacteria in dairy foods. *Milchwissenschaft. Milk Science International*, *55*(8), 441–445.

Goff, H. D., & Hartel, R. W. (2013). *Ice cream*. New York, NY: Springer. doi:10.1007/978-1-4614-6096-1

Gökmen, S., Palamutoğlu, R., & Sariçoban, C. (2012). Gıda Endüstrisinde Enkapsülasyon Uygulamaları. *Electronic Journal of Food Technologies Gıda Teknolojileri Elektronik Dergisi Electronic Journal of Food Technologies*, *7*(71), 36–5036. Retrieved from http://www.teknolojikarastirmalar.com/pdf/tr/02_12_7_1_121_723.pdf

Goldin, B. R. (2011). Probiotics and Health: From History to Future. In W. Kneifel & S. Salminen (Eds.), *Probiotics and Health Claims* (pp. 1–16). Blackwell Publishing. doi:10.1002/9781444329384.ch1

Golowczyc, M. A., Mobili, P., Garrote, G. L., Abraham, A. G., & de Antoni, G. L. (2007). Protective action of *Lactobacillus kefir* carrying S-layer protein against *Salmonella enterica* serovar *enteritidis*. *International Journal of Food Microbiology*, *118*(3), 264–273. doi:10.1016/j.ijfoodmicro.2007.07.042 PMID:17719671

Gomes, A. A., Braga, S. P., Cruz, A. G., Cadena, R. S., Lollo, P. C. B., Carvalho, C., ... Bolini, H. M. A. (2011). Effect of the inoculation level of Lactobacillus acidophilus in probiotic cheese on the physicochemical features and sensory performance compared with commercial cheeses. *Journal of Dairy Science*, *94*(10), 4777–4786. doi:10.3168/jds.2011-4175 PMID:21943729

Gomes, A. M., & Malcata, F. X. (1999). Bifidobacterium spp. and *Lactobacillus acidophilus*: Biological, biochemical, technological and therapeutical properties relevant for use as probiotics. *Trends in Food Science & Technology*, *10*(4), 139–157. doi:10.1016/S0924-2244(99)00033-3

Gómez-Guillén, M. C., Giménez, B., López-Caballero, M. E., & Montero, M. P. (2011). Functional and bioactive properties of collagen and gelatin from alternative sources: A review. *Food Hydrocolloids*, *25*(8), 1813–1827. doi:10.1016/j.foodhyd.2011.02.007

Göncüoğlu, M., Ormancı, F. S. B., & Doğru, A. K. (2009). Beyaz peynir üretiminde Enterococcus faecium un starter kültür olarak kullanılması. *Ankara Üniversitesi Veteriner Fakültesi Dergisi*, *56*(1), 249–254. doi:10.1501/Vetfak_0000002290

Gonzales De Llano, D., Ramos, M., Polo, C., Sanz, J., & Martinez-Castro, I. (1995). Evolution of the Volatile Components of an Artisanal Blue Cheese During Ripening. *Journal of Dairy Science*, *73*(7), 1676–1683. doi:10.3168/jds.S0022-0302(90)78842-X

Gonzalez, B., Arca, P., Mayo, B., & Suarez, J. E. (1994). Detection, purification, and partial characterization of plantaricin C, a bacteriocin produced by a *Lactobacillus plantarum* strain of dairy origin. *Applied and Environmental Microbiology*, *60*(6), 2158–2163. PMID:8080262

Gopal, P. K., Prasad, J., Smart, J., & Gill, H. S. (2001). In vitro adherence properties of *Lactobacillus rhamnosus* DR20 and *Bifidobacterium lactis* DR10 strains and their antagonistic activity against an enterotoxigenic *Escherichia coli*. *International Journal of Food Microbiology*, *67*(3), 207–216. doi:10.1016/S0168-1605(01)00440-8 PMID:11518430

Gori, K., Ryssel, M., Arneborg, N., & Jespersen, L. (2013). Isolation and identification of the microbiota of Danish farmhouse and industrially produced surface-ripened cheeses. *Microbial Ecology*, *65*(3), 602–615. doi:10.100700248-012-0138-3 PMID:23224222

Gote, M., Umalkar, H., Khan, I., & Khire, J. (2004). Thermostable α-galactosidase from Bacillus stearothermophilus (NCIM 5146) and its application in the removal of flatulence causing factors from soymilk. *Process Biochemistry*, *39*(11), 1723–1729. doi:10.1016/j.procbio.2003.07.008

Gouin, S. (2004). Microencapsulation. *Trends in Food Science & Technology*, *15*(7–8), 330–347. doi:10.1016/j.tifs.2003.10.005

Granato, D., Branco, G. F., Nazzaro, F., Cruz, A. G., & Faria, A. F. (2010). Functional foods and nondairy probiotic food development: Trends, concepts, and products. *Comprehensive Reviews in Food Science and Food Safety*, *9*(3), 292–303. doi:10.1111/j.1541-4337.2010.00110.x

Grande, M. J., Lucas, R., Abriouel, H., Valdivia, E., Ben Omar, N., Maqueda, M., ... G'alvez, A. (2006). Inhibition of toxicogenic *Bacillus cereus* in rice based foods by enterocin AS-48. *International Journal of Food Microbiology*, *106*(2), 185–194. doi:10.1016/j.ijfoodmicro.2005.08.003 PMID:16225949

Grattepanche, F., Miescher-Schwenninger, S., Meile, L., & Lacroix, C. (2008). Recent developments in cheese cultures with protective and probiotic functionalities. *Dairy Science & Technology*, *88*(4-5), 421–444. doi:10.1051/dst:2008013

Green, M. L. (1985). Cheddar cheesemaking with recombinant calf chymosin synthesized in *Escherichia coli*. *The Journal of Dairy Research*, *52*(2), 281–286. doi:10.1017/S0022029900024158

Griebeler, N., Polloni, A. E., Remonatto, D., Arbter, F., Vardanega, R., Cechet, J. L., ... Ninow, J. L. (2011). Isolation and screening of lipase-producing fungi with hydrolytic activity. *Food and Bioprocess Technology*, *4*(4), 578–586. doi:10.100711947-008-0176-5

Gripon, J. C. (1993). Mould-ripened cheeses. In. F. Fox, (Eds.), Cheese: Chemistry, Physics and Microbiology: Vol. 2. Major Cheese Groups (2nd ed.; pp. 111-136). London, UK: Chapman & Hall. doi:10.1007/978-1-4615-2648-3_4

Gripon, J.-C. (1977). The proteolytic system of *Penicillium roqueforti*. *Biochimie*, *59*(8), 679–686. doi:10.1016/S0300-9084(77)80246-0 PMID:23180

Gripon, J.-C., & Hermier, J. (1975). Le système protéolytique de *Penicillium roqueforti. Biochimie, 56*(10), 1323–1332. doi:10.1016/S0300-9084(75)80017-4 PMID:4219726

Gripon, J.-C., & Nardi, M. (1977). Le système protéolytique de *Penicillium roqueforti.* IV. Propriétés d'une carboxypeptidase acide. *Annales de Biologie Animale, Biochimie, Biophysique, 17*(3A), 283–298. doi:10.1051/rnd:19770304

Groenewald, M., Boekhout, T., Neuveglise, C., Caillardin, C., van Dijck, P. W. M., & Wyss, M. (2014). Yarrowia lipolytica: Safety assestment of an oleaginaus yeast with a great industrial potential. *Critical Reviews in Microbiology, 40*(3), 187–206. doi:10.3109/1040841X.2013.770386 PMID:23488872

Grossmann, L., Wefers, D., Bunzel, M., Weiss, J., & Zeeb, B. (2017). Accessibility of transglutaminase to induce protein crosslinking in gelled food matrices-influence of network structure. *Lebensmittel-Wissenschaft + Technologie, 75*, 271–278. doi:10.1016/j.lwt.2016.09.005

Guan, Y., & Zhong, Q. (2017). Encapsulation of ferulic acid ethyl ester in caseinate to suppress off-flavor formation in UHT milk. *Food Chemistry, 237*, 532–537. doi:10.1016/j.foodchem.2017.05.140 PMID:28764031

Guérin, D., Vuillemard, J.-C., & Subirade, M. (2003). Protection of bifidobacteria encapsulated in polysaccharide-protein gel beads against gastric juice and bile. *Journal of Food Protection, 66*(11), 2076–2084. doi:10.4315/0362-028X-66.11.2076 PMID:14627286

Guerin-Danan, C., Chabanet, C., Pedone, C., Popot, F., Vaissade, P., Bouley, C., … Andrieux, C. (1998). Milk fermented with yogurt cultures and Lactobacillus casei compared with yogurt and gelled milk: influence on intestinal microflora in healthy infants. *The American Journal of Clinical Nutrition, 67*(1), 111–7. Retrieved from http://www.ncbi.nlm.nih.gov/pubmed/9440384

Guerrero, C., Vera, C., Serna, N., & Ilanes, A. (2017). Immobilization of *Aspergillus oryzae* β-galactosidase in an agarose matrix functionalized by four different methods and application to the synthesis of lactulose. *Bioresource Technology, 232*, 53–63. doi:10.1016/j.biortech.2017.02.003 PMID:28214445

Guidone, A., Ianniello, R. G., Ricciardi, A., Zotta, T., & Parente, E. (2013). Aerobic metabolism and oxidative stress tolerance in the *Lactobacillus plantarum* group. *World Journal of Microbiology & Biotechnology, 29*(9), 1713–1722. doi:10.100711274-013-1334-0 PMID:23543191

Guignon, B., Duquenoy, A., & Dumoulin, E. D. (2002). Fluid Bed Encapsulation of Particles: Principles and Practice. *Drying Technology, 20*(2), 419–447. doi:10.1081/DRT-120002550

Güler, Z., Gün, I., & Uraz, T. (2004). The use of fungal lipase to accelerate the ripening of Kasar cheese. *Milchwissenschaft. Milk Science International, 59*(5-6), 277–279.

Güllüce, M., Karaday, M., & Barış, Ö. (2013). Bacteriocins: promising natural antimicrobials. In *Microbial pathogens and strategies for combating them: science, technology and education* (pp. 1016–1027). Badajoz, Spain: FORMATEX.

Gunasekera, T. S., Attfield, P. V., & Veal, D. A. (2000). A flow cytometry method for rapid detection and enumeration of fungal spores in the atmosphere. *Applied and Environmental Microbiology, 66*(3), 1228–1232. doi:10.1128/AEM.66.3.1228-1232.2000 PMID:10698799

Guo, J., Jin, Y. C., Yang, X. Q., Yu, S. J., Yin, S. W., & Qi, J. R. (2013). Computed microtomography and mechanical property analysis of soy protein porous hydrogel prepared by homogenizing and microbial transglutaminase crosslinking. *Food Hydrocolloids, 31*(2), 220–226. doi:10.1016/j.foodhyd.2012.10.023

Guo, X. H., Kim, J. M., Nam, H. M., Park, S. Y., & Kim, J. M. (2010). Screening lactic acid bacteria from swine origins for multistrain probiotics based on *in vitro* functional properties. *Anaerobe*, *16*(4), 321–326. doi:10.1016/j.anaerobe.2010.03.006 PMID:20304081

Gupta, R., Gupta, N., & Rathi, P. (2004). Bacterial lipases: An overview of production, purification and biochemical properties. *Applied Microbiology and Biotechnology*, *64*(6), 763–781. doi:10.100700253-004-1568-8 PMID:14966663

Gurr, M. I. (1987). Nutritional aspects of fermented milk products. *FEMS Microbiology Reviews*, *3*(3), 337–342. doi:10.1111/j.1574-6968.1987.tb02470.x

Gursoy, O., & Kinik, O. (2010). Incorporation of adjunct cultures of *Enterococcus faecium, Lactobacillus paracasei* subsp. *paracasei* and *Bifidobacterium bifidum* into White cheese. *Journal of Food Agriculture and Environment*, *2*(8), 107–112.

Gustaffson, C., Govindarajan, S., & Minshull, J. (2004). Codon bias and heterologous protein expression. *Trends in Biotechnology*, *22*(7), 346–353. doi:10.1016/j.tibtech.2004.04.006 PMID:15245907

Gustaw, W., Mleko, S., & Tomczynska-Mleko, M. (2008). Production of UF-milk cheeses: Effect of rennet and transglutami-nase on coagulation time, texture and syneresis. *Milchwissenschaft. Milk Science International*, *63*(2), 141–144.

Guyot, C., & Kulozik, U. (2011). Effect of transglutaminase-treated milk powders on the properties of skim milk yoghurt. *International Dairy Journal*, *21*(9), 628–635. doi:10.1016/j.idairyj.2010.10.010

Gyorgy, P., Mello, M. I., Torres, F. E., & Barness, L. A. (1953). Growth promotion in rats by crude concentrates of the bifidus factor. *Proceedings of the Society for Experimental Biology and Medicine. Society for Experimental Biology and Medicine, 84*(2), 464–7. Retrieved from http://www.ncbi.nlm.nih.gov/pubmed/13121074

Hagen, M., & Narvhus, J. (1999). Production of ice cream containing probiotic bacteria. *Milchwissenschaft. Milk Science International*, *54*(5), 265–268.

Hagens, S., & Loessner, M. J. (2007). Application of bacteriophages for detection and control of foodborne pathogens. *Applied Microbiology and Biotechnology*, *76*, 513–519.

Ha, J. K., & Lindsay, R. C. (1993). Release of volatile branched-chain and other fatty acids from ruminant milk fats by various lipases. *Journal of Dairy Science*, *76*(3), 677–690. doi:10.3168/jds.S0022-0302(93)77391-9 PMID:8463483

Hamlyn, P. F., Bradshaw, R. E., Mellon, F. M., Santiago, C. M., Wilson, J. M., & Peberdy, J. F. (1981). Efficient protoplast isolation from fungi using commercial enzymes. *Enzyme and Microbial Technology*, *3*(4), 321–325. doi:10.1016/0141-0229(81)90007-7

Han, M., Zhang, Y., Fei, Y., Xu, X., & Zhou, G. (2009). Effect of microbial transglutaminase on NMR relaxometry and microstructure of pork myofibrillar protein gel. *European Food Research and Technology*, *228*(4), 665–670. doi:10.100700217-008-0976-x

Han, X. Q., & Damodaran, S. (1996). Thermodynamic compatibility of substrate protiens affects their cross-linking by transglutaminase. *Journal of Agricultural and Food Chemistry*, *44*(5), 1211–1217. doi:10.1021/jf950569x

Haque, E., Chand, R., & Kapila, S. (2008). Biofunctional properties of bioactive peptides of milk origin. *Food Reviews International*, *25*(1), 28–43. doi:10.1080/87559120802458198

Harbutt, J. (2009). *World Cheese Book*. London: Dorling Kindersley Limited.

Hassouna, M., & Guizani, N. (1995). Evolution de la flore microbienne et des caractéristiques physico-chimiques au cours de la maturation du fromage Tunisien de type camembert fabriqué avec du lait pasteurisé. *Microbiolgy et Hygiene Alimentaire.*, *7*(18), 14–23.

Hatakka, K., Ahola, A. J., Yli-Knutila, H., Richardson, M., Poussa, T., Meurman, J. H., & Karpela, R. (2007). Probiotics reduce the prevalence, of oral Candida in the elderly-a randomized controlled Trial. *Journal of Dental Research*, *86*(2), 125–130. doi:10.1177/154405910708600204 PMID:17251510

Hati, S., Mandal, S., & Prajapati, J. B. (2013). Novel starters for value added fermented dairy products. *Current Research in Nutrition and Food Science*, *1*(1), 83–91. doi:10.12944/CRNFSJ.1.1.09

Håvarstein, H., Holo, H., & Nes, I. F. (1994). he leader peptide of colicin V shares consensus sequences with leader peptides that are common among peptide bacteriocins produced by gram-positive bacteria. *Microbiology*, *140*(9), 2383–2389. doi:10.1099/13500872-140-9-2383 PMID:7952189

Hayaloglu, A. A., Brechany, E. Y., Deegan, K. C., & McSweeney, P. L. H. (2008). Characterization of the chemistry, biochemistry and volatile profile of Kuflu cheese, a mould-ripened variety. *Lebensmittel-Wissenschaft + Technologie*, *41*(7), 1323–1334. doi:10.1016/j.lwt.2007.08.020

Hayaloglu, A. A., Fox, P. F., Guven, M., & Cakmakci, S. (2007). Cheeses of Turkey: 1. Varieties ripened in goat-skin bags. *Le Lait*, *87*(2), 79–95. doi:10.1051/lait:2007006

Hayaloglu, A. A., & Kirbag, S. (2007). Microbial quality and presence of moulds in Kuflu cheese. *International Journal of Food Microbiology*, *115*(3), 376–380. doi:10.1016/j.ijfoodmicro.2006.12.002 PMID:17258341

Hayashi, K. (1996). Accelerated maturation of cheese by proteolytic enzymes produced by *Brevibacterium linens*. *Japan Agricultural Research Quarterly*, *30*(2), 129–137.

Hayashi, K., Revell, D. F., & Law, B. A. (1990). Effect of Partially Purified Extracellular Serine Proteinases Produced by *Brevibacterium linens* on the Accelerated Ripening of Cheddar Cheese. *Journal of Dairy Science*, *73*(3), 579–583. doi:10.3168/jds.S0022-0302(90)78705-X

Hayek, S. A., & Ibrahim, S. A. (2013). Current limitations and challenges with lactic acid bacteria: A review. *Food and Nutrition Sciences*, *4*(11), 73–87. doi:10.4236/fns.2013.411A010

Hayes, M. G., Oliveira, J. C., McSweeney, P. L. H., & Kelly, A. L. (2002). Thermal inactivation of chymosin during cheese manufacture. *The Journal of Dairy Research*, *69*(02), 269–279. doi:10.1017/S0022029902005472 PMID:12222804

Hayes, M., Ross, R. P., Fitzgerald, G. F., & Stanton, C. (2007). Putting microbes to work: dairy fermentation, cell factories and bioactive peptides. Part I: overview. *Biotechnology Journal*, *2*(4), 426–434. doi:10.1002/biot.200600246 PMID:17407210

Haynes, I., & Playne, M. (2002). Survival of probiotic cultures in low-fat ice-cream. *Australian Journal of Dairy Technology*, *57*(1), 10–14.

Hechard, Y., Derijard, B., Letellier, F., & Cenatiempo, Y. (1992). Characterization and purification of mesentericin Y 105, an anti-listeria bacteriocin from *Leuconostoc mesenteroides*. *Journal of General Microbiology*, *138*(12), 2725–2731. doi:10.1099/00221287-138-12-2725 PMID:1487737

Heidabach, T., Först, P., & Kulozik, U. (2009). Transglutamşnase-induced caseinate gelation for themicroencapsulation of probiotic cells. *International Dairy Journal*, *19*(2), 77–84. doi:10.1016/j.idairyj.2008.08.003

Heidabach, T., Först, P., & Kulozik, U. (2010). Influence of casein-based microencapsulation on freeze-drying and storage of probiotic cells. *Journal of Food Engineering*, *98*(3), 309–316. doi:10.1016/j.jfoodeng.2010.01.003

Heinzelmann, K., Franke, K., Jensen, B., & Haahr, A.-M. (2000). Protection of fish oil from oxidation by microencapsulation using freeze-drying techniques. *European Journal of Lipid Science and Technology*, *102*(2), 114–121. doi:

Hekmat, S., & McMahon, D. J. (1992). Survival of *Lactobacillus acidophilus* and *Bifidobacterium bifidum* in ice cream for use as a probiotic food. *Journal of Dairy Science*, *75*(6), 1415–1422. doi:10.3168/jds.S0022-0302(92)77895-3 PMID:1500547

Hekmat, S., Soltani, H., & Reid, G. (2009). Growth and survival of *Lactobacillus reuteri* RC-14 and *Lactobacillus rhamnosus* GR-1 in yogurt for use as a functional food. *Innovative Food Science & Emerging Technologies*, *10*(2), 293–296. doi:10.1016/j.ifset.2008.10.007

Heller, K. J. (2001). Probiotic bacteria in fermented foods: Product characteristics and starter organisms. *The American Journal of Clinical Nutrition*, *73*, 374–379. PMID:11157344

Hempel, S., Newberry, A. J., Maher, A. R., Wang, A., Miles, J. N. V., Shanman, R., ... Shekelle, P. G. (2012). Probiotics for the Prevention and Treatment of Antibiotic-Associated Diarrhea. *Journal of the American Medical Association*, *307*(18), 1959–1969. doi:10.1001/jama.2012.3507 PMID:22570464

Hermet, A., Méheust, D., Mounier, J., Barbier, G., & Jany, J.-L. (2012). Molecular systematics in the genus Mucor with special regards to species encountered in cheese. *Fungal Biology*, *116*(6), 692–705. doi:10.1016/j.funbio.2012.04.002 PMID:22658314

Hernández, I., De Renobales, M., Virto, M., Pérez-Elortondo, F. J., Barron, L. J. R., Flanagan, C., & Albisu, M. (2005). Assessment of industrial lipases for flavour development in commercial Idiazabal (ewe's raw milk) cheese. *Enzyme and Microbial Technology*, *36*(7), 870–879. doi:10.1016/j.enzmictec.2004.12.032

Hill, A. (2017). *Cheese technology. University of Guelph.* Dairy Science and Technology Education Series.

Hinz, K., Huppertz, T., Kulozik, U., & Kelly, A. L. (2007). Influence of enzymatic cross-linking on milk fat globules and emulsifying properties of milk proteins. *International Dairy Journal*, *17*(4), 289–293. doi:10.1016/j.idairyj.2006.05.001

Hirayama, K., & Rafter, J. (2000). The role of probiotic bacteria in cancer prevention. *Microbes and Infection*, *2*(6), 681–686. doi:10.1016/S1286-4579(00)00357-9 PMID:10884619

Hocking, A. D. (2007). Current methods for the detection of molds and yeasts. In *Food Mycology 2007: Emerging mold problems and spoilage in food and beverages. Proceedings of Food Mycology*, *2007*, 4–8.

Holo, H., Faye, T., Brede, D. A., Nilsen, T., Odegard, I., Langsrud, T., ... Nes, I. F. (2002). Bacteriocins of propionic acid bacteria. *Le Lait*, *82*(1), 59–68. doi:10.1051/lait:2001005

Holo, H., Nilssen, O., & Nes, I. F. (1991). Lactococcin A, a new bacteriocin from *Lactococcus lactis* subsp. *cremoris*: Isolation and characterization of the protein and its gene. *Journal of Bacteriology*, *173*(12), 3879–3887. doi:10.1128/jb.173.12.3879-3887.1991 PMID:1904860

Holzapfel, W. H., & Schillinger, U. (2002). Introduction to pre and probiotics. *Food Research International*, *35*(2-3), 109–116. doi:10.1016/S0963-9969(01)00171-5

Homayouni, A., Ehsani, M. R., Azizi, A., Yarmand, M. S., & Razavi, S. H. (2008). Effect of microencapsulation and resistant starch on the probiotic survival and sensory properties of synbiotic ice cream. *Food Chemistry*, *111*(1), 50–55. doi:10.1016/j.foodchem.2008.03.036

Houbraken, J., Frisvad, J. C., & Samson, R. A. (2010). Sex in *Penicillium* series *Roqueforti. IMA Fungus: The Global Mycological Journal*, *1*(2), 171–180.

Houbraken, J., & Samson, R. (2011). Phylogeny of *Penicillium* and the segregation of *Trichocomaceae* into three families. *Studies in Mycology*, *70*(1), 1–51. doi:10.3114im.2011.70.01 PMID:22308045

Huang, J., Lacroix, C., Daba, H., & Simard, R. E. (1994). Growth of *Listeria monocytogenes* in milk and its control by pediocin 5 produced by *Pediococcus acidilactici* UL5. *International Dairy Journal*, *4*(5), 429–443. doi:10.1016/0958-6946(94)90057-4

Hugas, M. (1998). Bacteriocinogenic lactic acid bacteria for the biopreservation of meat and meat products. *Meat Science*, *49*, S139–S150. doi:10.1016/S0309-1740(98)90044-4 PMID:22060706

Hugenholtz, J. (2008). The lactic acid bacterium as a cell factory for food ingredient production. *International Dairy Journal*, *18*(5), 466–475. doi:10.1016/j.idairyj.2007.11.015

Hugenholtz, J., & Smid, E. J. (2002). Nutraceutical production with food-grade microorganisms. *Current Opinion in Biotechnology*, *13*(5), 497–507. doi:10.1016/S0958-1669(02)00367-1 PMID:12459344

Hung, M. N., & Lee, B. H. (2002). Purification and characterization of a recombinant beta-galactosidase with transgalactosylation activity from *Bifidobacterium infantis* HL96. *Applied Microbiology and Biotechnology*, *58*(4), 439–445. doi:10.100700253-001-0911-6 PMID:11954789

Huppertz, T., & de Kruif, C. D. (2007a). Ethanol stability of casein micelles cross-linked with transglutaminase. *International Dairy Journal*, *17*(5), 436–441. doi:10.1016/j.idairyj.2006.05.005

Huppertz, T., & de Kruif, C. D. (2008). Structure and stability of nanogel particles prepared by internal crosslinking of casein micelles. *International Dairy Journal*, *18*(5), 556–565. doi:10.1016/j.idairyj.2007.10.009

Huppertz, T., & de Kruif, C. G. (2007b). Rennet-induced coagulation of enzymatically cross-linked casein micelles. *International Dairy Journal*, *17*(5), 442–447. doi:10.1016/j.idairyj.2006.06.026

Huppertz, T., & Smiddy, M. A. (2008). Behaviour of partially cross-linked casein micelles under high pressure. *International Journal of Dairy Technology*, *61*(1), 51–55. doi:10.1111/j.1471-0307.2008.00370.x

Hutkins, R. W. (2006). Meat fermentation. In R. W. Hutkins (Ed.), *Microbiology and technology of fermented foods* (pp. 207–232). Blackwell Publishing. doi:10.1002/9780470277515.ch6

Hu, Y., Ge, K., Guo, H., Luo, J., Wang, F., & Ren, F. (2013). Effect of transglutaminase on yield, compositional and functional properties of low-fat Cheddar cheese. *Food Science and Technology Research*, *19*(3), 359–367. doi:10.3136/fstr.19.359

Ianniello, R. G., Ricciardi, A., Parente, E., Tramutola, A., Reale, A., & Zotta, T. (2015). Aeration and supplementation with heme and menaquinone affect survival to stresses and antioxidant capability of *Lactobacillus casei* strains. *Lebensmittel-Wissenschaft + Technologie*, *60*(2), 817–824. doi:10.1016/j.lwt.2014.10.020

Ianniello, R. G., Zotta, T., Matera, A., Genovese, F., Parente, E., & Ricciardi, A. (2016). Investigation of factors affecting aerobic and respiratory growth in the oxygen-tolerant strain *Lactobacillus casei* N87. *PLoS One*, *11*(11), e0164065. doi:10.1371/journal.pone.0164065 PMID:27812097

Ichishima, E., Takeuchi, M., Yamamoto, K., Sano, Y., & Kikuchi, T. (1978). Intracellular acid carboxypeptidase of *Penicillium roqueforti* isolated from blue cheese. *Current Microbiology*, *1*(2), 95–98. doi:10.1007/BF02605424

Ikura, K., Kometani, T., Sasaki, R., & Chiba, H. (1980). Cross-linking of soy bean 7S and 11S proteins by transglutaminase. *Agricultural and Biological Chemistry*, *44*, 2979–2984.

Ikura, K., Sasaki, R., & Motoki, M. (1992). Use of transglutaminase in quality-improvement and processing of food proteins. *Comments in Agricultural and Food Chemistry*, *2*(6), 389–407.

Iličić, M. D., Milanović, S. D., Carić, M. D., Dokić, L. P., & Kanurić, K. G. (2014). Effect of transglutaminase on texture and flow properties of stirred probiotic yoghurt during storage. *Journal of Texture Studies*, *45*(1), 13–19. doi:10.1111/jtxs.12038

Imm, J. Y., Lian, P., & Lee, C. M. N. (2000). Gelation and water binding properties of transglutaminase-treated skim milk powder. *Journal of Food Science*, *65*(2), 200–205. doi:10.1111/j.1365-2621.2000.tb15979.x

Iqbal, M. Z., Qadir, M. I., Hussain, T., Janbaz, K. H., Khan, Y. H., & Ahmad, B. (2014). Probiotics and their beneficial effects against various diseases. *Pakistan Journal of Pharmaceutical Sciences*, *27*(2), 405–415. PMID:24577933

Iwata, T., Tanaka, R., Suetsugu, M., Ishibashi, M., Tokunaga, H., Kikuchi, M., & Tokunaga, M. (2004). Efficient secretion of human lysozyme from the yeast, *Kluyveromyces lactis*. *Biotechnology Letters*, *26*(23), 1803–1808. doi:10.100710529-004-4614-9 PMID:15672218

Iyer, R., Tomar, S. K., Maheswari, T. U., & Singh, R. (2010). *Streptococcus thermophilus* strains: Multifunctional lactic acid bacteria. *International Dairy Journal*, *20*(3), 133–141. doi:10.1016/j.idairyj.2009.10.005

Izco, J. M., Irigoyen, A., Torre, P., & Barcina, Y. (2000). Effect of added enzymes on the free amino acids and sensory characteristics in Ossau-Iraty cheese. *Food Control*, *11*(3), 201–207. doi:10.1016/S0956-7135(99)00100-0

Izco, J. M., & Torre, P. (2000). Characterisation of volatile flavour compounds in Roncal cheese extracted by the 'purge and trap' method and analysed by GC–MS. *Food Chemistry*, *70*(3), 409–417. doi:10.1016/S0308-8146(00)00100-X

Izquierdo, E., Marchioni, E., Aoude-Werner, D., Hasselmann, C., & Ennahar, S. (2009). Smearing of soft cheese with *Enterococcus faecium* WHE 81, a multi-bacteriocin producer, against *Listeria monocytogenes*. *Food Microbiology*, *26*(1), 16–20. doi:10.1016/j.fm.2008.08.002 PMID:19028299

Jacob, M., Nöbel, S., Jaros, D., & Rohm, H. (2011). Physical properties of acid milk gels: Acidification rate significantly interacts with cross-linking and heat treatment of milk. *Food Hydrocolloids*, *25*(5), 928–934. doi:10.1016/j.foodhyd.2010.09.003

Jacops, K., & Botha, A. (2008). *Mucor renispora* sp. nov., a new coprophilous species from Southern Africa. *Fungal Diversity*, *29*, 27–35.

Jafari, S. M., Assadpoor, E., He, Y., & Bhandari, B. (2008). Encapsulation Efficiency of Food Flavours and Oils during Spray Drying. *Drying Technology*, *26*(7), 816–835. doi:10.1080/07373930802135972

Jagielski, T., Sandoval-Denis, M., Yu, J., Yao, L., Bakuła, Z., Kalita, J., ... Gené, J. (2016). Molecular taxonomy of *Scopulariopsis*-like fungi with description of new clinical and environmental species. *Fungal Biology*, *120*(4), 586–602. doi:10.1016/j.funbio.2016.01.014 PMID:27020159

Jakobsen, M., & Narvhus, J. (1996). Yeast and their possible beneficial and negativeeffects on the quality of dairy products. *International Dairy Journal*, *6*(8-9), 755–768. doi:10.1016/0958-6946(95)00071-2

James, J., Simpson, B. K., & Marshall, M. R. (1996). Application of enzymes in food processing. *Critical Reviews in Food Science and Nutrition*, *36*(5), 437–463. doi:10.1080/10408399609527735 PMID:8725673

James, T. Y., Kauff, F., Schoch, C. L., Matheny, P. B., Hofstetter, V., Cox, C. J., ... Vilgalys, R. (2006). Reconstructing the early evolution of Fungi using a six-gene phylogeny. *Nature*, *443*(7113), 818–822. doi:10.1038/nature05110 PMID:17051209

Jany, J., & Barbier, G. (2008). Culture-independent communities in cheese methods for identifying microbial communities in cheese. *Food Microbiology*, *25*(7), 839–848. doi:10.1016/j.fm.2008.06.003 PMID:18721671

Jaros, D., Heidig, C., & Rohm, H. (2007). Enzymatic modification through microbial transglutaminase enhances the viscosity of stirred yogurt. *Journal of Texture Studies*, *38*(2), 179–198. doi:10.1111/j.1745-4603.2007.00093.x

Jaros, D., Jacob, M., Otto, C., & Rohm, H. (2010). Excessive cross-linking of caseins by microbial transglutaminase and its impact on physical properties of acidified milk gels. *International Dairy Journal*, *20*(5), 321–327. doi:10.1016/j.idairyj.2009.11.021

Jaros, D., Partschefeld, C., Henle, T., & Rohm, H. (2006a). Transglutaminase in dairy products: Chemistry, physics and applications. *Journal of Texture Studies*, *37*(2), 13–155. doi:10.1111/j.1745-4603.2006.00042.x

Jaros, D., Pätzold, J., Schwarzenbolz, U., & Rohm, H. (2006b). Small and large deformation rheology of acid gels from transglutaminase treated milks. *FOBI*, *1*, 124–132.

Jaros, D., Schwarzenbolz, U., Raak, N., Löbner, J., Henle, T., & Rohm, H. (2014). Cross-linking with microbial transglutaminase: Relationship between polymerisation degree and stiffness of acid casein gels. *International Dairy Journal*, *38*(2), 174–178. doi:10.1016/j.idairyj.2013.10.011

Jay, J. M., Loessner, M. J., & Golden, D. A. (2008). *Modern food microbiology*. New York, NY: Springer.

Jaziri, F., Parisot, N., Abid, A., Denonfoux, J., Ribière, C., & Gasc, C. (2014). … Peyret, P. (2014). PhylOPDb: A 16S rRNA oligonucleotide probe database for prokaryotic identification. *Database*, (1): 1–7.

Jeong, D., Kim, D. H., Kang, B., Kim, H., Song, K. Y., Kim, H. S., & Seo, K. H. (2017). Characterization and antibacterial activity of a novel exopolysaccharide produced by *Lactobacillus kefiranofaciens* DN1 isolated from kefir. *Food Control*, *78*, 436–442. doi:10.1016/j.foodcont.2017.02.033

Jiang, J., & Fonden, R. (1998). Production of CLA by dairy starter cultures. *Journal of Applied Microbiology*, *85*, 95–102. doi:10.1046/j.1365-2672.1998.00481.x PMID:9721660

Jiang, S. J., Zhang, X., Ma, Y., Tuo, Y., Qian, F., Fu, W., & Mu, G. (2016). Characterization of whey protein-carboxymethylated chitosan composite films with and without transglutaminase treatment. *Carbohydrate Polymers*, *153*, 153–159. doi:10.1016/j.carbpol.2016.07.094 PMID:27561482

Jiang, S. J., & Zhao, X. H. (2011). Transglutaminase-induced cross-linking and glucosamine conjugation of casein and some functional properties of the modified product. *International Dairy Journal*, *21*(4), 198–205. doi:10.1016/j.idairyj.2010.12.004

Jiang, S. J., & Zhao, X. H. (2012). Cross-linking and glucosamine conjugation of casein by transglutaminase and the emulsifying property and digestibility in vitro of the modified product. *International Journal of Food Properties*, *15*(6), 1286–1299. doi:10.1080/10942912.2010.521274

Jiang, X. P., Yin, M. L., Chen, P., & Yang, Q. (2012). Constitutive expression, purification, and characterization of bovine prochymosin in *Pichia pastoris* GS115. *World Journal of Microbiology & Biotechnology*, *28*(5), 2087–2093. doi:10.100711274-012-1012-7 PMID:22806030

Jiao, Y. H., Zhang, L. W., & Liu, F. (2013). *Screening of lactic acid bacteria strains with respiration ability in the present of heme*. Paper presented at the Advanced Materials Research. 10.4028/www.scientific.net/AMR.726-731.448

Jiménez-Martín, E., Gharsallaoui, A., Pérez-Palacios, T., Ruiz Carrascal, J., & Antequera Rojas, T. (2015). Volatile compounds and physicochemical characteristics during storage of microcapsules from different fish oil emulsions. *Food and Bioproducts Processing*, *96*, 52–64. doi:10.1016/j.fbp.2015.07.005

Jimeno, J., Lazaro, M. J., & Sollberger, H. (1995). Antagonistic interactions between propionic acid bacteria and non-starter lactic acid bacteria. *Le Lait, 75*(4-5), 401–413. doi:10.1051/lait:19954-530

Johnson, B., Selle, K., O'Flaherty, S., Goh, Y. J., & Klaenhammer, T. (2013). Identification of extracellular surface-layer associated proteins in *Lactobacillus acidophilus* NCFM. *Microbiology, 159*(Pt_11), 2269–2282. doi:10.1099/mic.0.070755-0 PMID:24002751

Jones, B. V., Begley, M., Hill, C., Gahan, C. G., & Marchesi, J. R. (2008). Functional and comparative metagenomic analysis of bile salt hydrolase activity in the human gut microbiome. *Proceedings of the National Academy of Sciences of the United States of America, 105*(36), 13580–13585. doi:10.1073/pnas.0804437105 PMID:18757757

Jones, M. L., Martoni, C. J., & Prakash, S. (2012). Cholesterol lowering and inhibition of sterol absorption by *Lactobacillus reuteri* NCIMB 30242: A randomized controlled trial. *European Journal of Clinical Nutrition, 66*(11), 1234–1241. doi:10.1038/ejcn.2012.126 PMID:22990854

Jooyandeh, H., Kaur, A., & Minhas, K. S. (2009). Lipases in dairy industry: A review. *Journal of Food Science and Technology, 46*(3), 181–189.

Joubert, L., Derré-Bobillot, A., Gaudu, P., Gruss, A., & Lechardeur, D. (2014). HrtBA and menaquinones control haem homeostasis in *Lactococcus lactis. Molecular Microbiology, 93*(4), 823–833. doi:10.1111/mmi.12705 PMID:25040434

Joyce, S. A., MacSharry, J., Casey, P. G., Kinsella, M., Murphy, E. F., Shanahan, F., ... Gahan, C. G. (2014). Regulation of host weight gain and lipid metabolism by bacterial bile acid modification in the gut. *Proceedings of the National Academy of Sciences of the United States of America, 111*(20), 7421–7426. doi:10.1073/pnas.1323599111 PMID:24799697

Juarez-Tomas, M. S., Duhart, C. I. S., De Gregorio, P. R., Pingitore, E. V., & Nader-Macias, M. E. (2011). Urogenital pathogen inhibition and compatibility between vaginal *Lactobacillus* strains to be considered as probiotic candidates. *European Journal of Obstetrics, Gynecology, and Reproductive Biology, 159*(2), 399–406. doi:10.1016/j.ejogrb.2011.07.010 PMID:21862199

Julendra, H., Suryani, A. E., Istiqomah, L., Damayanti, E., Anwar, M., & Fitriani, N. (2017). Isolation of Lactic Acid Bacteria with cholesterol-lowering activity from digestive tracts of Indonesian native chickens. *Media Peternakan, 40*(1), 35–41. doi:10.5398/medpet.2017.40.1.35

Jyothi, N. V. N., Prasanna, P. M., Sakarkar, S. N., Prabha, K. S., Ramaiah, P. S., & Srawan, G. Y. (2010). Microencapsulation techniques, factors influencing encapsulation efficiency. *Journal of Microencapsulation, 27*(3), 187–197. doi:10.3109/02652040903131301 PMID:20406093

Kabak, B., & Dobson, D. W. (2011). An Introduction to the traditional fermented foods and beverages of Turkey. *Critical Reviews in Food Science and Nutrition, 51*(3), 248–260. doi:10.1080/10408390903569640 PMID:21390945

Kailasapathy, K. (2002). Microencapsulation of probiotic bacteria: Technology and potential applications. *Current Issues in Molecular Biology, 3*, 39–48. PMID:12400637

Kailasapathy, K. (2013). Commercial sources of probiotic strains and their validated and potential health benefits - a review. *International Journal of Fermented Foods, 2*(1), 1–17.

Kailasapathy, K., Anjani, K., & Seneweera, S. (2006). Recent trends in accelerated cheese ripening using microencapsulated enzymes. *Australian Journal of Dairy Technology, 61*(2), 78–80.

Kailasapathy, K., & Phillips, H. M. (2008). Survival of *Lactobacillus acidophilus* and *Bifidobacterium animalis* ssp. *lactis* in stirred fruit yogurts. *Lebensmittel-Wissenschaft + Technologie, 41*(7), 1317–1322. doi:10.1016/j.lwt.2007.08.009

Kailasapathy, K., & Sultana, K. (2003). Survival and [beta]-D-galactosidase activity of encapsulated and free *Lactobacillus acidophilus* and *Bifidobacterium lactis* in ice-cream. *Australian Journal of Dairy Technology*, *58*(3), 223–227.

Kailasapathy, K., & Sultana, K. (2003). Survival and [beta]-D-galactosidase activity of encapsulated and free. *Journal of Dairy Technology*, *58*(223). Retrieved from https://search.proquest.com/docview/199499876/fulltextPDF/7DEDFF93D6784502PQ/1?accountid=25088

Kalliomäki, M., Salminen, S., Poussa, T., Arvilommi, H., & Isolauri, E. (2003). Probiotics and prevention of atopic disease: 4-year follow-up of a randomised placebo-controlled trial. *Lancet*, *361*(9372), 1869–1871. doi:10.1016/S0140-6736(03)13490-3 PMID:12788576

Kalvrouzioti, I., Hatzıkamari, M., Litopoulou-Tzanetaki, E., & Tzanetakis, N. (2005). Production of hard cheese from caprine milk by the use of two types of probiotic cultures as adjunct. *International Journal of Dairy Technology*, *1*(58), 30–38. doi:10.1111/j.1471-0307.2005.00179.x

Kanamarlapudi, S. L. R., & Muddada, S. (2017). Characterization of Exopolysaccharide Produced by *Streptococcus thermophilus* CC30. *BioMed Research International*, *2017*, 1–12. doi:10.1155/2017/4201809 PMID:28815181

Kandylis, P., Pissaridi, K., Bekatorou, A., Kanellaki, M., & Koutinas, A. A. (2016). Dairy and non-dairy probiotic beverages. *Current Opinion in Food Science*, *7*, 58–63. doi:10.1016/j.cofs.2015.11.012

Kaneko, T., Takahashi, M., & Suzuki, H. (1990). Acetoin fermentation by citrate-positive *Lactococcus lactis* subsp. *lactis* 3022 grown aerobically in the presence of hemin or Cu2+. *Applied and Environmental Microbiology*, *56*(9), 2644–2649. PMID:16348274

Kang, Y. N., Kim, H., Shin, W. S., Woo, G., & Moon, T. W. (2003). Effect of disulfide bond reduction on bovine serum albumin-stabilized emulsion gel formed by microbial transglutaminase. *Journal of Food Science*, *68*(7), 2215–2220. doi:10.1111/j.1365-2621.2003.tb05749.x

Kanmani, P., Kumar, R. S., Yuvaraj, N., Paari, K. A., Pattukumar, V., & Arul, V. (2013). Probiotics and its functionally valuable products: A review. *Critical Reviews in Food Science and Nutrition*, *53*(6), 641–658. doi:10.1080/10408398.2011.553752 PMID:23627505

Kappeler, S. R., van den Brink, H. J. M., Rahbek-Nielsen, H., Farah, Z., Puhan, Z., Hansen, E. B., & Johansen, E. (2006). Characterization of recombinant camel chymosin reveals superior properties for the coagulation of bovine and camel milk. *Biocehmical and Biophysical Reserach Communications*, *342*(2), 647–654. doi:10.1016/j.bbrc.2006.02.014 PMID:16488399

Karlsson, M. (2013). *Evaluation of Lactobacillus reuteri DSM17938 as starter in cheese production* (Unpublished master's thesis). Department of Microbiology, Institutionen för mikrobiologi, Uppsala, Sweden.

Karthikeyan, N., Elango, A., Kumaresan, G., Golapakrishnamurty, T., & Raghunath, B. (2014). Enhancement of probiotic viability in ice cream by microencapsulation. *International Journal of Science. Environmental Technology*, *3*(1), 339–347.

Karzan, T. M., Nawal, H. S., & Ashna, T. A. (2016). The effect of microbial transglutaminase enzyme on some physicochemical and sensory properties of goat's whey cheese. *International Food Research Journal*, *23*, 688–693.

Kasimoglu, A., Goncuoglu, M., & Akgun, S. (2004). Probiotic white cheese with *Lactobacillus acidophilus*. *International Dairy Journal*, *14*(12), 1067–1073. doi:10.1016/j.idairyj.2004.04.006

Kaur, M. R. K., & Singh, T. P. (2014). Bacteriocins and their potential applications. In Microbes in the service of mankind: tiny bugs with huge impact (pp. 309-345). JBC Press.

Kaya, Y., Kök, M. Ş., & Öztürk, M. (2017). Molecular cloning, expression and characterization of bile salt hydrolase from *Lactobacillus rhamnosus* E9 strain. *Food Biotechnology*, *31*(2), 128–140. doi:10.1080/08905436.2017.1303778

Keersmaecker, S. C. J., Verhoeven, T. L. A., Desair, J., Marchal, K., Vanderleyden, J., & Nagy, I. (2006). Strong antimicrobial activity of *Lactobacillus rhamnosus* GG against *Salmonella typhimurium* is due to accumulation of lactic acid. *FEMS Microbiology Letters*, *259*(1), 89–96. doi:10.1111/j.1574-6968.2006.00250.x PMID:16684107

Kellerby, S. S., Gu, Y. S., McClements, D. J., & Decker, E. A. (2006). Lipid oxidation in a menhaden oil-in-water emulsion stabilized by sodium caseinate cross-linked with transglutaminase. *Journal of Agricultural and Food Chemistry*, *54*(26), 10222–10227. doi:10.1021/jf062143w PMID:17177563

Kent, A. D., Smith, D. J., Benson, B. J., Triplett, W., & Triplett, E. W. (2003). Web-Based phylogenetic assignment tool for analysis of terminal restriction fragment length polymorphism profiles of microbial communities. *Applied and Environmental Microbiology*, *69*(11), 6798. doi:10.1128/AEM.69.11.6768-6776.2003 PMID:14602639

Khademi, F., Abachi, S., Mortazavi, A., Ehsani, M. A., Tabatabaei, M. R., & Malekzadeh, F. A. (2013). Optimization of fungal rennet production by local isolate of *Rhizomucor nainitalensis* under solid substrate fermentation system. *IOSR Journal of Pharmacological and Biological Sciences*, *5*(2), 115–121. doi:10.9790/3008-052115121

Kheadr, E. E., Vuillemard, J. C., & El-Deeb, S. A. (2003). Impact of liposome-encapsulated enzyme cocktails on Cheddar cheese ripening. *Food Research International*, *36*(3), 241–252. doi:10.1016/S0963-9969(02)00166-7

Kheadr, E. E., Vuillemard, J.-C., & El Deeb, S. A. (2000). Accelerated Cheddar cheese ripening with encapsulated proteinases. *International Journal of Food Science & Technology*, *35*(5), 483–495. doi:10.1046/j.1365-2621.2000.00398.x

Kheadr, E. E., Vuillemard, J.-C., & El-Deeb, S. A. (2002). Acceleration of Cheddar cheese lipolysis by using liposome-entrapped lipases. *Journal of Food Science*, *67*(2), 485–492. doi:10.1111/j.1365-2621.2002.tb10624.x

Khosravi-Darani, K., Karamadi, D., Nikoopour, H., & Reyhani, S. F. S. (2016). Comparison of conjugated linoleic acid production in *Bifidobacterium* sp. and *Lactobacillus acidophilus* probiotic yoghurt. *Journal of Chemical and Pharmaceutical Research*, *8*(8), 45–53.

Khosravi-Darani, K., Reihani, F. S., & Feili, R. (2014). Bioproduction of conjugated linoleic acid in yogurt by probiotic bacteria. *International Journal of Biotechnology for Wellness Industries*, *3*(2), 62–68. doi:10.6000/1927-3037.2014.03.02.4

Kieliszek, M., & Misiewicz, A. (2014). Microbial transglutaminase and its application in the food industry. A review. *Folia Microbiologica*, *59*(3), 241–250. doi:10.100712223-013-0287-x PMID:24198201

Kilara, A., & Chandan, R. C. (2011). Enzyme-modified Dairy Ingredients. In R. Chandan & A. Kilara (Eds.), *Dairy Ingredients for Food Processing* (pp. 317–333). Wiley-Blackwell. doi:10.1002/9780470959169.ch12

Kilcawley, K. N., Nongonierma, A. B., Hannon, J. A., Doolan, I. A., & Wilkinson, M. G. (2012). Evaluation of commercial enzyme systems to accelerate Cheddar cheese ripening. *International Dairy Journal*, *26*(1), 50–57. doi:10.1016/j.idairyj.2012.03.015

Kim, D. H., Chon, J. W., Kim, H., & Seo, K. H. (2015). Modulation of intestinal microbiota in mice by kefir administration. *Food Science and Biotechnology*, *24*(4), 1397–1403. doi:10.100710068-015-0179-8

Kim, K. P., Klumpp, J., & Loessner, M. J. (2007b). *Enterobacter sakazakii* bacteriophages can prevent bacterial growth in reconstituted infant formula. *International Journal of Food Microbiology*, *115*(2), 195–203. doi:10.1016/j.ijfoodmicro.2006.10.029 PMID:17196280

Kim, K., Jang, S. S., Kim, S. K., Park, J. H., Heu, S., & Ryu, S. (2007a). Prevalence and genetic diversity of *Enterobacter sakazakii* in ingredients of infant foods. *International Journal of Food Microbiology, 122*(1-2), 196–203. doi:10.1016/j.ijfoodmicro.2007.11.072 PMID:18177966

Kim, Y., Oh, S., & Kim, S. H. (2009). Released exopolysaccharide (r-EPS) produced from probiotic bacteria reduce biofilm formation of enterohemorrhagic Escherichia coli O157:H7. *Biochemical and Biophysical Research Communications, 379*(2), 324–329. doi:10.1016/j.bbrc.2008.12.053 PMID:19103165

Kim, Y., Oh, S., Yun, H. S., Oh, S., & Kim, S. H. (2010). Cell-bound exopolysaccharide from probiotic bacteria induces autophagic cell death of tumour cells. *Letters in Applied Microbiology, 51*, 123–130. PMID:20536712

King, H., & Alexander, H. (1948). The mechanical destruction of bacteria. *Microbiology, 2*(3), 315–324.

Klaenhammer, T. R. (1993). Genetics of bacteriocins produced by lactic acid bacteria. *FEMS Microbiology Reviews, 12*(1-3), 39–86. doi:10.1111/j.1574-6976.1993.tb00012.x PMID:8398217

Klaubauf, S. (2008). *The role of fungal nitrogen cycling in soil systems: assessment of fungal diversity and development of a sequence set of fungal nitrate reductases*. Universität Wien.

Klaubauf, S., Inselsbacher, E., Zechmeister-Boltenstern, S., Wanek, W., Gottsberger, R., Strauss, J., & Gorfer, M. (2010). Molecular diversity of fungal communities in agricultural soils from Lower Austria. *Fungal Diversity*, 44. PMID:23794962

Klein, J. D., Guzman, E., & Kuehen, G. D. (1992). Purification and partial characterization of transglutaminase from *Physarum polycephalum. Journal of Bacteriology, 174*(8), 2599–2605. doi:10.1128/jb.174.8.2599-2605.1992 PMID:1348244

Klein, N., & Lortal, S. (1999). Attenuated starters: An efficient means to influence cheese ripening—a review. *International Dairy Journal, 9*(11), 751–762. doi:10.1016/S0958-6946(99)00155-7

Kobayashi, E., Hashiguchi, K., Yikozeki, K., & Yamanaka, S. (1998). Molecular cloning of the transglutaminase gene from *Bacillus subtilis* and its expression in *Escherichia coli. Bioscience, Biotechnology, and Biochemistry, 62*(6), 1109–1114. doi:10.1271/bbb.62.1109 PMID:9692191

Koç, M., Sakín, M., & Kaymak-Ertekin, F. (2010). Mikroenkapsülasyon ve Gıda Teknolojisinde Kullanımı Microencapsulation and its Applications in Food Technology. *Pamukkale Üniversitesi Mühendislik Bilimleri Dergisi, 16*(1), 77–86. Retrieved from https://www.journalagent.com/pajes/pdfs/PAJES_16_1_77_86.pdf

Koçak, C., & Güzel-Seydim, Z. B. (2011). Kazein Kimyası ve Sütün Pıhtılaşma Mekanizması. In Peynir Biliminin Temelleri (pp. 53-78). İzmir-Türkiye: Sidas.

Koçak, C., Bítlís, A., Gürsel, A., & Avsar, Y. K. (1996). Effect of added fungal lipase on the ripening of Kasar cheese. *Milchwissenschaft. Milk Science International, 51*(1), 13–16.

Kocaman, E. (2016). *Effects of ph and transglutaminase enzyme on gelation of caseinomacropeptide* (M.Sc. Thesis). Istanbul Technical University, Istanbul, Turkey.

Koebmann, B., Blank, L. M., Solem, C., Petranovic, D., Nielsen, L. K., & Jensen, P. R. (2008). Increased biomass yield of *Lactococcus lactis* during energetically limited growth and respiratory conditions. *Biotechnology and Applied Biochemistry, 50*(1), 25–33. doi:10.1042/BA20070132 PMID:17824842

Kok, J., Buist, G., Zomer, A. L., van Hijum, S. A., & Kuipers, O. P. (2005). Comparative and functional genomics of lactococci. *FEMS Microbiology Reviews, 29*(3), 411–433. doi:10.1016/j.fmrre.2005.04.004 PMID:15936843

Kołodziewska, I., & Piotrowska, B. (2007). The water vapour permeability, mechanical properties and solubility of fish gelatin-chitosan films modified with transglutamşnase or 1-ethyl-3-(3 –dimethylaminopropyl) carbodiimide (EDC) and plasticized with glycerol. *Food Chemistry*, *103*(2), 295–300. doi:10.1016/j.foodchem.2006.07.049

Konishi, K., Iida, A., Kaneko, M., Tomioka, K., Tokuda, H., Nishino, H., & Kumeda, Y. (2003). Cancer preventive potential of trichothecenes from *Trichothecium roseum. Bioorganic & Medicinal Chemistry*, *11*(12), 2511–2518. doi:10.1016/S0968-0896(03)00215-3 PMID:12757719

Konkit, M., Choi, W. J., & Kim, W. (2016). Aldehyde dehydrogenase activity in Lactococcus chungangensis: Application in cream cheese to reduce aldehyde in alcohol metabolism. *Journal of Dairy Science*, *99*(3), 1–7. doi:10.3168/jds.2015-10549 PMID:26723119

Kördikanlıoğlu, B., Şimşek, Ö., & Saris, P. E. J. (2015). Nisin production of *Lactococcus lactis* N8 with hemin-stimulated cell respiration in fed-batch fermentation system. *Biotechnology Progress*, *31*(3), 678–685. doi:10.1002/btpr.2075 PMID:25826783

Kovačević, G., Blažić, M., Draganić, B., Ostafe, R., Gavrović-Jankulović, M., Fischer, R., & Prodanović, R. (2014). Cloning, heterologous expression, purification and characterization of M12 mutant of *Aspergillus niger* glucose oxidase in yeast *Pichia pastoris* KM71H. *Molecular Biotechnology*, *56*(4), 305–311. doi:10.100712033-013-9709-x PMID:24122283

Kraggerud, H., Skeie, S., Høy, M., Røkke, L., & Abrahamsen, R. K. (2008). Season and ripening temperature influence fatty acid composition and sensory properties of semi-hard cheese during maturation. *International Dairy Journal*, *18*(8), 801–810. doi:10.1016/j.idairyj.2007.11.003

Krasaekoopt, W., Bhandari, B., & Deeth, H. (2004). The influence of coating materials on some properties of alginate beads and survivability of microencapsulated probiotic bacteria. *International Dairy Journal*, *14*(8), 737–743. doi:10.1016/j.idairyj.2004.01.004

Kringelum, B. W., Sorensen, N. M., Garrigues, C., Pedersen, M. B., & Gron, S. (2006). *Use of compounds involved in biosynthesis of nucleic acids to increase yield of bacterial cultures*. United States Patent No. US 20080171028 A1.

Krogsgård Nielsen, C., Kjems, J., Mygind, T., Snabe, T., Schwarz, K., Serfert, Y., & Meyer, R. L. (2016). Enhancing the antibacterial efficacy of isoeugenol by emulsion encapsulation. *International Journal of Food Microbiology*, *229*, 7–14. doi:10.1016/j.ijfoodmicro.2016.04.002 PMID:27089032

Kukkonen, K., Savilahti, E., Haahtela, T., Juntunen-Backman, K., Korpela, R., Poussa, T., ... Kuitunen, M. (2007). Probiotics and prebiotic galactooligosaccharides in the prevention of allergic diseases: A randomized, double-blind, placebo-controlled trial. *The Journal of Allergy and Clinical Immunology*, *119*(1), 192–198. doi:10.1016/j.jaci.2006.09.009 PMID:17208601

Kumar, A., Grover, S., Sharma, J., & Batish, V. K. (2010). Chymosin and other milk coagulants: Sources and biotechnological interventions. *Critical Reviews in Biotechnology*, *30*(4), 243–258. doi:10.3109/07388551.2010.483459 PMID:20524840

Kumar, A., Sharma, J., Grover, S., Mohanty, A. K., & Batish, V. K. (2007). Molecular cloning and expression of goat (*Capra hircus*) prochymosin in *E. coli. Food Biotechnology*, *21*(1), 57–69. doi:10.1080/08905430701191163

Kumar, B. V., Vijayendra, S. V. N., & Reddy, O. V. S. (2015). Trends in dairy and non-dairy probiotic products - a review. *Journal of Food Science and Technology*, *52*(10), 6112–6124. doi:10.100713197-015-1795-2 PMID:26396359

Kumar, D. S., & Ray, S. (2014). Fungal lipase production by Solid State Fermentation- An overview. *Journal of Analytical & Bioanalytical Techniques*, *6*(1), 230.

Kumar, L., Gangadharan, V. P., Rao, D. R., Saikia, T., Shah, S., Malhotra, H., ... Rao, R. (2005). Safety and efficacy of an indigenous recombinant interferonalpha-2b in patients with chronic myelogenous leukaemia: Results of a multicentre trial from India. *The National Medical Journal of India*, *18*(2), 66–70. PMID:15981440

Kumar, M., Kumar, A., Nagpal, R., Mohania, D., Behare, P., Verma, V., ... Yadav, H. (2010). Cancer-preventing attributes of probiotics: An update. *International Journal of Food Sciences and Nutrition*, *61*(5), 473–496. doi:10.3109/09637480903455971 PMID:20187714

Kumar, R., Grover, S., & Batish, V. K. (2012). Bile Salt Hydrolase (Bsh) activity screening of lactobacilli*: In vitro* selection of indigenous *Lactobacillus* strains with potential bile salt hydrolysing and cholesterol-lowering ability. *Probiotics and Antimicrobial Proteins*, *4*(3), 162–172. doi:10.100712602-012-9101-3 PMID:26782042

Kuraishi, C., Yamazaki, K., & Susa, Y. (2001). Transglutaminase: Its utilization in the food industry. *Food Reviews International*, *17*(2), 221–246. doi:10.1081/FRI-100001258

Kurmann, J. A., Rasic, J. L., & Kroger, M. (1992). *Encyclopedia of fermented fresh milk products: an international inventory of fermented milk, cream, buttermilk, whey, and related products*. New York: Van Nostrand Reinhold.

Kuru, A. (Photographer). (2011, February 13). *Yillar sonra Konya'da "kuflu peynir" in pazarda hala serbestce satilabildigini gormek sasirtici* [digital image]. Retrieved from https://www.flickr.com/photos/alikuru/5440805421

Kutahya, O. E., Starrenburg, M. J. C., Rademaker, J. L. W., Klaassen, C. H. W., van Hylckama Vlieg, J. E. T., Smid, E. J., & Kleerebezem, M. (2011). High-Resolution Amplified Fragment Length Polymorphism Typing of Lactococcus lactis Strains Enables Identification of Genetic Markers for Subspecies-Related Phenotypes. *Applied and Environmental Microbiology*, *77*(15), 5192–5198. doi:10.1128/AEM.00518-11 PMID:21666014

Kwak, H. S., K. M. Yang, & Ahn, J. (2003). *Microencapsulated Iron for Milk Fortification*. Academic Press. 10.1021/JF030199+

Ladero, V., Martínez, N., Martín, M. C., Fernández, M., & Alvarez, M. A. (2010). qPCR for quantitative detection of tyramine-producing bacteria in dairy products. *Food Research International*, *43*(1), 289–295. doi:10.1016/j.foodres.2009.10.007

Lafarge, V., Ogier, J., Girard, V., Maladen, V., Leveau, J., Gruss, A., & Delacroix-Buchet, A. (2004). Raw cow milk bacterial population shifts attributable to refrigeration raw cow milk bacterial population shifts attributable to refrigeration. *Applied and Environmental Microbiology*, *70*(9), 5644–5650. doi:10.1128/AEM.70.9.5644-5650.2004 PMID:15345453

Lamberet, G., & Lenoir, J. (1976). Les caractères du système lipolytique de l'espèce *Penicillium caseicolum*. Purification et propriétes de la lipase majeure. *Le Lait*, *56*(559–560), 622–644. doi:10.1051/lait:1976559-56033

Lamberet, G., & Menassa, A. (1983). Purification and properties of an acid lipase from *Penicillium roqueforti. The Journal of Dairy Research*, *50*(4), 459–468. doi:10.1017/S0022029900032696

Lam, E., Holt, C., Edwards, P., McKinnon, I., Otter, D., Li, N., & Hemar, Y. (2017). Effect of transglutaminase treatment on the physico-chemical properties of skim milk with added ethylendiaminetetraacetic acid. *Food Hydrocolloids*, *69*, 329–340. doi:10.1016/j.foodhyd.2017.02.002

Lan, C. Q., Oddone, G., Mills, D. A., & Block, D. E. (2006). Kinetics of *Lactococcus lactis* growth and metabolite formation under aerobic and anaerobic conditions in the presence or absence of hemin. *Biotechnology and Bioengineering*, *95*(6), 1070–1080. doi:10.1002/bit.21070 PMID:16807924

Land, M., Hauser, L., Jun, S.-R., Nookaew, I., Leuze, M. R., Ahn, T.-H., ... Ussery, D. W. (2015). Insights from 20 years of bacterial genome sequencing. *Functional & Integrative Genomics*, *15*(2), 141–161. doi:10.100710142-015-0433-4 PMID:25722247

Lane, C. N., & Fox, P. F. (1996). Contribution of starter and adjunct Lactobacilli to proteolysis in cheddar cheese during ripening. *International Dairy Journal*, *6*(7), 715–728. doi:10.1016/0958-6946(95)00067-4

Lanella, G. (2015). Equid milk renneting through pure camel chymosin and cheese manufacturing. [DRJAFS]. *Direct Research Journal of Agriculture and Food Science*, *3*(9), 167–172.

Lane, M. M., & Morrissey, J. P. (2010). *Kluyveromyces marxianus*: A yeast emerging from its sister's shadow. *Fungal Biology Reviews*, *24*(1-2), 17–26. doi:10.1016/j.fbr.2010.01.001

Langsrud, T., Reinbold, G. W., & Hammond, E. G. (1977). Proline production by Propionibacterium shermanii P59. *Journal of Dairy Science*, *1*(60), 16–23. doi:10.3168/jds.S0022-0302(77)83822-8

Laroute, V., Tormo, H., Couderc, C., Mercier-Bonin, M., Le Bourgeois, P., Cocaign-Bousquet, M., & Daveran-Mingot, M. L. (2017). From Genome to Phenotype: An Integrative Approach to Evaluate the Biodiversity of Lactococcus lactis. *Microorganisms*, 5. PMID:28534821

Larsen, M. D., Kristiansen, K. R., & Hansen, T. K. (1998). Characterization of the proteolytic activity of starter cultures of *Penicillium roqueforti* for production of blue veined cheeses. *International Journal of Food Microbiology*, *43*(3), 215–221. doi:10.1016/S0168-1605(98)00114-7 PMID:9801197

Lauber, S., Henle, T., & Klostermeyer, H. (2000). Relationship between the crosslinking of caseins by transglutaminase and the gel strength of yoghurt. *European Food Research and Technology*, *210*(5), 305–309. doi:10.1007002170050554

Lauber, S., Noack, I., Klostermeyer, H., & Henle, T. (2001). Oligomerization of β-lactoglobulin by microbial transglutaminase during high pressure treatment. *European Food Research and Technology*, *213*(3), 246–247. doi:10.1007002170100368

Law, B. A., & Wigmore, A. S. (1982a). Accelerated cheese ripening with food grade proteinases. *The Journal of Dairy Research*, *49*(1), 137–146. doi:10.1017/S0022029900022214

Law, B. A., & Wigmore, A. S. (1982b). Microbial proteinases as agents for accelerated cheese ripening. *International Journal of Dairy Technology*, *35*(2), 75–76. doi:10.1111/j.1471-0307.1982.tb02718.x

Le Bourhis, A.-G., Saunier, K., Dore, J., Carlier, J.-P., Chamba, J.-F., Popoff, M.-R., & Tholozan, J.-L. (2005). Development and validation of PCR primers to assess the diversity of Clostridium spp. in cheese by temporal temperature gradient gel electrophoresis. *Applied and Environmental Microbiology*, *71*(1), 29–38. doi:10.1128/AEM.71.1.29-38.2005 PMID:15640166

Le brie de Melun est une appellation d'origine désignant un fromage au lait cru de la région de la Brie et de Melun. (2015, May 15). Retrieved from https://commons.wikimedia.org/wiki/File:Wikicheese_-_Brie_de_Melun_-_20150515_-_015.jpg

Le Loir, Y., Baron, F., & Gautier, M. (2003). *Staphylococcus aureus* and food poisoning. *Genetics and Molecular Research*, *31*, 63–76. PMID:12917803

Le Rodez est un fromage produit dans la ville du même nom. (2017, February 5). Retrieved from https://commons.wikimedia.org/wiki/File:Wikicheese-XX_-_Rodez_-_20170905_-_011.jpg

LeBlanc, J. G., Laiño, J. E., del Valle, M. J., de Giori, G. S., Sesma, F. & Taranto, M. P. (2015). B-Group Vitamins Production by Probiotic Lactic Acid Bacteria. *Biotechnology of Lactic Acid Bacteria: Novel Applications,* 279.

LeBlanc, J. G., Milani, C., de Giori, G. S., Sesma, F., van Sinderen, D., & Ventura, M. (2013). Bacteria as vitamin suppliers to their host: A gut microbiota perspective. *Current Opinion in Biotechnology*, *24*(2), 160–168. doi:10.1016/j.copbio.2012.08.005 PMID:22940212

Lechardeur, D., Cesselin, B., Fernandez, A., Lamberet, G., Garrigues, C., Pedersen, M., ... Gruss, A. (2011). Using heme as an energy boost for lactic acid bacteria. *Current Opinion in Biotechnology*, *22*(2), 143–149. doi:10.1016/j.copbio.2010.12.001 PMID:21211959

Ledenbach, L. H., & Marshall, R. T. (2009). Microbiological Spoilage of Dairy Products. In W. H. Sperber & M. P. Doyle (Eds.), *Compendium of the Microbiological Spoilage of Foods and Beverages, Food Microbiology and Food Safety. Springer Science+Business Media, LLC*. doi:10.1007/978-1-4419-0826-1_2

Lee, Y. K., Yuan, K., & Salminen, S. (2009). Handbook of probiotics and prebiotics (2nd ed.). John Wiley & Sons.

Lee, H., Yoon, H., Ji, Y., Kim, H., Park, H., Lee, J., ... Holzapfel, W. (2011). Functional properties of *Lactobacillus* strains isolated from kimchi. *International Journal of Food Microbiology*, *145*(1), 155–161. doi:10.1016/j.ijfoodmicro.2010.12.003 PMID:21215484

Lee, J., Kim, Y., Yun, H. S., Kim, J. G., Oh, S., & Kim, S. H. (2010). Genetic and Proteomic Analysis of Factors Affecting Serum Cholesterol Reduction by *Lactobacillus acidophilus* A4. *Applied and Environmental Microbiology*, *76*(14), 4829–4835. doi:10.1128/AEM.02892-09 PMID:20495044

Lee, S. J., Choi, S. J., Li, Y., Decker, E. A., & McClements, D. J. (2011). Protein-stabilized nanoemulsions and emulsions: Comparison of physicochemical stability, lipid oxidation, and lipase digestibility. *Journal of Agricultural and Food Chemistry*, *59*(1), 415–427. doi:10.1021/jf103511v PMID:21133433

Lee, Y. K. (2009). Probiotic Microorganisms. In Y. K. Lee & S. Salminen (Eds.), *Handbook of Probiotics and Prebiotics* (2nd ed.; pp. 3–176). Wiley & Sons.

Lee, Y. K. (2009). Probiotic microorganisms. In Y. K. Lee & S. Salminen (Eds.), *Handbook of probiotics and prebiotics* (pp. 1–139). Hoboken, NJ: John Wiley & Sons.

Lee, Y. K., & Salminen, S. (2009). Effects on Human Health and Diseases. In Y. K. Lee & S. Salminen (Eds.), *Handbook of Probiotics and Prebiotics* (2nd ed.; pp. 267–321). Wiley & Sons.

Leistner, L. (1978). Hurdle effect and energy saving. In W. K. Downey (Ed.), *Food Quality and Nutrition* (pp. 553–557). London, UK: Applied Science Publishers.

Leistner, L. (2000). Basic aspects of food preservation by hurdle technology. *International Journal of Food Microbiology*, *55*(1-3), 181–186. doi:10.1016/S0168-1605(00)00161-6 PMID:10791741

Leistner, L., & Gorris, L. G. M. (1995). Food preservation by hurdle technology. *Trends in Food Science & Technology*, *6*(2), 41–46. doi:10.1016/S0924-2244(00)88941-4

Leite, A. M. O., Mayo, B., Rachid, C. T. C. C., Peixoto, R. S., Silva, J. T., Paschoalin, V. M. F., & Delgado, S. (2012). Assessment of the microbial diversity of Brazilian kefir grains by PCR-DGGE and pyrosequencing analysis. *Food Microbiology*, *31*(2), 215–221. doi:10.1016/j.fm.2012.03.011 PMID:22608226

Leite, A. M. O., Miguel, M. A. L., Peixoto, R. S., Ruas-Madiedo, P., Paschoalin, V. M. F., Mayo, B., & Delgado, S. (2015). Probiotic potential of selected lactic acid bacteria strains isolated from Brazilian kefir grains. *Journal of Dairy Science*, *98*(6), 3622–3632. doi:10.3168/jds.2014-9265 PMID:25841972

Lejeune, R., Callewaert, R., Crabbé, & De Vuyst. (1998). Modelling the growth and bacteriocin production by *Lactobacillus amylovorus* DCE 471 in batch cultivation. *Journal of Applied Microbiology*, *84*(2), 159–168. doi:10.1046/j.1365-2672.1998.00266.x

Lemay, M. J., Choquette, J., Delaquis, P. J., Claude, G., Rodrigue, N., & Saucier, L. (2002). Antimicrobial effect of natural preservatives in a cooked and acidified chicken meat model. *International Journal of Food Microbiology*, *78*(3), 217–226. doi:10.1016/S0168-1605(02)00014-4 PMID:12227640

Lemieux, L., & Simard, R. E. (1991). Bitter flavour in dairy products. 1. A review of the factors likely to influence its development, mainly in cheese manufacture. *Le Lait*, *71*(6), 599–636. doi:10.1051/lait:1991647

Lenoir, J. (1984). The surface flora and its role in the ripening of cheese. In *Federation Internationale de Laiterie Bulletin Annuel* (Vol. 171, pp. 3–20). Brussels: International Dairy Federation.

Lerner, A., & Matthias, T. (2015). Possible association between celiac diseaase and bacterial transglutaminase in food processing: A hypothesis. *Nutrition Reviews*, *73*(8), 544–552. doi:10.1093/nutrit/nuv011 PMID:26084478

Leroy, F., & De Vuyst, L. (2004). Lactic acid bacteria as functional starter cultures for the food fermentation industry. *Trends in Food Science & Technology*, *15*(2), 67–78. doi:10.1016/j.tifs.2003.09.004

Lessard, M.-H., Viel, C., Boyle, B., St-Gelais, D., & Labrie, S. (2014). Metatranscriptome analysis of fungal strains Penicillium camemberti and Geotrichum candidum reveal cheese matrix breakdown and potential development of sensory properties of ripened Camembert-type cheese. *BMC Genomics*, *15*(1), 235. doi:10.1186/1471-2164-15-235 PMID:24670012

Leverentz, B., Conway, W. S., Camp, M. J., Janisiewicz, W. J., Abuladze, T., Yang, M., ... Sulakvelidze, A. (2003). Biocontrol of *Listeria monocytogenes* on fresh-cut produce by treatment with lytic bacteriophages and a bacteriocin. *Applied and Environmental Microbiology*, *69*(8), 4519–4526. doi:10.1128/AEM.69.8.4519-4526.2003 PMID:12902237

Levic, S., Dordevic, V., Knezevic-Jugovic, Z., Kalusevic, A., Milasinovic, N., Bugarski, A., & Nedovic, V. (2017). Enzyme encapsulation. Technologies and their applications in food processing. In R. C. Ray & C. M. Rosell (Eds.), *Microbial enzyme technology in food applications* (pp. 469–502). Boca Raton, FL: CRC Press.

Li, C., Wang, C. L., Sun, Y., Li, A. L., Liu, F., & Meng, X. C. (2016). Microencapsulation of *Lactobacillus rhamnosus* GG by transglutaminase cross-linked soyprotein isolate to improve survival in simulated gastrointestinal conditions and yoghurt. *Journal of Food Science*, *81*(7), M1726–M1734. doi:10.1111/1750-3841.13337 PMID:27228279

Licitra, G., Ogier, J. C., Parayre, S., Pediliggieri, C., Carnemolla, T. M., Falentin, H., ... Lortal, S. (2007). Variability of bacterial biofilms of the "tina" wood vats used in the Ragusano cheese-making process. *Applied and Environmental Microbiology*, *73*(21), 6980–6987. doi:10.1128/AEM.00835-07 PMID:17720831

Lidbeck, A., Övervik, E., Rafter, J., Nord, C. E., & Gustafsson, J. A. (1992). Effect of *Lactobacillus acidophilus* Supplements on Mutagen Excretion in Faeces and Urine in Humans. *Microbial Ecology in Health and Disease*, *5*(1), 59–67. doi:10.3109/08910609209141305

Liepinsh, E., Andersson, M., Ruysschaert, J. M., & Otting, G. (1997). Saposin fold revealed by the NMR structure of NK-lysin. *Nature Structural Biology*, *4*(10), 793–795. doi:10.1038/nsb1097-793 PMID:9334742

Li, H., Cui, Y., Zhang, L., Luo, X. L., Fan, R., Xue, C., ... Han, X. (2015). Production of a transglutaminase from *Zea mays* in *Eschericia coli* and its impact on yoghurt properties. *International Journal of Dairy Technology*, *68*(1), 54–61. doi:10.1111/1471-0307.12165

Li, H., Cui, Y., Zhang, L., Zhang, L., & Yu, J. (2017). Optimization of recombinant *Zea mays* transglutaminase production and its influence on the functional properties of yoghurt. *Food Science and Biotechnology*, *26*(3), 723–730. doi:10.100710068-017-0083-5

Li, H., Zhang, L., Cui, Y., Luo, X. L., Xue, C., Wang, S., ... Han, X. (2014). Characterization of recombinant *Zea mays* transglutaminase expressed in Pichia pastoris and its impact on full and non-fat yoghurts. *Journal of the Science of Food and Agriculture, 94*(6), 1225–1230. doi:10.1002/jsfa.6402 PMID:24105803

Li, H., Zhang, S., Lu, J., Liu, L., Uluko, H., Pang, X., ... Lv, J. (2014). Antifungal activities and effect of *Lactobacillus casei* AST18 on the mycelia morphology and ultrastructure of *Penicillium chrysogenum. Food Control, 43*, 57–64. doi:10.1016/j.foodcont.2014.02.045

Likotrafiti, E., Valavani, P., Argiriou, A., & Rhoades, J. (2015). In vitro evaluation of potential antimicrobial synbiotics using *Lactobacillus kefiri* isolated from kefir grains. *International Dairy Journal, 45*, 23–30. doi:10.1016/j.idairyj.2015.01.013

Lim, T. J., Easa, A. M., Karim, A. A., Bhat, R., & Liong, M. T. (2011). Development of soy-based cream cheese via the addition of microbial transglutamianse, soy protein isolate and maltodextrin. *British Food Journal, 113*(9), 1147–1172. doi:10.1108/00070701111174587

Li, N., Wang, Y., Zhu, P., Liu, Z., Guo, B., & Ren, J. (2015). Improvement of exopolysaccharide production in *Lactobacillus casei* LC2W by overexpression of NADH oxidase gene. *Microbiological Research, 171*, 73–77. doi:10.1016/j.micres.2014.12.006 PMID:25644955

Lin, T. Y. (2000). Conjugated linoleic acid concentration as affected by lactic cultures and additives. *Food Chemistry, 69*(1), 27–31. doi:10.1016/S0308-8146(99)00218-6

Liong, M. T., & Shah, N. P. (2005). Bile salt deconjugation and BSH activity of five bifidobacterial strains and their cholesterol co-precipitating properties. *Food Research International, 38*(2), 135–142. doi:10.1016/j.foodres.2004.08.003

Liong, M. T., & Shah, N. P. (2005). Optimization of cholesterol removal by probiotics in prebiotics by using a response surface method. *Applied and Environmental Microbiology, 71*(4), 1745–1753. doi:10.1128/AEM.71.4.1745-1753.2005 PMID:15811997

Liptakova, D., Valik, L., & Bajusova, B. (2006). Effect of protective culture on the growth of *Candida maltosa* YP1 in yoghurt. *Journal of Food and Nutrition Research, 45*, 147–151.

Liu, P., Shen, S., Ruan, H., Zhou, Q., Ma, L., & He, G. (2011). Production of conjugated linoleic acids by *Lactobacillus plantarum* strains isolated from naturally fermented Chinese pickles. *Journal of Zheijang University Science B - Biomedicine & Biotechnology, 12*(11), 923-930.

Liu, F., Chiou, B. S., Avena-Bustillos, R. J., Zhang, Y., Li, Y., McHugh, T., & Zhong, F. (2017). Study of combined effects of glycerol and transglutaminase on properties of gelatin films. *Food Hydrocolloids, 65*, 1–9. doi:10.1016/j.foodhyd.2016.10.004

Liu, J. A. P., & Moon, N. J. (1982). CommensalisticInteractionBetweenLactobacillusacidophilus and Propionibacterium shermanii. *Applied and Environmental Microbiology*, 715–722. PMID:16346098

Liu, L., Chen, P., Zhao, W., Li, X., Wang, H., & Qu, X. (2017). Effect of microencapsulation with the Maillard reaction products of whey proteins and isomaltooligosaccharide on the survival rate of Lactobacillus rhamnosus in white brined cheese. *Food Control, 79*, 44–49. doi:10.1016/j.foodcont.2017.03.016

Liu, M., & Damodaran, S. (1999). Effect of Transglutaminase-catalysed polymerization of beta-casein on its emulsifying properties. *Journal of Agricultural and Food Chemistry, 47*(4), 1514–1519. doi:10.1021/jf981030c PMID:10564008

Liu, W., Zheng, Y., Kwok, L.-Y., Sun, Z., Zhang, J., Guo, Z., ... Zhang, H. (2015). High-throughput sequencing for the detection of the bacterial and fungal diversity in Mongolian naturally fermented cow's milk in Russia. *BMC Microbiology, 15*(1), 45. doi:10.118612866-015-0385-9 PMID:25887414

Liu, Y., Chen, Z., Jiang, Z., Yan, Q., & Yang, S. (2017). Biochemical characterization of a novel β-galactosidase from *Paenibacillus barengoltzii* suitable for lactose hydrolysis and galactooligosaccharides synthesis. *International Journal of Biological Macromolecules, 104*, 1055–1063. doi:10.1016/j.ijbiomac.2017.06.073 PMID:28652150

Liu, Y., Zhao, F., Liu, J., Wang, H., Han, X., Zhang, Y., & Yang, Z. (2017). Selection of cholesterol-lowering Lactic Acid Bacteria and its effects on rats fed with high-cholesterol diet. *Current Microbiology, 74*(5), 623–631. doi:10.100700284-017-1230-1 PMID:28286906

Ljungh, A., & Wadstrom, T. (2006). Lactic acid bacteria as probiotics. *Current Issues in Intestinal Microbiology, 7*(2), 73–90. PMID:16875422

Llorente, B., Malpertuy, A., Blandin, G., Artiguenave, F., Wincker, P., & Dujon, B. (2000). Genomic Exploration of the Hemiascomycetous Yeasts: 12. *Kluyveromyces marxianus* var. *marxianus. FEBS Letters, 487*(1), 71–75. doi:10.1016/S0014-5793(00)02283-3 PMID:11152887

Lollo, P. C. B., Morato, P. N., Moura, C. S., Almada, C. N., Felicio, T. L., Esmerino, E. A., ... Cruz, A. G. (2015). Hypertension parameters are attenuated by the continuous consumption of probiotic Minas cheese. *Food Research International, 76*, 611–617. doi:10.1016/j.foodres.2015.07.015 PMID:28455044

London, L. E. E., Ross, R. P., Fitzgerald, G. F., Shanahan, F., Caplice, N. M., & Stanton, C. (2016). Probiotics as Cell Factories for Bioactive Ingredients: Focus on Microbial Polysaccharides and Health Beneficial Effects. In P. Foerst & C. Santivarangkna (Eds.), *Advances in Probiotic Technology* (pp. 77–102). New York: CRC Press.

Lorenzen, P. C. (2000). Renneting properties of transglutaminase-treated milk. *Milchwissenschaft. Milk Science International, 54*, 563–566.

Lorenzen, P. C., Neve, H., Mautner, A., & Schlimme, E. (2002). Effect of enzymatic cross-linking of milk proteins on functional properties of set-style yoghurt. *International Journal of Dairy Technology, 55*(3), 152–157. doi:10.1046/j.1471-0307.2002.00065.x

Lorenzen, P. C., & Schlimme, E. (1998). Properties and potential fields of application of transglutamianse preparations in dairying. In *IDF Bulletin 332* (pp. 47–53). Brussels: International Dairy Federation.

Loveday, S. M., Su, J., Rao, M. A., Anema, S. G., & Singh, H. (2011). Effect of calcium on the morphology and functionality of whey protein nanofibrils. *Biomacromolecules, 12*(10), 3780–3788. doi:10.1021/bm201013b PMID:21894942

Low, D., Ahlgren, J. A., Horne, D., McMahon, D. J., Oberg, C. J., & Broadbent, J. R. (1998). Role of *Streptococcus thermophilus* MR- 1C capsular exopolysaccharide in cheese moisture retention. *Applied and Environmental Microbiology, 64*(6), 2147–2151. PMID:9603827

Lucas, A., Sodini, I., Monnet, C., Jolivet, P., & Corrieu, G. (2004). Probiotic cell counts and acidification in fermented milks supplemented with milk protein hydrolysates. *International Dairy Journal, 14*(1), 47–53. doi:10.1016/S0958-6946(03)00147-X

Luerce, T. D., Azevedo, M. S. P., LeBlanc, J. G., Azevedo, V., Miyoshi, A., & Pontes, D. S. (2014). Recombinant *Lactococcus lactis* fails to secrete bovine chymosine. *Bioengineered, 5*(6), 363–370. doi:10.4161/bioe.36327 PMID:25482140

Lule, V. K., Singh, R., Pophaly, S. D., & Tomar, S. K. (2016). Production and structural characterisation of dextran from an indigenous strain of *Leuconostoc mesenteroides* BA08 in Whey. *International Journal of Dairy Technology*, *69*(4), 520–531. doi:10.1111/1471-0307.12271

Lund, F., Filtenborg, O., & Frisvad, J. C. (1995). Associated mycoflora of cheese. *Food Microbiology*, *12*, 173–180. doi:10.1016/S0740-0020(95)80094-8

Luo, F., Jiang, W. H., Yang, Y. X., Li, J., & Jiang, M. F. (2016). Cloning and expression of yak active chymosin in *Pichia pastoris*. *Asian-Australasian Journal of Animal Sciences*, *29*(9), 1363–1370. doi:10.5713/ajas.16.0038 PMID:27004812

Lye, H.-S., Rahmat-Ali, G. R., & Liong, M.-T. (2010). Mechanisms of cholesterol removal by lactobacilli under conditions that mimic the human gastrointestinal tract. *International Dairy Journal*, *20*(3), 169–175. doi:10.1016/j.idairyj.2009.10.003

Lynch, C. M., McSweeney, P. L. H., Fox, P. F., Cogad, T. M., & Drinan, F. D. (1996). Manufacture of cheddar cheese with and without adjunct Lactobacilli under controlled microbiological conditions. *International Dairy Journal*, *6*(8-9), 851–867. doi:10.1016/0958-6946(96)00019-2

Lynch, C. M., McSweeney, P. L. H., Fox, P. F., Cogan, T. M., & Drinan, F. D. (1997). Contribution of starter lactococci and non-starter lactobacilli to proteolysis in Cheddar cheese with a controlled microflora. *Le Lait*, *77*(4), 441–459. doi:10.1051/lait:1997431

Lyon, W. J., Sethi, J. K., & Glatz, B. A. (1993). Inhibition of psychrotrophic organisms by propionicin PLG-1, a bacteriocin produced by *Propionibacterium thoenii*. *Journal of Dairy Science*, *76*(6), 1506–1513. doi:10.3168/jds.S0022-0302(93)77482-2 PMID:8326023

Maassen, C. B., van Holten-Neelen, C., Balk, F., den Bak-Glashouwer, M. J., Leer, R. J., Laman, J. D., … Claassen, E. (2000). Strain-dependent induction of cytokine profiles in the gut by orally administered Lactobacillus strains. *Vaccine*, *18*(23), 2613–23. Retrieved from http://www.ncbi.nlm.nih.gov/pubmed/10775795

Macedo, J., & Sato, H. (2005). Propriedades e aplicações da transglutaminase microbiana em alimentos. *Alimentos e Nutrição*, *16*(4), 413–419.

Macierzanka, A., Bordron, F., Rigby, N., Mills, E. N. C., Lille, M., Poutanen, K., & Mackie, A. R. (2011). Transglutaminase cross-linking kinetics of sodium caseinate is changed after emulsification. *Food Hydrocolloids*, *23*(5), 843–850. doi:10.1016/j.foodhyd.2010.07.010

Madene, A., Jacquot, M., Scher, J., & Desobry, S. (2006). Flavour encapsulation and controlled release - a review. *International Journal of Food Science & Technology*, *41*(1), 1–21. doi:10.1111/j.1365-2621.2005.00980.x

Madkor, S. A., Tong, P. S., & El Soda, M. (2000). Ripening of Cheddar cheese with added attenuated adjunct cultures of Lactobacilli. *Journal of Dairy Science*, *83*(8), 1684–1691. doi:10.3168/jds.S0022-0302(00)75037-5 PMID:10984143

Maedler, K., Dharmadhikari, G., Schumann, D. M., & Størling, J. (2009). Interleukin-1 beta targeted therapy for type 2 diabetes. *Expert Opinion on Biological Therapy*, *9*(9), 1177–1188. doi:10.1517/14712590903136688 PMID:19604125

Magarinos, H., Selaive, S., Costa, M., Flores, M., & Pizarro, O. (2007). Viability of probiotic micro-organisms (*Lactobacillus acidophilus* la-5 and *Bifidobacterium animalis* subsp. *lactis* bb-12) in ice cream. *International Journal of Dairy Technology*, *60*(2), 128–134. doi:10.1111/j.1471-0307.2007.00307.x

Mahmood, W. A., & Sebo, N. H. (2012). Improvement of yogurt properties by microbial transglutaminase. *Jordan Journal of Agricultural Sciences*, *8*, 333–341.

Maisnier-Patin, S., Deschamps, N., Tatini, S. R., & Richard, J. (1992). Inhibition of *Listeria monocytogenes* in Camembert cheese made with a nisin-producing starter. *Le Lait*, *72*(3), 249–263. doi:10.1051/lait:1992318

Maitan-Alfenas, G. P., Oliveira, M. B., Nagem, R. A., de Vries, R. P., & Guimarães, V. M. (2016). Characterization and biotechnological application of recombinant xylanases from *Aspergillus nidulans. International Journal of Biological Macromolecules*, *91*, 60–67. doi:10.1016/j.ijbiomac.2016.05.065 PMID:27235731

Maitan-Alfenas, G. P., Visser, E. M., & Guimarães, V. M. (2015). Enzymatic hydrolysis of lignocellulosic biomass: Converting food waste in valuable products. *Current Opinion in Food Science*, *1*, 44–49. doi:10.1016/j.cofs.2014.10.001

Makarova, K., Slesarev, A., Wolf, Y., Sorokin, A., Mirkin, B., Koonin, E., ... Mills, D. (2006). Comparative genomics of the lactic acid bacteria. *Proceedings of the National Academy of Sciences of the United States of America*, *103*(42), 15611–15616. doi:10.1073/pnas.0607117103 PMID:17030793

Malek, R. A., Hamdan, S., El Enshasy, H. A., Othman, N. Z., Zainol, N. A., Sarmidi, M. R., & Aziz, R. A. (2010). Production of *Lactobacillus salivarius*, a new probiotic strain isolated from human breast milk, in semi-industrial scale and studies on its functional characterization. In A. Médez-Vilas (Ed.), *Current research, technology and education topics in applied microbiology and microbial biotechnology* (Vol. 2, pp. 1196–1204). Badajoz, Spain: Formatex Research Centre.

Malheiros, P. S., Cuccovia, I. M., & Franco, B. D. G. M. (2016). Inhibition of Listeria monocytogenes in vitro and in goat milk by liposomal nanovesicles containing bacteriocins produced by Lactobacillus sakei subsp. sakei 2a. *Food Control*, *63*, 158–164. doi:10.1016/j.foodcont.2015.11.037

Malik, R. K., Rao, K. N., Bandhopadhyay, P., & Kumar, N. (2005). Bacteriocins: Natural and safe anti microbial peptides for food preservation. *Indian Food Industry*, *24*(1), 69–70.

Mandal, S., Puniya, A. K., & Singh, K. (2006). Effect of alginate concentrations on survival of microencapsulated Lactobacillus casei NCDC-298. *International Dairy Journal*, *16*(10), 1190–1195. doi:10.1016/j.idairyj.2005.10.005

Mandujano-Gonzales, V., Villa-Tanaca, L., Anducho-Reyes, M. A., & Mercado-Flores, Y. (2016). Secreted fungal aspartic proteases: A review. *Revista Iberoamericana de Micologia*, *33*(2), 76–82. doi:10.1016/j.riam.2015.10.003 PMID:27137097

Manning, T. S., & Gibson, G. R. (2004). Prebiotics. *Best Practice & Research. Clinical Gastroenterology*, *18*(2), 287–298. doi:10.1016/j.bpg.2003.10.008 PMID:15123070

Mantere-Alhonen, S. (1995). Propiobacteria used as probiotics-A review. *Le Lait*, *75*(4-5), 447–452. doi:10.1051/lait:19954-534

Manuel, P. M., Elena, B., Carolina, M. G., & Gabriela, P. (2017). Oral probiotics supplementation can stimulate the immune system in a stress process. *Journal of Nutrition & Intermediary Metabolism*, *8*, 29–40. doi:10.1016/j.jnim.2017.06.001

Maoz, A., Mayr, R., & Scherer, S. (2003). Temporal stability and Biodiversity of two complex antiristerial cheese-ripening microbial consortia. *Applied and Environmental Microbiology*, *69*(7), 4012–4018. doi:10.1128/AEM.69.7.4012-4018.2003 PMID:12839776

Marcellino, N., & Benson, D. R. (1992). Scanning electron and light microscopic study of microbial succession on Bethlehem St. Nectaire cheese. *Applied and Environmental Microbiology*, *58*(11), 3448–3454. PMID:16348797

Marcellino, S. N. M., & Benson, D. R. (2013). The good, the bad, and the ugly: Tales of mold-ripened cheese. *Microbiology Spectrum*, *1*(1), CM-0005–CM-0012. PMID:26184810

Margulies, M., Egholm, M., Altman, W. E., Attiya, S., Bader, J. S., Bemben, L. A., ... Rothberg, J. M. (2005). Genome Sequencing in Open Microfabricated High Density Picoliter Reactors. *Nature*, *437*(7057), 376–380. doi:10.1038/nature03959 PMID:16056220

Marilley, L., & Casey, M. G. (2004). Flavours of cheese products: Metabolic pathways, analytical tools and identification of producing strains. *International Journal of Food Microbiology*, *90*(2), 139–159. doi:10.1016/S0168-1605(03)00304-0 PMID:14698096

Marquez, R. G., Di Pierro, P., Esposito, M., Mariniello, L., & Porta, R. (2014). Application of transglutaminase-crosslinked whey protein /pectin films as water barrier coatings in fried and baked foods. *Food and Bioprocess Technology*, *7*, 47–455.

Marsh, A. J., Hill, C., Ross, R. P., & Cotter, P. D. (2014). Fermented beverages with health-promoting potential: Past and future perspectives. *Trends in Food Science & Technology*, *38*(2), 113–124. doi:10.1016/j.tifs.2014.05.002

Marteu, P., Minekus, M., Havenaar, R., & Veld, J. H. J. (1997). Survival of Lactic Acid Bacteria in a Dynamic Model of the Stomach and Small Intestine: Validation and the Effects of Bile. *Journal of Dairy Science*, *80*(6), 1031–1037. doi:10.3168/jds.S0022-0302(97)76027-2 PMID:9201571

Martı'nez, B., Obeso, J. M., Rodrı'guez, A., & Garcı'a, P. (2008). Nisin-bacteriophage crossresistance in *Staphylococcus aureus*. *International Journal of Food Microbiology*, *122*(3), 253–258. doi:10.1016/j.ijfoodmicro.2008.01.011 PMID:18281118

Martín, J. F., & Liras, P. (2017). Secondary Metabolites in Cheese Fungi. In J. M. Mérillon & K. G. Ramawat (Eds.), Fungal Metabolites (pp. 293–315). Cham: Springer International Publishing.

Martinis, E. C. P., & Franco, B. D. G. M. (1998). Inhibition of *Listeria monocytogenes* in a pork product by a *Lactobacillus sake* strain. *International Journal of Food Microbiology*, *42*(1-2), 119–126. doi:10.1016/S0168-1605(98)00059-2 PMID:9706804

Martín, J. F., & Coton, M. (2017). Blue Cheese: Microbiota and Fungal Metabolites. In J. Frias, C. Martinez-Villaluenga, & E. Peñas (Eds.), *Fermented Foods in Health and Disease Prevention* (pp. 275–303). Boston: Academic Press. doi:10.1016/B978-0-12-802309-9.00012-1

Martín-Platero, A. M., Valdivia, E., Maqueda, M., Martín-Sánchez, I., & Martínez-Bueno, M. (2008). Polyphasic approach to bacterial dynamics during the ripening of Spanish farmhouse cheese, using culture-dependent and -independent methods. *Applied and Environmental Microbiology*, *74*(18), 5662–5673. doi:10.1128/AEM.00418-08 PMID:18658288

Martins, F. S., Nardi, R. M. D., Arantes, R. M. E., Rosa, C. A., Neves, M. J., & Nicoli, J. R. (2005). Screening of yeasts as probiotic based on capacities to colonize the gastrointestinal tract and to protect against enteropathogen challenge in mice. *The Journal of General and Applied Microbiology*, *51*(2), 83–92. doi:10.2323/jgam.51.83 PMID:15942869

Mase, T., Asano, K., Ikeda, Y., Kato, Y., Esaki, H., & Isshiki, S. (2013). Characterization and application of Lipase 39-A from *Cryptococcus flavescens* for cheese flavoring. *Food Science and Technology Research*, *19*(1), 89–95. doi:10.3136/fstr.19.89

Mase, T., Sonoda, M., Morita, M., & Hirose, E. (2011). Characterization of a lipase from *Sporidiobolus pararoseus* 25-A which produces cheese flavor. *Food Science and Technology Research*, *17*(1), 17–20. doi:10.3136/fstr.17.17

Masotti, F., Cattaneo, S., Stuknytė, M., & De Noni, I. (2017). Technological tools to include whey proteins in cheese: Current status and perspectives. *Trends in Food Science & Technology*, *64*, 102–114. doi:10.1016/j.tifs.2017.04.007

Matalanis, A., Jones, O. G., & McClements, D. J. (2011). Structured biopolymer-based delivery systems for encapsulation, protection, and release of lipophilic compounds. *Food Hydrocolloids*, *25*(8), 1865–1880. doi:10.1016/j.foodhyd.2011.04.014

Mateo, C., Palomo, J. M., Fernandez-Lorente, G., Guisan, J. M., & Fernandez-Lafuente, R. (2007). Improvement of enzyme activity, stability and selectivity via immobilization techniques. *Enzyme and Microbial Technology*, *40*(6), 1451–1463. doi:10.1016/j.enzmictec.2007.01.018

Mathieu, F., Michel, M., Lebrihi, A., & Lefebvre, G. (1994). Effect of the bacteriocin carnocin CP5 and of the producing strain *Carnobacterium piscicola* CP5 on the viability of Listeria monocytogenes ATCC 15313 in salt solution, broth and skimmed milk, at various incubation temperatures. *International Journal of Food Microbiology*, *22*(2-3), 155–172. doi:10.1016/0168-1605(94)90139-2 PMID:8074969

Matsuaki, H., Endo, N., Sonomoto, K., & Ishikazi, A. (1996). Lantibiotic nisin Z fermentative production by *Lactococcus lactis* IO-1: Relationship between production of the lantibiotic and lactate and cell growth. *Applied Microbiology and Biotechnology*, *45*(1-2), 36–40. doi:10.1007002530050645 PMID:8920177

Matsuoka, H., Fuke, Y., Kaminogawa, S., & Yamauchi, K. (1991). Purification and debittering effect of aminopeptidase II from *Penicillium caseicolum*. *Journal of Agricultural and Food Chemistry*, *39*(8), 1392–1395. doi:10.1021/jf00008a007

Matsuzaki, C., Takagaki, C., Tomabechi, Y., Forsbeg, L. S., Heiss, C., Azadi, P., ... Hisa, K. (2017). Structural characterization of the immunostimulatory exopolysaccharide produced by *Leuconostoc mesenteroides* strain NTM048. *Carbohydrate Research*, *448*, 95–102. doi:10.1016/j.carres.2017.06.004 PMID:28633071

Mattila-Sandholm, T., Myllärinen, P., Crittenden, R., Mogensen, G., Fondén, R., & Saarela, M. (2002). Technological challenges for future probiotic foods. *International Dairy Journal*, *12*(2-3), 173–182. doi:10.1016/S0958-6946(01)00099-1

Mattila-Sandholm, T., & Saarela, M. (2000). *Functional dairy products*. Boca Raton, FL: CRC Press.

Mautner, A., Meisel, H., Lorenzen, P. C., & Schlimme, E. (1999). Determination of the dipeptide ε-(γ-glutamyl)lysine in transglutaminase crosslinked proteins by amino acid analysis. *Kieler Milchwirtschaftliche Fosrhungsberichte*, *51*, 155–163.

Mazuknaite, I., Guyot, C., Leskauskaite, D., & Kulozik, U. (2013). Influence of transglutaminase on the physical and chemical properties of acid milk gel and cottage type cheese. *Journal of Food Agriculture and Environment*, *11*(3-4), 119–124.

McClelland, R. G., & Pinder, A. C. (1994). Detection of Salmonella typhimurium in dairy products with flow cytometry and monoclonal antibodies. *Applied and Environmental Microbiology*, *60*(12), 4255–4262. PMID:7811064

McFarland, L. V. (2007). Meta-analysis of probiotics for the prevention of traveler's diarrhea. *Travel Medicine and Infectious Disease*, *5*(2), 97–105. doi:10.1016/j.tmaid.2005.10.003 PMID:17298915

McGee, H. (2004). *On food and cooking: The science and lore of the kitchen*. New York: Scribner.

McGovern, P. E., Zhang, J., Tang, J., Zhang, Z., Hall, G. R., Moreau, R. A., ... Wang, C. (2004). Fermented beverages of pre- and proto-historic China. *Proceedings of the National Academy of Sciences of the United States of America*, *101*(51), 17593–17598. doi:10.1073/pnas.0407921102 PMID:15590771

McSweeney, P. L. H. (2004). Biochemistry of cheese ripening: Introduction and overview. In Cheese: Chemistry, Physics and Microbiology: Vol 1. General Aspects, (pp. 347-360). London: Elsevier.

McSweeney, P. L. H., & Fox, P. F. (2004). Metabolism of residual lactose and of lactate and citrate. In Cheese: Chemistry, Physics and Microbiology: Vol 1. General Aspects (pp. 361-372). London: Elsevier. doi:10.1016/S1874-558X(04)80074-5

McSweeney, P. L. H. (2004). Biochemistry of cheese ripening. *International Journal of Dairy Technology*, *57*(2/3), 127–144. doi:10.1111/j.1471-0307.2004.00147.x

McSweeney, P. L. H., & Sousa, M. J. (2000). Biochemical pathways for the production of flavour compounds in cheeses during ripening: A review. *Le Lait*, *80*(3), 293–324. doi:10.1051/lait:2000127

Meijerink, M., van Hemert, S., Taverne, N., Wels, M., de Vos, P., Bron, P. A., ... Wells, J. M. (2010). Identification of genetic loci in *Lactobacillus plantarum* that modulate the immune response of dendritic cells using comparative genome hybridization. *PLoS One*, *5*(5), e10632. doi:10.1371/journal.pone.0010632 PMID:20498715

Meira, Q. G. S., Magnani, M., Júnior, F. C. M., Queiroga, R. C. R. E., Madruga, M. S., Gullóne, B., ... de Souza, E. L. (2015). Effects of added Lactobacillus acidophilus and Bifidobacterium lactis probiotics on the quality characteristics of goat ricotta and their survival under simulated gastrointestinal conditions. *Food Research International*, *76*, 828–838. doi:10.1016/j.foodres.2015.08.002 PMID:28455069

Menassa, A., & Lamberet, G. (1982). Contribution à l'étude du système lipolytique de *Penicillium roqueforti*. Caractères comparés de deux activités exocellulaires. *Le Lait*, *62*(611–612), 32–43. doi:10.1051/lait:1982611-6122

Meng, J., Gao, S. M., Zhang, Q. X., & Lu, R. R. (2015). Murein hydrolase activity of surface layer proteins from *Lactobacillus acidophilus* against *Escherichia coli*. *International Journal of Biological Macromolecules*, *79*, 527–532. doi:10.1016/j.ijbiomac.2015.03.057 PMID:25841372

Meng, Y., Zhao, M., Yang, M., Zhang, Q., Hao, J., & Meng, Y. (2014). Production and characterization of recombinant glucose oxidase from *Aspergillus niger* expressed in *Pichia pastoris*. *Letters in Applied Microbiology*, *58*(4), 393–400. doi:10.1111/lam.12202 PMID:24283586

Metwalli, A. M., De Jongh, H. H. J., & Van Boekel, M. J. S. (1998). Heat inactivation of bovine plasmin. *International Dairy Journal*, *8*(1), 47–56. doi:10.1016/S0958-6946(98)00017-X

Michael, D. R., Davies, T. S., Moss, J. W. E., Calvente, D. L., Ramji, D. P., Marchesi, J. R., ... Hughes, T. R. (2017). The anti-cholesterolaemic effect of a consortium of probiotics: An acute study in C57BL/6J mice. *Scientific Reports*, *7*(1), 2883. doi:10.103841598-017-02889-5 PMID:28588193

Miller, C. W., Nguyen, M. H., Rooney, M., & Kailasapathy, K. (2003). The control of dissolved oxygen content in probiotic yoghurts by alternative packaging materials. *Packaging Technology & Science*, *16*(2), 61–67. doi:10.1002/pts.612

Mishra, V., & Prasad, D. N. (2005). Application of in vitro methods for selection of *Lactobacillus casei* strains as potential probiotics. *International Journal of Food Microbiology*, *103*, 109–115. doi:10.1016/j.ijfoodmicro.2004.10.047 PMID:16040148

Miwa, N., Kumazawa, Y., Nakagoshi, H., & Sakaguchi, S. (2002). *Methods for modifying raw material milk and dairy product prepared by using the modified raw material milk*. European patent 1 197 152 A2.

Mizuno, A., Mitsuiki, M., & Motoki, M. (1999). Glass transition of casein as affected by transglutaminase. *Journal of Food Science*, *64*(5), 796–799. doi:10.1111/j.1365-2621.1999.tb15914.x

Modi, R., Hirvi, Y., Hill, A., & Griffiths, M. W. (2001). Effect of phage on survival of *Salmonella enteritidis* during manufacture and storage of Cheddar cheese made from raw and pasteurized milk. *Journal of Food Protection*, *64*(7), 927–933. doi:10.4315/0362-028X-64.7.927 PMID:11456198

Modler, H. W., Brunner, J. R., & Stine, C. M. (1974a). Extracellular protease of *Penicillium roqueforti*. I. Production and characteristics of crude enzyme preparation. *Journal of Dairy Science*, *57*(5), 523–527. doi:10.3168/jds.S0022-0302(74)84927-1 PMID:4858149

Modler, H. W., Brunner, J. R., & Stine, C. M. (1974b). Extracellular protease of *Penicillium roqueforti*. II. Characterization of a purified enzyme preparation. *Journal of Dairy Science*, *57*(5), 528–534. doi:10.3168/jds.S0022-0302(74)84928-3 PMID:4835412

Mohammadi, R., Mortazavian, A. M., Khosrokhavar, R., & Da Cruz, A. G. (2011). Probiotic ice cream: Viability of probiotic bacteria and sensory properties. *Annals of Microbiology*, *61*(3), 411–424. doi:10.100713213-010-0188-z

Mohan, M. S., Hopkinson, J., & Harte, F. (2014). Milk and ice cream processing. Food Processing: Principles and Applications (2nd ed.). *John Wiley & Sons, Ltd.*

Mohedano, A. F., Fernández, J., Garde, S., Medina, M., Gaya, P., & Nuñez, M. (1998). The effect of the cysteine proteinase from *Micrococcus sp. INIA 528* on the ripening process of Manchego cheese. *Enzyme and Microbial Technology*, *22*(5), 391–396. doi:10.1016/S0141-0229(97)00211-1

Moineau, S. (1999). *Applications of phage resistance in lactic acid bacteria. In Lactic Acid Bacteria: Genetics, Metabolism and Applications* (pp. 377–382). Springer. doi:10.1007/978-94-017-2027-4_23

Molimard, P., & Spinnler, H. E. (1996). Review: Compounds involved in the flavor of surface mold-ripened cheeses: Origins and properties. *Journal of Dairy Science*, *79*(2), 169–184. doi:10.3168/jds.S0022-0302(96)76348-8

Montel, M.-C., Buchin, S., Mallet, A., Delbes-Paus, C., Vuitton, D. A., Desmasures, N., & Berthier, F. (2014). Traditional cheeses: Rich and diverse microbiota with associated benefits. *International Journal of Food Microbiology*, *177*, 136–154. doi:10.1016/j.ijfoodmicro.2014.02.019 PMID:24642348

Moon, J. H., Hong, Y. H., Huppertz, T., Fox, P. F., & Kelly, A. L. (2009). Properties of casein micelles cross-linked by transglutaminase. *International Journal of Dairy Technology*, *62*(1), 27–32. doi:10.1111/j.1471-0307.2008.00442.x

Moreau, C. (1980). Le Penicillium roqueforti, morphologie, physiologie, intérêt en industrie fromagère, mycotoxines. (Révision bibliographique). *Le Lait*, *60*(595–596), 254–271. doi:10.1051/lait:1980595-59615

Moreira, S. R., Schwan, R. F., De Carvalho, E. P., & Wheals, A. E. (2001). Isolation and identification of yeasts and filamentous fungi from yoghurts in Brazil. *Brazilian Journal of Microbiology*, *32*(2), 117–122. doi:10.1590/S1517-83822001000200009

Moreno, M. R. F., Sarantinopoulos, P., Tsakalidou, E., & De Vuyst, L. (2006). The role and application of enterococci in food and health. *International Journal of Food Microbiology*, *106*(1), 1–24. doi:10.1016/j.ijfoodmicro.2005.06.026 PMID:16216368

Morita, H., He, F., Fuse, T., Ouwehand, A. C., Hashimoto, H., & Hosoda, M. … Kurisaki, J. (2002). Adhesion of lactic acid bacteria to caco-2 cells and their effect on cytokine secretion. *Microbiology and Immunology*, *46*(4), 293–7. Retrieved from http://www.ncbi.nlm.nih.gov/pubmed/12061632

Mortazavian, A. M., Ehsani, M. R., Mousavi, S. M., Reinheimer, J. A., Emamdjomeh, Z., Sohrabvandi, S., & Rezaei, K. (2006). Preliminary investigation of the combined effect of heat treatment and incubation temperature on the viability of the probiotic micro-organisms in freshly made yogurt. *International Journal of Dairy Technology*, *59*(1), 8–11. doi:10.1111/j.1471-0307.2006.00216.x

Mortazavian, A. M., Khosrokhvar, R., Rastegar, H., & Mortazaei, G. R. (2010). Effects of dry matter standardization order on biochemical and microbiological characteristics of freshly made probiotic Doogh (Iranian fermented milk drink). *Italian Journal of Food Science*, *22*, 98–102.

Moschopoulou, E. (2017). Microbial milk coagulants. In Microbial enzyme technology in food applications (pp. 199-210). CRC Press, Taylor and Francis Group.

Moschopoulou, E. (2017). Microbial milk coagulants. In R. C. Ray & C. M. Rosell (Eds.), *Microbial enzyme technology in food applications* (pp. 199–312). Boca Raton, FL: CRC Press.

Moslehi-Jenabian, S., Pedersen, L. L., & Jespersen, L. (2010). Beneficial Effects of Probiotic and Food Borne Yeasts on Human Health. *Nutrients*, *2*(4), 449–473. doi:10.3390/nu2040449 PMID:22254033

Moss, C. W., & Speck, M. L. (1963). Injury and death of *Streptococcus lactis* due to freezing and frozen storage. *Applied Microbiology*, *11*(4), 326–329. PMID:13936199

Motoki, M., Aso, H., Seguro, K., & Nio, N. (1987a). α_{s1}-casein film prepared using transglutaminase. *Agricultural and Biological Chemistry*, *51*, 993–996.

Motoki, M., & Nio, N. (1983). Crosslinking between different proteins by transglutaminase. *Journal of Food Science*, *48*(2), 561–566. doi:10.1111/j.1365-2621.1983.tb10790.x

Motoki, M., Nio, N., & Takinami, K. (1984). Functional properties of food proteins polymerized by transglutaminase. *Agricultural and Biological Chemistry*, *48*, 1257–1261.

Motoki, M., Nio, N., & Takinami, K. (1987b). Functional properties of heterologous polymer prepared by Transglutaminase between milk casein and soybean globulin. *Agricultural and Biological Chemistry*, *51*, 237–239.

Motoki, M., & Seguro, K. (1998). Transglutaminase and its use for food processing. *Trends in Food Science & Technology*, *9*(5), 204–210. doi:10.1016/S0924-2244(98)00038-7

Motoki, M., Seguro, K., Nio, N., & Takinami, K. (1986). Glutamine-specific deamidation of α_{s1}-casein by transglutaminase. *Agricultural and Biological Chemistry*, *50*, 3025–3030.

Motta, A. S., & Brandelli, A. (2002). Characterization of an antibacterial peptideproduced by *Brevibacterium linens*. *Journal of Applied Microbiology*, *92*(1), 63–70. doi:10.1046/j.1365-2672.2002.01490.x PMID:11849329

Mounsey, J. S., O'Kennedy, B. T., & Kelly, P. M. (2005). Influence of transglutaminase treatment on properties of micellar casein and products made therefrom. *Le Lait*, *85*(4-5), 405–418. doi:10.1051/lait:2005028

Mozafari, M. R., Flanagan, J., Matia-Merino, L., Awati, A., Omri, A., Suntres, Z. E., & Singh, H. (2006). Recent trends in the lipid-based nanoencapsulation of antioxidants and their role in foods. *Journal of the Science of Food and Agriculture*, *86*(13), 2038–2045. doi:10.1002/jsfa.2576

Mrvčić, J., Stanzer, D., Šolić, E., & Stehlik-Tomas, V. (2012). Interaction of lactic acid bacteria with metal ions: Opportunities for improving food safety and quality. *World Journal of Microbiology & Biotechnology*, *28*(9), 2771–2782. doi:10.100711274-012-1094-2 PMID:22806724

Mulder, J., Wels, M., Kuipers, O. P., Kleerebezem, M., & Bron, P. A. (2017). Unleashing natural competence in *Lactococcus lactis* by induction of the competence regulator ComX. *Applied and Environmental Microbiology*, *83*(20), e01320-17. doi:10.1128/AEM.01320-17 PMID:28778888

Mullan, W. M. A. (2014). Importance of Selected Genera. Encyclopedia of Food Microbiology, 3, 515-521.

Mullan, W. M. A. (2014). Importance of Selected Genera. In C. A. Batt & M. L. Tortorello (Eds.), *Encyclopedia of Food Microbiology, Volume:2* (pp. 515–521). USA: Elsevier. doi:10.1016/B978-0-12-384730-0.00321-9

Muriana, P. M., & Klaenhamer, T. R. (1991). Purification and partial characterization of lactacin F, a bacteriocin produced by *Lactobacillus acidophilus* 11088. *Applied and Environmental Microbiology*, *57*, 114–121. PMID:1903624

Mushtaq, M., Gani, A., Masoodi, F. A., & Ahmad, M. (2016). Himalayan cheese (Kalari/Kradi) - Effect of different probiotic strains on oxidative stability, microbiological, sensory and nutraceutical properties during storage. *Lebensmittel-Wissenschaft + Technologie*, *67*, 74–81. doi:10.1016/j.lwt.2015.11.039

Myllärinen, P., Buchert, J., & Autio, K. (2007). Effect of transglutaminase on rheological properties and microstructure of chemically acidified sodium caseinate gels. *International Dairy Journal*, *17*(7), 800–807. doi:10.1016/j.idairyj.2005.10.031

Nagayasu, M., Wardani, A. K., Nagahisa, K., Shimizu, H., & Shioya, S. (2007). Analysis of hemin effect on lactate reduction in *Lactococcus lactis*. *Journal of Bioscience and Bioengineering*, *103*(6), 529–534. doi:10.1263/jbb.103.529 PMID:17630124

Nagpal, R., Kumar, A., Kumar, M., Behare, P. V., Jain, S., & Yadav, H. (2012). Probiotics, their health benefits and applications for developing healthier foods: A review. *FEMS Microbiology Letters*, *334*(1), 1–15. doi:10.1111/j.1574-6968.2012.02593.x PMID:22568660

Nase, L., Hatakka, K., Savilahti, E., Saxelin, M., Ponka, A., Poussa, T., ... Meurman, J. H. (2001). Effect of Long-Term Consumption of a Probiotic Bacterium, *Lactobacillus rhamnosus* GG, in Milk on Dental Caries and Caries Risk in Children. *Caries Research*, *35*(6), 412–420. doi:10.1159/000047484 PMID:11799281

National Research Council (US) Committee on Metagenomics Challenges and Functional Applications. (2007). *The New Science of Metagenomics. The New Science of Metagenomics: Revealing the Secrets of Our Microbial Planet*. National Academies Press (US). Retrieved from http://www.ncbi.nlm.nih.gov/pubmed/21678629

Nayir, S. M. (2008). *Sütün yoğurda dönüşümü sırasında içerdiği feneolik antioksidan maddelere probiyotik bakteri etkisinin incelenmesi. Yüksek Lisans Tezi*. Trakya Üniversitesi. Retrieved from http://www.teknolojikarastirmalar.com/pdf/tr/02_2012_7_1_121_721.pdf

Nedovic, V., Kalusevic, A., Manojlovic, V., Levic, S., & Bugarski, B. (2011). An overview of encapsulation technologies for food applications. *Procedia Food Science*, *1*, 1806–1815. doi:10.1016/j.profoo.2011.09.265

Neelakantan, S., Mohanty, A. K., & Kaushik, J. K. (1999). Production and use of microbial enzymes for dairy processing. *Current Science*, *77*(1), 143–148.

Neish, A. S. (2017). Probiotics of the Acidophilus Group: *Lactobacillus acidophilus, delbrueckii* subsp. *bulgaricus* and *johnsonii*. In M. H. Floch, Y. Ringel, & A. Walker (Eds.), *The Microbiota in Gastrointestinal Pathophysiology* (pp. 71–78). Elsevier. doi:10.1016/B978-0-12-804024-9.00006-9

Nes, I. F., Diep, D. B., Håvarstein, L. S., & Brurberg, M. B. (1996). Biosynthesis of bacteriocins in lactic acid bacteria. *Antonie van Leeuwenhoek*, *70*, 113–128. doi:10.1007/BF00395929 PMID:8879403

Neves, A. R., Pool, W. A., Solopova, A., Kok, J., Santos, H., & Kuipers, O. P. (2010). Towards enhanced galactose utilization by Lactococcus lactis. *Applied and Environmental Microbiology*, *76*(21), 7048–7060. doi:10.1128/AEM.01195-10 PMID:20817811

Neviani, E., Bottari, B., Lazzi, C., & Gatti, M. (2013). New developments in the study of the microbiota of raw-milk, long-ripened cheeses by molecular methods: The case of Grana Padano and Parmigiano Reggiano. *Frontiers in Microbiology*, *4*(FEB), 1–14. PMID:23450500

Newman, M., Safro, M., Frazao, C., Khan, G., Zdanov, A., Tickle, I. J., ... Andreeva, N. (1991). X-ray analysis of aspartic proteinases IV. Structure and refinement at 2.2 Å resolution of bovine chymosin. *Journal of Molecular Biology*, *221*, 1295–1309. PMID:1942052

Ng, E. W., Yeung, M., & Tong, P. S. (2010). Effects of yogurt starter cultures on the survival of *Lactobacillus acidophilus*. *International Journal of Food Microbiology*, *145*(1), 169–175. doi:10.1016/j.ijfoodmicro.2010.12.006 PMID:21196060

Nielsen, K. F., Dalsgaard, P. W., Smedsgaard, J., & Larsen, T. O. (2005). Andrastins A–D, P*enicillium roqueforti* metabolites consistently produced in blue-mold-ripened cheese. *Journal of Agricultural and Food Chemistry, 53*(8), 2908–2913. doi:10.1021/jf047983u PMID:15826038

Nielsen, P. H., Malmos, H., Damhus, T., Diderichsen, B., Nielsen, H. K., Simonsen, M., ... Nielsen, T. K. (1994). Enzyme applications (industrial). In K. Othmer (Ed.), *Encyclopedia of Chemical Technology* (pp. 567–620). New York: John Wiley & Sons, Inc.

Nieuwenhuizen, W. F., Dekker, H. L., de Koning, L. J., Gröneveld, T., de Koster, C. G., & de Jong, G. A. (2003). Modification of glutamine and lysine residues in holo and apo alpha-lactalbumin with microbial transglutaminase. *Journal of Agricultural and Food Chemistry, 51*(24), 7132–7139. doi:10.1021/jf0300644 PMID:14611184

Nieuwenhuizen, W. F., Dekker, H. L., Gröneveld, T., Koster, C. G., & Jong, G. A. H. (2004). Transglutaminase-mediated modification of glutamine and lysine residues in native bovine β-lactoglobulin. *Biotechnology and Bioengineering, 85*(3), 248–258. doi:10.1002/bit.10898 PMID:14748079

Nieuwland, M., Bouwman, W. G., Bennik, M. L., Silletti, E., & de Jongh, H. H. J. (2014). Characterizing length scales that determine the mechanical behavior of gels from crosslinked casein micelles. *Food Biophysics, 10*(4), 416–427. doi:10.100711483-015-9399-y

Nio, N., Motoki, M., & Takinami, K. (1985). Gelation of casein and soybean globulins by transglutaminase. *Agricultural and Biological Chemistry, 49*, 2283–2286.

Nio, N., Motoki, M., & Takinami, K. (1986). Gelation mechanisms of protein solutions by transglutamianse. *Agricultural and Biological Chemistry, 48*, 851–855.

Nishimori, K., Kawagushi, Y., Hidaka, M., & Beppu, T. (1982). Expression of cloned calf prochymosin gene sequence in *Escherichia coli. Gene, 19*(3), 337–344. doi:10.1016/0378-1119(82)90024-5 PMID:6295885

Nocker, A., Sossa-Fernandez, P., Burr, M. D., & Camper, A. K. (2007). Use of propidium monoazide for live/dead distinction in microbial ecology. *Applied and Environmental Microbiology, 73*(16), 5111–5117. doi:10.1128/AEM.02987-06 PMID:17586667

Nonaka, M., Matsuura, Y., & Motoki, M. (1996). Incorporation of lysine dipeptides into α_{s1}-casein by Ca^{+2}-independent microbial transglutaminase. *Bioscience, Biotechnology, and Biochemistry, 60*(1), 131–133. doi:10.1271/bbb.60.131

Nonaka, M., Sakamoto, H., Toiguchi, S., Kawajiri, H., Soeda, T., & Motoki, M. (1992). Sodium caseinate and skim milk gels formed by incubation with microbial transglutaminase. *Journal of Food Science, 57*(5), 1214–1218. doi:10.1111/j.1365-2621.1992.tb11302.x

Nonaka, M., Tanaka, H., Okiyama, A., Motoki, M., Ando, H., Umeda, K., & Matsuura, A. (1989). Polymerization of several proteins by Ca^{+2}-independent transglutaminase derived from microorganisms. *Agricultural and Biological Chemistry, 53*(10), 2619–2623.

Norazman, N., Wu, W., Li, H., Wasinger, V., Zhang, H., & Chen, V. (2013). Evaluation of chemical cleaning of UF membranes fouled with whey protein isolates via analysis of residual protein components on membrane surface. *Separation and Purification Technology, 103*, 241–250. doi:10.1016/j.seppur.2012.10.039

Normand, A.-C., Cassagne, C., Ranque, S., L'ollivier, C., Fourquet, P., Roesems, S., ... Piarroux, R. (2013). Assessment of various parameters to improve MALDI-TOF MS reference spectra libraries constructed for the routine identification of filamentous fungi. *BMC Microbiology, 13*(1), 76. doi:10.1186/1471-2180-13-76 PMID:23565856

Nousia, F. G., Androulakis, P. I., & Fletouris, D. J. (2011). Survival of *Lactobacillus acidophilus* LMGP-21381 in probiotic ice cream and its influence on sensory acceptability. *International Journal of Dairy Technology*, *64*(1), 130–136. doi:10.1111/j.1471-0307.2010.00645.x

Novozymes. (2017). *What are enzymes?* Retrieved July 2nd, 2017, from http://www.novozymes.com/en/about-us/our-business/industrial-biotechnology/basic-technologies/recombinant-expression/for-the-experts-

Nuñez, M., Guillen, A. M., Rodriguez-Marin, M. A., Marcilla, A. M., Gaya, P., & Medina, M. (1991). Accelerated Ripening of Ewes' Milk Manchego Cheese: The Effect of Neutral Proteinases. *Journal of Dairy Science*, *74*(12), 4108–4118. doi:10.3168/jds.S0022-0302(91)78605-0

O'Callaghan, Y. C., O'Connor, T. P., & O'Brien, N. M. (2017). Nutritional Aspects of Cheese. In P. F. Fox, T. P. Guinee, T. M. Cogan, & P. L. H. McSweeney (Eds.), *Fundamentals of Cheese Science* (pp. 715–730). Boston, MA: Springer. doi:10.1007/978-1-4899-7681-9_20

O'Flaherty, S., Coffey, A., Meaney, W. J., Fitzgerald, G. F., & Ross, R. P. (2005). Inhibition of bacteriophage K proliferation on *Staphylococcus aureus* in raw bovine milk. *Letters in Applied Microbiology*, *41*(3), 274–279. doi:10.1111/j.1472-765X.2005.01762.x PMID:16108920

O'Mahony, T., Rekhif, N., Cavadini, C., & Fitzgerald, G. F. (2001). The application of a fermented food ingredient containing 'variacin', a novel antimicrobial produced by *Kocuria varians*, to control the growth of *Bacillus cereus* in chilled dairy products. *Journal of Applied Microbiology*, *90*(1), 106–114. doi:10.1046/j.1365-2672.2001.01222.x PMID:11155129

O'Sullivan, E., & Condon, S. (1997). Intracellular pH is a major factor in the induction of tolerance to acid and other stresses in *Lactococcus lactis*. *Applied and Environmental Microbiology*, *63*, 4210–4215. PMID:9361406

O'Sullivan, M. M., Kelly, A. L., & Fox, P. F. (2002). Effect of transglutamianse on the heat stability of milk: A possible mechanism. *Journal of Dairy Science*, *85*(1), 1–7. doi:10.3168/jds.S0022-0302(02)74045-9 PMID:11860101

Oberg, C. J., Moyes, L. V., Domek, M. J., Brothersen, C., & McMahon, D. J. (2011). Survival of probiotic adjunct cultures in cheese and challenges in their enumeration using selective media. *Journal of Dairy Science*, *94*(5), 2220–2230. doi:10.3168/jds.2010-3934 PMID:21524512

Oberhelman, R. A., Gilman, R. H., Sheen, P., Taylor, D., Black, R. E., Cabrera, L., ... Madico, G. (1999). A placebo-controlled trial of Lactobacillus GG to prevent diarrhea in undernourished Peruvian children. *The Journal of Pediatrics*, *134*(1), 15–20. doi:10.1016/S0022-3476(99)70366-5 PMID:9880443

Oberman, H., & Libudzisz, Z. (1998). Fermented milks. In B. J. B. Wood (Ed.), *Microbiology of fermented foods* (pp. 308–350). London: Blackie Academic and Professional. doi:10.1007/978-1-4613-0309-1_11

Ogel, Z. B. (1993). *Molecular characterization of a galactose oxidase gene* (Ph.D. thesis). The University of Leeds, UK.

Ogier, J.-C., Son, O., Gruss, A., Tailliez, P., & Delacroix-Buchet, A. (2002). Identification of the bacterial microflora in dairy products by temporal temperature gradient gel electrophoresis. *Applied and Environmental Microbiology*, *68*(8), 3691–3701. doi:10.1128/AEM.68.8.3691-3701.2002 PMID:12147461

Oh, J. H., Wang, B., Field, D. P., & Aglan, H. A. (2004). Characteristics of edible films made from dairy proteins and zein hydrolysate cross-linked with transglutaminase. *International Journal of Food Science & Technology*, *39*(3), 287–294. doi:10.1111/j.1365-2621.2004.00783.x

Ohmomo, S., Oguma, K., Ohashi, T., & Abe, M. (1978). Isolation of a new indole alkaloid, roquefortine D, from the cultures of *Penicillium roqueforti*. *Agricultural and Biological Chemistry*, *42*(12), 2387–2389.

Ohmomo, S., Sato, T., Utagawa, T., & Abe, M. (1975). Isolation of festuclavine and three new indole alkaloids, roquefortine A, B and C from the cultures of *Penicillium roqueforti. Agricultural and Biological Chemistry*, *39*(6), 1333–1334.

Oh, Y., Osato, M. S., Han, X., Bennett, G., & Hong, W. K. (2002). Folk yoghurt kills *Helicobacter pylori. Journal of Applied Microbiology*, *93*(6), 1083–1088. doi:10.1046/j.1365-2672.2002.01779.x PMID:12452966

Olempska-Beer, Z. S., Merker, R. I., Ditto, M. D., & DiNovi, M. J. (2006). Food-processing enzymes from recombinant microorganisms - a review. *Regulatory Toxicology and Pharmacology*, *45*(2), 144–158. doi:10.1016/j.yrtph.2006.05.001 PMID:16769167

Oliver, C. M., & Augustin, M. A. (2009). Using dairy ingredients for encapsulation. In Dairy-Derived Ingredients (pp. 565–588). Academic Press. doi:10.1533/9781845697198.3.565

Olsen, K. N., Brockmann, E., & Molin, S. (2007). Quantification of Leuconostoc populations in mixed dairy starter cultures using fluorescence in situ hybridization. *Journal of Applied Microbiology*, *103*(4), 855–863. doi:10.1111/j.1365-2672.2007.03298.x PMID:17897187

Olson, A. C., & Korus, R. A. (1976). Immobilized enzymes. In R. L. Ory & A. J. St. Angelo (Eds.), Enzymes in Food and Beverage Processing (pp. 100-131). Washington, DC: American Chemical Society.

Önal, D., Beyatlı, Y., & Aslım, B. (2005). Probiyotik Bakterilerin Epitel Yüzeylere Yapışması. *Orlab On-Line Mikrobiyoloji Dergisi, 3*(9), 1–10. Retrieved from www.mikrobiyoloji.org/pdf/702050901.pdf

Öner, O., Aslim, B., & Aydas, S. B. (2014). Mechanisms of cholesterol-lowering effects of lactobacilli and bifidobacteria strains as potential probiotics with their bsh gene analysis. *Journal of Molecular Microbiology and Biotechnology*, *24*(1), 12–18. doi:10.1159/000354316 PMID:24158048

Öner, Z., Karahan, A. G., Aydemir, S., & Sanlidere-Aloglu, H. (2008). Effect of transglutaminase on physicochemical properties of set-style yogurt. *International Journal of Food Properties*, *11*(1), 196–205. doi:10.1080/10942910701286346

Ong, L., Henriksson, A., & Shah, N. P. (2006). Development of probiotic Cheddar cheese containing *Lactobacillus acidophilus, Lb. casei, Lb. paracasei* and *Bifidobacterium* spp. and the influence of these bacteria on proteolytic patterns and production of organic acid. *International Dairy Journal*, *16*(5), 446–456. doi:10.1016/j.idairyj.2005.05.008

Ong, L., Henriksson, S., & Shah, N. P. (2007). Proteolytic pattern and organic acid profiles of probiotic Cheddar cheese as influenced by probiotic strains of *Lactobacillus acidophilus, Lb. paracasei, Lb. casei* or *Bifidobacterium* sp. *International Dairy Journal*, *17*(1), 67–78. doi:10.1016/j.idairyj.2005.12.009

Ong, L., & Shah, N. P. (2008). Release and identification of angiotensin-converting enzyme-inhibitory peptides as influenced by ripening temperatures and probiotic adjuncts in Cheddar cheeses. *Lebensmittel-Wissenschaft + Technologie*, *41*(9), 1555–1566. doi:10.1016/j.lwt.2007.11.026

Ong, L., & Shah, N. P. (2009). Probiotic Cheddar cheese: Influence of ripening temperatures on survival of probiotic microorganisms, cheese composition and organic acid profiles. *Lebensmittel-Wissenschaft + Technologie*, *42*(7), 1260–1268. doi:10.1016/j.lwt.2009.01.011

Ouwehand, A. (2004). The probiotic potential of Propionibacteria. In S. Salminen, A. Wright, & A. Ouwehand (Eds.), *Lactic Acid Bacteria: Microbiological and Functional Aspects* (pp. 125–159). London: Maecel and Dekker Inc. doi:10.1201/9780824752033.ch4

Ouwehand, A. C., Kirjavainen, P. V., Shortt, C., & Salminen, S. (1999). Probiotics: Mechanisms and established effects. *International Dairy Journal*, *9*(1), 43–52. doi:10.1016/S0958-6946(99)00043-6

Ouwehand, A. C., & Salminen, S. J. (1998). The health effects of cultured milk products with viable and non-viable bacteria. *International Dairy Journal*, *8*(9), 749–758. doi:10.1016/S0958-6946(98)00114-9

Ovsejevi, K., Grazu, V., & Batista-Viera, F. (1998). Beta-Galactosidase from *Kluyveromyces lactis* immobilized on to thiolsulfinate/thiolsulfonate supports for lactose hydrolysis in milk and dairy by-products. *Biotechnology Techniques*, *12*(2), 143–148. doi:10.1023/A:1008892518473

Oyeleke, S. B. (2009). Microbial assessment of some commercially prepared yoghurt retailed in Minna, Niger State. *African Journal of Microbiological Research*, *3*(5), 245–248.

Ozcan, T., & Kurdal, E. (2012). The effects of using a starter culture, lipase, and protease enzymes on ripening of Mihalic cheese. *International Journal of Dairy Technology*, *65*(4), 585–593. doi:10.1111/j.1471-0307.2012.00868.x

Özdemir, Ö. (2010). Various effects of different probiotic strains in allergic disorders: An update from laboratory and clinical data. *Clinical and Experimental Immunology*, *160*(3), 295–304. doi:10.1111/j.1365-2249.2010.04109.x PMID:20345982

Özer, B., Guyot, C., & Kulozik, U. (2012). Simultaneous use of transglutaminase and rennet in milk coagulation: Effect of initial milk pH and renneting temperature. *International Dairy Journal*, *24*(1), 1–7. doi:10.1016/j.idairyj.2011.10.002

Özer, B., Hayaloğlu, A. A., Yaman, H., Gürsoy, A., & Şener, L. (2013). Simultaneous use transglutaminase and rennet in white-brined cheese production. *International Dairy Journal*, *33*(2), 129–134. doi:10.1016/j.idairyj.2013.02.001

Özer, B., & Kırmacı, H. A. (2011). Technological and health aspects of probiotic cheese. In R. D. Foster (Ed.), *Cheese: Types, Nutrition and Consumption* (pp. 1–42). Nova Science Publishers, Inc.

Özer, B., Kirmaci, H. A., Oztekin, S., Hayaloglu, A., & Atamer, M. (2007). Incorporation of microbial transglutaminase into non-fat yogurt production. *International Dairy Journal*, *17*(3), 199–207. doi:10.1016/j.idairyj.2006.02.007

Özer, B., Kirmaci, H. A., Şenel, E., Atamer, M., & Hayaloğlu, A. (2009). Improving the viability of Bifidobacterium bifidum BB-12 and Lactobacillus acidophilus LA-5 in white-brined cheese by microencapsulation. *International Dairy Journal*, *19*(1), 22–29. doi:10.1016/j.idairyj.2008.07.001

Özer, C. O., Kılıç, B., & Başyiğit Kılıç, G. (2016). *In-vitro* microbial production of conjugated linoleic acid by probiotic *L. plantarum* strains: Utilization as a functional starter culture in sucuk fermentation. *Meat Science*, *114*, 24–31. doi:10.1016/j.meatsci.2015.12.005 PMID:26720888

Ozrenk, E. (2006). The use of transglutaminase in dairy products. *International Journal of Dairy Technology*, *59*(1), 1–7. doi:10.1111/j.1471-0307.2006.00220.x

Öztürk, H. İ., Demirci, T., & Akın, N. (2017). Production of functional probiotic ice creams with white and dark blue fruits of Myrtus communis: The comparison of the prebiotic potentials on Lactobacillus casei 431 and functional characteristics. *Lebensmittel-Wissenschaft + Technologie*, *90*, 339–345.

Ozturkoglu Budak, S. (2016). *Dynamics of complex microbiota and enzymes in Divle Cave cheese and their biochemical consequences*. Utrecht University.

Ozturkoglu Budak, S., Figge, M. J., Houbraken, J., & de Vries, R. P. (2016). The diversity and evolution of microbiota in traditional Turkish Divle Cave cheese during ripening. *International Dairy Journal*, *58*, 50–53. doi:10.1016/j.idairyj.2015.09.011

Ozturkoglu-Budak, S. (2016d). *Dynamics of complex microbiota and enzymes in Divle Cave cheese and their biochemical consequences* (PhD thesis). Utrecht University, The Netherlands.

Ozturkoglu-Budak, S., Gursoy, A., Aykas, D. P., Koçak, C., Dönmez, S., de Vries, R. P., & Bron, P. A. (2016b). Volatile compound profiling of Turkish Divle Cave cheese during production and ripening. *Journal of Dairy Science*, *99*(7), 1–12. doi:10.3168/jds.2015-10828 PMID:27108178

Ozturkoglu-Budak, S., Wiebenga, A., Bron, P. A., & de Vries, R. P. (2016c). Protease and lipase activities of fungal and bacterial strains derived from an artisanal raw ewe's milk cheese. *International Journal of Food Microbiology*, *237*, 17–27. doi:10.1016/j.ijfoodmicro.2016.08.007 PMID:27541978

Pakseresht, S., Tehrani, M. M., & Razavi, S. A. M. (2017). Optimization of low-fat set-type yoghurt: Effect of altered whey protein to casein ratio, fat content and microbial transglutaminase on rheological and sensorial properties. *Journal of Food Science and Technology*, *54*(8), 2351–2360. doi:10.100713197-017-2675-8 PMID:28740292

Pant, A. R., Graham, S. M., Allen, S. J., Harikul, S., Sabchareon, A., Cuevas, L., ... Hart, C. A. (1996). *Lactobacillus* GG and Acute Diarrhoea in Young Children in the Tropics. *Journal of Tropical Pediatrics*, *42*(3), 162–165. doi:10.1093/tropej/42.3.162 PMID:8699584

Papadimitriou, K., Alegria, A., Bron, P. A., de Angelis, M., Gobbetti, M., Kleerebezem, M., ... Kok, J. (2016). Stress Physiology of Lactic Acid Bacteria. *Microbiology and Molecular Biology Reviews*, *80*(3), 837–890. doi:10.1128/MMBR.00076-15 PMID:27466284

Papayannakos, N., Markas, G., & Kekos, D. (1993). Studies on modeling and simulation of lactose hydrolysis by free and immobilized β-galactosidase from *Aspergillus niger. Chemical Engineering Journal*, *52*(1), B1–B12. doi:10.1016/0300-9467(93)80044-O

Parente, E., & Cogan, T. M. (2004). Starter cultures: General aspects. In Cheese: Chemistry, Physics and Microbiology (Vol. 1, pp. 123-147): Academic Press.

Parente, E., & Cogan, T. M. (2004). Starter cultures: General aspects. In P. F. Fox, P. L. H. McSweeney, T. M. Cogan & T. P. Guinee (Eds.), Cheese: Chemistry, Physics and Microbiology: Vol. 1. General Aspects (3rd ed.; pp. 123-147). London, UK: Academic Press.

Parente, D., de Ferra, F., Galli, G., & Grandi, G. (1991). Prochymosin expression in *Bacillus subtilis. FEMS Microbiology Letters*, *61*(2-3), 243–249. doi:10.1111/j.1574-6968.1991.tb04356.x PMID:1903749

Parente, E., Giglio, M. A., Riccardi, A., & Clementi, F. (1998). The combined effect of nisin, leucocin F10, pH, NaCl and EDTA on the survival of *Listeria monocytogenes* in broth. *International Journal of Food Microbiology*, *40*(1-2), 65–75. doi:10.1016/S0168-1605(98)00021-X PMID:9600612

Pariza, M. W., & Johnson, E. A. (2001). Evaluating the safety of microbial enzyme preparations used in food processing: Update for a new century. *Regulatory Toxicology and Pharmacology*, *33*(2), 173–186. doi:10.1006/rtph.2001.1466 PMID:11350200

Pariza, M. W., Park, Y., & Cook, M. E. (2001). The biologically active isomers of conjugated linoleic acid. *Progress in Lipid Research*, *40*(4), 283–298. doi:10.1016/S0163-7827(01)00008-X PMID:11412893

Park, H. Y., Kim, H. J., Lee, J. K., Kim, D., & Oh, D. K. (2008). Galactooligosaccharide production by a thermostable β-galactosidase from *Sulfolobus solfataricus. World Journal of Microbiology & Biotechnology*, *24*(8), 1553–1558. doi:10.100711274-007-9642-x

Park, Y. M., Juarez, M., Ramos, M., & Haenlein, G. F. W. (2007). Physico-chemical characteristics of goat and sheep milk. *Small Ruminant Research*, *68*(1-2), 88–113. doi:10.1016/j.smallrumres.2006.09.013

Parussolo, G., Busatto, R. T., Schmitt, J., Pauletto, R., Schons, P. F., & Ries, E. F. (2017). Synbiotic ice cream containing yacon flour and *Lactobacillus acidophylus* NCFM. *Lebensmittel-Wissenschaft + Technologie*, *82*, 192–198. doi:10.1016/j.lwt.2017.04.049

Patel, A. K., Michaud, P., Singhania, R. R., Soccol, C. R., & Pandey, A. (2010). Polysaccharides from probiotics: New developments as food additives. *Food Technology and Biotechnology*, *48*(4), 451–463.

Patel, A. K., Singhania, R. R., Pandey, A., & Chincholkar, S. B. (2010). Probiotic bile salt hydrolase: Current developments and perspectives. *Applied Biochemistry and Biotechnology*, *162*(1), 166–180. doi:10.100712010-009-8738-1 PMID:19669939

Patel, A., Prajapati, J. B., Holst, O., & Ljungh, A. (2014). Determining probiotic potential of exopolysaccharide producing lactic acid bacteria isolated from vegetables and traditional Indian fermented food products. *Food Bioscience*, *5*, 27–33. doi:10.1016/j.fbio.2013.10.002

Patel, R. (2015). MALDI-TOF MS for the diagnosis of infectious diseases. *Clinical Chemistry*, *61*(1), 100–111. doi:10.1373/clinchem.2014.221770 PMID:25278500

Patra, F., Tomar, S. K., & Arora, S. (2009). Technological and functional applications of low-calorie sweeteners from lactic acid bacteria. *Journal of Food Science*, *74*(1), R16–R23. doi:10.1111/j.1750-3841.2008.01005.x PMID:19200114

Patra, F., Tomar, S. K., Rajput, Y. S., & Singh, R. (2011). Characterization of mannitol producing strains of Leuconostoc species. *World Journal of Microbiology & Biotechnology*, *27*(4), 933–939. doi:10.100711274-010-0536-y

Patterson, J., & Burkholder, K. (2003). Application of prebiotics and probiotics in poultry production. *Poultry Science*, *82*(4), 627–631. doi:10.1093/ps/82.4.627 PMID:12710484

Pavlovic, M., Huber, I., Konrad, R., & Busch, U. (2013). Application of MALDI-TOF MS for the Identification of Food Borne Bacteria. *The Open Microbiology Journal*, *7*(1), 135–141. doi:10.2174/1874285801307010135 PMID:24358065

Pavunc, A. L., Beganović, J., Kos, B., Buneta, A., Beluhan, S., & Šušković, J. (2011). Influence of microencapsulation and transglutaminase on viability of probiotic strain *Lactobacillus helveticus* M92 and consistency of set yoghurt. *International Journal of Dairy Technology*, *64*(2), 254–261. doi:10.1111/j.1471-0307.2010.00647.x

Pedersen, M. B., Garrigues, C., Tuphile, K., Brun, C., Vido, K., Bennedsen, M., ... Gruss, A. (2008). Impact of aeration and heme-activated respiration on *Lactococcus lactis* gene expression: Identification of a heme-responsive operon. *Journal of Bacteriology*, *190*(14), 4903–4911. doi:10.1128/JB.00447-08 PMID:18487342

Pedersen, M. B., Gaudu, P., Lechardeur, D., Petit, M.-A., & Gruss, A. (2012). Aerobic respiration metabolism in lactic acid bacteria and uses in biotechnology. *Annual Review of Food Science and Technology*, *3*(1), 37–58. doi:10.1146/annurev-food-022811-101255 PMID:22385163

Pedersen, M. B., Iversen, S. L., Sørensen, K. I., & Johansen, E. (2005). The long and winding road from the research laboratory to industrial applications of lactic acid bacteria. *FEMS Microbiology Reviews*, *29*(3), 611–624. doi:10.1016/j.fmrre.2005.04.001 PMID:15935510

Peker, H., & Arslan, S. (2011). Mikroenkapsülasyon ve süt teknolojisinde kullanım alanları. *Akademik Gıda*, *9*(6), 70–80.

Peltonen, K., El-Nezami, H., Haskard, C., Ahokas, J., & Salminen, S. (2001). Aflatoxin B1 binding by dairy strains of lactic acid bacteria and bifidobacteria. *Journal of Dairy Science*, *84*(10), 2152–2156. doi:10.3168/jds.S0022-0302(01)74660-7 PMID:11699445

Pennacchia, C., Ercolini, D., Blaiotta, G., Pepe, O., Mauriello, G., & Villani, F. (2004). Selection of *Lactobacillus* strains from fermented sausages for their potential use as probiotics. *Meat Science*, *67*(2), 309–317. doi:10.1016/j.meatsci.2003.11.003 PMID:22061328

Perdigon, G., Alvarez, S., de Macias, M. E., Roux, M., & de Ruiz Holgado, A. P. (1990). The Oral Administration of Lactic Acid Bacteria Increase the Mucosal Intestinal Immunity in Response to Enteropathogens. *Journal of Food Protection*, *53*(5), 404–410. doi:10.4315/0362-028X-53.5.404

Pereira, E. P., Cavalcanti, R. N., Esmerino, E. A., Silva, R., Guerreiro, L. R., Cunha, R. L., ... Cruz, A. G. (2016). Effect of incorporation of antioxidants on the chemical, rheological, and sensory properties of probiotic petit suisse cheese. *Journal of Dairy Science*, *99*(3), 1762–1772. doi:10.3168/jds.2015-9701 PMID:26805976

Pessela, B. C. C., Mateo, C., Fuentes, M., Vian, A., García, J. L., Carrascosa, A. V., ... Fernández-Lafuente, R. (2003). The immobilization of a thermophilic β-galactosidase on Sepabeads supports decreases product inhibition complete hydrolysis of lactose in dairy products. *Enzyme and Microbial Technology*, *33*(2-3), 199–205. doi:10.1016/S0141-0229(03)00120-0

Petersen, B. L., Dave, R. I., McMahon, D. J., Oberg, C. J., & Broadbent, J. R. (2000). Influence of Capsular and Ropy Exopolysaccharide-Producing *Streptococcus thermophilus* on Mozzarella Cheese and Cheese Whey. *Journal of Dairy Science*, *83*(9), 1952–1956. doi:10.3168/jds.S0022-0302(00)75071-5 PMID:11003223

Peterson, S. W. (2008). Phylogenetic analysis of Aspergillus species using DNA sequences from four loci. *Mycologia*, *100*(2), 205–226. doi:10.1080/15572536.2008.11832477 PMID:18595197

Petry, S., Furlan, S., Crepeau, M.-J., Cerning, J., & Desmazeaud, M. (2000). Factors affecting exocellular polysaccharide production by *Lactobacillus delbrueckii* subsp. *bulgaricus* grown in a chemically defined medium. *Applied and Environmental Microbiology*, *66*(8), 3427–3431. doi:10.1128/AEM.66.8.3427-3431.2000 PMID:10919802

Picón, A., Gaya, P., Medina, M., & Nuñez, M. (1997). Proteinases encapsulated in stimulated release liposomes for cheese ripening. *Biotechnology Letters*, *19*(4), 345–348. doi:10.1023/A:1018350901132

Pihlanto, A., & Korhonen, H. (2014). Bioactive peptides from fermented foods and health promotion. In W. Holzapfel (Ed.), *Advances in Fermented Foods and Beverages: Improving Quality, Technologies and Health Benefits* (p. 39). Elsevier.

Pimentel, T. C., Madrona, G. S., Garcia, S., & Prudencio, S. H. (2015). Probiotic viability, physicochemical characteristics and acceptability during refrigerated storage of clarified apple juice supplemented with *Lactobacillus paracasei* ssp. *paracasei* and oligofructose in different package type. *Lebensmittel-Wissenschaft + Technologie*, *63*(1), 415–422. doi:10.1016/j.lwt.2015.03.009

Pinho, O., Peres, C., & Ferreira, I. M. (2003). Solid-phase microextraction of volatile compounds in "Terrincho" ewe cheese. Comparison of different fibers. *Journal of Chromatography. A*, *1011*(1-2), 1–9. doi:10.1016/S0021-9673(03)01066-5 PMID:14518757

Pinilla, C. M. B., Noreña, C. P. Z., & Brandelli, A. (2017). Development and characterization of phosphatidylcholine nanovesicles, containing garlic extract, with antilisterial activity in milk. *Food Chemistry*, *220*, 470–476. doi:10.1016/j.foodchem.2016.10.027 PMID:27855927

Pino, A., Hoorde, K. V., Pitino, I., Russo, N., Carpino, S., Caggia, C., & Randazzo, C. L. (2017). Survival of potential probiotic lactobacilli used as adjunct cultures on Pecorino Siciliano cheese ripening and passage through the gastrointestinal tract of healthy volunteers. *International Journal of Food Microbiology*, *252*, 42–52. doi:10.1016/j.ijfoodmicro.2017.04.012 PMID:28458191

Pintado, M. E., Da Silva, L., Fernandes, P B., Malcata, F. X. & Hogg, T. A. (1996). Microbiological and rheological studies on Portuguese kefir grains. *International of Food Science + Technology, 31*(1), 15-26.

Pinto, M. G., Franz, C. M. A. P., Schillinger, U., & Holzapfel, W. H. (2006). *Lactobacillus* spp. with *in vitro* probiotic properties from human faeces and traditional fermented products. *International Journal of Food Microbiology, 109*(3), 205–214. doi:10.1016/j.ijfoodmicro.2006.01.029 PMID:16503361

Pitino, I., Randazzo, C. L., Mandalari, G., Curto, A. L., Faulks, R. M., Marc, Y. L., ... Wickham, M. S. J. (2010). Survival of *Lactobacillus rhamnosus* strains in the upper gastrointestinal tract. *Food Microbiology, 27*(8), 1121–1127. doi:10.1016/j.fm.2010.07.019 PMID:20832693

Pitt, J. I., Cruickshank, R. H., & Leistner, L. (1986). *Penicillium commune, P. camembertii*, the origin of white cheese moulds, and the production of cyclopiazonic acid. *Food Microbiology, 3*(4), 363–371. doi:10.1016/0740-0020(86)90022-5

Pitt, J. I., & Hocking, A. D. (2009). Zygomycetes. In *Fungi and Food Spoilage* (pp. 145–168). Boston, MA: Springer. doi:10.1007/978-0-387-92207-2_6

Pitt, J. I., & Hocking, A. D. (2009a). *Penicillium* and Related Genera. In J. I. Pitt & A. D. Hocking (Eds.), *Fungi and Food Spoilage* (pp. 169–273). Boston, MA: Springer. doi:10.1007/978-0-387-92207-2_7

Pitt, J. I., & Hocking, A. D. (2009b). Primary Keys and Miscellaneous Fungi. In J. I. Pitt & A. D. Hocking (Eds.), *Fungi and Food Spoilage* (pp. 53–143). Boston, MA: Springer. doi:10.1007/978-0-387-92207-2_5

Piveteau, P. G., Condon, S., & Cogan, T. M. (1995). Interactions between lactic and propionic acid bacteria. *Le Lait, 75*(4-5), 331–343. doi:10.1051/lait:19954-525

Placier, G., Watzlawick, H., Rabiller, C., & Mattes, R. (2009). Evolved β-galactosidases from *Geobacillus stearothermophilus* with improved transgalactosylation yield for galacto-oligosaccharides production. *Applied and Environmental Microbiology, 7*(19), 6312–6321. doi:10.1128/AEM.00714-09 PMID:19666723

Pogačić, T., Kelava, N., Zamberlin, Š., Dolenčić-Špehar, I., & Samaržija, D. (2010). Methods for culture-independent identification of lactic acid bacteria in dairy products. *Food Technology and Biotechnology, 48*(1), 3–10.

Poncelet, D. (n.d.). Microencapsulation: fundamentals, methods and applications. In Surface Chemistry in Biomedical and Environmental Science (pp. 23–34). Springer Netherlands. doi:10.1007/1-4020-4741-X_3

Poonam, S., Pophaly, S. D., Tomar, S. K., De, S., & Singh, R. (2012). Multifaceted attributes of dairy propionibacteria: A review. *World Journal of Microbiology & Biotechnology, 28*(11), 3081–3095. doi:10.100711274-012-1117-z PMID:22806746

Pophaly, S. D., Pophaly, S. D., Kapila, S., Nanda, D. K., Tomar, S. K., & Singh, R. (2017). Glutathione biosynthesis and activity of dependent enzymes in food grade lactic acid bacteria harboring multidomain bifunctional fusion gene (gshF). *Journal of Applied Microbiology, 123*(1), 194–203. doi:10.1111/jam.13471 PMID:28403558

Pophaly, S. D., Singh, P., Kumar, H., Tomar, S. K., & Singh, R. (2014). Selenium enrichment of lactic acid bacteria and bifidobacteria: A functional food perspective. *Trends in Food Science & Technology, 39*(2), 135–145. doi:10.1016/j.tifs.2014.07.006

Pophaly, S. D., Singh, R., Pophaly, S. D., Kaushik, J. K., & Tomar, S. K. (2012). Current status and emerging role of glutathione in food grade lactic acid bacteria. *Microbial Cell Factories, 11*(1), 114. doi:10.1186/1475-2859-11-114 PMID:22920585

Portaccio, M., Stellato, S., Rossi, S., Bencivenga, U., Mohy Eldin, M. S., Gaeta, F. S., & Mita, D. G. (1998). Galactose competitive inhibition of β-galactosidase (*Aspergillus oryzae*) immobilized on chitosan and nylon supports. *Enzyme and Microbial Technology*, *23*(1-2), 101–106. doi:10.1016/S0141-0229(98)00018-0

Postollec, F., Falentin, H., Pavan, S., Combrisson, J., & Sohier, D. (2011). Recent advances in quantitative PCR (qPCR) applications in food microbiology. *Food Microbiology*, *28*(5), 848–861. doi:10.1016/j.fm.2011.02.008 PMID:21569926

Powell, I. B., Broome, M. C., & Limsowtin, G. K. Y. (2003). Starter cultures: General aspects. In H. Roginski, J. W. Fuquay, & P. F. Fox (Eds.), *Encyclopedia of Dairy Sciences* (pp. 261–268). London, UK: Academic Press.

Prado, M. R., Blandon, L. M., Vandenberghe, P. S., Rodrigues, C., Castro, G. R., Thomaz-Soccol, V., & Soccol, C. R. (2015). Milk kefir: Composition, microbial cultures, biological activities, and related products. *Frontiers in Microbiology*, *6*, 1–10. doi:10.3389/fmicb.2015.01177 PMID:26579086

Prescott, L. M., Harley, J. P., & Klein, D. A. (2001). Microbiology (5th ed.). McGraw-Hill Higher Education.

Pretzer, G., Snel, J., Molenaar, D., Wiersma, A., Bron, P. A., Lambert, J., ... Kleerebezem, M. (2005). Biodiversity-based identification and functional characterization of the mannose-specific adhesin of *Lactobacillus plantarum. Journal of Bacteriology*, *187*(17), 6128–6136. doi:10.1128/JB.187.17.6128-6136.2005 PMID:16109954

Punt, P. J., Dingemanse, M. A., Jacobs-Meijsing, B. J. M., Pouwels, P. H., & van den Hondel, C. A. M. J. J. (1988). Isolation and characterization of the glyceraldehyde-3-phosphate dehydrogenase gene of *Aspergillus nidulans. Gene*, *69*(1), 49–57. doi:10.1016/0378-1119(88)90377-0 PMID:3066699

Qi, M., Gu, Y., Sakata, N., Kim, D., Shirouzu, Y., Yamamoto, C., ... Inoue, K. (2004). PVA hydrogel sheet macroencapsulation for the bioartificial pancreas. *Biomaterials*, *25*(27), 5885–5892. doi:10.1016/j.biomaterials.2004.01.050 PMID:15172501

Qi, W., Chen, C., Liu, M., Yu, G., Cai, X., Guo, P., ... Mei, S. (2015). Whey protein isolate modified by transglutaminase aggregation and emulsion gel properties. *Global Conference on Polymer and Composite Materials*, 1-11. 10.1088/1757-899X/87/1/012034

Qi, W., Ma, J., Yu, W., Xie, Y., Wang, W., & Ma, X. (2006). Behavior of microbial growth and metabolism in alginate–chitosan–alginate (ACA) microcapsules. *Enzyme and Microbial Technology*, *38*(5), 697–704. doi:10.1016/j.enzmictec.2005.10.003

Quintanilla-Carvajal, M. X., Camacho-Díaz, B. H., Meraz-Torres, L. S., Chanona-Pérez, J. J., Alamilla-Beltrán, L., Jimenéz-Aparicio, A., & Gutiérrez-López, G. F. (2010). Nanoencapsulation: A New Trend in Food Engineering Processing. *Food Engineering Reviews*, *2*(1), 39–50. doi:10.100712393-009-9012-6

Quintilla, R., Kolecka, A., Casaregola, S., Daniel, H. M., Houbraken, J., Kostrzewa, M., ... Groenewald, M. (2018). MALDI-TOF MS as a tool to identify foodborne yeasts and yeast-like fungi. *International Journal of Food Microbiology*, *266*, 109–118. doi:10.1016/j.ijfoodmicro.2017.11.016 PMID:29202340

Raak, N., Jaros, D., & Rohm, H. (2017a). Cross-linking with microbial transglutaminase: Isopeptide bonds and polymer size as drivers for acid casein gel stiffness. *International Dairy Journal*, *66*, 49–55. doi:10.1016/j.idairyj.2016.10.015

Raak, N., Rohm, H., & Jaros, D. (2017b). Enzymatic cross-linking of casein facilitates gel structure weakening ınduced by overacidification. *Food Biophysics*, *12*(2), 261–268. doi:10.100711483-017-9483-6

Radošević, V., Tonković, K., Gregurek, L., Kos, B., & Šušković, J. (2007). Production of fresh probiotic cheese with addition of transglutaminase [Proizvodnja svježeg probiotičkog sira s dodatkom transglutaminaze]. *Mljekarstvo*, *57*(1), 15–29.

Ramos, C. L., Thorsen, L., Schwan, R. F., & Jespersen, L. (2013). Strain-specific probiotics properties of *Lactobacillus fermentum, Lactobacillus plantarum* and *Lactobacillus brevis* isolates from Brazilian food products. *Food Microbiology, 36*(1), 22–29. doi:10.1016/j.fm.2013.03.010 PMID:23764216

Ranadheera, C. S., Evans, C., Adams, M., & Baines, S. (2013). Production of probiotic ice cream from goat's milk and effect of packaging materials on product quality. *Small Ruminant Research, 112*(1), 174–180. doi:10.1016/j.smallrumres.2012.12.020

Ranadheera, R. D. C. S., Baines, S. K., & Adams, M. C. (2010). Importance of food in probiotic efficacy. *Food Research International, 43*(1), 1–7. doi:10.1016/j.foodres.2009.09.009

Randazzo, C. L., Torriani, S., Akkermans, A. D. L., De Vos, W. M., & Vaughan, E. E. (2002). Diversity, dynamics, and activity of bacterial communities during production of an artisanal sicilian cheese as evaluated by 16S rRNA analysis. *Applied and Environmental Microbiology, 68*(4), 1882–1892. doi:10.1128/AEM.68.4.1882-1892.2002 PMID:11916708

Rao, J., & McClements, D. J. (2011). Food-grade microemulsions, nanoemulsions and emulsions: Fabrication from sucrose monopalmitate & lemon oil. *Food Hydrocolloids, 25*(6), 1413–1423. doi:10.1016/j.foodhyd.2011.02.004

Rao, M. B., Tanksale, A. M., Ghatge, S., & Deshpande, V. V. (1998). Molecular and biotechnological aspects of microbial proteases. *Microbiology and Molecular Biology Reviews, 62*(3), 597–635. PMID:9729602

Rastall, R. A., Gibson, G. R., Gill, H. S., Guarner, F., Klaenhammer, T. R., Pot, B., ... Sanders, M. E. (2005). Modulation of the microbial ecology of the human colon by probiotics, prebiotics and synbiotics to enhance human health: An overview of enabling science and potential applications. *FEMS Microbiology Ecology, 52*(2), 145–152. doi:10.1016/j.femsec.2005.01.003 PMID:16329901

Ratomahenina, R., Chabalier, C., & Galzy, P. (1994). Concerning *Sporendonema casei* Desmazieres. *Latte, 19*(6), 616–617.

Ratomahenina, R., Chabalier, C., Galzy, P., & Dieu, B. (1995). Study of *Chrysosporium sulfureum*, the mould responsible for Fleur jaune on Saint-Nectaire cheese. *Milchwissenschaft. Milk Science International, 50*(5), 266–267.

Ratomahenina, R., van den Boom, S., Galzy, P., & Dieu, B. (1995). Study of growth parameters of *Cylindrocarpon* sp., a mould isolated from Saint-Nectaire cheese. *Chemie, Mikrobiologie, Technologie. Der Lebensmittel, 17*, 169–171.

Rattray, F. P. (2003). Secondary cultures. In H. Roginski, J. W. Fuquay, & P. F. Fox (Eds.), *Encyclopedia of Dairy Sciences* (pp. 275–281). London, UK: Academic Press.

Rattray, F. P., & Fox, P. F. (1997). Purification and characterization of an intracellular esterase from Brevibacterium linens ATCC 9174. *International Dairy Journal, 7*(4), 273–278. doi:10.1016/S0958-6946(97)00013-7

Rattray, F. P., & Fox, P. F. (1999). Aspects of Enzymology and Biochemical Properties of Brevibacterium linens Relevant to Cheese Ripening: A Review1. *Journal of Dairy Science, 82*(5), 891–909. doi:10.3168/jds.S0022-0302(99)75308-7 PMID:10342227

Raya, R. R., Varey, P., Oot, R. A., Dyen, M. R., Callaway, T. R., Edrington, T. S., ... Brabban, A. D. (2006). Isolation and characterization of a new T-even bacteriophage, CEV1, and determination of its potential to reduce *Escherichia coli* O157:H7 levels in sheep. *Applied and Environmental Microbiology, 72*(9), 6405–6410. doi:10.1128/AEM.03011-05 PMID:16957272

Razvi, A. (2007). *A study of the hemin-stimulated respirative growth kinetics of the lactis using batch cultivation* (Master's Thesis). University of Ottawa (Canada). Masters Abstracts International, Volume: 47-06, page: 3670. database.

Reale, A., Di Renzo, T., Zotta, T., Preziuso, M., Boscaino, F., Ianniello, R., ... Coppola, R. (2016a). Effect of respirative cultures of *Lactobacillus casei* on model sourdough fermentation. *Lebensmittel-Wissenschaft + Technologie, 73*, 622–629. doi:10.1016/j.lwt.2016.06.065

Reale, A., Ianniello, R. G., Ciocia, F., Di Renzo, T., Boscaino, F., Ricciardi, A., ... McSweeney, P. L. H. (2016b). Effect of respirative and catalase-positive *Lactobacillus casei* adjuncts on the production and quality of Cheddar-type cheese. *International Dairy Journal, 63*, 78–87. doi:10.1016/j.idairyj.2016.08.005

Refstie, S., Sahlström, S., Bråthen, E., Baeverfjord, G., & Krogedal, P. (2005). Lactic acid fermentation eliminates indigestible carbohydrates and antinutritional factors in soybean meal for Atlantic salmon (Salmo salar). *Aquaculture (Amsterdam, Netherlands), 246*(1), 331–345. doi:10.1016/j.aquaculture.2005.01.001

Reid, G. (2008). Probiotics and prebiotics e progress and challenges. *International Dairy Journal, 18*(10-11), 969–975. doi:10.1016/j.idairyj.2007.11.025

Reid, G. (2015). The growth potential for dairy probiotics. *International Dairy Journal, 49*, 16–22. doi:10.1016/j.idairyj.2015.04.004

Reisinger, H., Seidel, H., Tschesche, P., & Hammes, W. P. (1980). The effect of nisin on murein synthesis. *Archives of Microbiology, 127*(3), 187–193. doi:10.1007/BF00427192 PMID:6255884

Reps, A. (1993). Bacterial surface-ripened cheeses. In P. F. Fox (Ed.), *Cheese: Chemistry, Physics and Microbiology* (2nd ed.; Vol. 2, pp. 137–172). Major Cheese Groups. doi:10.1007/978-1-4615-2648-3_5

Research and Markets. (2014). *Global functional food and nutraceuticals market (2014-2020) - by type (foods, beverages, supplements); benefits (health and wellness, disease prevention, fitness, beauty); origin & ingredient.* Author.

Retteger, L. F., Levy, M. N., Weinstein, L., & Weiss, J. E. (1935). *Lactobacillus acidophilus and its therapeutic Application.* New Haven, CT: Yale University Press.

Rezaïki, L., Cesselin, B., Yamamoto, Y., Vido, K., Van West, E., Gaudu, P., & Gruss, A. (2004). Respiration metabolism reduces oxidative and acid stress to improve long-term survival of *Lactococcus lactis. Molecular Microbiology, 53*(5), 1331–1342. doi:10.1111/j.1365-2958.2004.04217.x PMID:15387813

Rivera Espinoza, Y., & Gallardo Navarro, Y. (2010). Non-dairy probiotic products. *Food Microbiology, 27*(1), 1–11. doi:10.1016/j.fm.2008.06.008 PMID:19913684

Roberfroid, M. B. (2007). Concepts and strategy of functional food science: The European perspective. *The American Journal of Clinical Nutrition, 71*, 1660–1664. PMID:10837311

Robin, S., Petrov, K., Dintinger, T., Kujumdzvieva, A., Tellier, C., & Dion, M. (2003). Comparison of three microbial hosts for the expression of an active catalytic scFv. *Molecular Immunology, 39*(12), 729–738. doi:10.1016/S0161-5890(02)00253-5 PMID:12531284

Rocha, D. M. U. P., Martins, J. F. L., Santos, T. S. S., & Moreira, A. V. B. (2014). Labneh with probiotic properties produced from kefir: Development and sensory evaluation. *Food Science and Technology (Campinas), 34*(4), 694–700. doi:10.1590/1678-457x.6394

Rodriguez-Nogales, J. M. (2005). Enzymatic cross-linking of ewe's milk proteins by transglutaminase. *European Food Research and Technology, 221*(5), 692–699. doi:10.100700217-005-0041-y

Rodriguez-Nogales, J. M. (2006). Enhancement of transglutaminase-induced protein cross-linking by preheat treatment of cows' milk: A statistical approach. *International Dairy Journal, 16*(1), 26–32. doi:10.1016/j.idairyj.2005.01.003

Rodríguez-Sánchez, B., Alcalá, L., Marín, M., Ruiz, A., Alonso, E., & Bouza, E. (2016). Evaluation of MALDI-TOF MS (Matrix-Assisted Laser Desorption-Ionization Time-of-Flight Mass Spectrometry) for routine identification of anaerobic bacteria. *Anaerobe*, *42*, 101–107. doi:10.1016/j.anaerobe.2016.09.009 PMID:27702604

Rodtong, S., & Tannock, G. W. (1993). Differentiation of Lactobacillus strains by ribotyping. *Applied and Environmental Microbiology*, *59*(10), 3480–3484. PMID:7504432

Roger, T., Ngouné Léopold, T., & Carl Moses Funtong, M. (2015). Nutritional Properties and Antinutritional Factors of Corn Paste (Kutukutu) Fermented by Different Strains of Lactic Acid Bacteria. *International Journal of Food Sciences*. PMID:26904660

Rolim, F. R. L., dos Santos, K. M. O., de Barcelos, S. C., do Egito, A. S., Ribeiro, T. S., da Conceiçao, M. L., ... Queiroga, R. C. R. E. (2015). Survival of *Lactobacillus rhamnosus* EM1107 in simulated gastrointestinal conditions and its inhibitory effect against pathogenic bacteria in semi-hard goat cheese. *Lebensmittel-Wissenschaft + Technologie*, *63*(2), 807–813. doi:10.1016/j.lwt.2015.05.004

Roller, S. (2003). Introduction. In S. Roller (Ed.), *Natural antimicrobials for the minimal processing of foods* (pp. 1–10). Cambridge, UK: Woodhead Publishing Ltd.

Romeih, E., Abdel-Hamid, M., & Awad, A. A. (2014). The addition of buttermilk powder and transglutaminase improves textural and organoleptic properties of fat-free buffalo yogurt. *Dairy Science & Technology*, *94*(3), 297–309. doi:10.100713594-014-0163-8

Romeih, E., & Walker, G. (2017). Recent advances on microbial transglutaminase and dairy application. *Trends in Food Science & Technology*, *62*, 133–140. doi:10.1016/j.tifs.2017.02.015

Ronaghi, M., Karamohamed, S., Pettersson, B., Uhlén, M., & Nyrén, P. (1996). Real-Time DNA Sequencing Using Detection of Pyrophosphate Release. *Analytical Biochemistry*, *242*(1), 84–89. doi:10.1006/abio.1996.0432 PMID:8923969

Rönka, E., Malinen, E., Saarela, M., Rinta-Koski, M., Aarnikunnas, J., & Palva, A. (2003). Probiotic and milk technological properties of *Lactobacillus brevis*. *International Journal of Food Microbiology*, *83*(1), 63–74. doi:10.1016/S0168-1605(02)00315-X PMID:12672593

Ropars, J., Cruaud, C., Lacoste, S., & Dupont, J. (2012). A taxonomic and ecological overview of cheese fungi. *International Journal of Food Microbiology*, *155*(3), 199–210. doi:10.1016/j.ijfoodmicro.2012.02.005 PMID:22381457

Ropars, J., López-Villavicencio, M., Dupont, J., Snirc, A., Gillot, G., Coton, M., ... Giraud, T. (2014). Induction of sexual reproduction and genetic diversity in the cheese fungus *Penicillium roqueforti*. *Evolutionary Applications*, *7*(4), 433–441. doi:10.1111/eva.12140 PMID:24822078

Ropars, J., Rodríguez de la Vega, R. C., López-Villavicencio, M., Gouzy, J., Sallet, E., Dumas, É., ... Giraud, T. (2015). Adaptive horizontal gene transfers between multiple cheese-associated fungi. *Current Biology*, *25*(19), 2562–2569. doi:10.1016/j.cub.2015.08.025 PMID:26412136

Rösch, C., & Bothe, H. (2005). Improved assessment of denitrifying, N2 -fixing, and total-community bacteria by terminal restrictionfragment length polymorphism analysis using multiple restriction enzymes. *Applied and Environmental Microbiology*, *71*(4), 2026–2035. doi:10.1128/AEM.71.4.2026-2035.2005 PMID:15812035

Rose, R. P., Sporns, P., Dodd, H. M., Gasson, M. J., Mellon, F. A., & McMullen, L. M. (2003). Involvement of dehydroalanine and dehydrobutyrine in the addition of glutathione to nisin. *Journal of Agricultural and Food Chemistry*, *51*(10), 3174–3178. doi:10.1021/jf026022h PMID:12720411

Rossa, P. N., De Sá, E. M. F., Burin, V. M., & Bordignon-Luiz, T. M. (2011). Optimization of microbial transglutaminase activity in ice cream using response surface methodology. *LWT-Food Sceince and Technology, 44*, 29-34.

Rossello-Mora, R., & Amann, R. (2001). The species concept for procaryotes. *FEMS Microbiology Ecology, 25*(1), 39–67. doi:10.1111/j.1574-6976.2001.tb00571.x PMID:11152940

Ross, R. P., Desmond, C., Fitzgerald, G. F., & Stanton, C. (2005). Overcoming the technological hurdles in the development of probiotic foods. *Journal of Applied Microbiology, 98*(6), 1410–1417. doi:10.1111/j.1365-2672.2005.02654.x PMID:15916653

Routray, W., & Mishra, H. N. (2011). Scientific and technical aspects of yogurt aroma and taste: A Review. *Comprehensive Reviews in Food Science and Food Safety, 10*(4), 208–220. doi:10.1111/j.1541-4337.2011.00151.x

Ruas-Madiedo, P., Hugenholtz, J., & Zoon, P. (2002). An overview of the functionality of exopolysaccharides produced by lactic acid bacteria. *International Dairy Journal, 12*(2), 163–171. doi:10.1016/S0958-6946(01)00160-1

Ruijschop, R. M. A. J., Boelrijk, A. E. M., & te Giffel, M. C. (2008). Satiety effects of a dairy beverage fermented with propionic acid bacteria. *International Dairy Journal, 18*(9), 945–950. doi:10.1016/j.idairyj.2008.01.004

Ruiz-Barba, J. L., Cathcart, D. P., Warner, P. J., & Jimenez-Diaz, R. (1994). Use of Lactobacillus plantarum LPCO10, a bacteriocin producer, as a starter culture in Spanish-style green olive fermentations. *Applied and Environmental Microbiology, 60*(6), 2059. PMID:16349291

Russell, N. J. (2002). Bacterial membranes: The effects of chill storage and food processing. An overview. *International Journal of Food Microbiology, 79*(1), 27–34. doi:10.1016/S0168-1605(02)00176-9 PMID:12382682

Saavedra, L., & Taranto, M. P., Sesma, F., & Front de Valdez, G. (2003). Homade traditional cheeses fort he isolation of probiotic Enterococcus faecium strains. *International Journal of Food Microbiology, 2-3*(88), 242–245. PMID:14596996

Sagdic, O., Ozturk, I., Cankurt, H., & Tornuk, F. (2012). Interaction between some phenolic compounds and probiotic bacterium in functional ice cream production. *Food and Bioprocess Technology, 5*(8), 2964–2971. doi:10.100711947-011-0611-x

Sakamoto, H., Kumizawa, Y., & Motoki, M. (1994). Stregth of protein gels prepared with microbial transglutaminase as related to reaction conditions. *Journal of Food Science, 59*(4), 866–871. doi:10.1111/j.1365-2621.1994.tb08146.x

Salazar, N., Ruas-Madiedo, P., Kolida, S., Collins, M., Rastall, R., Gibson, G., & Clara, G. (2009). Exopolysaccharides produced by *Bifidobacterium longum* IPLA E44 and *Bifidobacterium animalis* subsp. *lactis* IPLA R1 modify the composition and metabolic activity of human faecal microbiota in pH-controlled batch cultures. *International Journal of Food Microbiology, 135*(3), 260–267. doi:10.1016/j.ijfoodmicro.2009.08.017 PMID:19735956

Salem, M., Fathi, F., & Awad, R. (2005). Production of probiotic ice cream. *Polish Journal of Food and Nutrition Sciences, 14*(3), 267–271.

Salers. (2015, February 20). Retrieved from https://commons.wikimedia.org/wiki/File:Salers_(fromage)_01.jpg

Salinas-Valdés, A., Millán, J. D. R., Serna-Saldívar, S. O., & Chuck-Hernández, C. (2015). Yield and textural characteristics of Panela cheeses produced with dairy-vegetable protein (soybean or peanut) blends supplemented with transglutaminase. *Journal of Food Science, 80*(12), S2950–S2956. doi:10.1111/1750-3841.13126 PMID:26550775

Samaržija, D., Tudor, M., Prtilo, T., Špehar, I. D., Zamberlin, S., & Havranek, J. (2009). Probiotic bacteria in prevention and treatment of diarrhea. *Mljekarstvo, 59*(1), 28–32.

Samson, R. A., Eckardt, C., & Orth, R. (1977). The taxonomy of *Penicillium* species from fermented cheeses. *Antonie van Leeuwenhoek*, *43*(3), 341–350. doi:10.1007/BF02313761 PMID:603235

Samson, R. A., Houbraken, J., Thrane, U., Frisvad, J. C., & Andersen, B. (2010). *Food and indoor fungi - CBS laboratory manual series*. Utrecht, The Netherlands: CBS-KNAW Fungal Biodiversity Centre.

Sander, J. D., & Joung, J. K. (2014). CRISPR-Cas systems for editing, regulating and targeting genomes. *Nature Biotechnology*, *32*(4), 347–355. doi:10.1038/nbt.2842 PMID:24584096

Sanger, F., Nicklen, S., & Coulson, A. R. (1977). DNA sequencing with chain-terminating inhibitors. *Proceedings of the National Academy of Sciences of the United States of America*, *74*(12), 5463–5467. doi:10.1073/pnas.74.12.5463 PMID:271968

Sani, A. M., Ehsani, M. R., & Asadi, M. M. (2005). Effect of *Propionibacterium shermanii* metabolites on sensory properties and shelf life of UF-Feta cheese. *Nutrition & Food Science*, *35*(2), 88–94. doi:10.1108/00346650510585877

Şanlıdere-Aloğlu, H., & Öner, Z. (2013). The effect of treating goat's milk with transglutaminase on chemical, structural and sensory properties of labneh. *Small Ruminant Research*, *109*(1), 31–37. doi:10.1016/j.smallrumres.2012.10.005

Şanlı, T., Sezgin, E., Deveci, O., Şenel, E., & Benli, M. (2011). Effect of using transglutaminase on physical, chemical and sensory properties of set-type yoghurt. *Food Hydrocolloids*, *25*(6), 1477–1481. doi:10.1016/j.foodhyd.2010.09.028

Şanlı, T., Sezgin, E., Şenel, E., & Benli, M. (2013). The effect of transglutaminase on some physicochemical and sensory properties of the Turkish drinking yoghurt Ayran. *International Journal of Dairy Technology*, *66*(3), 410–415. doi:10.1111/1471-0307.12045

Şanlı, T., Sezgin, E., Şenel, E., & Benli, M. (2014). The effects of transglutaminase, exopolysaccharide-producing starter culture, sensory and texture properties of low-fat set yoghurt. *International Journal of Dairy Technology*, *67*, 237–244. doi:10.1111/1471-0307.12127

Santivarangkna, C., Kulozik, U., & Foerst, P. (2007). Alternative drying processes for the industrial preservation of lactic acid starter cultures. *Biotechnology Progress*, *23*(2), 302–315. doi:10.1021/bp060268f PMID:17305363

Santivarangkna, C., Kulozik, U., & Foerst, P. (2008). Inactivation mechanisms of lactic acid starter cultures preserved by drying processes. *Journal of Applied Microbiology*, *105*(1), 1–13. doi:10.1111/j.1365-2672.2008.03744.x PMID:18266696

Saravanan, C., Gopu, V., & Shetty, P. H. (2015). Diversity and functional characterization of microflora isolated from traditional fermented food idli. *Journal of Food Science and Technology*, *52*(11), 7425–7432. doi:10.100713197-015-1791-6

Sarkar, N. K., Clarke, D. D., & Waelsch, H. (1957). An enzymatically catalyzed incorporation of amines into proteins. *Biochimica et Biophysica Acta*, *25*, 451–452. doi:10.1016/0006-3002(57)90512-7 PMID:13471608

Sarvari, F., Mortazavian, A. M., & Fazeli, M. R. (2014). Biochemical characteristics and viability of probiotic and yogurt bacteria in yogurt during the fermentation and refrigerated storage. *Applied Food Biotechnology*, *1*(1), 55–61.

Sayadi, A., Madadlou, A., & Khosrowshahi, A. (2013). Enzymatic cross-linking of whey proteins in low fat Iranian white cheese. *International Dairy Journal*, *29*(2), 88–92. doi:10.1016/j.idairyj.2012.10.006

Scheinbach, S. (1998). Probiotics: Functionality and Commercial Status. *Biotechnology Advances*, *16*(3), 581–608. doi:10.1016/S0734-9750(98)00002-0 PMID:14538145

Schmid, M., Sängerlaub, S., Wege, L., & Stäbler, A. (2014). Properties of transglutaminase crosslinked whey protein isolate coatings and cast films. *Packaging Technology & Science*, *27*(10), 799–817. doi:10.1002/pts.2071

Schokker, E. P., & van Boekel, M. A. J. S. (1997). Production, purification and partial characterization of extracellular proteinase from Pseudomonas fluorescens 22 F. *International Dairy Journal*, *7*(4), 265–271. doi:10.1016/S0958-6946(97)00008-3

Schornsteiner, E., Mann, E., Bereuter, O., Wagner, M., & Schmitz-Esser, S. (2014). Cultivation-independent analysis of microbial communities on Austrian raw milk hard cheese rinds. *International Journal of Food Microbiology*, *180*, 88–97. doi:10.1016/j.ijfoodmicro.2014.04.010 PMID:24794620

Schorsch, C., Carrie, H., & Norton, I. T. (2000). Cross-linking casein micelles by a microbial transglutaminase: Influence of cross-links in acid-induced gelation. *International Dairy Journal*, *11*, 785–793.

Schuller, F., Benz, R., & Sahl, H. G. (1989). The peptide antibiotic subtilin acts by formation of voltage-dependent multi-state pores in bacterial and artificial membranes. *European Journal of Biochemistry*, *182*(1), 181–186. doi:10.1111/j.1432-1033.1989.tb14815.x PMID:2471644

Schweiger, M., Eichmann, T. O., Taschler, U., Zimmermann, R., Zechner, R., & Lass, A. (2014). Measurement of Lipolysis. *Methods in Enzymology*, *538*, 171–193. doi:10.1016/B978-0-12-800280-3.00010-4 PMID:24529439

Scott, P. M., & Kennedy, B. P. C. (1976). Analysis of blue cheese for roquefortine and other alkaloids from *Penicillium roqueforti*. *Journal of Agricultural and Food Chemistry*, *24*(4), 865–868. doi:10.1021/jf60206a028 PMID:956545

Scott, P., & Kanhere, S. (1979). Instability of PR toxin in blue cheese. *Journal - Association of Official Analytical Chemists*, *62*(1), 141–147. PMID:422496

Seibert, T. M. (2010). Protective culture eliminates residual oxygen. *Die Fleischwirtschaft (Frankfurt)*, *90*, 59–61.

Selle, K. M., Klaenhammer, T. R., & Russell, W. M. (2014). Lactobacillus acidophilus. In C. A. Batt & M. L. Tortorello (Eds.), *Encyclopedia of Food Microbiology* (vol. 2, pp. 412–417). Elsevier. doi:10.1016/B978-0-12-384730-0.00179-8

Selmer-Olsen, E., Birkeland, S. E., & Sørhaug, T. (1999). Effect of protective solutes on leakage from and survival of immobilized *Lactobacillus* subjected to drying, storage and rehydration. *Journal of Applied Microbiology*, *87*(3), 429–437. doi:10.1046/j.1365-2672.1999.00839.x PMID:10540246

Senanayake, S. A., Fernando, S., Bamunuarachchi, A., & Arsekularatne, M. (2013). Application of *Lactobacillus acidophilus* (LA 5) strain in fruit-based ice cream. *Food Science & Nutrition*, *1*(6), 428–431. doi:10.1002/fsn3.66 PMID:24804052

Seneweera, S., & Kailasapathy, K. (2011). Impact of microencapsulated peptidase (*Aspergillus oryzae*) on Cheddar cheese proteolysis and its biologically active peptide profile. *Protein and Peptide Letters*, *18*(7), 741–746. doi:10.2174/092986611795446076 PMID:21342096

Seo, B. J., Rather, I. A., Kumar, V. J. R., Choi, U. H., Moon, M. R., Lim, J. H., & Park, Y. H. (2012). Evaluation of *Leuconostoc mesenteroides* YML003 as a probiotic against low-pathogenic avian influenza (H9N2) virus in chickens. *Journal of Applied Microbiology*, *113*(1), 163–171. doi:10.1111/j.1365-2672.2012.05326.x PMID:22548634

Serafini, M., Stanzione, A., & Foddai, S. (2012). Functional foods: Traditional use and European legislation. *International Journal of Food Sciences and Nutrition*, *63*(sup1), 7–9. doi:10.3109/09637486.2011.637488 PMID:22117621

Servin, A. L., & Coconnier, M.-H. (2003). Adhesion of probiotic strains to the intestinal mucosa and interaction with pathogens. *Best Practice & Research. Clinical Gastroenterology*, *17*(5), 741–754. doi:10.1016/S1521-6918(03)00052-0 PMID:14507585

Seth, S. D., & Maulik, M. (2011). Probiotics: A Pharmacologist's Perspective. In G. B. Nair & Y. Takeda (Eds.), *Probiotic Foods In Health And Disease* (pp. 42–48). CRC Press. doi:10.1201/b10770-6

Settanni, L., & Corsetti, A. (2007). The use of multiplex PCR to detect and differentiate food- and beverage-associated microorganisms: A review. *Journal of Microbiological Methods*, *69*(1), 1–22. doi:10.1016/j.mimet.2006.12.008 PMID:17280731

Settanni, L., & Moschetti, G. (2010). Non-starter lactic acid bacteria used to improve cheese quality and provide health benefits. *Food Microbiology*, *27*(6), 691–697. doi:10.1016/j.fm.2010.05.023 PMID:20630311

Shahani, K. M., & Ayebo, A. D. (1980). Role of dietary lactobacilli in gastrointestinal microecology. *The American Journal of Clinical Nutrition*, *33*(11), 2448–2457. doi:10.1093/ajcn/33.11.2448 PMID:6449143

Shah, N. P. (2007). Functional cultures and health benefits. *International Dairy Journal*, *17*(11), 1262–1277. doi:10.1016/j.idairyj.2007.01.014

Shah, N. P., & Lankaputhra, W. E. V. (1997). Improving viability of Lactobacillus acidophilus and Bifidobacterium spp. in yogurt. *International Dairy Journal*, *7*(5), 349–356. doi:10.1016/S0958-6946(97)00023-X

Shakila Begam, M., Pradeep, S. F., & Pradeep, B. V. (2012). Production, purification, characterization and applications of lipase from *Serratia marcescens MBB05. Asian Journal of Pharmaceutical and Clinical Research*, *5*(suppl.4), 237–245.

Sharma, R., Lorenzen, P. C., & Qvist, K. B. (2001). Influence of transglutaminase treatment of skim milk on the formation of ε-(γ-glutamyl)lysine and the susceptibility of individual proteins towards crosslinking. *International Dairy Journal*, *11*(10), 785–793. doi:10.1016/S0958-6946(01)00096-6

Sharma, R., Zakora, M., & Qvist, K. B. (2002). Susceptibility of an industrial α-lactalbumin concentrate to cross-linking by microbial transglutaminase. *International Dairy Journal*, *12*(12), 1005–1012. doi:10.1016/S0958-6946(02)00122-X

Sheehan, J. J., Wilkinson, M. G., & McSweeney, P. L. H. (2008). Influence of processing and ripening parameters on starter, non-starter and propionic acid bacteria and on the ripening characteristics of semi-hard cheeses. *International Dairy Journal*, *18*(9), 905–917. doi:10.1016/j.idairyj.2007.11.024

Sheehan, V. M., Ross, P., & Fitzgerald, G. F. (2007). Assessing the acid tolerance and the technological robustness of probiotic cultures for fortification in fruit juices. *Innovative Food Science & Emerging Technologies*, *8*(2), 279–284. doi:10.1016/j.ifset.2007.01.007

Sheehan, V. M., Sleator, R. D., Fitzgerald, G. F., & Hill, C. (2006). Heterologous expression of BetL, a betaine uptake system, enhances the stress tolerance of *Lactobacillus salivarius* UCC118. *Applied and Environmental Microbiology*, *72*(3), 2170–2177. doi:10.1128/AEM.72.3.2170-2177.2006 PMID:16517668

Shekh, S. L., Dave, J. M., & Vyas, B. R. M. (2016). Characterization of *Lactobacillus plantarum* strains for functionality, safety and γ-amino butyric acid production. *Lebensmittel-Wissenschaft + Technologie*, *74*, 234–241. doi:10.1016/j.lwt.2016.07.052

Shendure, J. (2005). Accurate Multiplex Polony Sequencing of an Evolved Bacterial Genome. *Science*, *309*(5741), 1728–1732. doi:10.1126cience.1117389 PMID:16081699

Sheng, H., Knecht, H. J., Kudva, I. T., & Hovde, C. J. (2006). Application of bacteriophages to control intestinal *Escherichia coli* O157:H7 levels in ruminants. *Applied and Environmental Microbiology*, *72*(8), 5359–5366. doi:10.1128/AEM.00099-06 PMID:16885287

Sheu, T.-Y., Marshall, R. T., & Heymann, H. (1993). Improving Survival of Culture Bacteria in Frozen Desserts by Microentrapment. *Journal of Dairy Science*, *76*(7), 1902–1907. doi:10.3168/jds.S0022-0302(93)77523-2 PMID:8345127

Shi, P., He, P., Teh, T. K. H., Morsi, Y. S., & Goh, J. C. H. (2011). Parametric analysis of shape changes of alginate beads. *Powder Technology*, *210*(1), 60–66. doi:10.1016/j.powtec.2011.02.023

Shi, W., Li, Y., Gao, X., & Fu, R. (2016). Improvement of the respiration efficiency of *Lactococcus lactis* by decreasing the culture pH. *Biotechnology Letters*, *38*(3), 495–501. doi:10.100710529-015-1999-6 PMID:26585330

Sieber, R., Collomb, M., Aeschlimann, A., & Eyer, H. (2004). Impact of microbial cultures on conjugated linoleic acid in dairy products – a review. *International Dairy Journal*, *14*(1), 1–15. doi:10.1016/S0958-6946(03)00151-1

Siezen, R. J., Bayjanov, J. R., Felis, G. E., van der Sijde, M. R., Starrenburg, M., Molenaar, D., ... van Hylckama Vlieg, J. E. (2011). Genome-scale diversity and niche adaptation analysis of *Lactococcus lactis* by comparative genome hybridization using multi-strain arrays. *Microbial Biotechnology*, *4*(3), 383–402. doi:10.1111/j.1751-7915.2011.00247.x PMID:21338475

Siezen, R. J., Bayjanov, J., Renckens, B., Wels, M., van Hijum, S. A., Molenaar, D., & van Hylckama Vlieg, J. E. (2010). Complete genome sequence of *Lactococcus lactis* subsp. *lactis* KF147, a plant-associated lactic acid bacterium. *Journal of Bacteriology*, *192*(10), 2649–2650. doi:10.1128/JB.00276-10 PMID:20348266

Silva, H. L. A., Balthazar, C. F., Esmerino, E. A., Vieira, A. H., Cappato, L. P., Neto, R. P. C., ... Cruz, A. G. (2017). Effect of sodium reduction and flavor enhancer addition on probiotic Prato cheese processing. *Food Research International*, *99*, 247–255. doi:10.1016/j.foodres.2017.05.018 PMID:28784481

Silva, J., Freixo, R., Gibbs, P., & Teixeira, P. (2011). Spray-drying for the production of dried cultures. *International Journal of Dairy Technology*, *64*(3), 321–335. doi:10.1111/j.1471-0307.2011.00677.x

Simon, S. M., & Blobel, G. (1992). Signal peptides open protein-conducting channels in *Escherichia coli*. *Cell*, *69*(4), 677–684. doi:10.1016/0092-8674(92)90231-Z PMID:1375130

Simons, G., Rutten, G., Homes, M., & de Vos, W. M. (1988). Production of bovine prochymosin by lactic acid bacteria. In *Proceedings 2nd Netherlands Biotechnology Congress*, 183-187.

Simova, E., Beshkova, D., Angelov, A., Hristozov, T., Frengova, G., & Spasov, Z. (2002). Lactic acid bacteria and yeasts in kefir grains and kefir made from them. *Journal of Industrial Microbiology & Biotechnology*, *28*(1), 1–6. doi:10.1038j/jim/7000186 PMID:11938463

Simpson, P., Stanton, C., Fitzgerald, G., & Ross, R. (2005). Intrinsic tolerance of *Bifidobacterium* species to heat and oxygen and survival following spray drying and storage. *Journal of Applied Microbiology*, *99*(3), 493–501. doi:10.1111/j.1365-2672.2005.02648.x PMID:16108790

Singhal, N., Kumar, M., Kanaujia, P. K., & Virdi, J. S. (2015). MALDI-TOF mass spectrometry: An emerging technology for microbial identification and diagnosis. *Frontiers in Microbiology*, *6*(AUG), 1–16. PMID:26300860

Singh, T. K., Drake, M. A., & Cadwallader, K. R. (2003). Flavor of Cheddar Cheese: A Chemical and Sensory Perspective. *Comprehensive Reviews in Food Science and Food Safety*, *2*(4), 166–189. doi:10.1111/j.1541-4337.2003.tb00021.x

Sipahioglu, H. N., & Heperkan, D. (2000). Lipolytic activity of *Trichothecium roseum* on hazelnut. *Food Microbiology*, *17*(4), 401–405. doi:10.1006/fmic.2000.0328

Sir'o, I., Kapolna, E., Kapolna, B., & Lugasi, A. (2008). Functional food. Product development, marketing and consumer acceptance-a review. *Appetite*, *51*(3), 456–467. doi:10.1016/j.appet.2008.05.060 PMID:18582508

Sisto, A., & Lavermicocca, P. (2012). Suitability of a probiotic *Lactobacillus paracasei* strain as a starter culture in olive fermentation and development of the innovative patented product "probiotic table olives". *Frontiers in Microbiology*, *3*(174), 1–5. PMID:22586426

Smiddy, M. A., Martin, J. E. G. H., Kelly, A. L., de Kruif, C. G., & Huppertz, T. (2006). Stability of casein micelles cross-linked by transglutaminase. *Journal of Dairy Science*, *89*(6), 1906–1914. doi:10.3168/jds.S0022-0302(06)72258-5 PMID:16702254

Smid, E. J., Erkus, O., Spus, M., Wolkers-Rooijackers, J. C., Alexeeva, S., & Kleerebezem, M. (2014). Functional implications of the microbial community structure of undefined mesophilic starter cultures. *Microbial Cell Factories*, *13*(Suppl 1), S2. doi:10.1186/1475-2859-13-S1-S2 PMID:25185941

Smid, E. J., & Hugenholtz, J. (2010). Functional genomics for food fermentation processes. *Annual Review of Food Science and Technology*, *1*(1), 497–519. doi:10.1146/annurev.food.102308.124143 PMID:22129346

Smit, G., Smit, B. A., & Engels, W. J. M. (2005). Flavour formation by lactic acid bacteria and biochemical flavour profiling of cheese products. *FEMS Microbiology Reviews*, *29*(3), 591–610. doi:10.1016/j.fmrre.2005.04.002 PMID:15935512

Smith, W. M. (2010). Molecular Mechanisms of Stress Resistance in Lactococcus lactis. Current Research, Technology and Education Topics in Applied Microbiology and Microbial Biotechnology, 2, 1106-1118.

Snow, D. (1949). The Germination of mould spores at controlled humidities. *Annals of Applied Biology*, *36*(1), 1–13. doi:10.1111/j.1744-7348.1949.tb06395.x PMID:18116920

Solomons, N. W. (2002). Fermentation, fermented foods and lactose intolerance. *European Journal of Clinical Nutrition*, *56*(S4), S50–S55. doi:10.1038j.ejcn.1601663 PMID:12556948

Solopova, A., Bachmann, H., Teusink, B., Kok, J., Neves, A. R., & Kuipers, O. P. (2012). A specific mutation in the promoter region of the silent cel cluster accounts for the appearance of lactose-utilizing *Lactococcus lactis* MG1363. *Applied and Environmental Microbiology*, *78*(16), 5612–5621. doi:10.1128/AEM.00455-12 PMID:22660716

Song, C. L., & Zhao, X. H. (2013). Rheological, gelling and emulsifying properties of a glycosylated and cross-linked caseinate generated by transglutaminase. *International Journal of Food Science & Technology*, *48*(12), 2595–2602. doi:10.1111/ijfs.12255

Songisepp, E., Kullisaar, T., Hütt, P., Elias, P., Brilene, T., Zilmer, M., & Mikelsaar, M. (2004). A new probiotic cheese with antioxidative and antimicrobial activity. *Journal of Dairy Science*, *87*(7), 2017–2023. doi:10.3168/jds.S0022-0302(04)70019-3 PMID:15328213

Song, M., Park, S., Lee, H., Min, B., Jung, S., Park, S., ... Oh, S. (2015). Effect of *Lactobacillus acidophilus* NS1 on plasma cholesterol levels in diet-induced obese mice. *Journal of Dairy Science*, *98*(3), 1492–1501. doi:10.3168/jds.2014-8586 PMID:25468703

Song, S., Lee, S. J., Park, D. J., Oh, S., & Lim, K. T. (2016). The anti-allergic activity of *Lactobacillus plantarum* L67 and its application to yogurt. *Journal of Dairy Science*, *99*(12), 9372–9382. doi:10.3168/jds.2016-11809 PMID:27743673

Soukoulis, C., Fisk, I. D., & Bohn, T. (2014). Ice cream as a vehicle for incorporating health-promoting ingredients: Conceptualization and overview of quality and storage stability. *Comprehensive Reviews in Food Science and Food Safety*, *13*(4), 627–655. doi:10.1111/1541-4337.12083

Sousa, M. J., & Malcata, F. X. (1997). Effects of processing conditions on the caseinolytic activity of crude extracts of Cynara cardunculus L. *Food Science & Technology International*, *2*(4), 255–263. doi:10.1177/108201329600200408

Spinnier, H. E., & Gripon, J. C. (2004). Surface mould-ripened cheeses. In P. F. Fox, P. L. H. McSweeney, T. M. Cogan, & T. P. Guinee (Eds.), *Cheese: Chemistry, Physics and Microbiology* (3rd ed.; Vol. 2, pp. 157–174). Major Cheese Groups.

Spohner, S. C., Schaum, V., Quitmann, H., & Czernak, P. (2016). Kluyveromyces lactis: An emerging tool in biotechnology. *Journal of Biotechnology*, *222*, 104–116. doi:10.1016/j.jbiotec.2016.02.023 PMID:26912289

Sridevi, N., Vishwe, P., & Prabhune, A. (2009). Hypocholesteremic effect of bile salt hydrolase from *Lactobacillus buchneri* ATCC 4005. *Food Research International*, *42*(4), 516–520. doi:10.1016/j.foodres.2009.02.016

Stadhouders, J. (1986). The control of cheese starter activity. *Netherlands Milk and Dairy Journal*, *40*, 155–173.

Stadhouders, J., & Leenders, G. J. M. (1984). Spontaneously developed mixed-strain cheese: Their behaviour toward phages and their use in the Dutch cheese industry. *Netherlands Milk and Dairy Journal*, *38*, 157–181.

Staley, C., Weingarden, A. R., Khoruts, A., & Sadowsky, M. J. (2017). Interaction of gut microbiota with bile acid metabolism and its influence on disease states. *Applied Microbiology and Biotechnology*, *101*(1), 47–64. doi:10.100700253-016-8006-6 PMID:27888332

Stanley, T., & Wilson, I. G. (2003). Multilocus enzyme electrophoresis: A practical guide. *Molecular Biotechnology*, *24*(2), 203–220. doi:10.1385/MB:24:2:203 PMID:12746557

Stanton, C., Desmond, C., Coakley, M., Collins, J. K., Fitzgerald, G., & Ross, R. P. (2003). Challenges facing development of probiotic-containing functional foods. In F. Er (Ed.), *Handbook of fermented functional foods* (pp. 50–79). Boca Raton, FL: CRC Press. doi:10.1201/9780203009727.ch2

Stanton, C., Gardiner, G., Lynch, P. B., Collins, J. K., Fitzgerald, G., & Ross, R. P. (1998). Probiotic cheese. *International Dairy Journal*, *8*(5-6), 491–496. doi:10.1016/S0958-6946(98)00080-6

Stark, J. (2007). Cheese and fermented sausages. In J. Dijksterhuis & R. A. Samson (Eds.), *Food Mycology* (pp. 319–331). Boca Raton, FL: CRC Press.

Steele, J., Broadbent, J., & Kok, J. (2013). Perspectives on the contribution of lactic acid bacteria to cheese flavor development. *Current Opinion in Biotechnology*, *24*(2), 135–141. doi:10.1016/j.copbio.2012.12.001 PMID:23279928

Steffen, C., Eberhard, P., Bosset, J. O., & Rüegg, M. (1993). Swiss-Type Variedties. In P. F. Fox (Ed.), *Cheese: Chemistry, Physics and Microbiology* (2nd ed.; Vol. 2, pp. 83–110). Major Cheese Groups. doi:10.1007/978-1-4615-2648-3_3

Steinkraus, K. H. (1994). Nutritional significance of fermented foods. *Food Research International*, *27*(3), 259–267. doi:10.1016/0963-9969(94)90094-9

Steinkraus, K. H. (1997). Bio-enrichment: production of vitamins in fermented foods. In B. J. B. Wood (Ed.), *Microbiology of Fermented Foods* (pp. 603–621). Springer.

Stepaniak, L., Kornacki, K., Grabska, J., Rymaszewski, J., & Cichosz, G. (1980). Lipolytic and proteolytic activity of *Penicillium roqueforti, Penicillium candidum* and *Penicillium camemberti* strains. *Acta Alimentaria Polonica*, *6*(3), 155–165.

Stilton cheese made by Cropwell Bishop sitting on waxed paper. (2013, October 11). Retrieved from https://commons.wikimedia.org/wiki/File:Stilton_Cheese_02.png

Strauch, E., Hammerl, J. A., & Hertwig, S. (2007). Bacteriophages: New tools for safer food? *Journal für Verbraucherschutz und Lebensmittelsicherheit*, *2*(2), 138–143. doi:10.100700003-007-0188-5

Streit, F., Delettre, J., Corrieu, G., & Béal, C. (2008). Acid adaptation of *Lactobacillus delbrueckii* subsp. *bulgaricus* induces physiological responses at membrane and cytosolic levels that improves cryotolerance. *Journal of Applied Microbiology*, *105*(4), 1071–1080. doi:10.1111/j.1365-2672.2008.03848.x PMID:18498349

Sudhamani, M., Ismaiel, E., Geis, A., Batish, V., & Heller, K. J. (2008). Characterisation of pSMA23, a 3.5 kbp plasmid of *Lactobacillus casei*, and application for heterologous expression in *Lactobacillus*. *Plasmid*, *59*(1), 11–19. doi:10.1016/j.plasmid.2007.09.001 PMID:17961648

Sultana, K., Godward, G., Reynolds, N., Arumugaswamy, R., Peiris, P., & Kailasapathy, K. (2000). Encapsulation of probiotic bacteria with alginate–starch and evaluation of survival in simulated gastrointestinal conditions and in yoghurt. *International Journal of Food Microbiology*, *62*(1–2), 47–55. doi:10.1016/S0168-1605(00)00380-9 PMID:11139021

Sumantha, A., Larroche, C., & Pandey, A. (2006). Microbiology and industrial biotechnology of food-grade proteases: A perspective. *Food Technology and Biotechnology*, *44*, 211–220.

Sumarah, M. W., Miller, J. D., & Blackwell, B. A. (2005). Isolation and metabolite production by *Penicillium roqueforti, P. paneum* and *P. crustosum* isolated in Canada. *Mycopathologia*, *159*(4), 571–577. doi:10.100711046-005-5257-7 PMID:15983744

Summerbell, R., Gueidan, C., Schroers, H.-J., de Hoog, G., Starink, M., Rosete, Y. A., ... Scott, J. (2011). *Acremonium* phylogenetic overview and revision of *Gliomastix, Sarocladium*, and *Trichothecium. Studies in Mycology*, *68*, 139–162. doi:10.3114im.2011.68.06 PMID:21523192

Sunanliganon, C., Thong-Ngam, D., Tumwasorn, S., & Klaikeaw, N. (2012). *Lactobacillus plantarum* B7 inhibits *Helicobacter pylori* growth and attenuates gastric inflammation. *World Journal of Gastroenterology*, *18*(20), 2472–2480. doi:10.3748/wjg.v18.i20.2472 PMID:22654444

Suomalainen, T. H., & Mäyrä-Makinen, A. M. (1999). Propionic acid bacteria as protective cultures in fermented milks and breads. *Le Lait*, *79*(1), 165–174. doi:10.1051/lait:1999113

Swaisgood, H. E. (1982). Chemistry of milk proteins. In P. F. Fox (Ed.), *Developments in dairy chemistry* (Vol. 1, pp. 1–59). London: Applied Science Publishers.

Sybesma, W., Starrenburg, M., Tijsseling, L., Hoefnagel, M. H. N., & Hugenholtz, J. (2003). Effects of cultivation conditions on folate production by lactic acid bacteria. *Applied and Environmental Microbiology*, *69*(8), 4542–4548. doi:10.1128/AEM.69.8.4542-4548.2003 PMID:12902240

Szecsi, P. B., & Harboe, M. (2013). Chymosin. In N. D. Rawlings & G. S. Salvesen (Eds.), *Handbook of proteolytic enzymes* (3rd ed.; Vol. 1). Academic Press. doi:10.1016/B978-0-12-382219-2.00005-3

Tabbers, M. M., Chmielewska, A., Roseboom, M. G., Crastes, N., Perrin, C., Reitsma, J. B., ... Benninga, M. A. (2011). Fermented milk containing *Bifidobacterium lactis* DN-173 010 in childhood constipation: A randomized, double-blind, controlled trial. *Pediatrics*, *127*(6), e1392–e1399. doi:10.1542/peds.2010-2590 PMID:21606153

Tachedjian, G., Aldunate, M., Bradshaw, C. S., & Cone, R. A. (2017). The role of lactic acid production by probiotic *Lactobacillus* species in vaginal health. *Research in Microbiology*, 1–11. PMID:28435139

Takehana, S., Washizu, K., Ando, K., Koikeda, S., Takeuchi, K., Matsui, H., ... Takagi, H. (1994). Chemical synthesis of the gene for microbial transglutaminase from *Streptoverticillium* and its expression in *Escherica coli. Bioscience, Biotechnology, and Biochemistry*, *58*(1), 88–92. doi:10.1271/bbb.58.88 PMID:7765335

Takó, M., Kotogán, A., Krisch, J., Vágvölgyi, C., Mondal, K. C., & Papp, T. (2015). Enhanced production of industrial enzymes in *Mucoromycotina* fungi during solid-state fermentation of agricultural wastes/by-products. *Acta Biologica Hungarica*, *66*(3), 348–360. doi:10.1556/018.66.2015.3.10 PMID:26344030

Talwalkar, A., & Kailasapathy, K. (2003). Effect of microencapsulation on oxygen toxicity in probiotic bacteria. *Australian Journal of Dairy Technology*, *58*(1), 36–39.

Talwalkar, A., Miller, C. W., Kailasapathy, K., & Nguyen, M. H. (2004). Effect of packaging materials and dissolved oxygen on the survival of probiotic bacteria in yoghurt. *International Journal of Food Science & Technology*, *39*(6), 605–611. doi:10.1111/j.1365-2621.2004.00820.x

Tamime, A. Y., & Robinson, R. K. (2007). Yoghurt science and technology (3rd ed.). Abington, UK: Woodhead Publishing Ltd. LLC.

Tamime, A. Y., & Marshall, V. M. E. (1997). Microbiology and technology of fermented milks. In B. A. Law (Ed.), *Microbiology and Biochemistry of Cheese and Fermented Milk* (pp. 153–192). Boston, MA: Springer US; doi:10.1007/978-1-4613-1121-8_3

Tang, C.-H., & Ma, C.-H. (2007). Modulation thermal stability of β-lactoglobulin by transglutaminase treatment. *European Food Research and Technology*, *225*(5-6), 649–652. doi:10.100700217-006-0460-4

Tang, W., Dong, M., Wang, W., Han, S., Rui, X., Chen, X., ... Li, W. (2017). Structural characterization and antioxidant property of released exopolysaccharides from *Lactobacillus delbrueckii* ssp. *bulgaricus* SRFM-1. *Carbohydrate Polymers*, *173*, 654–664. doi:10.1016/j.carbpol.2017.06.039 PMID:28732910

Tang, Y., Xue, H., Bi, Y., Li, Y., Wang, Y., Zhao, Y., & Shen, K. (2015). A method of analysis for T-2 toxin and neosolaniol by UPLC-MS/MS in apple fruit inoculated with *Trichothecium roseum*. *Food Additives & Contaminants: Part A*, *32*(4), 480–487. doi:10.1080/19440049.2014.968884 PMID:25254921

Tan, H. A. D., Dizon, E. I., Barraquio, V. L., & Mercado, S. M. (2016). Efficacy of Alginate-Taro (Colocasia esculenta L. Schott) Starch for Yoghurt Starter Culture Encapsulation. *Journal of Science, Engineering and Technology*, *4*(0), 1–10. Retrieved from http://ijterm.org/index.php/jset/article/view/159

Tanimoto, S.-Y., & Kinsella, J. E. (1988). Enzymatic modifications of proteins: Effects of transglutaminase cross-linking on some properties of β-lactoglobulin. *Journal of Agricultural and Food Chemistry*, *36*(2), 281–285. doi:10.1021/jf00080a010

Tannock, G. A. (1999). Fresh look at the intestinal microflora. In G. Tannock (Ed.), *Probiotics: A Critical Review* (p. 514). Norfolk, UK: Horizon Scientific Press.

Tappler, P., Twrdik, F., Reinthaler, F. F., Haas, D., Galler, H., Gorfer, M. … Samson, R. A. (2013). Beurteilung, Messmethoden, Identifizierung. In Pilze in Innenräumen und am Arbeitsplatz (pp. 195–422). Vienna: Springer Vienna.

Taşdemir, A. (2017). Probiyotikler, Prebiyotikler ve Sinbiyotikler. *Kastamonu Sağlık Akademisi*, *2*(1), 71–88. Retrieved from http://dergipark.gov.tr/download/article-file/305079

Tavano, O. L. (2017). Protesaes. In R. C. Ray & C. M. Rosell (Eds.), *Microbial enzyme technology in food applications* (pp. 162–179). Boca Raton, FL: CRC Press.

Tavaria, F. K., Silva Ferreira, A. C., & Malcata, F. X. (2004). Volatile free fatty acids as ripening indicators for Serra da Estrela cheese. *Journal of Dairy Science*, *87*(12), 4064–4072. doi:10.3168/jds.S0022-0302(04)73548-1 PMID:15545367

Teixeira, P. (2014). *Lactobacillus delbrueckii* ssp. *bulgaricus*. In C. A. Batt & M. L. Tortorello (Eds.), *Encyclopedia of Food Microbiology* (vol. 2, pp. 425–431). Elsevier. doi:10.1016/B978-0-12-384730-0.00177-4

Temmerman, R., Scheirlinck, I., Huys, G., & Swings, J. (2003). Culture-independent analysis of probiotic products by denaturing gradient gel electrophoresis culture-independent analysis of probiotic products by denaturing gradient gel electrophoresis. *Applied and Environmental Microbiology*, *69*(1), 220–226. doi:10.1128/AEM.69.1.220-226.2003 PMID:12513998

Teran, C. G., Teran-Escalera, C. N., & Villarroel, P. (2009). Nitazoxanide vs. probiotics for the treatment of acute rotavirus diarrhea in children: A randomized, single-blind, controlled trial in Bolivian children. *International Journal of Infectious Diseases*, *13*(4), 518–523. doi:10.1016/j.ijid.2008.09.014 PMID:19070525

Teschke, O., & de Souza, E. F. (2002). Liposome Structure Imaging by Atomic Force Microscopy: Verification of Improved Liposome Stability during Adsorption of Multiple Aggregated Vesicles. *Langmuir, 18*(17), 6513–6520. doi:10.1021/la025689v

Thierry, A., Deutsch, S., Falentin, H., Dalmasso, M., Cousin, F. J., & Jan, G. (2011). New insight into physiology and metabolisms of *Propionibacterium freudenreichii. International Journal of Food Microbiology, 149*(1), 19–27. doi:10.1016/j.ijfoodmicro.2011.04.026 PMID:21620505

Thomas, L. V., Clarkson, M. R., & Delves-Broughton, J. (2000). Nisin. In A. S. Naidu (Ed.), *Natural food antimicrobial systems* (pp. 463–524). Boca-Raton, FL: CRC Press.

Thomas, L. V., & Delves-Broughton, J. (2001). New advances in the application of the food preservative nisin. *Advances in Food Sciences, 2*, 11–22.

Toledano, O., & Magdassi, S. (1997). Formation of surface active gelatin by covalent attachment of hydrophobic chains. *Journal of Colloid and Interface Science, 193*(2), 172–177. doi:10.1006/jcis.1997.5030 PMID:9344517

Tolkach, A., & Kulozik, U. (2005). Fractionation of whey proteins and caseinomacro-peptide by means of enzymatic crosslinking and membrane separation techniques. *Journal of Food Engineering, 67*(1-2), 13–20. doi:10.1016/j.jfoodeng.2004.05.058

Tolkach, A., & Kulozik, U. (2006). Transport of whey proteins through 0.1 mm ceramic membrane: Phenomena, modeling and consequences for concentration or diafiltration processes. *Desalination, 199*(1-3), 340–341. doi:10.1016/j.desal.2006.03.183

Tomme de Savoie. (2015, June 19). Retrieved from https://commons.wikimedia.org/wiki/File:WikiCheese_-_Tomme_de_Savoie_-_20150619_-_002.jpg

Tournas, V. H. (2009). Evaluation of the Hydrophobic Grid Membrane Filter for the Enumeration of Moulds and Yeasts in Naturally- Contaminated Foods. *Microbiology Insights, 2*, 31–37. doi:10.4137/MBI.S2906

Traorè, F., & Meunier, J. C. (1992). Crosslinking activity of placental F XIIIa on whey proteins and caseins. *Journal of Agricultural and Food Chemistry, 40*(3), 399–402. doi:10.1021/jf00015a007

Trépanier, G., Abboudi, M. E., Lee, B. H., & Simard, R. E. (1992). Accelerated Maturation of Cheddar Cheese: Influence of Added Lactobacilli and Commercial Protease on Composition and Texture. *Journal of Food Science, 57*(4), 898–902. doi:10.1111/j.1365-2621.1992.tb14318.x

Troise, A. D., Vitiello, D., Tsang, C., Fiore, A., Glomb, M. A., Fogliano, V., ... Eisenbrand, G. (2016). Encapsulation of ascorbic acid promotes the reduction of Maillard reaction products in UHT milk. *Food & Function, 7*(6), 2591–2602. doi:10.1039/C6FO00151C PMID:27240727

Truong, V., Clare, D. A., Catignani, G. L., & Swaisgood, H. (2004). Cross-linking and rheological changes of whey proteins treated with microbial transglutaminase. *Journal of Agricultural and Food Chemistry, 52*(5), 1170–1176. doi:10.1021/jf034397c PMID:14995116

Tsai, G.-J., Lin, S.-M., & Jiang, S.-T. (1996). Transglutaminase from *Streptoverticillium ladakanum* and application to minced fish product. *Journal of Food Science, 61*(6), 1234–1238. doi:10.1111/j.1365-2621.1996.tb10968.x

Tsevdou, M. S., Eleftheriou, E. G., & Taoukis, P. S. (2013). Transglutamianse treatment of thermally and high pressure processed milk: Effects on the properties and storage stability of set yoghurt. *Innovative Food Science & Emerging Technologies, 17*, 144–152. doi:10.1016/j.ifset.2012.11.004

Tsuda, H., Matsumoto, T., & Ishimi, Y. (2012). Selection of lactic acid bacteria as starter cultures for fermented meat products. *Food Science and Technology Research*, *18*(5), 713–721. doi:10.3136/fstr.18.713

Tu, N., Hoang Khue, H., & Ngoc, N. H. (2013). Exopolysaccharide in *Lactobacillus rhamnosus* Pn04 after co-culture with *Leuconostoc mesenteroides* Vtcc-B-643. *Journal of Applied Pharmaceutical Science, 3*(7), 14-17.

Tuomola, E., Crittenden, R., Playne, M., Isolauri, E., & Salminen, S. (2001). Quality assurance criteria for probiotic bacteria. *The American Journal of Clinical Nutrition*, *73*(2), 393–398. PMID:11157347

Turcatti, G., Romieu, A., Fedurco, M., & Tairi, A. P. (2008). A new class of cleavable fluorescent nucleotides: Synthesis and optimization as reversible terminators for DNA sequencing by synthesis. *Nucleic Acids Research*, *36*(4), e25. doi:10.1093/nar/gkn021 PMID:18263613

Turgut, T., & Cakmakci, S. (2009). Investigation of the possible use of probiotics in ice cream manufacture. *International Journal of Dairy Technology*, *62*(3), 444–451. doi:10.1111/j.1471-0307.2009.00494.x

Turnbaugh, P. J., Ley, R. E., Mahowald, M. A., Magrini, V., Mardis, E. R., & Gordon, J. I. (2006). An obesity-associated gut microbiome with increased capacity for energy harvest. *Nature*, *444*(7122), 1027–1131. doi:10.1038/nature05414 PMID:17183312

Tyagi, A., Kumar, A., Mohanty, A. K., Kaushik, J. K., Groveer, S., & Batish, V. K. (2017). Expression of buffalo chymosin in *Pichia pastoris* for application in mozarella cheese. *Lebensmittel-Wissenschaft + Technologie*, *84*, 733–739. doi:10.1016/j.lwt.2017.06.033

Tyagi, A., Kumar, A., Yadav, A. K., Saklani, A. C., Grover, S., & Batish, V. K. (2016). Functional expression of recombinant goat chymosin in *Pichia pastoris* bioreactor cultures: A commercially viable alternate. *Lebensmittel-Wissenschaft + Technologie*, *69*, 217–224. doi:10.1016/j.lwt.2016.01.015

Uniacke-Lowe, T., Huppertz, T., & Fox, P. (2010). Equine milk proteins: Chemistry, structure and nutritional significance. *International Dairy Journal*, *20*(9), 609–629. doi:10.1016/j.idairyj.2010.02.007

Upadhyay, V. K., McSweeney, P. L. H., Magboul, A. A. A., & Fox, P. F. (2004). Proteolysis in cheese during ripening. In Cheese: Chemistry, Physics and Microbiology: Vol 1. General Aspects (pp. 391-433). London: Elsevier. doi:10.1016/S1874-558X(04)80076-9

Upadhyay, V. K., Kelly, A. L., & Mcsweeney, P. L. H. (2006). Use of nattokinase, a subtilisin-like serine proteinase, to accelerate proteolysis in Cheddar cheese during ripening. *Le Lait*, *86*(3), 227–240. doi:10.1051/lait:2006004

USDA. (2010). *Food and Drug Administration Compliance Policy Guide Sec. 527.300 Dairy Products - Microbial Contaminants and Alkaline Phosphatase Activity is now available.* Retrieved from Federal Register notice of availability: http://origin.www.gpo.gov/fdsys/pkg/FR-2010-12-12/pdf/2010-32232.pdf

Ustunol, Z., & Hicks, C. L. (1990). Effect of milk-clotting enzymes on cheese yield. *Journal of Dairy Science*, *73*(1), 8–16. doi:10.3168/jds.S0022-0302(90)78639-0

Uzeh, R. E., Ohenhen, R. E., & Rojugbokan, A. K. (2006). Microbiological and nutritional qualities of dairy products : Nono and Wara. *Nature and Science*, *4*(3), 37–40.

Vafopoulou, A., Alichanidis, E., & Zerfiridis, G. (1989). Accelerated ripening of Feta cheese with heat-shocked cultures or microbial proteinases. *The Journal of Dairy Research*, *56*(2), 285–296. doi:10.1017/S0022029900026492

Vaillancourt, J. (2006). *Regulating pre- and probiotics: a U.S. FDA perspective. In Ending the war metaphor: The changing agenda for unraveling the host-microbe relationship* (pp. 229–237). Washington, DC: National Academies Press.

Valdes-Stauber, N., & Götz, H., & Busse, M. (1991). Antagonistic effect of coryneform bacteria from red smear cheese against Listeria species. *International Journal of Food Microbiology*, *2*(13), 131–130. PMID:1909545

Valdes-Stauber, N., & Scherer, S. (1994). Isolation and characterization of Linocin M18, A bacteriosin produced by *Brevibacterium linens. Applied and Environmental Microbiology*, *60*(10), 3809–3814. PMID:7986050

Valence, F., Richoux, R., Thierry, A., Palva, A., & Lortal, S. (1998). Autolysis of *Lactobacillus helveticus* and *Propionibacterium freudenreichii* in Swiss cheeses: First evidence by using species-specific lysis markers. *The Journal of Dairy Research*, *65*(4), 609–620. doi:10.1017/S0022029998003021

van Bokhorst-van de Veen, H., Lee, I. C., Marco, M. L., Wels, M., Bron, P. A., & Kleerebezem, M. (2012). Modulation of *Lactobacillus plantarum* gastrointestinal robustness by fermentation conditions enables identification of bacterial robustness markers. *PLoS One*, *7*(7), e39053. doi:10.1371/journal.pone.0039053 PMID:22802934

van Bokhorst-van de Veen, H., van Swam, I., Wels, M., Bron, P. A., & Kleerebezem, M. (2012). Congruent Strain Specific Intestinal Persistence of *Lactobacillus plantarum* in an Intestine-Mimicking In Vitro System and in Human Volunteers. *PLoS One*, *7*(9), e44588. doi:10.1371/journal.pone.0044588 PMID:22970257

Van De Guchte, M., Kodde, J., Van Der Vossen, J. M. B. M., Kok, J., & Venema, G. (1990). Heterologous gene expression in *Lactococcus lactis subsp. lactis*: Synthesis, secretion, and processing of the *Baccilius subtilis* neutral protease. *Applied and Environmental Microbiology*, *56*(9), 2606–2611. PMID:2125811

van den Tempel, T. (1998). Yeast associated with Danablu. *International Dairy Journal*, *8*(1), 25–31. doi:10.1016/S0958-6946(98)00013-2

van der Aa Kühle, A., Skovgaard, K., & Jespersen, L. (2005). In vitro screening of probiotic properties of *Saccharomyces cerevisiae* var. *boulardii* and food-borne *Saccharomyces cerevisiae* strains. *International Journal of Food Microbiology*, *101*(1), 29–39. doi:10.1016/j.ijfoodmicro.2004.10.039 PMID:15878404

van Kranenburg, R., Kleerebezema, M., van Hylckama Vliega, J., Ursing, B. M., Boekhorst, J., Smit, A. B., ... Siezen, R. J. (2002). Flavour formation from amino acids by lactic acid bacteria: Predictions from genome sequence analysis. *International Dairy Journal*, *12*(2-3), 111–121. doi:10.1016/S0958-6946(01)00132-7

van Ooyen, A. J. J., Dekker, P., Huang, M., Olsthoorn, M. M. A., Jacobs, D. I., Colussi, P. A., & Taron, C. H. (2006). Heterologous protein production in the yeast *Kluyveromyces lactis. FEMS Yeast Research*, *6*(3), 381–392. doi:10.1111/j.1567-1364.2006.00049.x PMID:16630278

Vanin, F. M., Sobral, P. J. A., Menegalli, F. C., Carvalho, R. A., & Habitante, A. M. Q. B. (2005). Effect of plasticizers and their concentrations on thermal and functional properties of gelatin-based films. *Food Hydrocolloids*, *19*(5), 899–907. doi:10.1016/j.foodhyd.2004.12.003

Vardar, N. B., & Öksüz, Ö. (2007). Artisan strawberry ice cream made with supplementation of Lactococci or *Lactobacillus acidophilus. Italian Journal of Food Science*, *19*(4), 403–412.

Vardjan, T., Lorbeg, P. M., Rogelj, I., & Majheniz, A. C. (2013). Characterization and stability of lactobacilli and yeast microbiota in kefir grains. *Journal of Dairy Science*, *96*(5), 2729–2736. doi:10.3168/jds.2012-5829 PMID:23497996

Verruck, S., Prudencio, E. S., Vieira, C. R. W., Amante, E. R., & Amboni, R. D. M. C. (2015). The buffalo Minas Frescal cheese as a protective matrix of *Bifidobacterium* BB-12 under in vitro simulated gastrointestinal conditions. *Lebensmittel-Wissenschaft + Technologie*, *63*(2), 1179–1183. doi:10.1016/j.lwt.2015.04.014

Vido, K., le Bars, D., Mistou, M.-Y., Anglade, P., Gruss, A., & Gaudu, P. (2004). Proteome analyses of heme-dependent respiration in *Lactococcus lactis*: Involvement of the proteolytic system. *Journal of Bacteriology*, *186*(6), 1648–1657. doi:10.1128/JB.186.6.1648-1657.2004 PMID:14996795

Vijayageetha, V., Begum, S. K., & Reddy, Y. K. (2011). Technology and quality attributes of probiotic ice cream. *Tamilnadu. Journal of Veterinary & Animal Sciences (Lahore)*, *7*(6), 299–302.

Villalva, F. J., Cravero Bruneri, A. P., Vinderola, G., Gonçalvez de Oliveira, E., Paz, N. F., & Ramón, A. N. (2017). Formulation of a peach ice cream as potential symbiotic food. *Food Science and Technology (Campinas)*, *37*(3), 456–461. doi:10.1590/1678-457x.19716

Vinderola, C. G., Prosello, W., Ghiberto, T. D., & Reinheimer, J. A. (2000). Viability of probiotic (Bifidobacterium, Lactobacillus acidophilus and Lactobacillus casei) and nonprobiotic microflora in Argentinian Fresco cheese. *Journal of Dairy Science*, *83*(9), 1905–1911. doi:10.3168/jds.S0022-0302(00)75065-X PMID:11003217

Vinderola, G., Binetti, A., Burns, P., & Reinheimer, J. (2011). Cell viability and functionality of probiotic bacteria in dairy products. *Frontiers in Microbiology*, *28*(70), 1–6. PMID:21833320

Vinderola, G., Perdigon, G., Duarte, J., Farnworth, E., & Matar, C. (2006). Effects of the oral administration of the exopolysaccharide produced by *Lactobacillus kefiranofaciens* on the gut mucosal immunity. *Cytokine*, *36*(5-6), 254–260. doi:10.1016/j.cyto.2007.01.003 PMID:17363262

Vinderola, G., Prosello, W., Molinari, F., Ghiberto, D., & Reinheimer, J. (2009). Growth of Lactobacillus paracasei A13 in Argentinian probiotic cheese and its impact on the characteristics of the product. *International Journal of Food Microbiology*, *135*(2), 171–174. doi:10.1016/j.ijfoodmicro.2009.08.021 PMID:19751954

Visser, S. (1993). Proteolytic Enzymes and Their Relation to Cheese Ripening and Flavor: An Overview. *Journal of Dairy Science*, *76*(1), 329–350. doi:10.3168/jds.S0022-0302(93)77354-3

von der Weid, T., Bulliard, C., & Schiffrin, E. J. (2001). Induction by a Lactic Acid Bacterium of a Population of CD4+ T Cells with Low Proliferative Capacity That Produce Transforming Growth Factor and Interleukin-10. *Clinical and Vaccine Immunology; CVI*, *8*(4), 695–701. doi:10.1128/CDLI.8.4.695-701.2001 PMID:11427413

von Staszewski, M., & Jagus, R. J. (2008). Natural antimicrobials: Effect of Microgard (TM) and nisin against *Listeria innocua* in liquid cheese whey. *International Dairy Journal*, *18*(3), 255–259. doi:10.1016/j.idairyj.2007.08.012

Vos, P., Hogers, R., Bleeker, M., Reijans, M., van de Lee, T., Hornes, M., ... Kuiper, M. (1995). AFLP: A new technique for DNA fingerprinting. *Nucleic Acids Research*, *23*(21), 4407–4414. doi:10.1093/nar/23.21.4407 PMID:7501463

Voulgari, K., Hatzikamari, M., Delepoglou, A., Georgakopoulos, P., Litopoulou-Tzanetaki, E., & Tzanetakis, N. (2010). Antifungal activity of non-starter lactic acid bacteria isolates from dairy products. *Food Control*, *21*(2), 136–142. doi:10.1016/j.foodcont.2009.04.007

Vuyst, L., & Degeest, B. (1999). Heteropolysaccharides from lactic acid bacteria. *FEMS Microbiology Reviews*, *23*(2), 153–177. doi:10.1111/j.1574-6976.1999.tb00395.x PMID:10234843

Walstra, P., Wouters, J. T. M., & Geurts, T. J. (2006). *Dairy science and technology (Second Edi)*. Boca Raton, FL: CRC Press, Taylor & Francis Group.

Wang, W. (2013). *Whey proteins cross-linked by transglutaminase or glycated with maltodextrin: physicochemical bases of the improved heat stability* (PhD Thesis). University of Tenesse.

Wang, N., Wang, K. Y., Li, G., Guo, W., & Liu, D. (2015). Expression and characterization of camel chymosin in *Pichia pastoris*. *Protein Expression and Purification*, *111*, 75–81. doi:10.1016/j.pep.2015.03.012 PMID:25837439

Wang, W., Zhong, Q., & Hu, Z. (2012). Nanoscale understanding of thermal aggregation of whey protein pretreated by transglutaminase. *Journal of Agricultural and Food Chemistry*, *61*(2), 435–446. doi:10.1021/jf304506n PMID:23252670

Wang, Y., Ahmed, Z., Feng, W., Li, C., & Song, S. (2008). Physicochemical properties of exopolysaccharide produced by *Lactobacillus kefiranofaciens* ZW3 isolated from Tibet kefir. *International Journal of Biological Macromolecules*, *43*(3), 283–288. doi:10.1016/j.ijbiomac.2008.06.011 PMID:18662712

Wang, Y., Liu, A., Ye, R., Wang, W., & Li, X. (2015). Transglutamianse-induced crosslinking of gelatin-calcium carbonate composite films. *Food Chemistry*, *166*, 414–422. doi:10.1016/j.foodchem.2014.06.062 PMID:25053075

Washizu, K., Ando, K., Koikeda, S., Hirose, S., Matsuura, A., Akagi, H., ... Takeuchi, K. (1994). Molecular cloning of the gene for microbial transglutamianse *Streptoverticillium* and its expression in *Streptomyces lividans*. *Bioscience, Biotechnology, and Biochemistry*, *58*(1), 82–87. doi:10.1271/bbb.58.82 PMID:7765334

Watanabe, M., van der Veen, S., & Abee, T. (2012a). Impact of respiration on resistance of *Lactobacillus plantarum* WCFS1 to acid stress. *Applied and Environmental Microbiology*, *78*(11), 4062–4064. doi:10.1128/AEM.00287-12 PMID:22447603

Watanabe, M., van der Veen, S., Nakajima, H., & Abee, T. (2012b). Effect of respiration and manganese on oxidative stress resistance of *Lactobacillus plantarum* WCFS1. *Microbiology*, *158*(1), 293–300. doi:10.1099/mic.0.051250-0 PMID:22016573

Watson, R. S., Preedy, V. R., Monedero, V., & Rodríguez-Díaz, J. (2016). Intestinal Microbiota and Susceptibility to Viral Infections: Role of Probiotics. In Probiotics, Prebiotics, and Synbiotics (pp. 813–826). Academic Press. doi:10.1016/B978-0-12-802189-7.00062-9

Watson, R. S., Preedy, V. R., Likotrafiti, E., & Rhoades, J. (2016). *Probiotics, Prebiotics, Synbiotics, and Foodborne Illness*. Probiotics, Prebiotics, and Synbiotics. doi:10.1016/B978-0-12-802189-7.00032-0

Wedajo, B. (2015). Lactic acid bacteria: Benefits, selection criteria and probiotic potential in fermented food. *Journal of Probiotics & Health*, *3*(02), 129. doi:10.4172/2329-8901.1000129

Wegmann, U., O'Connell-Motherway, M., Zomer, A., Buist, G., Shearman, C., Canchaya, C., ... Kok, J. (2007). Complete genome sequence of the prototype lactic acid bacterium *Lactococcus lactis* subsp. *cremoris* MG1363. *Journal of Bacteriology*, *189*(8), 3256–3270. doi:10.1128/JB.01768-06 PMID:17307855

Wegmuller, B., Luthy, J., & Candrian, U. (1993). Direct polymerase chain reaction detection of Campylobacter jejuni and Campylobacter coli in raw milk and dairy products. *Applied and Environmental Microbiology*, *59*(7), 2161–2165. PMID:8357250

Weinbrenner, D. R., Barefoot, S. F., & Grinstead, D. A. (1997). Inhibition of yogurt starter cultures by Jenseniin G, a *Propionibacterium* bacteriocin. *Journal of Dairy Science*, *80*(7), 1246–1253. doi:10.3168/jds.S0022-0302(97)76053-3

Welman, A. D., & Maddox, I. S. (2003). Exopolysaccharides from lactic acid bacteria: Perspectives and challenges. *Trends in Biotechnology*, *21*(6), 269–274. doi:10.1016/S0167-7799(03)00107-0 PMID:12788547

Wen-qiong, W., Lan-wei, Z., Xue, H., & Yi, L. (2017). Cheese whey protein recovery by ultrafiltration through transglutamianse (TG) catalysis whey protein cross-linking. *Food Chemistry*, *215*, 31–40. doi:10.1016/j.foodchem.2016.07.057 PMID:27542447

Werner, H., Nielsen, E. W., Ardö, Y., Rage, A., & Antila, V. (1999). North European varieties of cheese. In P. F. Fox (Ed.), *Cheese: Chemistry, Physics and Microbiology* (pp. 245–262). Boston, MA: Springer. doi:10.1007/978-1-4615-2800-5_8

Whitaker, R. D., & Batt, C. A. (1991). Characterization of the Heat Shock Response in *Lactococcus lactis* subsp. *lactis*. *Applied and Environmental Microbiology*, *57*, 1408–1412. PMID:16348482

WHO/FAO. (2002a). *Guidelines for the Evaluation of Probiotics in Food*. London: Author.

Wierzbicka-Woś, A., Cieśliński, H., Wanarska, M., Kozłowska-Tylingo, K., Hildebrandt, P., & Kur, J. (2011). A novel cold-active β-d-galactosidase from the *Paracoccus sp.* 32d - gene cloning, purification and characterization. *Microbial Cell Factories*, *49*(12), 2122–2133. PMID:22166118

Wilcox, C. P., Clare, D. A., Valentine, V. W., & Swaisgood, H. (2002). Immobilization and utilization of the recombinant fusion proteins trypsin-streptavidin and streptavidin-transglutaminase for modification of whey protein isolate functionality. *Journal of Agricultural and Food Chemistry*, *50*(13), 3723–3730. doi:10.1021/jf011603c PMID:12059150

Wilcox, C. P., & Swaisgood, H. E. (2002). Modification of the rheological properties of whey protein isolate through the use of an immobilized microbial transglutaminase. *Journal of Agricultural and Food Chemistry*, *50*(20), 5546–5551. doi:10.1021/jf0117154 PMID:12236677

Wilkinson, M. G. (1999). Acceleration of cheese ripening. In P.F. Fox (Ed.), Cheese: Chemistry, Physics and Microbiology: Vol. 1. General Aspects (pp. 523-556). Gaithersburg, MD: Aspen Publication.

Wilkinson, M. G., & Kilcawley, K. N. (2005). Mechanisms of incorporation and release of enzymes into cheese during ripening. *International Dairy Journal*, *15*(6-9), 817–830. doi:10.1016/j.idairyj.2004.08.021

Wisselink, H. W., Weusthuis, R. A., Eggink, G., Hugenholtz, J., & Grobben, G. J. (2002). Mannitol production by lactic acid bacteria: A review. *International Dairy Journal*, *12*(2), 151–161. doi:10.1016/S0958-6946(01)00153-4

Witthuhn, R. C., Schoema, T., & Britz, T. J. (2005). Characterisation of the microbial population at different stages of Kefir production and Kefir grain mass cultivation. *International Dairy Journal*, *15*(4), 383–389. doi:10.1016/j.idairyj.2004.07.016

Wright, A. V., & Axelsson, L. (2011). Lactic acid bacteria: an introduction. In S. Lahtinen, A. C. Ouwehand, S. Salminen, & A. V. Wright (Eds.), *Lactic acid bacteria: microbiology and functional aspects*. Boca Raton, FL: CRC Press. doi:10.1201/b11503

Wróblewska, B., Kaliszewska, A., Kołakowski, P., Pawlikowska, K., & Troszyńska, A. (2011). Impact of transglutaminase reaction on the immunoreactive and sensory quality of yoghurt starter. *World Journal of Microbiology & Biotechnology*, *27*(2), 215–227. doi:10.100711274-010-0446-z

Wróblewska, B., Kaliszewsk-Suchodoła, A., Kołakowski, P., Pawlikowska, K., & Troszyńska, A. (2013). The effect of microbial transglutaminase on the immunoreactive and sensory properties of fermented milk beverages. *International Journal of Food Science & Technology*, *48*(5), 1007–1017. doi:10.1111/ijfs.12054

Xiong, Z. Q., Wang, Q. H., Kong, L. H., Song, X., Wang, G. Q., Xia, Y. J., ... Ai, L. Z. (2017). Short communication: Improving the activity of bile salt hydrolases in *Lactobacillus casei* based on *in silico* molecular docking and heterologous expression. *Journal of Dairy Science*, *100*(2), 975–980. doi:10.3168/jds.2016-11720 PMID:28012620

Yamamoto, Y., Gaudu, P. & Gruss, A. (2011). Oxidative stress and oxygen metabolism in lactic acid bacteria. *Lactic Acid Bacteria and Bifidobacteria: Current Progress in Advanced Research,* 91-102.

Yang, Y.-X., He, M., Hu, G., Wei, J., Pages, P., Yang, X.-H., & Bourdu-Naturel, S. (2008). Effect of a fermented milk containing *Bifidobacterium lactis* DN-173010 on Chinese constipated women. *World Journal of Gastroenterology*, *14*(40), 6237. doi:10.3748/wjg.14.6237 PMID:18985817

Yasueda, H., Nakanishi, K., Kumazawa, Y., Nagase, K., Motoki, M., & Matsui, H. (1995). Tissue-type transglutaminase from red sea bream (*Pagrus major*) sequence analysis of the cDNA and functional expression in *Eschericia coli*. *European Journal of Biology*, *232*(2), 411–419. doi:10.1111/j.1432-1033.1995.tb20826.x PMID:7556189

Yavuzdurmaz, H. (2007). *Isolation*. İzmir: Characterization, Determination of Probiotic Properties of Lactic Acid Bacteria From Human Milk. Master of Science, Food Engineering, the Graduate School of Engineering and Sciences of Izmir Institute of Technology.

Yazdanpanah, S., Ehsani, M. R., & Mizani, M. (2015). Water mobility in accelerated ripening of UF Feta cheese. *Journal of Agricultural Science and Technology*, *17*(3), 663–674.

Yegin, S., & Dekker, P. (2013). Progress in the field of aspartic proteinases in cheese manufacturing: Structures, functions, catalytic machanism, inhibition, and engineering. *Dairy Science & Technology*, *93*(6), 565–594. doi:10.100713594-013-0137-2

Yegin, S., Fernandez-Lahore, M., Jose Gama Salgado, A., Guvenc, U., Goksungur, Y., & Tari, C. (2011). Aspartic proteinases from Mucor spp. in cheese manufacturing. *Applied Microbiology and Biotechnology*, *89*(4), 949–960. doi:10.100700253-010-3020-6 PMID:21127856

Yeo, S. K., & Liong, M. T. (2010). Angiotensin I-converting enzyme inhibitory activity and bioconversion of isoflavones by probiotics in soymilk supplemented with prebiotics. *International Journal of Food Sciences and Nutrition*, *61*(2), 161–181. doi:10.3109/09637480903348122 PMID:20085504

Yerlikaya, O. (2014). Starter cultures used in probiotic dairy product preparation and popular probiotic dairy drinks. *Food Science and Technology (Campinas)*, *34*(14), 221–229. doi:10.1590/fst.2014.0050

Yeşil Peynirin Faydaları- Erzurum. (2016, October 23). Retrieved from http://www.erzurumsayfasi.com/haber/yesil-peynirin-faydalari-erzurum-h1089.html

Yeşilova, Y., Çalka, Ö., Akdeniz, N., & Berktaş, M. (2012). Effect of Probiotics on the Treatment of Children with Atopic Dermatitis. *Annals of Dermatology*, *24*(2), 189–193. doi:10.5021/ad.2012.24.2.189 PMID:22577270

Yi, J. B., Kim, Y. T., Bae, H., Whiteside, W. S., & Park, H. J. (2006). Influence of transglutamianse-induced cross-linking on properties of fish gelatin films. *Journal of Food Science*, *71*(9), E376–E383. doi:10.1111/j.1750-3841.2006.00191.x

Yildirim, M., & Hettiarachchy, N. S. (1998). Properties of films produced by cross-linking whey proteins and 11S globulin using transglutaminase. *Journal of Food Science*, *63*(2), 248–252. doi:10.1111/j.1365-2621.1998.tb15719.x

Yilmaz-Ersan, L., Ozcan, T., Akpinar-Bayizit, A., Turan, M. A., & Taskin, M. B. (2017). Probiotic cream: Viability of probiotic bacteria and chemical characterization. *Journal of Food Processing and Preservation*, *41*(1), 1–9. doi:10.1111/jfpp.12797

Yilmaz, G., Ayar, A., & Akin, N. (2005). The effect of microbial lipase on the lipolysis during the ripening of Tulum cheese. *Journal of Food Engineering*, *69*(3), 269–274. doi:10.1016/j.jfoodeng.2004.08.017

Ying, D. Y., Phoon, M. C., Sanguansri, L., Weerakkody, R., Burgar, I., & Augustin, M. A. (2010). Microencapsulated Lactobacillus rhamnosus GG Powders: Relationship of Powder Physical Properties to Probiotic Survival during Storage. *Journal of Food Science*, *75*(9), E588–E595. doi:10.1111/j.1750-3841.2010.01838.x PMID:21535593

Yoder, J. A., Benoit, J. B., & Zettler, L. W. (2003). Effects of salt and temperature on the growth rate of a tick-associated fungus, *Scopulariopsis brevicaulis* Bainier (Deuteromycota). *International Journal of Acarology*, *29*(3), 265–269. doi:10.1080/01647950308684338

Yokoyama, K., Nio, N., & Kikuchi, Y. (2004). Properties and applications of microbial transglutaminase. *Applied Microbiology and Biotechnology*, *64*(4), 447–454. doi:10.100700253-003-1539-5 PMID:14740191

Yoon, K. Y., Woodams, E. E., & Hang, Y. D. (2006). Production of probiotic cabbage juice by lactic acid bacteria. *Bioresource Technology*, *97*(12), 1427–1430. doi:10.1016/j.biortech.2005.06.018 PMID:16125381

Yoon, Y., Lee, S., & Choi, K.-H. (2016). Microbial benefits and risks of raw milk cheese. *Food Control*, *63*, 201–215. doi:10.1016/j.foodcont.2015.11.013

Young, S. L., Sarda, X., & Rosenberg, M. (1993). Microencapsulating Properties of Whey Proteins. 1. Microencapsulation of Anhydrous Milk Fat. *Journal of Dairy Science*, *76*(10), 2868–2877. doi:10.3168/jds.S0022-0302(93)77625-0

Yüksel, Z., Avcı, E., & Erdem, Y. K. (2011). Modification of the rennet process in Berridge substrate by transglutamianse. *International Journal of Dairy Technology*, *64*(3), 365–371. doi:10.1111/j.1471-0307.2011.00667.x

Yüksel, Z., & Erdem, Y. K. (2010). The influence of transglutaminase on functional properties of set yoghurt. *International Journal of Dairy Technology*, *63*(1), 86–97. doi:10.1111/j.1471-0307.2009.00539.x

Yun, B., Oh, S., & Griffiths, M. W. (2014). Lactobacillus acidophilus modulates the virulence of *Clostridium difficile*. *Journal of Dairy Science*, *97*(8), 4745–4758. doi:10.3168/jds.2014-7921 PMID:24856984

Yu, S., Yan, L., Wu, X., Li, F., Wang, D., & Xu, H. (2017). Multiplex PCR coupled with propidium monoazide for the detection of viable Cronobacter sakazakii, Bacillus cereus, and Salmonella spp. in milk and milk products. *Journal of Dairy Science*, *100*(10), 7874–7882. doi:10.3168/jds.2017-13110 PMID:28755939

Yvon, M., & Rijnen, L. (2001). Cheese flavour formation by amino acid catabolism. *International Dairy Journal*, *11*(4), 185–201. doi:10.1016/S0958-6946(01)00049-8

Zago, M., Fornasari, M. E., Carminati, D., Burns, P., Suarez, V., Vinderola, G., ... Giraffa, G. (2011). Characterization and probiotic potential of *Lactobacillus plantarum* strains isolated from cheeses. *Food Microbiology*, *28*(5), 1033–1040. doi:10.1016/j.fm.2011.02.009 PMID:21569949

Zaharia, N. S. (2012). Research regarding accelerated ripening of pasta filata cheeses. *The Annals of the University Dunarea de Jos of Galati. Fascicle VI, Food Technology*, *36*(1), 26–38.

Zanirati, D. F., Abatemarco Jr, M., de Cicco Sandes, S. H., Nicoli, J. R., Nunes, A. C., & Neumann, E. (2015). Selection of lactic acid bacteria from Brazilian kefir grains for potential use as starter or probiotic cultures. *Anaerobe*, *32*, 70–76. doi:10.1016/j.anaerobe.2014.12.007 PMID:25542841

Zeeb, B., Grossmann, L., & Weiss, J. (2013). Accessibility of transglutaminase to induce protein crosslinking in gelled food matrices-Impact of membrane structure. *Food Biophysics*, 1–8.

Zhang, L., Zhang, L., Yi, H., Du, M., Ma, C., Han, X., ... Zhang, Y. (2012). Enzymatic characterization of transglutaminase from *Streptomyces mobaraensis* DSM 40587 in high salt and effect of enzymatic cross-linking of yak milk proteins on functional properties of stirred yogurt. *Journal of Dairy Science*, *95*(7), 3559–3568. doi:10.3168/jds.2011-5125 PMID:22720914

Zhang, N., & Zhao, X.-H. (2010). Study of *Mucor* spp. in semi-hard cheese ripening. *Journal of Food Science and Technology*, *47*(6), 613–619. doi:10.100713197-010-0108-z PMID:23572694

Zhang, W., Xiao, S., Samaraweera, H., Lee, E. J., & Ahn, D. U. (2010). Improving functional value of meat products. *Meat Science*, *86*(1), 15–31. doi:10.1016/j.meatsci.2010.04.018 PMID:20537806

Zhang, W., & Zhong, Q. (2009). Microemulsions as nanoreactors to produce whey protein nanoparticles with enhanced heat stability by sequential enzymaticc-linking and thermal pretreatments. *Journal of Agricultural and Food Chemistry*, *57*(19), 9181–9187. doi:10.1021/jf902045u PMID:19807163

Zhang, Y.-H., Liu, J.-Q., Xu, D., & Zhao, X.-H. (2016). Impacts of glucosamine/ oligochitosan glycation and cross-linking by transglutaminase on the structure and *in vitro* antigenicity of whey proteins. *International Journal of Dairy Technology*, *69*(2), 169–176. doi:10.1111/1471-0307.12246

Zhao, L., Jiang, Y., Zhang, T., Duan, C., Huang, C., Zhao, Y., ... Li, S. (2017). Protective effects of *Lactobacillus plantarum* C88 on chronic ethanolinduced liver injury in mice. *Journal of Functional Foods*, *35*, 97–104. doi:10.1016/j.jff.2017.05.017

Zhao, R., Zheng, S., Duan, C., Liu, F., Yang, L., & Huo, G. (2013). NAD-dependent lactate dehydrogenase catalyses the first step in respiratory utilization of lactate by *Lactococcus lactis*. *FEBS Open Bio*, *3*(1), 379–386. doi:10.1016/j.fob.2013.08.005 PMID:24251099

Zhong, Q., Wang, W., Hu, Z., & Ikeda, S. (2013). Sequential preheating and transglutaminase pretreatments improve stability of whey protein isolate at pH 7.0 during thermal sterilization. *Food Hydrocolloids*, *31*(2), 306–316. doi:10.1016/j.foodhyd.2012.11.018

Zhou, J., Xiaoli, L., Hanhu, J., & Mingsheng, D. (2009). Analysis of the microflora in Tibetan kefir grains using denaturing gradient gel electrophoresis. *Food Microbiology*, *26*(8), 770–775. doi:10.1016/j.fm.2009.04.009 PMID:19835760

Zhu, Y., & Tramper, J. (2008). Novel applications for microbial transglutaminase beyond food processing. *Trends in Biotechnology*, *26*(10), 559–565. doi:10.1016/j.tibtech.2008.06.006 PMID:18706723

Zotta, T., Guidone, A., Ianniello, R. G., Parente, E., & Ricciardi, A. (2013). Temperature and respiration affect the growth and stress resistance of *Lactobacillus plantarum* C17. *Journal of Applied Microbiology*, *115*(3), 848–858. doi:10.1111/jam.12285 PMID:23782242

Zotta, T., Parente, E., & Ricciardi, A. (2017a). Aerobic metabolism in the genus Lactobacillus: Impact on stress response and potential applications in the food industry. *Journal of Applied Microbiology*, *122*(4), 857–869. doi:10.1111/jam.13399 PMID:28063197

Zotta, T., Ricciardi, A., Ianniello, R. G., Parente, E., Reale, A., Rossi, F., ... Coppola, R. (2014). Assessment of aerobic and respiratory growth in the *Lactobacillus casei* group. *PLoS One*, *9*(6), e99189. doi:10.1371/journal.pone.0099189 PMID:24918811

Zotta, T., Tabanelli, G., Montanari, C., Ianniello, R. G., Parente, E., Gardini, F., & Ricciardi, A. (2017b). Tween 80 and respiratory growth affect metabolite production and membrane fatty acids in *Lactobacillus casei* N87. *Journal of Applied Microbiology*, *122*(3), 759–769. doi:10.1111/jam.13373 PMID:27981716

Zuidam, N. J., & Shimoni, E. (2010). Overview of Microencapsulates for Use in Food Products or Processes and Methods to Make Them. In *Encapsulation Technologies for Active Food Ingredients and Food Processing* (pp. 3–29). New York, NY: Springer New York; doi:10.1007/978-1-4419-1008-0_2

About the Contributors

Şebnem Öztürkoğlu Budak was born in Adana, Turkey on March 18th 1978. She received her BSc diploma in Food Engineering from Ankara University (Turkey) in 2001. She obtained her Master of Science Degree in Food Engineering from Middle East Technical University (Turkey) in 2004. In September 2007, she started her PhD in Food Engineering Department of Ankara University (Turkey), and graduated in March 2014. In 2011, she began her second PhD in Utrecht University, and performed her laboratory studies in CBS-KNAW Fungal Biodiversity Center (recently named Westerdijk Biodiversity Center, The Netherlands). She obtained her second PhD degree in September 2016. In September 2010, Dr. Öztürkoğlu Budak joined the Dairy Technology in Faculty of Agriculture in Ankara University, where she contributes the teaching and research program on the dairy technology program. She also continued her research on the molecular biology, molecular fungal physiology and enzyme technology in Westerdijk Biodiversity Center (The Netherlands). Dr. Öztürkoğlu Budak has published more than 20 journal articles, chapters and reviews, 50 abstracts, she have been coordinator or researcher in about 20 projects and has an h-index of 6. Her areas of expertise include fermented dairy products, enzyme technology and dairy microbiology including starter and adjunct cultures.

H. Ceren Akal graduated from Food Engineering Department of Selçuk University, in 2008. She received her Masters Degree in Dairy Technology from Ankara University in 2011 and her PhD in Dairy Technology from Ankara University in 2017. She has been working Department of Dairy Technology for seven years. Her areas of expertise include fermented dairy products, enzyme technology and dairy microbiology.

***

Nihat Akın has completed his PhD in Chemical Engineering Department at Loughborough University of Technology, United Kingdom. He has been working as a Professor for 15 years and been working as Head of Department of Food Engineering at Selcuk University. He has published more than 100 scientific papers about Milk Products and Microbiology.

Harun Aksu was born in Istanbul in 1966. He graduated from Istanbul University Veterinary Faculty in 1987. He completed his Ph.D. program in 1994 at Istanbul University Health Sciences Institute Food Hygiene and Technology Department. In 2005 he took his Professor title. He worked in various organizations in food sector. ISO 9001, ISO 22000 and FSSC 22000. He has worked on various national and international projects. There are many studies on food.

Mehmet Alpaslan graduated from Trakya University and completed his Ph.D. program at Food Engineering. He has done a lot of research in various fields of food chemistry and technology. He was Head of the Department of Food Engineering and Dean of Faculty of Engineering in Inonu University. He is still Head of Department of Nutrition and Dietetics of Namık Kemal University.

Shekhar Banala Bashetty is holding a bachelor's degree in Dairy Technology and did his Masters in Dairy Microbiology from Chhattisgarh Kamdhenu Vishwavidyalaya, Durg, Chhattisgarh, India. His research interest spans from LAB, gut microbiology, food safety and probiotics. He has attended several national and international conferences and workshops in relevant areas.

Gulden Basyigit Kilic is an Associate Professor in the Department of Food Engineering at Mehmet Akif Ersoy University, Burdur, Turkey. She received her Master in 2004 and PhD in 2009 at Department of Food Engineering from Suleyman Demirel University, Isparta. She joined "Research & Development in Postharvest Biology and Technology" course conducted by the Ministry of Agriculture & Rural Development in the Volcani Center, Israel in 2004 with the Mashav scholarship. She was visiting scientist in Department of Food Science at the University of Wisconsin-Madison, USA for six months in 2007 with the scholarship from Danone Institute, Turkey. In 2012, she was a visiting scientist in the same university for three months with scholarship from Higher Education Council of Turkey. She also visited University of Ghent, Belgium and gave lecture with Erasmus Programme in 2010. Her research interests are lactic acid bacteria, probiotics, starter cultures, functional foods and food biotechnology. She has 13 refereed international publications, 7 refereed national publications and 55 other publications/proceedings/posters. She has lead 8 projects as coordinator and worked in 15 projects as a researcher. She was a member of Cost Action FA 1001: The application of innovative fundamental foodstructure-property relationships to the design of foods for health, wellness and pleasure health between 2010 and 2014.

Peter A. Bron initiated his research career in 1999 as a PhD student within TI Food & Nutrition (Wageningen, The Netherlands) on the molecular response of Lactobacillus plantarum to intestinal passage and conditions. He has been working in the scientific area of host-microbe interactions ever since and succeeded his career with a 2.5-year post-doc in Cork (Ireland) investigating Listeria monocytogenes temporal gene expression profiles during infection. Subsequently, he returned to TIFN in 2007 as a post-doctoral researcher in a project on fermentation-enhanced probiotic function. Nowadays, he holds a senior scientist position within the fermentation workgroup at NIZO (Ede, the Netherlands). He was also one of the principal investigators that in 2014 cofounded flagship 10 within the BE-Basic consortium that aims to develop natural methods for strain improvement for the food industry. His work has so far resulted in over 50 publications in peer reviewed journals (H-index = 28 according to google scholar) and 5 patent filings.

Sabrina Neves Casarotti is a professor at Nutrition Faculty of Federal University of Mato Grosso (UFMT) since July 2016. In 2009 she received her MS in Microbiology at São Paulo State University (UNESP), Brazil. In 2013 she got her PhD degree on Food Science and Technology at the same University. From February to July 2012 she did part of her PhD at Teagasc Agriculture and Food Development Authority, Fermoy, Ireland. Currently she is developing research and publishing papers on the fields of fermented products, probiotics and intestinal microbiota.

Manorama Chauhan is working as Professor and Head in Department of Dairy Microbiology, College of Dairy Science and Food Technology, Raipur (CG), India. She is actively engaged in UG, PG and Ph.D. teaching and research. She holds Master's and Doctorate in Dairy Microbiology from the premier National Dairy Research Institute, Karnal (India). Her research interest is mainly in the area of Dairy Biotechnology i.e. improvement of starter cultures and their utilization in fermented foods.

Ronald P. de Vries obtained his PhD degree from Wageningen University on Aspergillus enzymes involved in plant polysaccharide degradation. Since then he has continued and broadened this research through postdocs and appointments at Wageningen University and Utrecht University and since 2009 he is Group Leader Fungal Physiology at the Westerdijk Fungal Biodiversity Institute (formerly CBS-KNAW Fungal Biodiversity Centre) in Utrecht, The Netherlands, where he focusses mainly on conversion of natural substrates by fungi with a special emphasis on fungal diversity, regulatory systems, extracellular enzymes and carbon metabolism. While maintaining his position at the institute, he was also appointed Professor in Fungal Molecular Physiology at Utrecht University in 2015 and in 2017 as visiting professor at University of Helsinki (Finland).

Annereinou R. Dijkstra, after finishing her study in Biotechnology, worked for several years as a Research Specialist at NIZO, Ede, The Netherlands. In 2015 she successfully defended her PhD thesis entitled "Spray drying of starter cultures - Diverse solutions within Lactococcus lactis to improve robustness". Currently, she is working as a Project Manager in the Fermentation group of NIZO.

Ayşe Gürsoy is a Professor in Department of Dairy Technology, Faculty of Agriculture, Ankara University. She received Ph.D degree from Graduate School of Natural and Applied Sciences, Ankara University about White pickled cheese and salt penetration. Her research area low fat cheese, ice cream (with natural additives) and yoghurt.

Sylvia Klaubauf received her MSc degree in Molecular Biology from the University of Vienna, Austria in 2008. In 2010 she started her PhD project within the Fungal Physiology group at the CBS-KNAW Fungal Biodiversity Centre in Utrecht, the Netherlands. She received her doctoral degree in 2015 and pursued a postdoctoral training in enzymology at Chalmers University of Technology in Sweden. In June 2017 she started as a scientist in the enzyme program of Dutch DNA Biotech in the Netherlands.

Celalettin Koçak was graduated from Ankara University, Department of Dairy Technology with a bachelor degree in 1973. He earned his Ph.D. in the same department from Ankara University in 1979. He is currently an emeritus professor of the same department. His research interests are; Dairy Technology, Cheese Technology and Food Quality and Safety. Dr. Koçak has published over 100 articles in peer-reviewed journals.

Vaibhao K. Lule is B. Tech. (Dairy Technology) from MAFSU, Nagpur followed by M. Tech. and Ph. D. (Dairy Microbiology) from the National Dairy Research Institute (NDRI), Karnal (Haryana-132001), India. He has several peer reviewed national and international publications to his credit. He is actively involved in research on therapeutic benefits of probiotics and fermented milks. He attended several National and International conferences/seminars on probiotics, beneficial microbes, food safety and bio-preservation of milk and dairy foods at national and international platforms. He bestowed with several awards like, ICAR-Junior Research Fellow, ICAR-Senior Research Fellow, Recipient of Travel Fellowship award from Centre for International Co-operation in Science (CICS), Chennai and Department of Biotechnology (DBT), Government of India.

Gabriela Piccolo Maitan-Alfenas is a professor at Nutrition Faculty of Federal University of Mato Grosso (UFMT) since July 2016. In 2011 and 2014, respectively, she received her MS and PhD degrees in Agricultural Biochemistry at Federal University of Viçosa (UFV), Brazil. From May 2012 to May 2013 she did part of her PhD at CBS-KNAW- Fungal Biodiversity Centre (Westerdijk Fungal Biodiversity Institute), Utrecht, The Netherlands. Currently she is developing research and publishing papers on the fields of biotechnological application of microbial enzymes.

Banu Metin is an Assist. Prof. in the Department of Food Engineering at Istanbul Sabahattin Zaim University. She has been working at this university since 2014 giving undergraduate and graduate courses and conducting research on the mating type loci and sexual reproduction of fungi and microbial diversity of fermented foods. She completed her BSc degree in the Department of Food Engineering at Middle East Technical University in 1999. After that, she started to work in the same department as a research assistant that lasted until 2006. During this time, she received her MSc degree in 2002 from the same university studying the alpha-galactosidase gene of *Aspergillus fumigatus* under the supervision of Prof. Zümrüt B. Ögel. After finishing MSc, she started PhD at the same university studying a multicopper oxidase gene and the associated polyketide biosynthesis gene cluster of Aspergillus fumigatus again under the supervision of Prof. Zümrüt B. Ögel. During her PhD, just after getting married, she spent 4 months in Prof. Gillian Turgeon's lab in the Department of Plant Pathology at Cornell University (USA) as a visiting scholar where she studied the function of a polyketide synthase gene and on the evolution of the mating type genes of Cochliobolus heterostrophus. After coming back from the US, she received her PhD degree in 2007. After that, she worked as a post-doctoral researcher in Prof. Joseph Heitman's lab in the Department of Molecular Microbiology and Genetics at Duke University in 2007-2008. There, she worked on the mating type locus and sexual reproduction of Cryptococcus heveanensis, the phylogeny of pathogenic *Cryptococcus* species and the mating type locus of the dermatophyte Microsporum gypseum. Then she returned to Turkey and worked as an R&D scientist at ATQ Biyoteknoloji (Ankara, Turkey) in 2009-2011, where she designed and developed pyrosequencing assays for the detection of specific mutations and for microorganism genotyping. In 2012, she gave birth to a girl, Duru. Then they moved to Istanbul and she started working as an Assist. Prof. in the Department of Food Engineering at Istanbul Sabahattin Zaim University in 2014. Since then, she conducted one TUBITAK-funded project on the mating type locus and sexual reproduction of the opportunist black yeast Exophiala dermatitidis and one university funded project on the metagenomic profiling of the bacterial diversity of fermented foods. She still lives in Istanbul with her family.

Ekaterini Moschopoulou received the B.Sc. degree in Agricultural Science from the Agricultural University of Athens (AUA) in 1986, the M.Sc. degree in Dairy Science from Reading University of U.K. in 1992 and the Ph.D. degree in Food Science & Technology from AUA in 2003. In 2009 she started working as Lecturer in the Department of Food Science & Human Nutrition of AUA and from late 2017 she promoted to Assistant Professor position. Her research interests lie in the field of: milk-clotting enzymes; cheese, yoghurt and ice cream technology; modern technologies applied in milk and dairy products; analytical methods of dairy products. She has co-authored 26 journal papers, 2 books, 3 book chapters and 37 presentations at international and national symposia and seminars. She is referee in scientific journals in the area of Dairy Science & Technology and member of Scientific expert groups.

Zumrut Ogel completed her B.Sc. and M.Sc. in the Food Engineering Department of the Middle East Technical University, Turkey in 1985 and 1987, and her Ph.D. in the University of Leeds, UK, Department of Biochemistry and Molecular Biology in 1993. She was invited as a visiting academic to Oxford University for post-doctoral research on plant sciences for 7 months. In 1994 she became an academic in the Middle East Technical University, becoming Assistant Professor in 1994, Associate Professor in 1998 and Professor in 2003. Ogel received the prestigious TUBITAK Encouragement in Science award for young scientists in 2003 in the field of Engineering Sciences. Since 2015, Ogel is a member and Head of the Department of Food Engineering of the Konya Food and Agriculture University.

Barbaros Özer received his undergraduate and graduate degrees from Ankara University Department of Dairy Technology. He completed PhD studies at The University of Reading Department of Food Science and Technology and started working at Harran University Department of Food Engineering as Assistant Professor in 1998. In 2006, he was promoted to full Professorship position. Currently he is working at Ankara University Department of Dairy Technology. He has over 120 published scientific papers and congress proceedings. He also wrote 20 book chapters and 5 books in the field of dairy science and technology. Up to now, He has involved in more than 30 research projects supported by various institutions. He is currently a member of Turkish Milk, Meat, Food Industrialists and Manufacturers Union (SETBIR), Turkish Food Safety Association (GGD) and Turkish Food Codex Milk and Milk Products sub-committee. He is also an authorized trainer of European Hygienic Engineering and Design Group (EHEDG).

Hale İnci Öztürk has completed her master in Food Engineering Department at Selcuk University, Turkey and is continuing her PhD in the same department and university. She has been working on dairy technology and microbiology. She has published scientific papers about dairy products and microbiology.

Sarang Dilip Pophaly is working as Assistant Professor at College of Dairy Science and Food Technology, Raipur, CGKV, C.G., India since 2012 where he is actively engaged in UG, PG and PhD teaching and research. He is holding a bachelors degree in Dairy Technology and did his and Masters and PhD in Dairy Microbiology from National Dairy Research Institute, Karnal, India. He also holds a postgraduate diploma in Intellectual Property Rights (IPR) from Indira Gandhi National Open University (IGNOU), India. His research interest spans from LAB, gut microbiology, food safety and probiotics and has actively published several peer-reviewed papers and book chapters on varied aspects of dairy and food microbiology. He has attended several national and international conferences and workshops in relevant areas. He is also an awardee of Senior Research Fellowship Grant for pursuing PhD from ICAR, New Delhi.

Poonam Sarang holds a bachelor's degree in Sciences and Masters in food microbiology. She completed her PhD in Dairy Microbiology from National Dairy Research Institute, Karnal, India. She is working in the area of biomolecule production from Propionibacteria and Lactic Acid Bacteria. She has actively published several peer-reviewed papers and book chapters on varied aspects of dairy and food microbiology. She has attended several national and international conferences and workshops in relevant areas.

Frank J.J. Segers received his MSc degree in Biology from Wageningen University in the Netherlands in 2010. He started his PhD project in 2011 within the Applied and Industrial Mycology group at the CBS-KNAW Fungal Biodiversity Centre in Utrecht, the Netherlands. He received his doctoral degree in 2017 on his thesis entitled "Growth of indoor fungi under changing water conditions".

Osman Şimşek was born in Bayburt in 1962. He graduated from the Department of Agricultural Products Technology with BSc degree, Atatürk University Agricultural Faculty, in 1983. He graduated with a Master of Science degree in Agricultural Products Technology, Atatürk University Graduate School of Natural and Applied Sciences in 1986. He completed his PhD in the Department of Dairy Technology, Ege University Graduate School of Natural and Applied Sciences in 1989. Osman Şimşek became Associate Professor in 1995 and Professor in 2001. He worked at Trakya University between 1990-2006 and at Namık Kemal University after 2006. He is currently working at the Department of Food Engineering in the same university.

Tejinder Pal Singh has completed his B.Sc. in Industrial Microbiology from Kurukshetra University, Kurukshetra, India (2009, subsequently he graduated from his M.Sc. programme from Dairy Microbiology (2011) from National Dairy Research Institute, Karnal, India; and recently he completed his Ph.D. in Dairy Microbiology (2015) from National Dairy Research Institute, Karnal, India. Currently, he is working as Assistant Professor (Dairy Microbiology) in College of Dairy Science and Technology of Lala Lajpat Rai University of Veterinary and Animal Sciences.

Jitesh Tarak is a PG student at College of Dairy Science and Food Technology, Raipur, CGKV, C.G., India He is holding a bachelor's degree in Dairy Technology and pursuing his Masters Dairy Microbiology from College of Dairy Science and Food Technology, Raipur, CGKV, C.G., India. His research interest spans from Lactic Acid Bacteria, gut microbiology, food safety and probiotics. He has attended several national conferences and workshops in relevant areas.

Kiran Thakur received her BS (2007), MS (2009) and PhD (2015) degrees from Himachal Pradesh University, Lovely Professional University and National Dairy Research Institute, respectively. Currently, She is associate professor at HFUT. Dr. Kiran's research focuses on the functional potential of lactic acid bacteria and B-vitamin production

Sudhir Kumar Tomar has been working as Principal Scientist at Dairy Microbiology Division, ICAR-National Dairy Research Institute, Karnal (Haryana)-132001, India. He has a research and teaching experience of more than 3 decades in the realm of Dairy Microbiology and is in charge of Technofunctional Starters Lab and National Collection of Dairy Cultures at his institute. His research area has been bioprospecting Lactic Acid bacteria from dairy and non-dairy niche and their evaluation technofunctional attributes.

Nazli Turkmen has been working as a Research Assistant at the Department of Dairy Technology, Faculty of Agriculture, Ankara University since 2012. Her M.Sc. thesis was about using odourless fish oil in fruit yoghurt production. She is still continuing her Ph.D. and working on Maras ice cream which is a traditional product within the scope of doctorate thesis topic.

Atila Yetisemiyen was born in Ankara (Turkey) in 1954. He graduated from the Department of Dairy Technology, Faculty of Agriculture, University of Ankara, as a Agricultural Engineer (MSc). After working for Turkish Standarts Institution, Çankırı Milk Factory and General Directorate of Food Affairs which are public institutions for one year, He started as research assistant in January 1979, in the department of Dairy Technology which he graduated from. He assumed the title of PhD with the finishing of thesis with the subject of 'Some Physical and Chemical Properties of Milk Powders obtained from different kinds of milks'. He made researches on "Ultra-filtration Technique and Production of Production of White Brined Cheese through This Technique" in Hungary for 10 months in 1986 and for 14 months in Germany in 1989, by scholarships that he received, as well as cheese, yoghurt and milk powder intensive studies in Turkey. He also made scientific studies on "Milk and Milk Products Technology" 6 to 7 times in various institutes in Austria and Germany for 3 months at each time, at various occasions since 1994. He assumed the titles of "Associated Professor" in October 1990 and "Professor" in 1996. He currently works for above mentioned Department as academic member.

Seydi Yıkmış was born in 1982 in Adıyaman. In 2005, he graduated from Inönü University, Faculty of Engineering, Department of Food Engineering. In 2013, he graduated from Süleyman Demirel University, Institute of Science, Department of Food Engineering. In 2013, he started his PhD education at Istanbul University, Institute of Health Sciences, Department of Food Hygiene and Technology. He is still working as a lecturer in Nutrition and Dietetics Department of Namık Kemal University.

Index

A

adjunct 55, 58-59, 77, 106, 108, 184-185, 187-190, 234-244, 246
Aerobic Respiration 87-96, 100
Aflatoxins 68

B

Bacteriocins 54, 58-59, 71-77, 80, 101, 144, 195, 238, 243
Bacteriophages 70, 78-81, 183
Bile salt hydrolase 1-2, 8-10
Biomass Production 9, 87, 91, 93, 95
Biopreservation 58-59, 68-72, 79
Biopreservative 59, 95
Bioprotective Cultures 59, 77, 81

C

cheese 1-2, 25-26, 29, 31-32, 34-40, 49, 51, 55, 58, 60, 76-77, 94-95, 104-106, 123, 141-143, 159, 163-166, 182-197, 204-214, 221-226, 230-231, 234-246, 257-268, 283
Cheese Flavor 2, 58, 185, 192, 194, 197, 204, 207-209, 212, 260
Cheese Making 123, 204-205, 214
Cheese Ripening 2, 38-40, 58, 141, 182-183, 185-189, 191, 196, 204, 206-212, 221, 234, 237
Cheese Texture 204, 212-213
Cheese Yield 164-166, 204, 206, 210, 212-213
chymosin 2, 186-187, 209, 222-225, 227-230
Classical Microbiology 24-25
Cloning 1-2, 5, 7, 38, 224
Colony Forming Units 25, 315
Cryotolerance 303, 315
Cultures 13, 23, 26, 28-29, 33, 36-37, 39-40, 51, 54, 56-63, 68, 77-79, 87-88, 90-96, 101-113, 121-125, 127, 143, 160, 162, 182-187, 189, 192-193, 231, 234-236, 239-240, 242-244, 246, 257-261, 264, 267, 276, 297-300, 302-303, 306-309
Culturing Methods 24-26
Cytochrome Oxidase 88-90, 92, 100

D

Dairy 1-2, 4, 7-8, 11-14, 23-24, 27-28, 31-34, 37, 39-40, 49, 51, 54, 56, 58-62, 69, 74, 76-80, 87, 101, 105-106, 108, 111, 113, 123-124, 132, 141, 144, 152-153, 155, 159, 163, 165-166, 183-184, 186, 188, 193, 205-206, 214, 227, 236, 238, 244, 246, 261, 276-277, 279, 283, 286, 297-298, 306-307, 315
DNA 2, 5, 23, 28-39, 47, 51-52, 62, 92, 124-126, 136, 158, 222-223, 227, 230, 261, 279
DNA Sequencing 23, 29-30, 33, 261

E

edible films 152, 159-160, 206
Electron Transport Chain 88-89, 96, 100
Encapsulation 105, 108-109, 113, 132, 136-138, 140-141, 143-144, 207, 304
Enzymatic production 5-6
Enzymes 1-10, 12-14, 28-29, 31-35, 40, 47, 54, 58-59, 70, 73, 88-90, 93, 95, 100-101, 103, 106, 125, 135-136, 138, 141-142, 152-155, 157-158, 160, 163-164, 166, 182-183, 185-189, 193, 195-197, 204-214, 221, 223-229, 234-235, 238, 240, 259, 263-264, 278-279, 303
Exopolysaccharide 101, 281
Experimental Evolution 126-127

F

Fermentates 76-77, 80
Fermentation 1-3, 6, 28, 34, 39, 54-56, 58-60, 62-63, 68, 70, 76-78, 87-91, 93, 96, 100-101, 104-105, 107, 109-112, 121, 123, 125-127, 136, 160, 182-183, 188, 205-206, 210, 222, 227, 234-238, 246, 257, 282-283, 285-286, 299-300, 303-304, 306-309
Fermented Foods 55, 58-61, 95, 104, 113, 298
Fingerprinting Techniques 28-29, 31, 34
Flavor 2, 13, 55, 58, 93-94, 101, 108-109, 113, 124-125, 140-141, 182-188, 190-197, 204, 207-212, 229, 234-235, 237, 240, 257, 260, 262, 264-265, 268, 276, 306
Food Safety 3, 55, 69-72, 74, 77-80
Food Security 71
Fusarium domesticum 257-258, 265-266

G

Gastrointestinal Tract 9-10, 121, 140
Gel-Electrophoresis 47
Genotype-Phenotype Matching 124-125
Glucose oxidase 1-2, 8, 12-13
Glycolysis 58, 88-89, 94, 100-101, 182, 188-189, 204, 207

H

health 8-10, 13, 23-24, 34, 55, 59-63, 68, 71, 88, 93, 102-105, 107, 111, 113, 121, 132-134, 144, 185, 209, 228, 234, 241, 244-246, 276-280, 286, 298, 309, 315
Health Benefits 8, 24, 61, 113, 144, 209, 276-277, 298, 309, 315
Heme 88-89, 91-95, 100
Hypercholesterolemia 1, 10

I

Ice Cream 105, 108-109, 139, 142, 159, 166, 297-309, 315
Immobilization 3-4, 13, 143, 158

K

Kefir 58, 62, 103, 105, 109, 276-277, 280-281, 283-286
Kluyveromyces 185, 228-229, 241, 264, 276, 285-286

L

Lactic Acid Bacteria 11, 27-29, 34, 37, 54-55, 58-63, 70-71, 76, 87-90, 92, 95-96, 101-102, 123, 135, 142-143, 183-185, 188, 210, 224, 234-238, 246, 258-259, 277, 280, 300
Lactobacillus 8, 10, 28, 31, 60-61, 63, 90, 102, 104-108, 110-111, 113, 123-124, 134-136, 142-144, 162, 166, 183, 185, 210, 212, 235, 238-240, 244, 276-283, 298, 300, 303-305, 309
Leuconostoc 28, 58, 61, 90, 112, 123, 183, 189, 235, 240, 276, 284
Lipase 26, 186, 188-189, 193, 196-197, 204-207, 209-212, 221, 240, 260, 264
Locus 29-30, 37, 47
Low costs induction 1

M

Menaquinone 89-92, 95
Microbial Community Profiling 35, 47
Microbial Enzyme 207, 214, 221
microbial heterologous expression 226
Microbial Lipase 210-211
Microbial Protease 210
Microencapsulation 4, 104-105, 109, 132, 136, 138-139, 141-144, 166, 208, 304-305, 309, 315
Microorganism 3, 5, 7, 29, 37, 107, 134, 190, 208, 212, 221, 223, 242, 277, 282, 285, 304
Morphology 26-27, 29, 263
Mucor spp 242, 244, 257-258, 262-263, 265-267

O

Overrun 300, 303, 305, 307-309, 315

P

PCR 5, 23, 29-30, 32-33, 35, 37-39, 47, 52, 226
Penicillium camemberti 185, 196, 257, 263
Penicillium roqueforti 185, 211, 257-258
Phage 56, 60, 63, 68, 79-80, 103, 184, 227
Phage Resistance 60, 63, 103
Phylogenetic Marker 47
Polymerase Chain Reaction 5, 29, 47
Polyols 58, 68
Prebiotics 8, 107-109, 113, 121, 133-134, 144, 299, 301-302, 308-309, 315

Primary 59, 62, 72, 153, 164, 183-184, 186-188, 191, 207-209, 234-235, 240, 243, 258-259
Probiotic 1, 10, 12-14, 34, 54, 60-62, 87-88, 92-93, 101-113, 132, 134-136, 142-144, 162, 166, 213, 231, 244, 246, 276-283, 285-286, 297-300, 303-309, 315
Probiotics 8, 10, 13, 36, 61-63, 68, 88, 101-105, 108, 110-112, 121, 132-136, 142, 144, 244-246, 277-280, 286, 297-309, 315
Production 1-3, 5-6, 8-10, 12-14, 23, 28, 37, 55-60, 62-63, 72-73, 77, 87-89, 91, 93-96, 101-104, 106-113, 122-125, 132, 135-136, 138-139, 142-144, 152-153, 158-160, 163, 166, 182-190, 192-197, 204-205, 210-212, 214, 222-224, 226-231, 234-241, 243-244, 246, 257-262, 264-267, 279, 281, 283-286, 297-300, 302-309
Protease 62, 186-188, 192-193, 196-197, 207, 209-211, 221, 224-225, 259, 264
proteolysis 14, 58, 60, 62, 101, 165-166, 182, 186-189, 191-193, 195-197, 204, 207-211, 214, 224-225, 234, 238, 240, 243-244, 259-260, 264-265
Purification 1-4, 6-8, 12, 158, 229

R

Rennet 1-2, 58, 163-166, 183, 186-187, 189, 195, 204-205, 207-208, 211-213, 222-231, 259
Respiration 34, 87-96, 100
Restriction Enzyme 33, 47
Ribosomal RNA Operon 47
Ripening 2, 35, 37-40, 58, 106, 132, 141, 165, 182-191, 193-197, 204, 206-214, 221, 225, 234-244, 246, 257-260, 262-265, 267

S

Saccharomyces 102, 228, 241, 264, 276, 278, 280, 285
Scopulariopsis spp 266-267
Secondary 9, 30, 55, 76, 138, 159, 164, 184-190, 192-197, 207, 210, 234-235, 240, 243, 246, 258, 261, 265
Selective Media 26, 49
Species Identification 29, 38-39, 53
Starter Culture 54-55, 63, 68, 87, 92-94, 100, 104, 106-110, 112, 121-126, 160, 163, 184, 193, 224, 234-235, 241, 257-258, 265
Starter Functionality 68
Stress Resistance 88, 91-92, 95
Stress Survival 124-126
Super-expression 1
Symbiotics 315

T

Texture 14, 55-57, 77, 95, 102, 107-108, 123, 155, 160, 182, 186, 193, 204, 212-213, 234-236, 239, 246, 257, 260, 264-265, 268, 300, 307
Transcriptome-Phenotype Matching 125-127
Transglutaminase 153, 204-206, 212, 214, 221
Trichothecium roseum 257-258, 265

V

Viable Bacterial Counts 315

Y

yeasts 24-25, 27, 32-33, 38, 50, 153, 193, 196, 205, 226-229, 235, 240-243, 246, 258, 264, 276-277, 285-286
yield 6-7, 9, 12, 58, 72, 87-88, 93, 126, 152, 155, 163-166, 204, 206, 210, 212-214, 224, 228-229, 240
Yogurt 13-14, 55, 58, 62, 77, 107, 109, 143, 159-160, 162-163, 223, 226, 231

www.ingramcontent.com/pod-product-compliance
Lightning Source LLC
LaVergne TN
LVHW060402300926
849162LV00006B/1325
9781522553632